Numerical Linear Algebra, Digital Signal Processing and Parallel Algorithms

NATO ASI Series

Advanced Science Institutes Series

A series presenting the results of activities sponsored by the NATO Science Committee, which aims at the dissemination of advanced scientific and technological knowledge, with a view to strengthening links between scientific communities.

The Series is published by an international board of publishers in conjunction with the NATO Scientific Affairs Division

A Life Sciences B Physics	Plenum Publishing Corporation London and New York
C Mathematical and Physical Sciences D Behavioural and Social Sciences E Applied Sciences	Kluwer Academic Publishers Dordrecht, Boston and London
F Computer and Systems Sciences G Ecological Sciences H Cell Biology I Global Environmental Change	Springer-Verlag Berlin Heidelberg New York London Paris Tokyo Hong Kong Barcelona Budapest

NATO-PCO DATABASE

The electronic index to the NATO ASI Series provides full bibliographical references (with keywords and/or abstracts) to more than 30000 contributions from international scientists published in all sections of the NATO ASI Series. Access to the NATO-PCO DATABASE is possible in two ways:

– via online FILE 128 (NATO-PCO DATABASE) hosted by ESRIN, Via Galileo Galilei, I-00044 Frascati, Italy.

– via CD-ROM "NATO-PCO DATABASE" with user-friendly retrieval software in English, French and German (© WTV GmbH and DATAWARE Technologies Inc. 1989).

The CD-ROM can be ordered through any member of the Board of Publishers or through NATO-PCO, Overijse, Belgium.

Series F: Computer and Systems Sciences Vol. 70

Numerical Linear Algebra, Digital Signal Processing and Parallel Algorithms

Edited by

Gene H. Golub
Computer Science Department
Stanford University
Stanford, CA 94305, USA

Paul Van Dooren
Philips Research Laboratory
Av. Albert Einstein 4
B-1348 Louvain-la-Neuve, Belgium

Springer-Verlag
Berlin Heidelberg New York London Paris Tokyo
Hong Kong Barcelona Budapest
Published in cooperation with NATO Scientific Affairs Division

Proceedings of the NATO Advanced Study Institute on Numerical Linear Algebra, Digital Signal Processing and Parallel Algorithms, held in Leuven, Belgium, August 1–12, 1988.

ISBN-13: 978-3-642-75538-5 e-ISBN-13: 978-3-642-75536-1
DOI: 10.1007/978-3-642-75536-1

Softcover reprint of the hardcover 1st edition 1991

45/3140-543210 – Printed on acid-free-paper

Preface

The NATO Advanced Study Institute on **Linear Algebra, Digital Signal Processing and Parallel Algorithms** took place in Leuven from August 1 through August 12, 1988. There were over 90 participants (including the 15 invited speakers) from 16 countries and there was a significant industrial participation.

The invited speakers were: M. Bellanger (TRT, France), B. Bitmead (ANU, Australia), A. Björck (Linköping, Sweden), R. Brent (ANU, Australia), Y. Genin (PRLB, Belgium), S. Hammarling (NAG, UK), I. Ipsen (Yale, USA), T. Kailath (Stanford, USA), F. Luk (Cornell, USA), J. McWhirter (RSRE, UK), G. Meurant (CEL, France), D. Sorensen (Argonne, USA), J. Vandewalle (KUL, Belgium), and the codirectors G. Golub (Stanford, USA) and P. Van Dooren (PRLB, Belgium).

The goal of this meeting was to synthesize the three topics mentioned in the title as there is currently a great deal of activity in each one of these areas. It was felt that there were many interactions at the meeting, not only between participants who were familiar with each one of these areas but also between people working in different areas of interest.

The following areas emerged as major themes at the meeting:

1) *Singular value and eigenvalue decompositions, including applications*

The current techniques used for the computation of the singular values and eigenvalues of matrices were addressed by several speakers. Modified singular value and eigenvalue problems were also discussed, especially in view of their application in various signal processing problems (e.g., total least squares, generalized SVD, time varying eigenvalue problems, etc.).

2) *Toeplitz matrices, including special algorithms and architectures*

Several speakers focussed on so-called fast or $O(n^2)$ algorithms for computing $L.L^t$ and $Q.R$ decompositions of a Toeplitz matrix. Special attention was given here to Levinson and Schur type algorithms as well as to their split version. The fast algorithms based on displacement ranks and on updating and downdating were also addressed.

3) *Recursive least squares in linear algebra, digital signal processing and control*

The recursive least squares algorithms occurring in signal processing and control often have a specific structure and hence allow for fast solutions. Various of these fast algorithms were explained and their potential numerical deficiencies were pointed out. Issues such as error build-up, error feedback, exponential windowing, sliding windowing and so on, were addressed.

4) *Updating and downdating techniques in linear algebra and signal processing*

The two main techniques of updating and downdating are low rank corrections and low norm corrections. Low rank corrections are used in divide and conquer methods for eigenvalue and singular value computations and were shown to yield powerful parallel computational methods. They can also efficiently be used for computing new least squares solutions for modified data fitting problems using the theory of orthogonal polynomials and modified moments. Low norm corrections are used in slowly time varying problems as encountered in various signal processing problems.

5) *Error analysis and stability of algorithms and sensitivity analysis of special recursive least squares problems*

Error propagation in linear algebra is a well established discipline but insufficiently known to the signal processing community. Their basic principles were explained and their applications to specific problems in signal processing were addressed (fast recursive least squares, Toeplitz solvers, Kalman filtering, and so on). The relevance of special forms of stability (weak, strong, mixed stability) was also emphasized in problems of signal processing and linear algebra with special structure.

6) *Special architectures (including supercomputers and distributed processor arrays) for linear algebra and signal processing*

Exploiting the parallelism of present (and future) computer architectures has been an area of significant interest in the last few years. The state of the art in computational algorithms for supercomputers and distributed arrays of processors (such as systolic arrays) was widely covered. Also special problems in linear algebra and signal processing were given full attention.

Although most of the presentations can easily be classified in one of the above "themes", it became apparent during the meeting that there was a strong interconnection between several of the themes. This lead to lively discussions both during lectures and breaks. Several of the contributed talks focussed on specific applications of these topics. In particular we mention here radar technology, medical applications and robotics.

The feedback we had from the participants was very positive in general, for the talks as well as for the social contacts made possible by the meeting. For making all this possible we are in great debt to the organizing committee: Paul Van Dooren (Philips Research Lab. Belgium), Gene Golub (Stanford University, USA), Jack Dongarra (Argonne National Laboratory, USA), Sven Hammarling (NAG Ltd., Oxford, UK), Danny Sorensen (Argonne National Laboratory, USA) and Joos Vandewalle (Katholieke Universiteit Leuven, Belgium), to the North Atlantic Treaty Organization, Scientific Affairs Division and to the co-sponsors of the Advanced Study Institute: the National Science Foundation of the USA, N.V. Philips Belgium, SWIFT Belgium, Cray Research USA, IBM Belgium, and the Office of Naval Research London.

Finally, special thanks are also due to M. Vanbegin who helped edit the proceedings and to E. Moes who helped typeset most of the papers.

December 1990

P. Van Dooren (Director)
G. Golub (Co-director)

Table of Contents

(In case of several authors, ⋆ indicates who presented the paper)

I. Invited Lectures

II. Contributed Lectures

The Family of Fast Least Squares Algorithms for Adaptive Filtering

M. G. Bellanger

T.R.T.
5, av. Réaumur
92350 Le Plessis Robinson, France

Abstract

Three representative algorithms from the family of Fast Least Squares (FLS) algorithms are reviewed. They correspond to the transversal FIR filter, the lattice-ladder structure and the rotation approach respectively. They all solve the least squares problem recursively and are convenient for adaptive filtering.

They lead to three different levels of computational complexity and robustness to round-off errors. Each of them also has special implementation aspects and provides a particular set of signal parameters.

Overall, these three types of algorithms offer a wide and welcome choice to the adaptive filter designer.

1 Introduction

Fast Least Squares (FSL) algorithms have been worked out for adaptive filters with different kinds of specifications and various structures, and they can handle one-or multidimensional signals [1].

The filter structure may be dictated by the application, but it can also be chosen according to the computational complexity, the particular set of signal parameters which is of interest to the designer or on the basis of robustness against numerical limitations. It must be pointed out that an important aspect of least squares techniques is the round-off error accumulation, which takes place in the transversal filter for example, but apparently not in the lattice-ladder structure or in the rotation method. The purpose of this paper is to review these three major approaches and for each one, assess the computational complexity and comment on the operational organization, the wordlength limitation effects and the set of signal parameters available.

The principle of an adaptive filter is shown in Fig. 1, which gives the various signals involved namely the input $x(n)$, the reference $y(n)$ and the error $e(n)$, The transversal structure is dealt with first, since the order of growing complexity is retained.

2 FLS Algorithm for Transversal Structure

A typical FLS algorithm for the transversal structure is given in Fig. 2. A detailed derivation is available in reference [1] and will not be reproduced here. Earlier references include [2,3,4].

That approach has two strong points, which are its low computational complexity and the usefulness of the signal parameters involved.

If the filter order is N, the operations listed in Fig. 2 correspond to $8N+4$ multiplications and 2 divisions in the adaptation gain updating section and $2N$ multiplications in the filter section. The amount of memories is about $6N$, to store the coefficients and internal variables. Altogether, the complexity can be viewed as about five times that of the least mean squares (LMS) approach, which is just $2N$ multiplications.

The sequencing of the operations is reasonably favorable for implementation since weighted sums form the bulk of the processing. A difficulty may arise from the two divisions, in connection with the real time constraints. It is worth mentioning that some flexibility exists in the computation ordering [5].

The adaptation gain updating procedure consists of a real time analysis of the input signal, through linear prediction. The set of forward and backward coefficients obtained is useful in many technical areas, for example for spectral analysis purposes, extraction of information or system modelling and measurement.

The major problem with the FLS transversal adaptive filter is the round-off error accumulation, which takes place, mainly in the backward prediction coefficient updating loops, with a finite precision implementation. In most cases the consequence is instability. Although the problem is still a subject for research, significant progress has been made. First of all, several minor modifications can be made to the basic algorithm, which permit proper working in a number of situations ([6] and chapter 6 in [1]).

The forward prediction error energy updating equation can be modified as follows:

$$E_a(n+1) = W E_a(n) + e_a(n+1)\,\epsilon_a(n+1) + C \tag{1}$$

The positive constant C prevents the variable $E_a(n+1)$ from reaching zero, which is a crucial point since it is used subsequently as a divider. Moreover it can be shown that the backward prediction coefficient updating recursion is also modified and becomes:

$$B(n+1) = (1-\gamma)\,B(n) + G(n+1)e_b(n+1) \tag{2}$$

The expectation of the term γ is positive, which yields a leakage effect able to counter the round-off error accumulation. The impact on the forward prediction coefficient updating is difficult to investigate. It can be shown that a leakage effect is also obtained for some signals, for example sinusoids in noise. However further work is needed on this topic.

Another approach is to control the round-off error accumulation by exploiting redundancies in the algorithm, in connection with a recursive least squares procedure. To that purpose a reliable variable has to be chosen. The last element of the gain vector $G_1(n+1)$ is $m(n+1)$, which is related to the backward prediction-variables by the equation:

$$m(n+1) = \frac{\epsilon_b(n+1)}{E_b(n+1)} \tag{3}$$

The two variables $\epsilon_b(n+1)$ and $E_b(n+1)$ can be computed directly, which yields another value, say $m_0(n+1)$. In finite wordlength implementation, a least squares procedure can be worked out and the equations modified in such a way that the energy of the difference between $m(n+1)$ and $m_0(n+1)$

is minimized. That technique has been shown to be successful in a number of cases [7].

Another control variable can be derived from the prediction error ratio and the adaptation gain. Several control variables can also be combined [8].

As concerns the wordlength estimations, in connection with the filter performance specifications, they are difficult to derive and further work is needed. However, rough estimations can be derived from those already available for the LMS algorithm. Consider, for example the wordlength b_c of the prediction coefficients; a first guess can be:

$$b_c = log\left(\frac{N}{1-W}\right) + log_2(G_p) + log_2(a_{\max}) \tag{4}$$

Where G_p is the prediction gain and $a_{\max}$ is the magnitude of the largest coefficient [1]. According to that expression the transversal FLS algorithms require larger wordlengths than LMS algorithms and the difference is $log_2(N)$.

3 The Case of Updating and Downdating (Sliding Window)

In most applications of adaptive filters, least squares algorithms are based on cost functions involving infinite exponential weighting of the error sequence. However, in some specific cases, finite memory might yield better performance, for example, when sharp changes are present in the signal, on in image processing, due to the edges. Finite weighting is implemented through sliding windows, and Fast Least Squares (FLS) algorithms can be derived specifically for that case [9,10].

Given a reference sequence $y(n)$ and an input signal $x(n)$, the cost function considered is:

$$J_{SW}(n) = \sum_{p=n+1-N_0}^{n} [y(p) - X^t(p)H(n)] \tag{5}$$

Where N_0 is the length of the observation time window, which slides among the time axis. The autocorrelation matrix and cross-correlation vector estimations are:

$$R_N(n) = \sum_{p=n+1-N_0}^{n} X(p)X^t(p) \tag{6}$$

$$r_{yx}(n) = \sum_{p=n+1-N_0}^{n} y(p)X(p) \tag{7}$$

Using the recurrence relations:

$$\begin{aligned} R_N(n+1) &= R_N(n) + X(n+1)X^t(n+1) \\ &\quad -X(n+1-N_0)X^t(n+1-N_0) \end{aligned} \tag{8}$$

$$\begin{aligned} r_{yx}(n+1) &= r_{yx}(n) + y(n+1)X(n+1) \\ &\quad -y(n+1-N_0)X(n+1-N_0) \end{aligned} \tag{9}$$

the following recursive updating of the coefficients is obtained [1]:

$$H(n+1) = H(n) + G(n+1)e(n+1) - G_0(n+1)e_0(n+1) \tag{10}$$

Where $e_0(n+1)$ is the backward innovation error:

$$e_0(n+1) = y(n+1-N_0) - X^t(n+1-N_0)H(n) \tag{11}$$

and $G_0(n+1)$ is the backward adaptation gain:

$$G_0(n+1) = R_N^{-1}(n+1)\,X(n+1-N_0) \tag{12}$$

Both adaptation gains can be updated recursively, in a very similar manner, and the complete algorithm is given in Fig. 3.

Clearly, the computational complexity is about twice that of the exponential window algorithm of the previous section.

As concerns performance, the estimation of the mean residual error power can be derived from the same estimation for the exponential weighting case through the equivalence:

$$W = \frac{N_0 - 1}{N_0 + 1} \tag{13}$$

where W is the weighting factor.

In order to assess the convergence properties, let us assume that, at time n_0, a system to be identified undergoes a step change in coefficients from H_1 to H_2. Then, from the definition of $H(n)$ [10]:

$$H(n) = R_N^{-1}(n) \sum_{n+1-N_0}^{n_0} X(p)y(p) + \sum_{n_0+1}^{n} X(p)y(p) \tag{14}$$

One gets for the exponential window:

$$E[H(n) - H_2] = W^{n-n_0} [H_1 - H_2] \tag{15}$$

and for the sliding window:

$$E[H(n) - H_2] = \frac{N_0 - (n - n_0)}{N_0} (H_1 - H_2) \qquad n_0 \leq n \leq n_0 + N_0 \tag{16}$$

The results obtained for the prediction error energy in linear prediction, are shown in Fig. 4; the finite memory effect is clearly apparent.

As concerns wordlength limitation effects, they are difficult to investigate in that case. Simulations show that, with some stationary signals, the round-off error accumulation mentioned in the previous section does not occur. It is so for sinusoids in noise for example. However if the input signal is turned on and off regularly for example, then accumulation shows up again, leading to divergence. Further investigation is needed; however, it is clear that the sliding window algorithm is highly sensitive to round-off errors and some control techniques have to be worked out.

The algorithm in Fig. 2 belongs to a group which is based essentially on time recursions. Another group exploits both time and order recursions [11,12].

4 A Lattice-Ladder FLS Algorithm

Many different combinations of time and order recursions can lead to FLS algorithms for the lattice-ladder structure. The algorithm given in Fig. 5 is obtained when precedence is given to time recursions, which are known for their better robustness to wordlength limitations. The computational complexity amounts to $16N + 2$ multiplications and $3N$ divisions (inverse calculations). About $7N$ memories are required.

A major advantage of that approach is that it provides filters with all orders from 1 to N. Accordingly the operations are arranged in a loop configuration and the programming can be very efficient. For hardware realization a modular cascade structure can be taken, which is quite convenient. Great care has to be exercised in the operation organization, to preserve the indexes and save the relevant variable.

The signal parameters are the reflection or partial correlation coefficients. They can be exploited for signal analysis or feature extraction purposes, but not as widely and easily as the transversal coefficients. From the implementation standpoint, they have a nice property, namely the expectation of their magnitude is bounded by one; the same applies also to the variables $\phi_i(n)$, which are the ratios of a posteriori to a priori prediction errors. The errors being bounded by the signal values, it turns out that there are no scaling uncertainties in the algorithm of Fig. 5, which turns out to be very robust to round-off errors [13]. A first guess for the wordlength estimation can be obtained from expression (4), taking $a_{\max} = 1$ and $N = 1$:

$$b_{CL} = log\left(\frac{1}{1-w}\right) + log_2\left(G_p\right) \tag{17}$$

As concerns round-off error accumulation, it has been shown theoretically, on an approximate model and to some extent verified experimentally, that the lattice-ladder FLS algorithm is stable [14]. Indeed, it is a critical aspect, which can, in some circumstances, justify the increase in computational complexity

An important aspect for use in adaptive filtering applications is the availability in each lattice stage of the variable $\phi_i(n)$, the ratio of a posteriori to a priori prediction errors.

The proper functioning of each stage independently can be checked through that variable, which has to take its value in the interval between zero and unity.

A different type of algorithm, but with similar properties is provided by the rotation method.

5 The FLS-QR Algorithm

In the Q-R decomposition technique the solution to a least squares problem is obtained in two consecutive steps. First an orthogonal matrix is used to transform the data matrix into a matrix in which all the entries are zero except for a submatrix, which takes on a triangular shape. Then the desired coefficients are calculated by solving the new triangular system [15].

The approach can be implemented recursively and the triangularisation process can be performed through a set of rotation operations, which are known for their robustness against finite precision effects.

Let us first show that the rotation matrix $Q(n)$ can be computed recursively. The data matrix $X_N(n)$ is a $(n+1) \times N$ matrix defined by:

$$X_N(n) = \begin{bmatrix} x(n) & x(n-1) & \cdots & x(n+1-N) \\ 0 & 0 & \cdots & 0 \\ \vdots & \vdots & & \vdots \\ 0 & 0 & \cdots & 0 \\ w^{n/2} S_N(n-1) & \cdots & \cdots & 0 \end{bmatrix}$$

Now, the rotation matrix $Q(n)$ is such that:

$$Q_N(n) X_N(n) = \begin{bmatrix} 0 & \cdots & 0 \\ \vdots & & \vdots \\ 0 & \cdots & 0 \\ \hline \multicolumn{2}{|c}{S_N(n)} & \vdots \\ \cdots & \cdots & 0 \end{bmatrix} \tag{18}$$

Where $S_N(n)$ is a $N \times N$ triangular matrix:

$$S_N(n) = \begin{bmatrix} S_{N;N1}(n) & \cdots & \cdots & S_{N;NN}(n) \\ \vdots & & & \vdots \\ S_{N;11}(n) & \cdots & \cdots & 0 \end{bmatrix} \tag{19}$$

The data matrix can be rewritten as:

$$X_N(n) = \begin{bmatrix} x(n) & x(n-1) & \cdots & x(n+1-N) \\ & w^{1/2}X_N(n-1) & & \end{bmatrix}$$

Consider the following product:

$$\begin{bmatrix} 1 & 0 \\ 0 & Q_N(n-1) \end{bmatrix} X_N(n) = \begin{bmatrix} x(n) & x(n-1) & \cdots & x(n+1-N) \\ 0 & 0 & \cdots & 0 \\ \vdots & \vdots & & \vdots \\ 0 & 0 & \cdots & 0 \\ w^{1/2}S_N(n-1) & \cdots & \cdots & 0 \\ \cdots & \cdots & \cdots & 0 \end{bmatrix} \tag{20}$$

If the above matrix is multiplied by the rotation matrix:

$$\begin{bmatrix} \cos\theta_1 & \cdots & -\sin\theta_1 & 0 \\ \vdots & 1 & & \\ \sin\theta_1 & \cdots & \cos\theta_1 & 0 \\ \vdots & & & \vdots \\ 0 & \cdots & \cdots & 1 \end{bmatrix} \tag{21}$$

the term $x(n+1-N)$ is replaced by zero if:

$$\cos\theta_1 x(n+1-N) - \sin\theta_1 w^{1/2} S_{N,NN}(n-1) = 0 \tag{22}$$

which defines the rotation angle such that:

$$\cos\theta_1 = \frac{w^{1/2}S_{N;NN}(n-1)}{S_{N;NN}(n)}$$

$$\sin\theta_1 = \frac{x(n+1-N)}{S_{N;NN}(n)} \tag{23}$$

where

$$S_{N^2,NN}(n) = wS_{N^2,NN}(n-1) + x^2(n+1-N) \tag{24}$$

Repeating the above operations for $x(n+2-N),\ldots,x(n)$ leads to the recursion:

$$Q_N(n) = \hat{Q}_N(n) \begin{bmatrix} 1 & 0 \\ 0 & Q_N(n-1) \end{bmatrix} \tag{25}$$

where the matrix $Q_N(n)$ is a product of N rotations similar to (21).

Taking the identity matrix I_N as the initial matrix in the recursive procedure, it appears that the matrix $Q_N(n)$ is made of rotations and it is orthogonal:

$$Q_N(n)\, Q_N^t(n) = I_N \tag{26}$$

An important property of the rotation matrices is that they keep constant the norms of the vectors. Therefore, the definition equation (18) yields:

$$S_{N^2,NN}(n) = \sum_{p=0}^{n} w^{n-p}\, x^2(p+1-N) \tag{27}$$

It is the input signal power and the recursive form is provided by (24).

Once the data matrix has been triangularized, the least squares problem is solved by a set of N substitutions.

Following the principle of other fast least squares algorithms, the fast QR algorithm consists of updating the rotation matrix with extended dimension using forward linear prediction and then obtain the desired matrix with the help of backward linear prediction [16].

The sequence of operations is as follows:

$$\hat{Q}_N(n) \to \hat{Q}_{N+1}(n+1) \to \hat{Q}_N(n+1)$$

The corresponding algorithm is given in Fig. 6. In fact it is not $Q_N(n)$ which is involved, but a simplified $(N+1)\times(N+1)$ version, denoted $Q_a(n)$ and obtained by deleting all the lines and columns whose elements are only zeros and ones.

For the initialization, all the internal variables are set to zero except for the cosines which are set to one and the prediction error energy $E_a(0)$, which is set to a small positive value, as in the other types of FLS algorithms.

As concerns the operation count, the algorithm requires about $26N$ multiplications, $2N+1$ divisions and $2N+1$ square root calculations. Although some simplifications can be made, the computational load is significantly more important than that of the algorithms of the previous sections. However, it must be pointed out that special operators can be designed to carry out the rotations in hardware. Moreover gains can be expected from a reduced internal data wordlength, since most of the internal data are normalized. Further work is needed on that specific topic.

In fact the above algorithm is comparable to a lattice-ladder algorithm with normalization of the variables.

The fact that the operations are essentially rotations leads to a high level of robustness against round-off errors and there are apparently no such divergence effects as encountered in the transversal algorithm of section 2.

Like other FLS algorithms, the rotation algorithm performs a real time analysis of the input signal. It can be observed that the rotation procedure corresponds to a projection operations on the signal vectors and the number of large value elements in the transformed vector $XQ(n)$ is, at least for some classes of signals, related to the dimension of the signal space. That property can have interesting practical value and deserves further investigation.

6 Conclusion

The main features of the three algorithms which have been reviewed are presented in Fig. 7. Overall, it can be stated that they adequately reflect the characteristics of the family to which they belong. From a user point of view the choice for a given application will depend on what is considered as the most crucial property, computational complexity or numerical stability, operation accuracy or signal parameters. Whether the implementation is carried out in software or hardware may also affect the decision. Worth pointing out also is the need for a fine tuning of the adaptive filter parameters, in order to get a good efficiency. In that respect, the proper choice of the values of the weighting factor W and the initial prediction error energy E_0 is crucial.

As concerns research aspects, further work is needed for all three algorithm families, but the rotation method might deserve special attention.

References

[1] M. Bellanger, "Adaptive Digital Filters and Signal Analysis", Marcel Dekker Inc, New York, 1987.

[2] D. Falconer and L. Ljung, "Application of Fast Kalman Estimation to adaptive Equalization", IEEE Trans. COM-26, Oct. 1978, pp. 1439-1446.

[3] G. Carayannis, D. Manolakis and N. Kalouptsidis, "A Fast Alg. for LS Filtering and Prediction", IEEE Trans. ASSP-31, Dec. 1983, pp. 1394-1402.

[4] J. Cioffi and T. Kailath, "Fast Recursive Least Squares Transversal Filters for Adaptive Filtering", IEEE Trans. ASSP-32, April 1984, pp. 304-337.

[5] V.B. Lawrence and S.K. Tewkesbury, "Multiprocessor Implementation of Adaptive Digital Filters", IEEE Trans. COM-31, June 1983, pp. 826-835.

[6] P. Fabre and C. Gueguen, "Fast Recursive Least Squares : Preventing Divergence", Proc. IEEE-ICASSP-85, Tampa, Florida, 1985, pp. 1149-1152.

[7] J.L. Botto, "Stabilization of Fast Recursive Least Squares Transversal Filters", Proc. IEEE-ICASSP-87, Dallas, Texas, 1987, pp. 403-406.

[8] D.T.M. Slock and T. Kailath, "Numerically Stable Fast Recursive Least Squares Transversal Filters", Proceedings of IEEE-ICASSP-88, New York, April 1988, pp. 1365-1368.

[9] M.L. Honig and D.G. Messerschmitt, "Adaptive Filter Structures, Algorithms and Applications". Kluwer Academic, Boston, 1984.

[10] D. Manolakis, F. Ling and J. Proakis, "Efficient Time-Recursive Least Squares Algorithms for Finite Memory Adaptive Filtering", IEEE Transactions, Vol. CAS-34, No. 4, April 1987, pp. 400-407.

[11] B. Friedlander, "Lattice Filters for Adaptive Processing", Proc. IEEE, 70, August 1982, pp. 829-867.

[12] J.M. Turner, "Recursive Least Squares Estimation and Lattice Filters", Chapter 5 in "Adaptive Filters", Prentice-Hall, Englewood Cliffs, New Jersey, 1985.

[13] F. Ling, D. Manolakis and J. Proakis, "Numerically Robust LS Lattice-Ladder Algorithm with Direct Updating of the Coefficients", IEEE Trans. ASSP-34, August 1986, pp. 837-845.

[14] S. Ljung and L. Ljung, "Error Propagation Properties of Recursive Least Squares Adaptation Algorithms", Automatica 21, 1985, pp. 157-167.

[15] S. Haykin, "Adaptive Filter Theory", Prentice-Hall, Englewood Cliffs, New Jersey, 1985.

[16] J.M. Cioffi, "High Speed Systolic implementation of Fast Q.R. Adaptive Filters", Proceedings of IEEE ICASSP-88, New York, April 1988, pp. 1584-1587.

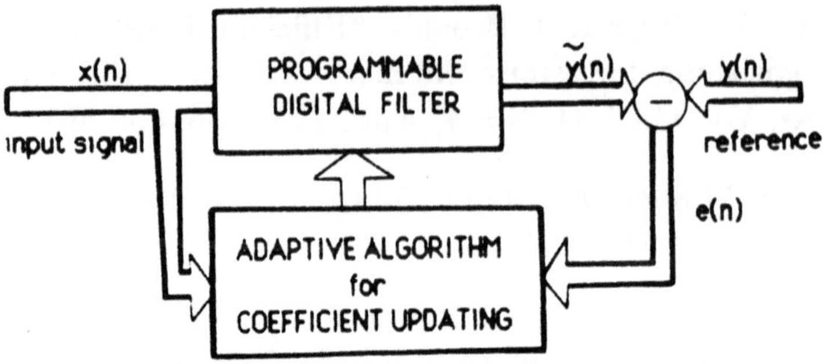

Figure 1 Principle of an adaptive filter.

AVAILABLE AT TIME n.

COEFFICIENTS OF ADAPTIVE FILTER	$H(n)$
" FORWARD PREDICTOR	$A(n)$
" BACKWARD PREDICTOR	$B(n)$
DATA VECTOR	$X(n)$
ADAPTATION GAIN	$G(n)$
FORWARD PREDICTION ERROR ENERGY	$E_a(n)$

NEW DATA AT TIME n:

Input Signal : $x(n+1)$; Reference $y(n+1)$

ADAPTATION GAIN UPDATING :

$$e_a(n+1) = x(n+1) - A^t(n)\,X(n)$$
$$A(n+1) = A(n) + G(n)\,e_a(n+1)$$
$$\varepsilon_a(n+1) = x(n+1) - A^t(n+1)\,X(n)$$
$$E_a(n+1) = W\,E_a(n) + e_a(n+1)\,\varepsilon_a(n+1)$$
$$G_1(n+1) = \begin{bmatrix} 0 \\ G(n) \end{bmatrix} + \frac{\varepsilon_a(n+1)}{E_a(n+1)} \begin{bmatrix} 1 \\ -A(n+1) \end{bmatrix} = \begin{bmatrix} M(n+1) \\ m(n+1) \end{bmatrix}$$
$$e_b(n+1) = x(n+1-N) - B^t(n)\,X(n+1)$$
$$G(n+1) = \frac{1}{1-m(n+1)\,e_b(n+1)}\,\big(M(n+1) + m(n+1)\,B(n)\big)$$
$$B(n+1) = B(n) + G(n+1)\,e_b(n+1)$$

ADAPTIVE FILTER :

$$e(n+1) = y(n+1) - H^t(n)\,X(n+1)$$
$$H(n+1) = H(n) + G(n+1)\,e(n+1)$$

Figure 2 FLS algorithm for the transversal structure

AVAILABLE AT TIME n:

COEFFICIENTS OF ADAPTIVE FILTER : $H(n)$
COEFF. OF FORWARD PREDICTOR : $A(n)$
COEFF. OF BACKWARD PREDICTOR : $B(n)$
DATA VECTOR : $X(n)$
WINDOW LENGTH : N_0
DELAYED DATA VECTOR : $X(n-N_0)$
ADAPTATION GAIN : $G(n)$
ADAPTATION GAIN FOR DELAYED DATA : $G_0(n)$
FORWARD PREDICTION ERROR ENERGY : $E_a(n)$

NEW DATA AT TIME n:

Input Signal : $x(n+1)$; Reference : $y(n+1)$

ADAPTATION GAIN UPDATING

$$e_a(n+1) = x(n+1) - A^t(n)\,X(n)$$
$$e_{0a}(n+1) = x(n+1-N_0) - A^t(n)\,X(n-N_0)$$
$$A(n+1) = A(n) + G(n)\,e_a(n+1) - G_0(n)\,e_{0a}(n+1)$$
$$\varepsilon_a(n+1) = x(n+1) - A^t(n+1)\,X(n)$$
$$\varepsilon_{0a}(n+1) = x(n+1-N_0) - A^t(n+1)\,X(n-N_0)$$
$$E_a(n+1) = E_a(n) + e_a(n+1)\,\varepsilon_a(n+1) - e_{0a}(n+1)\,\varepsilon_{0a}(n+1)$$

$$G_1(n+1) = \begin{bmatrix} 0 \\ G(n) \end{bmatrix} + \frac{e_a(n+1)}{E_a(n+1)} \begin{bmatrix} 1 \\ -A(n+1) \end{bmatrix} = \begin{bmatrix} M(n+1) \\ m(n+1) \end{bmatrix}$$

$$G_{01}(n+1) = \begin{bmatrix} 0 \\ G_0(n) \end{bmatrix} + \frac{\varepsilon_{0a}(n+1)}{E_a(n+1)} \begin{bmatrix} 1 \\ -A(n+1) \end{bmatrix} = \begin{bmatrix} M_0(n+1) \\ m_0(n+1) \end{bmatrix}$$

$$e_b(n+1) = x(n+1-N) - B^t(n)\,X(n+1)$$
$$e_{0b}(n+1) = x(n+1-N-N_0) - B^t(n)\,X(n+1-N_0)$$
$$k = m(n+1) \,/\, [\,1 + m_0(n+1)\,e_{0b}(n+1)\,]$$
$$k_0 = m_0(n+1) \,/\, [\,1 - m(n+1)\,e_b(n+1)\,]$$

$$G(n+1) = \frac{1}{1 - k\,e_b(n+1)} [\,M(n+1) + k\,B(n) - k\,e_{0b}(n+1)\,M_0(n+1)\,]$$

$$G_0(n+1) = \frac{1}{1 + k_0 e_{0b}(n+1)} [\,M_0(n+1) + k_0\,B(n) + k_0 e_b(n+1)\,M(n+1)\,]$$

$$B(n+1) = B(n) + G(n+1)\,e_b(n+1) - G_0(n+1)\,e_{0b}(n+1)$$

ADAPTIVE FILTER

$$e(n+1) = y(n+1) - H^t(n)\,X(n+1)$$
$$e_0(n+1) = y(n+1-N_0) - H^t(n)\,X(n+1-N_0)$$
$$H(n+1) = H(n) + G(n+1)\,e(n+1) - G_0(n+1)\,e_0(n+1)$$

Figure 3 FLS algorithm for adaptive filtering with a sliding window

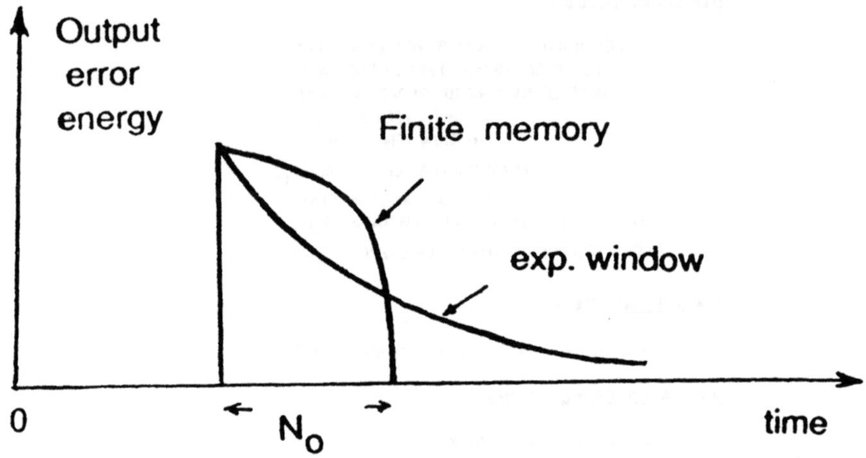

Figure 4 Responses of exponential and sliding window to step change in input signal

Algorithm	Comput. Complexity			Round-off Error Accumulation	Operation Accuracy	Signal Parameters
	Mult	Div	SqRt			
Transversal	10N	2	0	yes	moderate	very useful
Lattice/ Ladder	16N	3N	0	no	moderate - low	can be useful
Rotation Method	26N	2N+1	2N+1	no	low? **	may be useful **

** Further work is needed

Figure 7 Main features of the three algorithms

AVAILABLE AT TIME n:

REFLECTION COEFFICIENTS : $K_a(n)$, $K_b(n)$

FILTER COEFFICIENTS : $K_f(n)$

BACKWARD PREDICTION ERRORS : $[e_b(n)]$

PREDICTION ERROR ENERGIES : $[E_a(n)]$, $[E_b(n)]$

WEIGHTING FACTOR : W

NEW DATA AT TIME n:

Input signal : $x(n+1)$; Reference : $y(n+1)$

INITIALIZATIONS:

$$e_{a0}(n+1)=e_{b0}(n+1)=x(n+1) \; ; \; e_0(n+1)=y(n+1)$$

$$\varphi_0(n+1)=1 \; ; \; E_{a0}(n+1)=E_{b0}(n+1)=WE_{a0}(n)+x^2(n+1)$$

$0 \leq i \leq N-1$

PREDICTION SECTION:

$$e_{a(i+1)}(n+1) = e_{ai}(n+1) - k_{b(i+1)}(n)\, e_{bi}(n)$$

$$e_{b(i+1)}(n+1) = e_{bi}(n) - k_{a(i+1)}(n)\, e_{ai}(n+1)$$

$$k_{a(i+1)}(n+1) = k_{a(i+1)}(n) + e_{ai}(n+1)\varphi_i(n)\, e_{b(i+1)}(n+1)/E_{ai}(n+1)$$

$$k_{b(i+1)}(n+1) = k_{b(i+1)}(n) + e_{a(i+1)}(n+1)\, e_{bi}(n)\varphi_i(n)/E_{bi}(n)$$

$$E_{a(i+1)}(n+1) = W\, E_{a(i+1)}(n) + e_{a(i+1)}^2(n+1)\varphi_{i+1}(n)$$

$$\varphi_{i+1}(n+1) = \varphi_i(n+1) - \varphi_i^2(n+1)\, e_{bi}^2(n+1)/E_{bi}(n+1)$$

$$E_{b(i+1)}(n+1) = W\, E_{b(i+1)}(n) + e_{b(i+1)}^2(n+1)\varphi_{i+1}(n+1)$$

FILTER SECTION:

$$e_{i+1}(n+1) = e_i(n+1) - k_{fi}(n)\, e_{bi}(n+1)$$

$$k_{fi}(n+1) = k_{fi}(n) + e_{bi}(n+1)\, e_{i+1}(n+1)\varphi_i(n+1)/E_{bi}(n+1)$$

Figure 5 FLS algorithm for the lattice-ladder structure

AVAILABLE AT TIME n :

Input Data Rotation $(N+1)\times(N+1)$ Matrix $Q_a(n)$ $[\cos\theta(i) ; \sin\theta(i) ; i=1,N]$

Backward Data Rotation Matrix $Q_b(n)$ $[\cos\theta_b(i) , \sin\theta_b(i) ; i=1,N]$

Transformed Input Data N-element Vector : $XQ(n)$

Transformed Reference N-element Vector : $YQ(n)$

Forward Prediction Error Energy : $E_a(n)$

Weighting Factor : W

NEW DATA AT TIME n :

Input Signal : $x(n+1)$; Reference : $y(n+1)$

PREDICTION SECTION

$$\begin{bmatrix} \epsilon_{aq}(n+1) \\ XQ(n+1) \end{bmatrix} = Q_a(n) \begin{bmatrix} x(n+1) \\ W^{1/2}\, XQ(n) \end{bmatrix}$$

$$E_a(n+1) = W\, E_a(n) + \epsilon_{aq}^2(n+1)$$

$$\cos\theta_\alpha = [W\, E_a(n) / E_a(n+1)]^{1/2}$$

$$\sin\theta_\alpha = \epsilon_{aq}(n+1) / [E_a(n+1)]^{1/2}$$

$$Q_\alpha = \begin{bmatrix} \cos\theta_\alpha & 0 & -\sin\theta_\alpha \\ 0 & 1 & 0 \\ \sin\theta_\alpha & 0 & \cos\theta_\alpha \end{bmatrix}$$

$$B_1(n+1) = Q_b^t(n) \begin{bmatrix} 1 \\ 0 \\ \vdots \\ 0 \end{bmatrix}$$

$$\begin{bmatrix} U_0 \\ B_2(n+1) \end{bmatrix} = Q_\alpha \begin{bmatrix} Q_a(n) & 0 \\ 0 & 1 \end{bmatrix} \begin{bmatrix} 0 \\ B_1(n+1) \end{bmatrix}$$

$$Q_b^t(n+1) \begin{bmatrix} 1 \\ 0 \\ \vdots \\ 0 \end{bmatrix} = B_2(n+1)$$

$$\begin{bmatrix} \gamma_1(n+1) \\ G_{N+1}(n+1) \end{bmatrix} = Q_\alpha \begin{bmatrix} Q_a(n) & 0 \\ 0 & 1 \end{bmatrix} \begin{bmatrix} 1 \\ 0 \end{bmatrix}$$

$$\begin{bmatrix} V_0 \\ G_N(n+1) \end{bmatrix} = Q_b(n+1)\; G_{N+1}(n+1)$$

$$Q_a(n+1) \begin{bmatrix} 1 \\ 0 \\ \vdots \\ 0 \end{bmatrix} = \begin{bmatrix} \gamma(n+1) \\ G_N(n+1) \end{bmatrix}$$

$$\epsilon_a(n+1) = \epsilon_{aq}(n+1)\, \gamma(n+1)$$

FILTER SECTION :

$$\begin{bmatrix} \epsilon_q(n+1) \\ YQ(n+1) \end{bmatrix} = Q_a(n+1) \begin{bmatrix} y(n+1) \\ YQ(n) \end{bmatrix}$$

$$\epsilon(n+1) = \epsilon_q(n+1)\, \gamma(n+1)$$

Figure 6 FLS-QR algorithm

Adaptive Control Algorithms

Robert R. Bitmead

Department of Systems Engineering
Australian National University
Canberra, ACT 2601, Australia

Abstract

The thinking behind adaptive control is to combine the processes of system identification, to gain a system model for an unknown plant, and of feedback control design for a known plant model. This yields a design-while-identify control strategy otherwise known as self-tuning or self-adjusting. To be naive one would suppose that a good adaptive control procedure is able to be manufactured by cobbling together a good parameter identifier and a good control law, and all could be expected to perform. This is partially true in that a successful adaptive control law must incorporate both of these features, but this may not be sufficient for good performance. A third feature needs to be taken into account which addresses the marrying of the identification objective with the control objective through the design or selection of adaptation speed, model structure, signal conditioning, control design rule, reference signal properties, speed of plant variation etc.

The aim of this work is to present and develop a view of the salient aspects of adaptive control algorithms with the intent of encapsulating their dominant features and explaining the reasons for their specific numerically desirable properties. We shall concentrate upon the development of a coherent standpoint from which the practical aspects of adaptive controller design emerge.

1 Introduction

What is adaptive control and why?

Without becoming too involved in semantics, adaptive control is concerned with the coupling of system parameter identification methods with simultaneous controller design to produce a robust feedback controller capable of dealing with both initial plant uncertainties and slow plant time-variation. This description is best supported by considering some representative examples of applications of adaptive control, which will help to fix ideas.

Example 1 In the processing industries, such as paper manufacture, ore crushing, chemical refineries etc, it is frequently the goal to attempt to maintain product output quality at a desired level in the face of changing material properties of the inputs and/or changing operating points of the equipment. In such an environment, one often strives to regulate variables such as moisture content or particulate size in such a fashion that the product satisfies tolerances but does not oversatisfy them. Thus an adaptive control law may be useful to account for material variations better than could be achieved by, say, robust or insensitive control design. In very many such applications, the adaptive, or "self-tuning", controller is chosen to circumvent the explicit need to tune-up a robust generic controller.

Example 2 Ship steering systems must be designed to deal with the dramatic alterations arising through differing loads and differing sea and wind states. Because the dynamic response of a large ship to command signals alters significantly with the above changes, accurate steering demands the adjustment of the control procedure. This could be achieved off-line by measuring these variables and then selecting the appropriate control law, so called "gain scheduling", or by adaptive means where the control law self-adjusts in response to its current performance, thereby short circuiting much of the need for exceedingly complex control design.

Example 3 In future aircraft constructed of light and somewhat flexible materials, the fluttering of the wings is a serious problem, even in level flight, which can lead to complete material failure when the flutter becomes unstable. Consequently, active manipulation of control surfaces is necessary to suppress this flutter. The control problem arises in that the flutter commences as a stable oscillation of the wing which soon becomes unstable. The frequency of this oscillation is, however, very dependent upon environmental and flying conditions, fuel level, load etc and a robust controller capable of overcoming this uncertainty is essentially impossible to design, at least with a reasonable complexity. Consequently, adaptation to the frequency of oscillation is needed to stabilize these dynamics.

Numerous other examples of adaptive control applications exist in mechanical systems, such as robots and disk drives, chemical systems, such as bio-reactors, and electrical systems. There are several underlying common features, however. They are very frequently applied with linear plant models coupled with linear control laws and one operates under the assumption that the major form of plant uncertainty arises from parametric uncertainty in the model structure. The adaptive controller then is constructed based upon a parameter estimator connected to a control design procedure and a "design while identify" strategy is adopted.

While the underlying parametric model is assumed to be moderately accurate, it is never exact and, further, the parameter yielding the best plant model is usually (slowly) time-varying. Equally, the actual plant must operate in an environment which is both noisy and signal-dependent, with the adaptation mechanism itself injecting a nonlinear effect into the closed loop in spite of the possible linearity assumptions of its derivation.

The consequences of these modelling and operating environment uncertainties is that, in general, adaptive controllers need to incorporate the features of both robust identification and of robust control. Further, a level of caution and of scepticism needs to be asserted by both these components. To counterbalance this very tentative nature of adaptive control, it is frequently the case that a considerable amount of data is available online and that data efficiency is typically not a problem. In computational terms, this leads to the need to develop algorithms for adaptive control which implement these robustness features in their handling of data but which need not produce particularly accurate parameter estimates because of the implied scepticism with respect to the data. Similarly, the specification of the control objective does not benefit greatly from the inclusion of a large number of design variables because it is not clear how they should be chosen to best advantage. Thus adaptive designs typically do not possess many design parameters themselves.

The remainder of this article will proceed as follows. Because adaptive controllers need to contain both robust identifiers and robust controllers, we shall begin by briefly considering the basic algorithm types focussing on recursive linear regression style estimators and depending upon how the parametrization is linked to the control law design. Following this, in Section 3, we shall investigate how the standard Recursive Least Squares parameter identification algorithm is modified to account for the various effects and conditions attendant to adaptive control. Section 4 is devoted to the presentation of control reference signal conditions, so called persistence of excitation, which are sometimes required of the control law in order that the identification be able to proceed well. Section 5 treats the salient features of the choice of control law for the design stage of the adaptive controller, and Section 6 concludes with a brief description of the interplay between control and identification. Throughout, our goal will be to provide an overview of the aspects of adaptive control versus adaptive identification and/or nonadaptive control. More details and practical discussions can be found in the recent book of Åström and Wittenmark [1] and the book of Goodwin and Sin [2].

2 Basic Algorithm Types

Operating under the equality

Adaptive Control = Parametrized Control Law + On-line Parameter Estimator,

we may consider the archetypal adaptive control algorithm as consisting of a standard regression based estimator coupled to a controller design. These estimators have the familiar algorithmic form,

$$\hat{\theta}_{k+1} = \hat{\theta}_k + \alpha f(\phi_k) g(e_k), \tag{1}$$

where $\hat{\theta}_k$, α, ϕ_k and e_k represent the parameter estimate, the algorithm step size, the regressor vector and the error signal respectively, and $f(.)$ and $g(.)$ are functions to be specified.

A naïve approach to adaptive control design would be to propose that any convergent estimator plus any stabilizing control law should be capable of providing good adaptive control. A more sensible approach is to be more circumspect and to adjust, or fiddle, with both parts systematically so that the interplay between identification and control is robust. Before presenting some techniques for attempting this, however, we shall describe briefly the basic algorithm types subdivided according to the nature of the identified parameter and the associated error signal and regressor in (1).

2.1 Indirect Adaptive Control

In this scheme the parameter identifier focuses upon deriving an explicit model of the plant system and the controller is designed from this plant model. Here $\hat{\theta}_k$ is the estimate of the parameter entering in a parametric model of the plant, ϕ_k then is the standard regressor of lagged plant inputs and outputs, and e_k is the prediction error between plant model and measured plant output.

Thus a typical such configuration would take the plant model structure to be in ARMAX form,

$$\begin{aligned} y_k &= -a_1 y_{k-1} - a_2 y_{k-2} - \ldots - a_n y_{k-n} \\ &\qquad + b_1 u_{k-1} + b_2 u_{k-2} + \ldots + b_n u_{k-n} + \epsilon_k, \end{aligned}$$

and the model predicted output is generated by

$$\begin{aligned} \hat{y}_k &= -\hat{a}_{1,k} y_{k-1} - \ldots - \hat{a}_{n,k} y_{k-n} \\ &\qquad + \hat{b}_{1,k} u_{k-1} + \hat{b}_{2,k} u_{k-2} + \ldots + \hat{b}_{n,k} u_{k-n} \\ &= \begin{pmatrix} -y_{k-1} & -y_{k-2} & \cdots & -y_{k-n} & u_{k-1} & \cdots & u_{k-n} \end{pmatrix} \begin{pmatrix} \hat{a}_{1,k} \\ \hat{a}_{2,k} \\ \vdots \\ \hat{a}_{n,k} \\ \hat{b}_{1,k} \\ \vdots \\ \hat{b}_{n,k} \end{pmatrix} \\ &\triangleq \phi_k^T \hat{\theta}_k. \end{aligned}$$

The estimation error of (1) is then the (output) prediction error between this model and the actual plant,

$$e_k = y_k - \phi_k^T \hat{\theta}_k.$$

The parameter update using, say, Recursive Least Squares (RLS) may then be performed as,

$$\hat{\theta}_{k+1} = \hat{\theta}_k + \frac{P_{k-1}\phi_k}{1+\phi_k^T P_{k-1}\phi_k}\left\{y_k - \phi_k^T\hat{\theta}_k\right\} \tag{2}$$

$$P_k = P_{k-1} - \frac{P_{k-1}\phi_k\phi_k^T P_{k-1}}{1+\phi_k^T P_{k-1}\phi_k}. \tag{3}$$

Given a parametrized plant model, the control design may proceed by any available procedure based on the assumption that the estimated $\hat{\theta}_k$ is the actual plant parameter. This assumption, known as *Certainty Equivalence Adaptive Control*, demonstrates an implied confidence in the adaptive estimation procedure. Other approaches, such as called Cautious Adaptive Control, attempt the incorporate a further measure of scepticism into the controller by slowing the adoption of new data into the parameter calculation but this is roughly equivalent to other modifications to the RLS algorithm to be described shortly. Once a parameter is given, the controller design can proceed according to any stabilizing control law procedure such as LQG, Pole-positioning, H^∞ etc.

In particular, continuing the above example, pole-positioning can proceed as follows from the parameter estimate vector $\hat{\theta}_k$. Define the polynomials,

$$\hat{A}_k(z) \triangleq 1 + \hat{a}_{1,k}z + \ldots + \hat{a}_{n,k}z^n$$
$$\hat{B}_k(z) \triangleq \hat{b}_{1,k}z + \ldots + \hat{b}_{n,k}z^n,$$

and select a given stable polynomial $P(z)$ of degree $2n$. Then solve the Bezout equation,

$$\hat{A}_k(z)C_k(z) + \hat{B}_k(z)D_k(z) = P(z), \tag{4}$$

for polynomials $C_k(z)$ and $D_k(z)$, with decompositions conformant with $\hat{A}_k$ and $\hat{B}_k$ above, to yield the feedback controller as,

$$u_k = -c_{1,k}u_{k-1} - \ldots - c_{n,k}u_{k-n} + d_{0,k}y_k + \ldots + d_{n,k}y_{k-n}. \tag{5}$$

The idea behind such a pole-positioning scheme is that, if a plant described by transfer function $A^{-1}(z)B(z)$ is connected in feedback with a controller with transfer function $C^{-1}(z)D(z)$ then the closed loop input output transfer function becomes $(A(z)C(z)+B(z)D(z))^{-1}B(z)C(z)$. Thus, if the parametric model is exact, $P(z)$ becomes the closed loop characteristic equation and the closed loop poles are shifted to the zeros of $P(z)$.

The numerical issues arising directly from such an adaptive control method need to address such features as the speed and stability of the Least Squares iteration (2)-(3) through the use of fast implementations, square root algorithms etc, and numerical stability and speed for the solution of the Bezout equation (4). Since this latter equation is often solved by the inversion of a Sylvester matrix, difficulties can arise when $\hat{A}_k(z)$ and $\hat{B}_k(z)$ have closely located zeros which need not be zeros of $P(z)$, and measures (to be discussed later) may need to be taken to avoid such $\hat{\theta}$ parameter values. The feature of adaptive control different from straight off-line control design is that, since calculations of the control signal take place at each parameter update, speed of calculation is important. The accuracy of computation, however, often need not be so crucial because, by assumption, one expects the parameter estimates to change value with time and so one is not particularly confident in their particular current values. Thus insensitivity of the control law is a desired feature and so explicit computation need not be very exact. This aspect of adaptive control vis-à-vis off-line design is a significant distinguishing property of the former.

2.2 Direct Adaptive Control

While indirect adaptive control is based on the generation of an explicit plant model and subsequent computation of a control law, which may effectively be of any standard type, some choices of control law (notably pole-positioning) admit the explicit estimation of the controller parameters without the need to generate an intermediate plant model. These schemes are known as Direct Adaptive Control and hinge on the ability to construct a useful error signal e_k for such a parameter update (1). We shall describe the approach with a pole-positioning example.

As in (4), for the plant with transfer function $A^{-1}(z)B(z)$, write

$$A(z)C(z) + B(z)D(z) = P(z) \tag{6}$$
$$A(z)Q(z) + B(z)R(z) = -z^{2n}, \tag{7}$$

for polynomials $C(z)$, $D(z)$, $Q(z)$ qnd $R(z)$ of degree n. Such polynomials always exist, subject to coprimeness. Now define the controller parameter vector

$$\hat{\theta}_k = (\,\hat{c}_{1,k} \quad \ldots \quad \hat{c}_{n,k} \quad \hat{d}_{0,k} \quad \ldots \quad \hat{d}_{n,k} \quad \hat{q}_{1,k} \quad \ldots \quad \hat{q}_{n,k} \quad \hat{r}_{1,k} \quad \ldots \quad \hat{r}_{n,k}\,)^T,$$

and the associated regressor vector

$$\begin{aligned}\phi_k \;=\; (\,& u_{k-2n} \quad \ldots \quad u_{k-3n+1} \quad y_{k-2n} \quad \ldots \quad y_{k-3n+1} \\ & P(z)u_k \quad \ldots \quad P(z)u_{k-n+1} \quad P(z)y_k \quad \ldots \quad P(z)y_{k-n+1}\,)^T,\end{aligned}$$

to define the control error signal

$$e_k = u_{k-2n+1} - \phi_k^T \hat{\theta}_k$$

and the control itself through

$$u_k = -\hat{c}_{1,k}u_{k-1} - \ldots - \hat{c}_{n,k}u_{k-n} + \hat{d}_{0,k}y_k + \hat{d}_{n,k}y_{k-n}.$$

We note that, with this problem formulation, the controller parameter is explicitly identified and so the computation of the controller from the estimated plant is obviated. Hence, the attendant computational issues connected with the solution of (4) are avoided although other issues of parameter identifiability arise. Several authors, notably Kreisselmeier and Smith [3] and Giri *et al* [4], have proposed the joint implementation of an adaptive controller with both plant parameter estimation followed by (indirect) controller calculation and direct control computation, with the difference between the two computed controller parameters $C(z)$ and $D(z)$ by the respective methods giving a guide to the confidence in the accuracy of the current estimate. This confidence figure is then used to alter the adaptation environment to enhance the parameter identification.

2.3 Other Variations

In addition to the separation into indirect and direct adaptive control, which is rooted in the meaning of the identified parameter, other divisions of algorithm type are made based upon the specification of the control objective and hence of the error signal used in a direct adaptive scheme.

Specifically, Model Reference Adaptive Control, *MRAC*, uses as a control objective the following of a prespecified model output by the plant output. Hence, such adaptive schemes may be constructed in either the indirect form, using plant identification followed by model matching controller design, or, more usually, in direct form using the model following error as the algorithm error. In this way, the adaptation should take place to minimize the closed loop deviation in behaviour from the desired model.

Another familiar form for adaptive control algorithms is with a Minimum Output Variance control objective. In industrial control problems this frequently signifies a desire to have more uniform product and is connected with a closed loop which removes all but the completely unpredictable component of the plant output. Here the design differs from MRAC in that no ideal response characteristic is specified but the predictable part of the plant's deviation from its nominal value is removed. Because the plant identification also is based upon prediction error minimization, the modelling and performance errors are the same. There are strong algorithmic similarities between minimum variance adaptive control, adaptive data encoding schemes and adaptive channel equalization methods all of which attempt to produce a statistically white system output sequence.

Extensions of this idea of minimizing the plant output variance, $\mathrm{E}(\hat{y}^2_{k+1|k})$, lead naturally to Generalized Minimum Variance (GMV) adaptive control [5], which minimizes

$$\mathrm{E}\left(\hat{y}^2_{k+1|k} + \lambda u_k^2\right),$$

and to Generalized Predictive Control (GPC) [6], which minimizes

$$\mathrm{E}\left(\sum_{i=N_1}^{N_2} \hat{y}^2_{k+i|k} + \lambda \sum_{i=1}^{N_u} u^2_{k+i-1}\right),$$

on which we shall comment more later, since GPC currently is finding considerable success in industrial applications of adaptive control.

3 Identification in Adaptive Control

The common structure of all the above adaptive control schemes is that they involve a parameter estimator incorporating a regression vector of plant signals and an error signal. Further, they include, explicitly or implicitly, a control law selection embodying a closed loop objective. Here we shall investigate what are the types of difficulties which can arise with the identifier operating in closed loop and what is usually done to modify the identification algorithm to alleviate this. That is, "What can go wrong?" and "What 'fixes' can be made?".

3.1 The Need to Track Time-variations

Recursive Least Squares has the algorithmic form of (2) and (3)

$$\begin{aligned}\hat{\theta}_{k+1} &= \hat{\theta}_k + \frac{P_{k-1}\phi_k}{1+\phi_k^T P_{k-1}\phi_k}\left\{y_k - \phi_k^T\hat{\theta}_k\right\}\\ P_k &= P_{k-1} - \frac{P_{k-1}\phi_k\phi_k^T P_{k-1}}{1+\phi_k^T P_{k-1}\phi_k}.\end{aligned}$$

With a regressor sequence, $\{\phi_k\}$, which does not tend rapidly to zero and a finite nonzero initial P_0, this algorithm produces estimates, $\hat{\theta}_k$, which asymptotically approach the true off-line Least Squares fit. Further, rewriting (3) one sees that

$$P_k = \left\{P_0^{-1} + \sum_{i=1}^{k}\phi_i\phi_i^T\right\}^{-1}.$$

Thus, for nonvanishing $\{\phi_k\}$, one has $P_k \to 0$ as $k \to \infty$ and asymptotically the updating of $\hat{\theta}_k$ ceases because P_k dictates the step size of the algorithm. While this might be a desirable feature for statistically consistent estimation of a fixed parameter in a time-invariant system with noise, it is unacceptable in the adaptive context because of the paradigm of needing to track slow time variations of the system. How can this objective of keeping the identifier 'awake' be achieved while still preserving the structural identity and self scaling properties of the RLS method?

RLS with Forgetting Factor

This most familiar modification to the RLS algorithm may be derived by considering the exponential weighting of data in the least squares criterion. Thus an exponential discount factor, $0 < \lambda < 1$, is chosen to diminish the value attributed to past prediction errors relative to recent ones. The modified RLS algorithm with a forgetting factor (RLSFF) then takes the form of (2) plus covariance update,

$$P_k = \frac{1}{\lambda}\left\{P_{k-1} - \frac{P_{k-1}\phi_k\phi_k^T P_{k-1}}{\lambda + \phi_k^T P_{k-1}\phi_k}\right\} \tag{8}$$

$$= \left\{\lambda^k P_0^{-1} + \sum_{i=1}^{k} \lambda^{k-i}\phi_i\phi_i^T\right\}^{-1}. \tag{9}$$

Covariance Addition

A similar alternative is to use Covariance Addition, consisting of the usual P-update followed by a modification as follows,

$$\bar{P}_k = P_{k-1} - \frac{P_{k-1}\phi_k\phi_k^T P_{k-1}}{1 + \phi_k^T P_{k-1}\phi_k} \tag{10}$$

$$P_k = \bar{P}_k + Q. \tag{11}$$

From the above expressions (9) and (11) for P_k, we see that, P_k never tends to zero with finite $\{\phi_k\}$, and thus the algorithm remains alert to potential parameter movements (at the expense of consistency in the time-invariant case). However, another issue arises in that if $\phi_k \to 0$ as $k \to \infty$, which is often the goal of the adaptive control law, then $P_k \to \infty$ which indicates that the parameter estimator becomes very sensitive and wild. How can this be alleviated without losing the tracking property?

Covariance Reset

Here the normal RLS algorithm (3) is run until one reaches $P_k < \tilde{P}$ or $P_k > P'$ for some prespecified matrices $\tilde{P}$ and P', at which time the covariance matrix P_k is reset to P_0 and the algorithm recommenced.

Constant Trace Algorithm

Again the normal update is performed, but now is followed by a rescaling

$$\bar{P}_k = P_{k-1} - \frac{P_{k-1}\phi_k\phi_k^T P_{k-1}}{1 + \phi_k^T P_{k-1}\phi_k}$$

$$P_k = \frac{\operatorname{tr}(P_{k-1})}{\operatorname{tr}(\bar{P}_k)}\bar{P}_k.$$

Numerous other variants also exist; Directional Dependent Forgetting, Exponential Forgetting and Resetting, Fixed P, Condition Number Monitoring, etc. etc.

Åström has described the area of adaptive identification algorithm design as "a fiddler's paradise", which is an accurate statement given the plethora of modifications available and the lack of underlying guidelines. Nevertheless, practitioners persist with the RLS algorithm and its variants rather than gradient methods because they value the automatic scaling properties of RLS, which play a more important rôle in adaptive control than in, say adaptive signal processing where signal powers often remain relatively constant. Similarly, system orders and time constants and the economic benefit of plant control usually allow significant computation to be performed in adaptive control problems, so that computation of condition numbers etc can be contemplated. This feature should be compared to adaptive communications systems where orders can be very high, time scales very short and computational costs a significant factor in component pricing. These timescale and dynamical range properties of adaptive control distinguish it from other adaptive systems.

3.2 Integrity of Data

It is a fundamental feature of adaptive control that frequently the control objective of regulation is contradictory to the identification objective of system parameter estimation. For a nominally linear system with input-output model

$$A(z)y_k = B(z)u_k + \eta_k$$

the signal η_k embodies both measurement noise and unmodelled effects. If the closed loop control is effective then both $\{y_k\}$ and $\{u_k\}$ may be kept small and then $\{\eta_k\}$ really matters in deriving the input-output description of the plant. That is, just when good control is being achieved, the data may fail to have much integrity with respect to its providing information about a good model of the plant for control purposes. This is one of the mechanisms used by Rohr's *et al.* [7] to demonstrate instabilities of non-modified adaptive controllers. Solutions to this issue usually take the form of ignoring the error signal when it is suitably small.

Deadzones

The simplest way of rejecting small error signals is to threshold them before allowing an update of the parameter estimator. In this way small errors are associated with undue noise corruption and hence are neglected. The usual fashion for achieving this is to replace the error signal, e_k, in (1) by $g(e_k, d)$ defined as follows,

$$g(e,d) = \begin{cases} 0 & \text{if} |e| < d \\ e - d & e > d \\ e + d & e < -d \end{cases} \tag{12}$$

This simply discards small errors and reduces the notional value of other errors. If such steps are not taken, then, when the model fit is accurate, the identifier could operate simply as an integrator of measurement noise values, leading to a random walk of the parameters or a constant drift, which in turn can produce stability problems, [7]. Naturally the deadzone parameter, d, is a design variable.

Relative Deadzones

In adaptive control, compared to open loop adaptation, the closed loop signals are functions of the estimated parameters and large input signals can be caused by either strong deviation of the estimated plant from the best model or by large closed loop gains amplifying the measurement noise signal. To account for this effect, where the error signal is possibly large but still is dominated by measurement noise rather than information rich system parameter deviations, a relative deadzone may be introduced to deweight the error when it is small relative to other closed loop signals.

A normalization signal is defined as follows,

$$m_k = (1 - \rho)m_{k-1} + \max(\rho, |\phi_k|), \tag{13}$$

where $\rho > 0$ is another design variable and m_k is a signal whose magnitude reflects recent closed loop signal values. e_k in (1) is then replaced by $g(\frac{e_k}{m_k}, d)$. Such a procedure is used by [8] to enhance robustness.

Normalization

Taking the relative deadzone philosophy further, where errors are assessed for their importance according to recent signal values, the idea of normalization (due chiefly to Praly, [9]) is to replace the regressor and the plant output, ϕ_k and y_k, by their normalized values,

$$\bar{\phi}_k = \frac{\phi_k}{m_k},$$

$$\bar{y}_k = \frac{y_k}{m_k},$$

with m_k from (13) above, in the entire RLS iteration (2) and (3). The reason for this is that, in adaptive control, internal signals can temporarily become very large as parameters are identified and this has the effect of swamping the local data weighting too greatly with these signals, thereby affording parameter estimation only in directions associated with the mode of instability. Normalization has the property of protecting the RLS algorithm from short term large signal values by allowing concentration on the geometry of the error/regressor signals more or less independently from the local magnitude.

3.3 *A Priori* Data

In many circumstances for adaptive control, a *priori* data is available concerning possible localities of the space of parameters, $\Re^{2n}$, in which the best plant fit might lie. There are several ways in which this can be included into the RLS scheme and, indeed, there are many algorithm variations where convergence can only be guaranteed provided stability conditions are ensured on the estimated closed loop system. Further, for indirect controller computations, such as pole positioning via (4), a controller solution may only exist provided pole-zero cancellations in the estimated plant are avoided.

In such situations where the estimated plant parameters must, by a *priori* knowledge or by assumption, belong to a particular set, K say, one may alter the RLS algorithm either to enforce the residence of $\hat{\theta}_k$ in K by **Projection**,

$$\begin{aligned}\bar{\theta}_{k+1} &= \hat{\theta}_k + \frac{P_{k-1}\phi_k}{1+\phi_k^T P_{k-1}\phi_k}\left\{y_k - \phi_k^T\hat{\theta}_k\right\} \\ \hat{\theta}_k &= \mathrm{Proj}_K(\bar{\theta}_k),\end{aligned}$$

or to encourage the membership of K by adding a **Leakage** term

$$\hat{\theta}_{k+1} = \hat{\theta}_k + \frac{P_{k-1}\phi_k}{1+\phi_k^T P_{k-1}\phi_k}\left\{y_k - \phi_k^T\hat{\theta}_k\right\} + \sigma h(\hat{\theta}_k, K) \tag{14}$$

where $h(\hat{\theta}, K)$ is a function penalizing nonmembership of the set and persuading the estimator to tend towards member solutions and $\sigma > 0$.

3.4 Summary of the Identification Algorithm

We have presented and described how the adaptive control paradigm is reflected in the development of a considerable number of RLS algorithm 'fixes'. These modifications are part of the structural coercion of the parameter estimation algorithm to 'robustify' it against time-variations and classes of extraneous signals. Not all amendments to the RLS need be included in any particular application, but certainly covariance monitoring, deadzones and projection are fairly standard.

There are certain features of the adaptive control problem which directly affect the numerical issues associated with these identification methods. Typically, the sample rate is not very high and so a modest amount of computing is possible at each time instant. Further, the questionable nature of the data means that approximate solutions only need be obtained for estimation problems and similarly there is frequently little need to be efficient with the data. The major numerical features of the identification algorithms focus upon; achieving speed since the intersample time is reasonable but not great, guaranteeing stability via projection and methods for achieving this, and, issues of parametrization and scaling of the models and signals. It is for its self-scaling features that, inspite of all the modifications made to it, RLS is persisted with in applications.

4 Persistence of Excitation

The philosophical basis for adaptive control is that on-line system parameter identification is conducted in tandem with closed loop controller design. The overall control objective is typically to produce a closed loop independent of the particular plant and often to regulate the plant output to a constant value. The dichotomy of adaptive control is that this control objective may work counter to the requirements of system identification. Indeed, if a finite-time optimal regulation problem is posed for a plant with unknown parametrization described in a suitable Bayesian fashion, then the solution involves the compromise between excitation of the system via the input in order to learn the parameters well versus regulation of the system to keep the control criterion small.

In terms of the tracking of time-variations and the learning of plant parameters, the external signals applied to the input of the plant need to cause the regressor vector sequence, $\{\phi_k\}$, not to diminish to zero but continually to probe the plant and to excite all system modes. Usually, the vehicle for exerting such an exciting input is to manipulate the reference signal involved in the controller calculation. This artificial disturbance of the reference to yield good identification often will work counter to the enunciated control objective. To compound the difficulty, the level of external excitation frequently has a minimum threshold for achieving robustness.

The fundamental feature of linear regression parameter estimators to which we appeal here is that, to identify a linear parametric model of n-parameters requires that the n-dimensional regressor sequence, $\{\phi_k\}$, be persistently spanning in the sense that there exist positive constants T, α and β such that, for all times t,

$$0 < \alpha I \leq \sum_{i=t}^{t+T} \phi_i \phi_i^T \leq \beta I < \infty. \tag{15}$$

This condition is known as **Persistence of Excitation** and ensures the uniform identifiability of the parameter vector from the input-output measurements. Recall that we require the uniformity here because of the need to cope with time-variation of the parameters.

In terms of the input properties of the system, the regressor condition (15) can be assured for a linear time-invariant system of degree $n/2$ provided the input contains n complex exponentials or $n/2$ sinusoids of sufficiently distinct frequencies, see [10]. To see how this condition arises, one can contemplate the attempted identification of a linear system from input output data when the input is a constant value. That is, the input is constant, $u_k = u$, and the system output is (exponentially) also a constant, $y_k = y = G(1)u$. Thus only the dc gain $G(1)$ is discernible from signal measurements.

Because persistence of excitation is essentially necessary for accurate identification, one has two effective choices in the design of adaptive control algorithms. Either the input is rendered exciting by the injection of signals to the reference or these signals are monitored on-line and adaptation is only allowed to proceed when conditions suit. Recent novel self-exciting control laws have been proposed by Kreisselmeier and Smith [3] which run both direct adaptive control and an indirect adaptive identifier with controller calculation. The deviation between the controller parameters computed by both methods is used as a guide to the level of external system excitation needed.

5 Controller Design

5.1 Guiding Principles

Given the scepticism asserted earlier about the accuracy of parameter values and finite-dimensional linear system models in general, it is apparent that the controller component of an adaptive control scheme should be chosen to be robust to certain types of system uncertainty. In particular, because of time variations, it should be robust to parametric uncertainty but also, because of modelling questions, it needs to be robust to undermodelling of the system order and general unstructured uncertainty. Robust control system design, given reasonably complete characterization of the nominal open-loop plant, has been a subject area of considerable recent interest in much of control systems theory. Thus Loop Transfer Recovery techniques for robust Linear Quadratic Gaussian controller design, LQG/LTR methods, have recently been advocated [11] and the methodology of H^∞, or min-max frequency domain design, is a current 'hot topic' [12].

The central features of these robust control design schemes is that they provide a formalism with which to approach serious practical controller design issues which previously occupied a prohibitive proportion of design time, and that typically they are iterative and often computationally intensive. That is, they allow the control engineer to conduct a more focussed search of potential controllers through the manipulation of different variables.

In terms of adaptive control, the applicability of these robust design methodologies to provide the control design stage is tempered by the constraints imposed by the available computational power and the time window between samples and parameter updates. Further, the inadequacy of system knowledge can lead to difficulties in the specification of appropriate design objectives suitable for the use of sophisticated control laws. Thus a generically robust control law is needed. In addition, the normal iterative component of design should not rely too heavily upon the specification of too many nebulous design parameters. Recall from the introduction that many adaptive controllers are implemented to obviate the need for explicit robust loop design. The guiding principles for adaptive controller design then are,

- Generic Robustness of the control law both in control system robustness terms and numerical stability terms,
- Computational Feasibility within the time frame and given the range of concurrent identification, monitoring and safety jacketing tasks,
- Simplicity of the external design in terms of number of variables and their direct interpretation in terms of effects on achieved closed loop performance.

5.2 An Example - GPC

We present here a brief description of the Generalized Predictive Control law of Clarke *et al.* [6], which is representative of a currently promising class of practical adaptive control law design procedures. Our aim clearly is to demonstrate how the above principles have found their way into industrial algorithms.

From an estimated ARMAX plant model, involving inputs y_k outputs u_k and noise values ξ_k,

$$A(z)y_k = B(z)u_k + C(z)\xi_k, \tag{16}$$

derived from the identification component at time k, one computes the control sequence $\{u_{k+j}, \quad j = 0, \ldots, N_u - 1\}$ which minimizes the finite horizon quadratic cost criterion,

$$J(u, k, N) = \min \quad \mathrm{E}\left\{\sum_{j=1}^{N}(y_{k+j} - w_{k+j})^2 + \lambda \sum_{j=1}^{N_u} u_{k+j-1}^2\right\}, \tag{17}$$

subject to the constraint that $u_{t+j} = 0$ for $j = N_u, \ldots, N - 1$ and where $\{w_k\}$ is a sequence of reference values to be tracked.

This future control sequence may be solved by explicit minimization of the quadratic criterion by using the plant model (16) to generate predicted future plant output values, $\hat{y}_{k+j|k}$, given input and output data up to time k and assuming zero control signal from k onwards. Denoting this vector of predictions by $\mathbf{f}$, the vector of future control values by $\mathbf{u}$, the vector of future reference values by $\mathbf{w}$ and the matrix of impulse response coefficients of the plant derived from the ARMAX model by $\mathbf{G}$, we have the solution

$$\mathbf{u} = (\mathbf{G}^T\mathbf{G} + \lambda I)^{-1}\mathbf{G}^T(\mathbf{w} - \mathbf{f}). \tag{18}$$

Of the solution vector $\mathbf{u}$, consisting of future control values for times k to $k + N_u - 1$, only the first control value is applied. At the next time instant a complete new optimization is performed and another N_u future controls calculated using the updated system model, of which also only the first value will be applied. This strategy is known as receding horizon optimal control and is utilized here in order to make the adaptive control law computationally feasible in industrial situations.

Ideally, one would like to solve a full infinite horizon Linear Quadratic control problem at each time instant and then to utilize, say, loop transfer recovery design to make this solution robust. However, the compromise is struck that a receding horizon problem is solved and the design variables N, N_u and λ manipulated to achieve generic closed loop robustness relying only on a small manageable number of variables. Recent theoretical work [13] on this subject has shown how this particular choice of controller design somewhat serendipitously satisfies all the principles above.

6 Conclusion

We have presented a discussion of the salient features of the identification and control algorithmic components of an adaptive control scheme. In particular, we have tried to utilize the adaptive control paradigm of modelling whilst controlling to justify the peculiar aspects of the field. The remaining issues are really how to connect a suitably modified identifier to a generically robust control design stage to produce a complete adaptive controller.

6.1 Interplay between Identification and Control

Much of the current robust control design methodologies focusses upon the frequency domain characterization of the particular system robustness specified and achieved. Thus it is desirable to have an identification procedure which helps to satisfy the robustness requirements of particular control laws as expressed in the frequency domain. For an understanding of the performance of identifiers in frequency domain terms of models identified, one may turn to the work of Ljung [14] to ascertain what are the critical elements of the model fitting procedure, especially in closed loop.

The major result of Ljung is that the weighting in the frequency domain of the model fit is determined directly by the frequency content of the input signal fed to the system and by any signal prefiltering performed jointly upon the input and output measurements. Thus these become further design variables for the adaptive control engineer and bandpass filters are usually introduced into all the signal measurements to effect a suitable range of model fit and to emphasize areas of potential problem. By considering this issue in closed loop, it is possible to establish how the control law and the identifier interact in the complete adaptive controller.

6.2 Coda

At this final stage, one may consider what ingredients are available to the adaptive control system designer to carry out a recipe for an adaptive control algorithm. For the Recursive Least Squares identifier one has the option of including deadzones, leakage, normalization, forgetting, resetting, filtering. The control law design hinges upon choice of structure; model reference, direct or indirect, Linear Quadratic or pole-positioning, observers or filters and their roll-off rates, the use of integrators etc. Additionally, the reference signal supplied to the closed loop needs to be correctly selected in terms of spectral range, spectral content and potential cost to closed loop performance.

The choice of a particular adaptive control algorithm needs to take into account the design principles specified earlier as well as the operating constraints imposed by computational considerations and the multitude of other concurrent tasks scheduled. The above discussion has attempted to show how the specific conjunction of typical constraints and influences has led to the distinctive nature of adaptive control algorithms in response to the paradigm presented earlier.

References

[1] K.J. Åström and B. Wittenmark, *Adaptive Control*, Addison-Wesley, Reading MA, 1989.

[2] G.C. Goodwin and K.S. Sin, *Adaptive Filtering, Prediction and Control*, Prentice-Hall, Englewood Cliffs NJ, 1984.

[3] G. Kreisselmeier and M.C. Smith, 'Stable Adaptive Regulation of n-th Order Plants', *IEEE Trans. Automatic Control*, **AC-31**, 299-305, 1986.

[4] F. Giri, M. M'Saad, L. Dugard and J.M. Dion, 'Robust Pole Placement Indirect Adaptive Control', *International J. Adaptive Control and Signal Processing*, **2**, 33-47,1988.

[5] D.W. Clarke and P.J. Gawthrop, 'A Self-Tuning Controller', *IEE Proc Series D*, **122**, 929-934, 1975.

[6] D.W. Clarke, C. Mohtadi and P.S. Tuffs, 'Generalized Predictive Control Parts I and II', *Automatica*, **23**, 137-160, 1987.

[7] C. Rohrs, L.S. Valavani, M. Athans and G. Stein, 'Robustness of Continuous-time Adaptive Control Algorithms in the Presence of Unmodelled Dynamics', *IEEE Trans. Automatic Control*, **AC-30**, 881-889, 1985.

[8] G. Kreisselmeier and B.D.O Anderson, 'Robust Model Reference Adaptive Control', *IEEE Trans. Automatic Control*, **AC-31**, 127-133, 1986.

[9] L.Praly, 'Global Stability of a Direct Adaptive Control Scheme with Respect to a Graph Topology', in *Adaptive and Learning Systems - Theory and Applications* ed. K.S. Narendra, Plenum Press, New York, 1986.

[10] B.D.O. Anderson, R.R. Bitmead, C.R. Johnson Jr, P.V. Kokotovic, R.L. Kosut, L. Praly and B.D. Riedle, *Stability of Adaptive Systems: Passivity and Averaging Analysis*, MIT Press, Cambridge MA, 1986.

[11] G. Stein and M. Athans, 'The LQG/LTR Procedure for Multivariable Feedback Control Design', *IEEE Trans. Automatic Control*, **AC-32**, 105-114, 1987.

[12] B.A. Francis, *A Course in H^∞ Control Theory*, Springer-Verlag, New York, 1987.

[13] R.R. Bitmead, M.R. Gevers and V. Wertz, *Adaptive Predictive Control*, Snodfart Press, Palo Alto CA, 1989.

[14] L.Ljung, *System Identification: Theory for the User*, Prentice-Hall, Englewood Cliffs NJ, 1987.

Error Analysis of Least Squares Algorithms

Åke Björck

Department of Mathematics
University of Linköping
S-581 83 Linköping, Sweden

Abstract

A finite algebraic algorithm starts with a set of data $d_1, \ldots, d_r$, from which it computes via fundamental arithmetic operations a solution $f_1, \ldots, f_t$. In forward error analysis one attempts to bound $|\bar{f}_j - f_j|$, where $\bar{f}_j$ denotes the computed element f_j. In backward error analysis, pioneered by J.H. Wilkinson in the late fifties, one attempts to determine a modified set of data $\bar{d}_i$ such that the computed solution $\bar{f}_j$ is the *exact solution*. When it applies it tends to be very markedly superior to forward analysis. To yield error bounds for the solution, the backward error analysis has to be complemented with a perturbation analysis, which naturally leads to the concept of condition number of a problem.

There are several possible definitions of the stability of an algorithm related to different types of error analysis. The concepts of forward and backward stability and of weak and strong stability are discussed.

Many of the common problems in signal processing can be formulated as solutions to (a sequence of) linear least squares problems of the form $\min_x \|Xw - y\|_2$. We review the perturbation theory of such problems and discuss methods for the estimation of the corresponding condition numbers. We survey stability results for the method of normal equations and methods based on orthogonal reductions.

Very often it is required to recursively recalculate the solution x when equations are successively added to and/or deleted from the

NATO ASI Series, Vol. F 70
Numerical Linear Algebra, Digital Signal Processing and Parallel Algorithms
Edited by G. H. Golub and P. Van Dooren

least squares problem. Many different algorithms have been proposed to effectuate this. Most of these involve updating or downdating the Cholesky factor R of A^TA. This can be achieved using orthogonal and hyperbolic transformations. The numerical stability of such recursive algorithms is not yet completely analyzed. A new method, using iterative refinement, is suggested as a means of increasing the reliability of downdating algorithms.

1 Problems and Algorithms

We define a **numerical problem** to be a functional connection (explicit or implicit) between a vector of **input data** $(a_1, \ldots, a_r)$ and a vector of **output data** $(w_1, \ldots, w_t)$.

A finite algebraic **algorithm** for a numerical problem is a complete description of a sequence of elementary arithmetic and logical operations, which transforms each feasible set of input data into output data. For a given numerical problem one can consider many different algorithms. These can give approximate solutions of widely differing accuracy.

Example 1.1

To determine the roots of the equation

$$x^3 + a_1x^2 + a_2x + a_3 = 0,$$

is a numerical problem with input data a_1, a_2, a_3 and output data the roots x_1, x_2, x_3. An algorithm could be based on Newton's method supplemented with rules for chosing the initial approximations and a criterion for terminating the iterations.

1.1 Perturbation and Conditioning

In scientific computing the given input data $\hat{a}_1, \ldots, \hat{a}_r$ are usually imprecise. The difference from the exact input data can often be bounded in some norm

$$\|\hat{a} - a\|_{in} \leq \epsilon\|a\|_{in} \tag{1.1}.$$

We define the **condition number** for the problem to be

$$\kappa(a) = \limsup_{\epsilon\to 0} \frac{1}{\epsilon}\frac{\|\hat{w} - w\|_{out}}{\|w\|_{out}}.$$

Then for sufficiently small ϵ we have the estimate for the perturbations in the output

$$\|\hat{w} - w\|_{out} \le \epsilon\kappa(a)\|w\|_{out} + O(\epsilon^2). \qquad (1.2).$$

Hence $s = \log_{10}\kappa(a)$ roughly gives the number of significant decimal digits that may be lost in the accuracy of the solution. If $\kappa(a)$ is "large" the problem is said to be **ill-conditioned.** The definition of large may differ from problem to problem and depends on the accuracy of the data and the accuracy needed in the solution.

Note that $\kappa(a)$ is a function of the input data a and depends on the choice of norms in the data space and solution space. The most common norms are special cases, $p = 1, 2$ and ∞, of the family of L_p-norms,

$$\|x\|_p = (x_1|^p + |x_2|^p + \cdots + |x_n|^p)^{1/p}, \qquad 1 \le p < \infty$$

Sometimes it is more realistic to consider **elementwise relative bounds** on the perturbations in the data, i.e. instead of (1.1) we consider the smallest ϵ such that

$$|\hat{a}_i - a_i| \le \epsilon|a_i|, \qquad i = 1,\ldots,r \qquad (1.3).$$

Example 1.2

Consider the problem of solving the linear system $Xw = y$, where $X \in \mathbf{R}^{n\times n}$, $y \in \mathbf{R}^n$, are the given data and $w \in \mathbf{R}^n$ is the output. The condition number for this problem is

$$\kappa_p(X) = \|X\|_p\|X^{-1}\|_p. \qquad (1.4)$$

if the p–norm is used. If we use the elementwise relative bounds for the input and the ∞–norm for the output then (Skeel [1979])

$$\kappa_{|X|}(X) = \||X^{-1}||X|\|_\infty, \qquad (1.5)$$

where $|X|$ denotes the matrix with elements $|x_{ij}|$. It can be shown that $\kappa_{|X|} \le \kappa_\infty$, and it is possible for $\kappa_{|X|}$ to be much smaller than κ_∞. The conditioning of problems can to some degree be illustrated geometrically. A numerical problem P, means a mapping of a space X onto a space Y. In the figure below we picture a mapping in two dimensions. Since we are considering relative perturbations we can take the coordinate axis to be logarithmically scaled. A small circle of radius r is mapped onto an ellipse whose major axis is $R = \kappa r$.

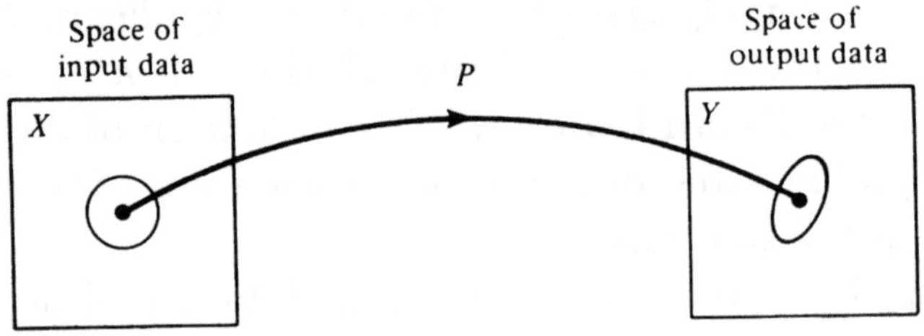

1.2 Unstable Algorithms

Poor accuracy in the computed solution can be expected when the problem is ill-conditioned. However, it can also be caused by a poor algorithm. In such a case we can say that the **algorithm** is unstable.

Example 1.3.
Compute $w = x + y$ from the system

$$\begin{pmatrix} 1 & \alpha \\ \alpha & 1 \end{pmatrix} \begin{pmatrix} x \\ y \end{pmatrix} = \begin{pmatrix} 1 \\ 0 \end{pmatrix},$$

where $\alpha \neq 1$ is the only input data. This problem is well-conditioned when $\alpha \approx 1$, since

$$w = x + y = 1/(1 - \alpha^2) - \alpha/(1 - \alpha^2) = 1/(1 + \alpha).$$

On the other hand, the problem of computing x (and y) is ill-conditioned when $\alpha \approx 1$. An algorithm which first solves the linear system for x and y and then adds to get w gives bad accuracy when $\alpha \approx 1$. For example, with $\alpha = 0.9900$, using four digit decimal arithmetic one gets $w = x + y = 50.25 - 49.75 = 0.5000$, while the correct answer is 0.5025, a loss of two digits. An algorithm often breaks down the problem into a chain of subproblems $P_1, P_2, \ldots, P_k$. As the previous example illustrated, if a subproblem is ill-conditioned this can cause a loss of accuracy in the final solution. In the figure below we illustrate, for k=2, a good and a poor algorithm.

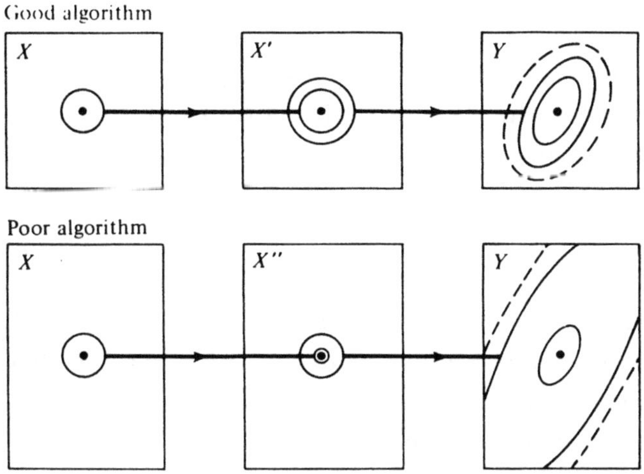

In the poor algorithm in the figure there is a strong contraction in the first mapping P_1 from X to X'', and then an equally strong expansion by P_2 from X'' to Y. Hence rounding errors in X'' can have a large effect on Y.

Example 1.4.

The eigenvalues of a symmetric matrix A are always well-conditioned functions of the elements (a_{ij}) in A. Consider an algorithm in which one first computes the coefficients of the characteristic polynomial of A, and then solves for the zeros of the polynomial. It is well known that the latter problem can be very ill-conditioned, and therefore this is not a stable algorithm.

2 Concepts of Stability

In this section we discuss several different definitions of stability, which are used for algorithms in numerical linear algebra. The discussion follows to a large extent that of Bunch [1987]. For a more formal approach see de Jong [1977].

2.1 Forward, Backward and Mixed Stability

We say that an algorithm is **forward stable** for the data a if the computed solution $\hat{w}$ is close to the exact solution,

$$\|\hat{w} - w\|/\|w\| \leq c_1 u, \tag{2.1}$$

where c_1 is a not too large constant. Similarly, we say that an algorithm is **backward stable** for the data a if the computed solution $\hat{w}$ is the exact solution of a slightly perturbed problem with data $\hat{a}$ such that

$$\|\hat{a} - a\|/\|a\| < c_2 u, \tag{2.2}$$

where c_2 is a not too large constant and u is a measure of the precision used in the computations. We would like these estimates to hold not only for a single problem but for some *class of input data.* We are usually satisfied if we can prove forward or backward stability for $\|\cdot\|_2$ or $\|\cdot\|_\infty$, although we may like the estimates to hold elementwise,

$$|\hat{a}_i - a_i|/|a_i| < c_2 u, \qquad i = 1, \ldots, r.$$

Forward stability guarantees an accurate solution. However, very few algorithms are forward stable. For a backward stable algorithm we are not guaranteed to get an accurate solution. We can only show that

$$\|\hat{w} - w\| \leq c_1 u \kappa(X) \|w\|.$$

However, if the perturbations $\hat{a} - a$ are within the uncertainties of the given data, *the computed solution is as good as our data warrants*! Therefore, backward stability is usually very satisfactory. As we shall see in the next section, although many important algorithms are not backward stable, they are **mixed stable.** By this we mean that the computed solution $\bar{w}$ is close to the exact solution $\hat{w}$ of a slightly perturbed problem with data $\hat{a}$, where $\hat{a}$ satisfies (2.2) and

$$\|\bar{w} - \hat{w}\|/\|w\| \leq c_1 u.$$

2.2 Stability of Algorithms for Linear Systems

We now discuss stability of algorithms for solving systems of linear equations $Xw = y$. The following definition of backward stability is the one introduced by Wilkinson [1965]. Forward and mixed stability are defined analogously.

Definition 2.1

A numerical algorithm for solving systems of linear equations is **backward stable** *for a class of matrices* $\mathcal{X}$ *if for each* $X \in \mathcal{X}$ *and for each* y *the computed solution* $\hat{w}$ *satisfies* $\hat{X}\hat{w} = \hat{y}$ *where* $\hat{X}$ *and* $\hat{y}$ *are close to* X *and* y.

Example 2.1

Gaussian elimination with partial pivoting and the Householder method is stable for $\mathcal{X}_1 =$ the class of nonsingular matrices. Similarly the Cholesky algorithm is stable for $\mathcal{X}_2 =$ the class of symmetric positive definite matrices.

An important property of backward stable algorithms is given in the following theorem.

Theorem 2.1

An algorithm for solving $Xw = y$ *is backward stable according to Definition 2.1 if and only if the computed solution has a small residual*

$$\|y - X\hat{w}\| \leq c_2 u \|X\| \|\hat{w}\|. \tag{2.3}$$

Proof: If (2.3) holds we define for the 2-norm

$$\delta X = r\hat{w}^T / \|\hat{w}\|_2^2, \qquad r = y - X\hat{w}.$$

Then it holds exactly that $(X + \delta X)\hat{w} = X\hat{w} + r = y$, where

$$\|\delta X\|_2 = \|r\|_2 / \|\hat{w}\|_2 \leq c_2 u \|X\|_2.$$

We can take $\delta y = 0$ and hence the algorithm is stable by Definition 2.1. On the other hand, if $\hat{X}\hat{w} = \hat{y}$, we have $y - X\hat{w} = (\hat{X} - X)\hat{w} + y - \hat{y}$, and if

$$\|\hat{X} - X\|_2 \leq c_3 u \|X\|_2, \qquad \|\hat{y} - y\|_2 \leq c_4 u \|y\|_2,$$

then an estimate of the form (2.3) for the norm of the residual holds. Note that in Definition 2.1 it is not required that $\hat{X}$ is in class $\mathcal{X}$. If the system comes from a physical problem such that X always belongs to some class

$\mathcal{X}$, then we have not solved a nearby physical problem unless also the perturbed $\hat{X}$ is in $\mathcal{X}$. For example, if X is Toeplitz we would like also $\hat{X}$ to be Toeplitz. Therefore, the following alternative definition of stability has been suggested.

Definition 2.2

A numerical algorithm for solving linear equations is **strongly backward stable** *for a class of matrices $\mathcal{X}$ if for each $X \in \mathcal{X}$ and for each y the computed solution $\hat{w}$ satisfies $\hat{X}\hat{w} = \hat{y}$ where $\hat{X}$ and $\hat{y}$ are close to X and y and $\hat{X}$ is in $\mathcal{X}$.*

Strong forward stability does not seem to be a useful concept, since any known structure of the solution can usually be imposed. Although the algorithms in example 2.1 are strongly backward stable, it may be difficult to prove strong stability for many other important algorithms. For example, only recently has it been proved that Gaussian elimination with partial pivoting is strongly backward stable for the class of symmetric positive definite systems, the difficulty being to show that $\hat{X}$ is symmetric. In this particular case one can also argue that since the condition number for unsymmetric perturbations of a symmetric system is the same as for symmetric perturbations backward stability is sufficient. Hence, the definition 2.1 often seems the more useful one.

Many important algorithms for solving linear systems, for example most iterative methods, are not backward stable. Therefore the following weaker form of stability is needed.

Definition 2.3

An algorithm for solving $Xw = y$ is **stable** *for a class of matrices $\mathcal{X}$ if the computed solution $\hat{w}$ satisfies*

$$\|\hat{w} - w\| \leq c_3 u \kappa(X) \|w\|.$$

It is straightforward to show that backward stability implies stability, but *the converse is not true.*

Many users are satisfied if the algorithm they use produces accurate solutions for well-conditioned problems. The following definition of weak stability has been suggested by Bunch [1987]. Weak stability may be sufficient for giving confidence that an algorithm may be used successfully.

Definition 2.4

An algorithm for solving $Xw = y$ is **weakly stable** *for a class of matrices $\mathcal{X}$ if for each well-conditioned X in $\mathcal{X}$ the computed solution $\hat{w}$ is such that $\|\hat{w} - w\|/\|w\|$ is small.*

For example, the Levinson algorithm is weakly stable for the class of symmetric positive definite Toeplitz matrices, see Cybenko [1980].

3 Error Analysis

3.1 Floating Point Number System

A floating point number system consists of numbers x which can be represented as

$$x = m \cdot \beta^e, \qquad m = \pm d_0.d_1 d_2 \dots d_t,$$

where m is the mantissa , e the exponent and $0 \le d_i < \beta$. The integer β is the base (usually $\beta = 2$). The mantissa is usually normalized so that $1 \le |m| < \beta$, and the exponent satisfies $L \le e \le U$. Hence, the floating point number system is characterized by the set (β, t, L, U).

The result of arithmetic operations on floating-point numbers can in general not be represented exactly as floating-point numbers. We denote by

$$fl(x + y), \quad fl(x - y), \quad fl(xy), \quad fl(x/y)$$

the result of floating-point operation (after rounding or chopping). In the standard model for floating point arithmetic, assuming that underflow does not occur,

$$fl(x \ op \ y) = (x \ op \ y)(1 + \epsilon), \qquad |\epsilon| \le u, \tag{3.1}$$

where u is the machine unit roundoff or **machine precision**. This model is true for most computers, but false on e.g. Cray for which

$$fl(x \ op \ y) = x(1 + \epsilon_1) \ op \ y(1 + \epsilon_2), \qquad |\epsilon_i| \le u.$$

This is not as satisfactory, and can cause inaccuracy in addition and subtraction.

3.2 Rounding Error Analysis

Consider a finite algebraic algorithm which from the data $(a_1, \ldots, a_r)$, through intermediate values $(c_1, \ldots, c_s)$ computes a solution $(w_1, \ldots, w_t)$. There are two basic forms of error analysis for such an algorithm, which both are useful:

(i) In **classical forward error analysis** one attempts to find bound for the errors in the intermediate values $|\hat{c}_i - c_i|$, and finally bounds for $|\hat{w}_i - w_i|$.

(ii) In **backward error analysis** one attempts to determine a modified set of data $\tilde{a}_i$ such that the computed solution $\hat{w}_i$ is the *exact solution*, and give bounds for $|\tilde{a}_i - a_i|$. There may be an infinite number of such sets; sometimes there is just one and it can happen , even for very simple algorithms, that no such set exists. Notice that no reference is made to the exact solution for the original data.

A great advantage of backward error analysis is that when it can be applied it tends to give much sharper results than a forward error analysis. Perhaps more important, it usually also gives a better insight into the stability (or lack of it) of the algorithm. It cannot be stressed too much that *the primary purpose of rounding error analysis is insight.*

Example 3.1

By repeatedly using the formula for floating point multiplication one can show that the computed product $fl(x_1 x_2 \cdots x_n)$ is exactly equal to

$$x_1 x_2 (1 + \epsilon_2) x_3 (1 + \epsilon_3) \cdots x_n (1 + \epsilon_n),$$

where $|\epsilon_i| \leq u, \quad i = 2, 3, \ldots, n$. This can be interpreted as a backward error analysis; we have shown that the computed product is the exact product of factors x_1, $\tilde{x}_i = x_i(1 + \epsilon_i)$, $i = 2, \ldots, n$. It also follows from this analysis that

$$|fl(x_1 x_2 \cdots x_n) - x_1 x_2 \cdots x_n| \leq \delta |x_1 x_2 \cdots x_n|$$

where

$$\delta = (1 + u)^{n-1} - 1 < 1.06(n - 1)u,$$

and the last inequality holds if the condition $(n-1)u < 0.1$ is satisfied. This bounds the forward error in the computed result.

Similar results can easily be derived for basic vector and matrix operations, see Wilkinson [1965, pp.114-118]. For an inner product $x^T y$ we have

$$fl(x^T y) = x_1 y_1(1+\delta_1) + x_2 y_2(1+\delta_2) + \ldots + x_n y_n(1+\delta_n)$$

where

$$|\delta_1| < 1.06nu, \qquad |\delta_r| < 1.06(n+2-r)u, \quad r = 2,\ldots,n.$$

Hence, the computed result is the exact inner product of x and $\hat{y}$, where $\hat{y}_i = y_i(1+\delta_i)$. Useful consequences of this result are

$$|fl(x^T y) - x^T y| \le \sum |x_i||y_i||\delta_i| \le 1.06nu|x^T||y| \le 1.06nu\|x\|_2\|y\|_2. \quad (3.2)$$

This result is easily generalized to yield a *forward* error analysis of matrix times matrix multiplication. However, for this case there is no backward error analysis, since the rows and columns of the two matrices participate in many inner products!

3.3 Running Error Analysis

A different approach to rounding error analysis is to perform the analysis automatically, for *each particular computation*. This gives an *a posteriori* error analysis as compared to the *a priori* error analysis discussed above. An example is the use of **interval analysis** for which special purpose hardware and software now exists.

A simple form of a posteriori analysis, called running error analysis, was used in the early days of computing by Wilkinson, see Wilkinson [1986]. To illustrate his idea we rewrite the basic model for floating point arithmetic as

$$x \; op \; y = fl(x \; op \; y)(1+\epsilon).$$

These are also satisfied for most implementations of floating point arithmetic. Then, the actual error can be estimated $|fl(x \; op \; y) - x op y| \le u|fl(x \; op \; y)|$. Note that the error is now given in terms of the *computed* result and is available in the computer at the time the operation is performed. This running error analysis can often be easily implemented. We just take an existing program and modify it, so that as each arithmetic operation is performed, the absolute value of the computed results is added into the accumulating error bound.

Example 3.2

The inner product $fl(x^T y)$ is computed by the program

$$
\begin{aligned}
&\hat{s}_0 = 0; \quad \eta = 0;\\
&\text{for } i = 1, 2, \ldots, n\\
&\text{begin}\\
&\quad \hat{t}_i = fl(x_i y_i); \quad \eta = \eta + |\hat{t}_i|;\\
&\quad \hat{s}_i = fl(\hat{s}_{i-1} + \hat{t}_i);\\
&\quad \eta = \eta + |\hat{s}_i|;\\
&\text{end}
\end{aligned}
$$

For the final error we have the estimate $|fl(x^T y) - x^T y| \leq \eta u$. Note that a running error analysis takes advantage of cancellations in the sum. This is in contrast to the previous estimates, which we call a priori error analysis, where the error estimate is the same for all distribution of signs of the elements x_i and y_i. For an application of running error analysis in signal processing, see Luk and Qiao[1988].

3.4 Rounding Errors in Normal Equations

As an application of the results in Sec. 3.2 we now consider rounding errors in the formation of the system of normal equations. For the computed elements in the matrix $M = X^T X$, where X is $p \times n$, we have

$$\hat{m}_{ij} = fl\Big(\sum_{k=1}^{p} x_{ik} x_{jk}\Big) = \sum_{k=1}^{p} x_{ik} x_{jk}(1 + \delta_k),$$

where $|\delta_k| < 1.06(p + 2 - k)u$. This yields

$$|\hat{m}_{ij} - m_{ij}| < 1.06u \sum_{k=1}^{p} (p + 2 - k)|x_{ik}||x_{jk}|.$$

It follows that the computed matrix satisfies

$$fl(X^TX) = X^TX + E, \qquad |e_{ij}| < 1.06pu\|x_i\|_2\|x_j\|_2. \tag{3.3}$$

A similar estimate holds for the rounding errors in the computed vector X^Ty.

Note that it is *not* possible to show that $\overline{X^TX} = (X+E)^T(X+E)$ for some error matrix E, i.e. the rounding errors in forming the matrix X^TX are not in general equivalent to small perturbations of the initial data matrix X. From this we can deduce that **no method based on forming the normal equations can be backward stable.** The following example illustrates the possible loss of information when X^TX is formed.

Example 3.3

Consider the system $Xw = y$, where

$$X = \begin{pmatrix} 1 & 1 \\ \epsilon & \\ & \epsilon \end{pmatrix}, \qquad y = \begin{pmatrix} 1 \\ 0 \\ 0 \end{pmatrix}, \qquad |\epsilon| \ll 1.$$

This may correspond to a problem where the sum $w_1 + w_2$ has been determined much more accurately than the individual components of w. We have, exactly

$$X^TX = \begin{pmatrix} 1+\epsilon^2 & 1 \\ 1 & 1+\epsilon^2 \end{pmatrix}, \qquad X^Ty = \begin{pmatrix} 1 \\ 1 \end{pmatrix},$$

and the exact solution and residual are

$$w = \frac{1}{2+\epsilon^2}(1,1)^T, \quad r = \frac{\epsilon}{2+\epsilon^2}(\epsilon,-1,-1)^T.$$

Now assume that $\epsilon = 10^{-4}$, and that we use eight-digit decimal floating point arithmetic. Then $1+\epsilon^2 = 1.00000001$ round to 1, and

$$\overline{X^TX} = \begin{pmatrix} 1 & 1 \\ 1 & 1 \end{pmatrix}$$

which is singular. Hence, $X+E$ must also be singular. We know that this implies that $\|E\|_2 \geq \sigma_2 = \epsilon \approx \sqrt{u}$. Note that we have lost all information contained in the last two rows of X.

4 Conditioning Of Least Squares Solutions

4.1 The Singular Value Decomposition

The singular value decomposition (SVD) is a powerful tool both for analyzing and solving linear least squares problems.

Theorem 4.1

Given the matrix $X \in \mathbf{R}^{p\times n}$ and vector $y \in \mathbf{R}^p$, consider the general linear least squares problem

$$\min_{w\in S} \|w\|_2, \qquad S = \{w \in \mathbf{R}^n | \|y - Xw\|_2 = \min\},$$

where rank $(X) = r \leq \min(p,n)$. This problem always has a unique solution, the pseudo-inverse solution, which can be written $w = X^I y$, where X^I is the unique **pseudo-inverse** *of X. Let the SVD of X be*

$$X = U \begin{pmatrix} \Sigma_r & 0 \\ 0 & 0 \end{pmatrix} V^T, \tag{4.1}$$

where U and V are orthogonal matrices, $\Sigma_r = diag(\sigma_1, \sigma_2, \ldots, \sigma_r)$ and $\sigma_1 \geq \sigma_2 \geq \ldots \geq \sigma_r > 0$ are the ordered nonzero singular values of X. Then the pseudo-inverse is

$$X^I = V \begin{pmatrix} \Sigma_r^{-1} & 0 \\ 0 & 0 \end{pmatrix} U^T \tag{4.2}$$

4.2 Perturbation Analysis

We now consider the effect of perturbations of X and y on the least squares solution w and the residual $r = y - Xw$. In this analysis the generalized condition number

$$\kappa(X) = \|X\| \|X^I\|, \tag{4.3}$$

plays a significant role. In particular for the 2-norm since $\|X\|_2 = \sigma_1$, $\|X^I\|_2 = \sigma_r^{-1}$, we have

$$\kappa_2(X) = \sigma_1/\sigma_r. \tag{4.4}$$

It is important to note that there are efficient methods to estimate $\kappa(X)$, see Sec. 5.2.

Theorem 4.2

Assume that rank$(X) = n$, *that the perturbations in the data satisfy*

$$\|\delta X\|_2/\|X\|_2 \leq \epsilon_X, \quad \|\delta y\|_2/\|y\|_2 \leq \epsilon_y, \tag{4.5}$$

and that $\eta = \|X^I\|_2\|\delta X\|_2 \leq \kappa_2(X)\epsilon_X < 1$. *Then rank*$(X + \delta X) = n$,

$$\|\delta w\|_2 \leq \frac{\kappa_2}{1-\eta}\Big(\epsilon_X\|w\|_2 + \frac{\|y\|_2}{\|X\|_2}\epsilon_y + \epsilon_X\kappa_2\frac{\|r\|_2}{\|X\|_2}\Big), \tag{4.6}$$

where $r = y - Xw$, *and*

$$\|\delta r\|_2 \leq \epsilon_X\|w\|_2\|X\|_2 + \|y\|_2\epsilon_y + \epsilon_X\kappa_2\|r\|_2 \tag{4.7}$$

Proof: See Wedin [1973].

It can be shown that for an arbitrary matrix X and vector y there exist perturbations δX and δy such that the estimates in Theorem 4.1 are almost attained. Golub and Wilkinson [1966] were the first to note that for problems where $r \neq 0$ a term proportional to κ^2 occurs in the bound for δw. A geometrical explanations for this term is that the least squares solution satisfies $Xw = \tilde{y}$, $\tilde{y} = P_X y$, where P_X is the orthogonal projector onto the range of X. When X is ill-conditioned the projection $\tilde{y}$ is sensitive to perturbations.

If we let $\epsilon_X = \epsilon_y = \epsilon$ in Eq. (4.6), then it follows that the condition number for the solution w is

$$\kappa_2(X, y) = 2\kappa_2\left(1 + \kappa_2\frac{\|r\|_2}{\|X\|_2\|w\|_2}\right). \tag{4.8}$$

In contrast to the linear equation case *the condition number for the least squares problem depends in an essential way on the right hand side* y. For small residual problems which are only moderately ill-conditioned, i.e. if

$$\kappa_2\|r\|_2 < \|X\|_2\|w\|_2, \tag{4.9}$$

the second term does not dominate the error and the condition number is 2κ. In particular, if $r = 0$, Eq.(4.6) reduces to the well known estimate for the linear equation case

$$\|\delta w\| \leq \frac{\kappa_2}{1-\eta}(\epsilon_X + \epsilon_y)\|w\|.$$

4.3 Scaling

It should be stressed that in order for the perturbation analysis above to be useful the matrix X and vector y should be scaled so that the class of perturbations defined by (4.6) is relevant. Note that we are free to scale columns of X, which corresponds to a scaling of the solution

$$Xw = (XD)(D^{-1}w) = \bar{X}\bar{w} = y.$$

It can be proved that choosing D so that the columns in X have unit l_2–norm gives a condition number which is within a factor of $\sqrt(n)$ from the minimum over all diagonal scalings with $D > 0$. Note that if no column pivoting is performed such a scaling will only effect the error bound, and *not the accuracy of the computed solution* $\bar{w}$.

However, unlike the linear system case, we are not allowed to scale the rows of (X, y), since this would change the norm which defines the solution. It may occur that the matrix (X, y) has some rows with large norms. (Such rows correspond to equations with small errors.) For such problems it is more realistic to consider componentwise bounds on the perturbations in the data. We define the norm $\|(X, y)\|_{rel}$ as the smallest number ω such that

$$|\delta X| \le \omega |X|, \qquad |\delta y| \le \omega |y| \qquad (4.10).$$

Here, if X is a matrix with elements x_{ij}, $|X|$ denotes the matrix with elements $|x_{ij}|$ and the inequalities should be interpreted componentwise. We now consider the class of perturbations $\|(X, y)\|_{rel} \le \epsilon$. Note that this class of perturbations is invariant under row and column scalings and that zero elements are not perturbed. If we use the maximum norm in the output data space then it can be shown that

$$\|\delta w\|_\infty \le \epsilon \kappa_{rel}(X, y)\|w\|_\infty + O(\epsilon^2),$$

where the condition number is

$$\kappa_{rel}(X, y) = \||X^I|(|X||w| + |y|)\|_\infty + \||(X^T X)^{-1}||X^T||r|\|_\infty / \|w\|_\infty. \quad (4.11)$$

For a method to estimate this condition number see Sec. 5.2.

5 Condition Estimation

To assess the accuracy using the bounds given in Section 4, it is necessary to compute a reliable estimate the condition number κ. If we compute the SVD of X, then we have $\kappa_2 = \sigma_1/\sigma_r$. However, usually we have only the QR decomposition of X, and we now discuss estimates of κ based on this decomposition. The most satisfactory estimates require that column pivoting is used in the QR decomposition, and we start by discussing the pivoting strategy.

5.1 Pivoting in the QR Decomposition

We modify the Householder method to include column pivoting using a pivoting strategy introduced by Golub [1965]. Assume that after $(k-1)$ steps in this algorithm we have computed

$$X^{(k)} = (P_{k-1}\cdots P_1)X(\Pi_1\cdots\Pi_{k-1}) = \begin{pmatrix} R_{11}^{(k)} & R_{12}^{(k)} \\ 0 & X_{22}^{(k)} \end{pmatrix}, \qquad (5.1)$$

where $R_{11}^{(k)} \in \mathbf{R}^{(k-1)\times(k-1)}$, P_i are Householder matrices and Π_i are elementary permutation matrices, $i = 1,\ldots,k-1$. Let

$$X_{22}^{(k)} = (\tilde{x}_k^{(k)},\ldots,\tilde{x}_n^{(k)}).$$

Then, in the next step the permutation matrix Π_k is chosen to interchange the columns p and k, where p is the smallest index that satisfies

$$s_p^{(k)} \geq s_j^{(k)}, \quad s_j^{(k)} = \|\tilde{x}_j^{(k)}\|_2^2, \quad j = k,\ldots,n.$$

This column pivoting strategy is equivalent to searching for the column of largest norm in the submatrix $X_{22}^{(k)}$.

This strategy can also be used with the Givens or modified Gram-Schmidt algorithms. At each step the diagonal element r_{kk} will be maximized. Hence, the diagonal elements in R form a non-increasing sequence. It is easily shown that the elements in R also satisfy the inequalities

$$r_{kk}^2 \geq \sum_{i=k}^{j} r_{ij}^2, \quad j = k+1,\ldots,n. \qquad (5.2)$$

In particular, if $r_{kk} = 0$, then $r_{ij} = 0$, $i, j \geq k$.

If column pivoting has been performed we can easily get upper and lower bounds for $\sigma_1(R)$ in terms of r_{11},

$$|r_{11}| \leq \sigma_1(R) \leq n^{1/2}|r_{11}|. \tag{5.3}$$

The diagonal elements of R^{-1} equal r_{ii}^{-1}, $i = 1, \ldots, n$. Hence, for the smallest singular value we have, assuming that $r_{nn} \neq 0$,

$$\sigma_n^{-1}(R) = \|R^{-1}\|_2 \geq |r_{nn}^{-1}|,$$

From these estimates we obtain a *lower bound* for the condition number $\kappa(X) = \kappa(R)$

$$\kappa(R) = \sigma_1/\sigma_n > |r_{11}/r_{nn}| \tag{5.4}$$

In practice this gives a fairly reliable estimate of $\kappa(X)$. Indeed, extensive numerical testing by Stewart [1980] has shown that it usually underestimates κ only by a factor of 2 – 3 and seldom by more than 10. However, there exist contrived examples for which this is a considerable underestimate of $\kappa(X)$, see Example 5.1.

Pivoting also simplifies rank determination. However, note that because of rounding errors the submatrix $X_{22}^{(k)}$ in (5.1) will usually not be exactly zero when X has rank $r = k - 1$. Therefore, we should terminate the Householder algorithm when $X_{22}^{(k)}$ is small in norm relative to X, say when for some small $\epsilon > 0$ we have $\|X_{22}^{(k)}\|_2 \leq \epsilon\|X\|_2$. It can be shown that this inequality holds provided

$$|r_{kk}| \leq= (n - k + 1)^{-1/2}\epsilon|r_{11}|, \tag{5.5}$$

which can be used to terminate the QR algorithm.

Unfortunately, the matrix X can be nearly rank deficient without any diagonal element in R being small. The best lower bound of σ_n in terms of the diagonal elements that can be proved is (Faddeev et al. [1968])

$$\sigma_n(R) \geq 3(4^n + 6n - 1)^{-1/2}|r_{nn}| \geq 2^{1-n}|r_{nn}|. \tag{5.6}$$

That this lower bound can almost be attained is shown by the following example due to Kahan [1966].

Example 5.1

It is easily verified that the upper triangular matrices

$$R_n = \mathrm{diag}(1, s, \dots, s^{n-1}) \begin{pmatrix} 1 & -c & -c & \dots & -c \\ & 1 & -c & \dots & -c \\ & & 1 & & \vdots \\ & & & \ddots & \vdots \\ & & & & 1 \end{pmatrix},$$

where $s^2 + c^2 = 1$, are invariant under QR decomposition with column pivoting. Taking, for example, $n = 100$, $c = 0.2$ we find that $r_{nn} = 0.133$ although the matrix is nearly singular since $\sigma_n = 0.368 \cdot 10^{-8}$. In case no column pivoting is done one can still try to use the estimate

$$\kappa(R) = \max_i |r_{ii}| / \min_i |r_{ii}|. \tag{5.7}$$

That this is less reliable is well known. Indeed, Golub and Wilkinson [1976] show that r_{min} can be bounded in terms of the singular values by

$$\sigma_n \le r_{min} \le \sigma_n(\sigma_1/\sigma_n)^{(1-\frac{1}{n})},$$

and these are in general best possible bounds. For example, the matrices

$$R_n = \begin{pmatrix} \alpha & 1 & & \\ & \alpha & 1 & \\ & & \ddots & 1 \\ & & & \alpha \end{pmatrix} \in \mathbf{R}^{n \times n},$$

with $|\alpha| \le 1$ has $\sigma_1 \approx 1$ and $\sigma_n \approx \alpha^n$. Hence, for $\alpha = 0.1$ and $n = 20$ the matrix R_n is singular to double precision accuracy on most computers although no diagonal element is small. Foster [1988] has made a probabilistic study of when r_{min} is much larger than σ_n. His conclusion is that although usually they are of the same magnitude, r_{min} is substantially larger often enough that one should be very cautious in using (5.7), when no pivoting has been performed.

5.2 Condition Estimators

We now describe a condition estimator, which is also effective when no column pivoting has been performed in the QR decomposition. The estimator is based on inverse iteration, and requires $0(n^2)$ extra operations.

Let d be a random vector and compute y and z from $R^T y = d$, $Rz = y$. Since

$$z = (R^T R)^{-1} d = (X^T X)^{-1} d$$

this is equivalent to one step of inverse iteration with $X^T X$. Let $R = U\Sigma V^T$ be the singular value decomposition of R. Note that $\sigma_i(X) = \sigma_i(R)$, $i = 1, \ldots, n$. If we write $d = \sum_{i=1}^n \alpha_i v_i$ then

$$y = \sum_{i=1}^{n} (\alpha_i/\sigma_i) v_i, \qquad z = \sum_{i=1}^{n} (\alpha_i/\sigma_i^2) v_i,$$

and provided α_n, the component of d along v_n, is not very small we have

$$\sigma_n^{-1} \approx \|z\|_2 / \|y\|_2. \tag{5.8}$$

To enhance this probability we take $d = (\pm 1, \pm 1, \ldots, \pm 1)^T$, where the sign of d_s being determined at the stage when y_s is computed so as to obtain a vector y of large norm.

A condition estimator based on Eq. (5.8) has been implemented in the LINPACK set of subroutines, and has proved to be very reliable in practice. It will detect near rank deficiency of the matrix X even in case this is not revealed by a small diagonal element in R. This is important since failure to detect near rank deficiency can lead to meaningless solutions of very large norm.

Suppose that R is numerically rank deficient. Then we would like to compute a **rank revealing QR decomposition**, i.e. find a column permutation such that in the corresponding QR decomposition r_{nn} is small. When $r = n-1$, we can always find such a column permutation by inspecting the elements of v_n, the right singular vector of R corresponding to the singular value σ_n.

Let the vector v, $\|v\|_2 = 1$, satisfy $\|Xv\|_2 = \epsilon$, and let Π be a permutation such that if $\Pi^T v = w$, then $|w_n| = \|w\|_\infty$. Then if $X\Pi = QR$ is the QR decomposition of X, then it can be shown that

$$|r_{nn}| \leq n^{1/2} \epsilon.$$

If we let $v = v_n$, the right singular vector corresponding to the smallest singular value $\sigma_n(X)$ we have

$$\sigma_n(X) \geq n^{-1/2}|r_{nn}|. \tag{5.9}$$

In practice an approximation to v_n can be determined by one or two steps of inverse iteration. This results leads to a two step procedure:

(i) Determine a QR decomposition of X.

(ii) Use the condition estimator to determine an elementary permutation matrix Π and compute the QR decomposition $R\Pi = \bar{Q}\bar{R}$, from which we get the QR decomposition of $A\Pi$.

Note that the QR decomposition of $R\Pi$ can be computed in less than $2n^2$ flops using the updating techniques. Foster [1983] introduced rank revealing QR decompositions for sparse matrices. In Chan [1987] the above procedure is extended to the case when A has numerical rank $r < n-1$.

We finally mention that it is possible to estimate the condition number (4.11) corresponding to componentwise relative perturbations using a recent method of Hager [1984]. This method estimates the ∞-norm (or 1-norm) of a matrix B using only matrix-vector products of the form Bx and B^Ty. We note that the terms in (4.11) are of the form $\||B|g\|_\infty$, where $g > 0$. Arioli, Demmel and Duff [1988] have noticed that using the equality

$$\||B|g\|_\infty = \|BG\|_\infty, \qquad G = diag(g).$$

Hager's method can be used to estimate $\||B|g\|_\infty$ provided matrix vector products BGx and G^TB^Ty can be computed. Now (4.11) can be written as

$$\kappa_r(X,y) \leq \left(\||B_1|g_1\|_\infty + \||B_2|g_2\|_\infty\right)/\|x\|_\infty, \tag{5.10}$$

where

$$B_1 = X^+, \quad g_1 = |X||x| + |b|, \qquad B_2 = (X^TX)^{-1}, \quad g_2 = |X^T||r|. \tag{5.11}$$

It follows that in order to apply Hager's algorithm it suffices that we can compute the matrix-vector products X^+x, $(X^+)^Ty$, and $(X^TX)^{-1}x$. When the QR-factorization of X is known this can be done cheaply.

6 Modified Least Squares Problems

6.1 Updating and Downdating

Consider the least squares problem

$$\min_w \|Xw - s\|_2, \quad X \in \mathbf{R}^{p\times n}, \quad p > n.$$

Let the QR decomposition of the augmented matrix (X, s) be

$$Q^T(X, s) = \begin{pmatrix} R & u \\ 0 & \rho \end{pmatrix}, \tag{6.1}$$

where R is the R factor of X. Then the least squares solution is obtained by solving the triangular linear system $Rw = u$, and the residual norm is equal to ρ.

Frequently one knows the R factor in (6.1), and wishes to find the solution to a modified problem

$$\min_w \|\bar{X}w - \bar{s}\|_2,$$

where either a new observation $y^T w = \sigma$ is added (**updating**),

$$\bar{X} = \begin{pmatrix} X \\ y^T \end{pmatrix}, \quad \bar{s} = \begin{pmatrix} s \\ \sigma \end{pmatrix},$$

or an old observation $z^T = \sigma$ is removed (**downdating**)

$$X = \begin{pmatrix} z^T \\ \bar{X} \end{pmatrix}, \quad s = \begin{pmatrix} \sigma \\ \bar{s} \end{pmatrix}.$$

Sometimes both modifications are carried out simultaneously. By Eq.(6.1), this can be reduced to modifying the R factor of the corresponding augmented matrix $(\bar{X}, \bar{s})$. In the following, for simplicity of notation, we discuss the updating and downdating of the R factor of X , but it is to be understood that actually the factor in Eq. (6.1) is modified.

If R and $\bar{R}$ are the R factors of X and $\bar{X}$ respectively, then we have

$$\bar{R}^T\bar{R} = R^T R + yy^T, \quad \bar{R}^T\bar{R} = R^T R - zz^T$$

for updating and downdating. From the relation $\sigma_i^2(A) = \lambda_i(A^TA)$ and classical perturbation theory for eigenvalues it follows that the singular values $\bar{\sigma}_i = \sigma_i(\bar{R})$ interleave with $\sigma_i = \sigma_i(R)$.

$$\bar{\sigma}_1 \geq \sigma_1 \geq \bar{\sigma}_2 \geq \cdots \geq \bar{\sigma}_n \geq \sigma_n \geq 0,$$

for updating and

$$\sigma_1 \geq \bar{\sigma}_1 \geq \sigma_2 \geq \cdots \geq \sigma_n \geq \bar{\sigma}_n \geq 0,$$

for downdating. Updating will not decrease the smallest singular value. In downdating it may decrease and we can have $\bar{\sigma}_n = 0$, even when R has full column rank. This shows that downdating can be a sensitive problem.

6.2 Updating the QR Decomposition

Assume that we have computed the QR decomposition of X,

$$Q^TX = \begin{pmatrix} R \\ 0 \end{pmatrix}, \quad Q \in \mathbf{R}^{p\times p}. \tag{6.2}$$

Then we have

$$\begin{pmatrix} Q^T & 0 \\ 0 & 1 \end{pmatrix}\begin{pmatrix} X \\ y^T \end{pmatrix} = \begin{pmatrix} R \\ 0 \\ y^T \end{pmatrix} = \Pi \begin{pmatrix} R \\ y^T \\ 0 \end{pmatrix},$$

where Π is a permutation matrix. The elements in y^T can now be eliminated and the updated factor $\bar{R}$ computed by a sequence of orthogonal rotations $U = J_1 \cdots J_n$ so that

$$U^T \begin{pmatrix} R \\ y^T \\ 0 \end{pmatrix} = \begin{pmatrix} \bar{R} \\ 0 \end{pmatrix}. \tag{6.3}$$

It then follows that

$$\bar{Q} = \begin{pmatrix} Q & 0 \\ 0 & 1 \end{pmatrix} \Pi U \tag{6.4}$$

is the updated factor $\bar{Q}$. *Note that Q is not needed for the construction of U and the updated factor $\bar{R}$.* It is a standard result that this algorithm for updating is backward stable. Indeed, if we construct R by a sequence of such modifications the resulting algorithm is equivalent to the sequential row orthogonalization method for computing the QR decomposition, see Björck [1989, Sec.17].

6.3 Downdating the QR Decomposition

Assume that we have computed

$$X = \begin{pmatrix} z^T \\ \bar{X} \end{pmatrix} = Q \begin{pmatrix} R \\ 0 \end{pmatrix}. \tag{6.5}$$

To remove the first row z^T we adjoin a column e_1 to the left of X,

$$(e_1, X) = \begin{pmatrix} 1 & z^T \\ 0 & \bar{X} \end{pmatrix}.$$

Using (6.5) it follows that

$$Q^T(e_1, X) = \begin{pmatrix} q_1 & R \\ q_2 & 0 \end{pmatrix},$$

where $q^T = (q_1^T, q_2^T)$ is the first row of Q. We can now determine a sequence of rotations

$$U^T = J_1 \cdots J_n J_{n+1} \cdots J_{p-1} \tag{6.6}$$

such that

$$U^T \begin{pmatrix} q_1 & R \\ q_2 & 0 \end{pmatrix} = \begin{pmatrix} 1 & v^T \\ 0 & \bar{R} \\ 0 & 0 \end{pmatrix}.$$

Here J_k, $k = 1, \ldots, p-1$, is a rotation in the plane $(k, k+1)$, chosen to zero the $(k+1)$st component in q. Then we have

$$\hat{Q}^T \begin{pmatrix} 1 & z^T \\ 0 & \bar{X} \end{pmatrix} = \begin{pmatrix} 1 & v^T \\ 0 & \bar{R} \\ 0 & 0 \end{pmatrix}, \tag{6.7}$$

where $\hat{Q} = QU$. Equating the first column on both sides we see that $\hat{Q}^T e_1 = e_1$, so the first row in $\hat{Q}$ equals e_1. Hence, $\hat{Q}$ must have the form

$$\begin{pmatrix} \pm 1 & 0 \\ 0 & \bar{Q} \end{pmatrix},$$

and it follows that $v^T = z^T$. Dropping the first row and column now gives the downdated QR decomposition of $\bar{X}$. *Note, that to construct the matrix U we need the first row of Q.*

We now show that this downdating algorithm is mixed stable, i.e. the computed $\bar{R}$ is close to the exact factor of $\bar{X} + E$, where E is small. By the standard error analysis for the QR decomposition (see Wilkinson [1965, pp.240-241]) there exists an *exactly orthogonal* $\hat{Q}$ such that the computed v and $\bar{R}$ satisfy

$$\hat{Q}^T(e_1 + f, X + E) = \begin{pmatrix} r_{11} & v^T \\ 0 & \bar{R} \end{pmatrix},$$

where

$$\|f\|_2 \leq c_1 u, \quad \|E\|_2 < c_2 u \|X\|_2, \tag{6.8}$$

c_1, c_2 are constants depending only on the dimensions p and n, and u is the machine precision. Hence,

$$\hat{Q}^T(e_1, X + E) = \begin{pmatrix} r_{11} - \delta & v^T \\ -d & \bar{R} \end{pmatrix},$$

where we have put

$$\hat{Q}^T f = \begin{pmatrix} \delta \\ d \end{pmatrix}.$$

This matrix can now be reduced to triangular form by an exact orthogonal matrix to give

$$\bar{Q}^T(e_1, X + E) = \begin{pmatrix} 1 & v^T + \Delta v \\ 0 & \bar{R} + \Delta \bar{R} \end{pmatrix},$$

where

$$\|\Delta \bar{R}\|_2 \leq c_3 u \|z\|_2. \tag{6.9}$$

Since $\bar{Q}$ is exactly orthogonal this again deflates and shows that $\bar{R} + \Delta \bar{R}$ is the exact factor of $\bar{X} + E$. Since the norms of $\Delta \bar{R}$ and E are bounded by (6.8) and (6.9) we have mixed stability.

A drawback with downdating the full QR decomposition is that we need to store and modify the full matrix $Q \in \mathbf{R}^{p \times p}$. In many cases this may not be feasible, for example when $p \gg n$ or when X is large and sparse. A more attractive alternative is then to update only the n first columns of Q, i.e., the Gram-Schmidt QR decomposition. Such methods were considered by Daniel et al. [1976], where Algol procedures also were given. Recently FORTRAN subroutines, which are modifications of these have been developed in Reichel and Gragg [1988]. These algorithms also allow the stable downdating of least squares problems.

6.4 The LINPACK Algorithm

We now show how to construct the downdating transformation (6.6), when the Q factor is not available. Note that the first $(p-n-1)$ transformations in (6.6) only affect the last $(p-n)$ rows and leave q_1 and R unchanged,

$$J_{n+1}\cdots J_{p-1}\begin{pmatrix} q_1 & R \\ q_2 & 0 \end{pmatrix} = \begin{pmatrix} q_1 & R \\ \gamma e_1 & 0 \end{pmatrix}, \quad \gamma = \|q_2\|_2. \tag{6.10}$$

From the first row of the QR decomposition of X we have

$$z^T = (\, q_1^T \quad q_2^T \,)\begin{pmatrix} R \\ 0 \end{pmatrix} = q_1^T R.$$

Since $1 = \|q\|_2^2 = \|q_1\|_2^2 + \|q_2\|_2^2$, it follows that q_1 and γ can be computed from

$$R^T q_1 = z, \quad \gamma = (1 - \|q_1\|_2^2)^{1/2}. \tag{6.11}$$

We can then construct the rotations $J_1, \ldots, J_n$ and apply them to (6.10). This is the algorithm implemented in LINPACK.

Although closely related to the downdating algorithm in Sec. 6.3 using Q, the error properties for the LINPACK algorithm are quite different. A detailed analysis by Stewart [1979] shows that the downdated factor $\bar{R}$ computed by the LINPACK algorithm satisfies

$$\hat{Q}\begin{pmatrix} R \\ 0 \end{pmatrix} = \begin{pmatrix} z^T + \delta z^T \\ \bar{R} + E \end{pmatrix}, \tag{6.12}$$

where $\hat{Q}$ is exactly orthogonal, and δz and E are small. Hence, $\bar{R}$ is close to the exact factor corresponding to downdating R with a slightly perturbed vector $z^T + \delta z^T$. The LINPACK method therefore can be said to be mixed stable *if R and z are considered as the data.* However, it is important to note that *there may be no matrix $\bar{X} + E$ close to $\bar{X}$ which has factor $\bar{R}$, and the method is not mixed stable with respect to the data $\bar{X}$ and z.*

We now consider the effect of a perturbation in z on the downdated $\bar{R}$. Following Stewart [1979] we write

$$\bar{R}^T\bar{R} = R^T R - zz^T, \quad \tilde{R}^T\tilde{R} = R^T R - (z+\delta z)(z+\delta z)^T, \tag{6.13}$$

for the exact and perturbed problems. Subtracting the equations in (6.13) and taking norms it follows that

$$\|\tilde{R}^T\tilde{R} - \bar{R}^T\bar{R}\|_2 \le 2\|\delta z\|_2\|z\|_2 + O(\|\delta z\|_2^2).$$

If we assume that

$$\|\delta z\|_2 \le u\|X\|_2 = u\sigma_1,$$

where $\sigma_1 = \sigma_1(R) = \sigma_1(X)$, and denote $\tilde{\sigma}_i = \sigma_i(\tilde{R})$, $\bar{\sigma}_i = \sigma_i(\bar{R})$, then applying classical perturbation results for eigenvalues gives

$$|\tilde{\sigma}_i^2 - \bar{\sigma}_i^2| \le 2u\sigma_1^2 + O(u^2), \quad i = 1, \dots, n.$$

Neglecting higher order terms and using the elementary inequality $|1 - \sqrt{1+x}| \le |x|$ we get

$$|\tilde{\sigma}_i - \bar{\sigma}_i| \le 2u\frac{\sigma_1^2}{\bar{\sigma}_i}, \quad i = 1, \dots, n. \tag{6.14}$$

In particular

$$|\tilde{\sigma}_n - \bar{\sigma}_n| \le 2u\sigma_1\left(\frac{\sigma_1}{\bar{\sigma}_n}\right),$$

and this bound can be attained. Hence, an *upper bound* for the condition number of the downdating problem with respect to perturbations in z is given by

$$\kappa \le 2\sigma_1/\bar{\sigma}_n.$$

Example 6.1 below shows that this bound can be realistic. Alternatively, one can show, see again Stewart [1979], that the condition number is large if any singular value σ_i is reduced by a significant amount in the downdating.

Since the object is to downdate least squares solutions, we are interested in obtaining an accurate solution w, and not in necessarily in an accurate factor R, see also Section 6.6. However,for any matrix $\tilde{X}$, whose R factor equals $\tilde{R}$ we have $\sigma_i(\tilde{X}) = \tilde{\sigma}_i$, and since

$$\|\tilde{X} - \bar{X}\|_2 \ge \max_i |\tilde{\sigma}_i - \bar{\sigma}_i|,$$

the above result shows that there may be no such matrix $\tilde{X}$ which is closer to $\bar{X}$ than $2u\sigma_1(\sigma_1/\bar{\sigma}_n)$.

We now illustrate this perturbation result in a simple case.

Example 6.1

For $p = 2$, $n = 1$, we have

$$X = \begin{pmatrix} \xi \\ \alpha \end{pmatrix} = Q \begin{pmatrix} \rho \\ 0 \end{pmatrix}, \quad \rho = \sqrt{\xi^2 + \alpha^2}, \quad Q = \frac{1}{\rho} \begin{pmatrix} \xi & -\alpha \\ \alpha & \xi \end{pmatrix},$$

After deleting the first row the downdated factors are

$$\bar{\rho} = \alpha, \quad \bar{Q} = 1.$$

Consider now a perturbation $\delta\xi$ in ξ. For the downdated R factor we get

$$\bar{\rho}^2 = \rho^2 - (\xi + \delta\xi)^2 = \alpha^2 - 2\xi\delta\xi - \delta\xi^2.$$

This computation can break down. Neglecting the term $\delta\xi^2$ we get $\bar{\rho} = 0$ provided that $2\delta\xi/\xi = (\alpha/\xi)^2$. Hence, a relative perturbation in ξ of order u will make $\bar{\rho} = 0$ if $\alpha/\xi = \sqrt{u}$.

The above perturbation analysis shows that downdating R may be a much more ill-conditioned problem than downdating the original matrix X. This is so because the original row in X is not perturbed in the same way as the vector z used to compute the downdating transformation. The LINPACK method may therefore fail to give good results when $\bar{\sigma}_n \ll \sigma_1$. To illustrate this we use again a simple 2 by 1 example.

Example 6.2

Take $\xi = 1$, $\alpha = \epsilon = \sqrt{u}$ in Example 6.1. Then the QR decomposition correctly rounded to single precision is

$$X = \begin{pmatrix} 1 \\ \epsilon \end{pmatrix} = \begin{pmatrix} 1 & -\epsilon \\ \epsilon & 1 \end{pmatrix} \begin{pmatrix} 1 \\ 0 \end{pmatrix}.$$

The LINPACK algorithm will compute

$$q_1 = 1/1 = 1, \quad \gamma^2 = 1 - 1 = 0, \quad J_1 = I,$$

and we obtain $\bar{\rho} = 0$ instead of the correct value $\rho = \epsilon$. Note that *no method working only from R can hope to do better*! Indeed, the information from the second row in X is not present in R, only in Q. It is left to the reader to verify that if we downdate using Q we get the correct result.

6.5 Downdating using Seminormal Equation

To solve the least squares problem

$$\min_w \|Xw - s\|_2,$$

when only the R factor of X is available one may use the **seminormal equations**

$$R^T Rw = X^T s. \tag{6.15}$$

Without modification this method is in general no more accurate than the method of normal equations. However, it has been showed in Björck [1987] that if combined with one step of iterative refinement, using single precision throughout, this method is conditionally stable in the following sense: Provided that a condition $cu\kappa^2 < 1$ holds, where $c = c(p, n)$ is a constant depending on the details of the arithmetic, the forward error will not be worse than for a backward stable method.

We now adapt this method to solve the downdating problem. We first need the following result.

Theorem 6.1

Let w be the solution to

$$\min_w \|e_1 - Xw\|_2.$$

and R the R factor of X. Then the R factor of (X, e_1) is

$$\bar{R} = \begin{pmatrix} R & q_1 \\ 0 & \gamma \end{pmatrix}, \quad q_1 = Rw, \quad \gamma = \|e_1 - Xw\|_2. \tag{6.16}$$

If we use the seminormal equations with iterative refinement to solve for w we get the following algorithm for downdating:

Algorithm 6.1

(i) $R^T q_1 = z;\ Rw = q_1;$

(ii) $r = e_1 - Xw;\ R^T \delta q_1 = X^T r;\ R\delta w = \delta q_1;$

(iii) $q_1 := q_1 + \delta q_1;\ w := w + \delta w;$

(iv) $$\gamma = \|e_1 - Xw\|_2;$$

(v)
$$U^T \begin{pmatrix} q_1 & R \\ \gamma & 0 \end{pmatrix} = \begin{pmatrix} 1 & v^T \\ 0 & \bar{R} \end{pmatrix}.$$

The first triangular system in step (i) and step (iv) are the same as in the LINPACK algorithm.

Example 6.3
Let X be as in Example 6.2. In the method of seminormal equations we compute

$$w = q_1 = 1, \quad \gamma = \| \begin{pmatrix} 1 \\ 0 \end{pmatrix} - \begin{pmatrix} 1 \\ \epsilon \end{pmatrix} 1\|_2 = \epsilon.$$

There is no need for the refinement steps (ii) and (iii) here, and we get

$$U^T \begin{pmatrix} 1 & 1 \\ \epsilon & 0 \end{pmatrix} = \begin{pmatrix} 1 & 1 \\ 0 & \epsilon \end{pmatrix}, \quad \bar{\rho} = \epsilon,$$

which is the correct result.

Algorithm 6.1 requires three more triangular solves than the LINPACK method, if the iterative refinement is carried out. Also, the data matrix X needs to be stored, and two extra matrix times vector multiplications with X and X^T are required. Hence, the added computational complexity is quite high. However, sometimes X may have a special structure so that X can be stored cheaply and the multiplications Xw and X^Tr performed by fast algorithms. This is the case, for example, when X is a Toeplitz matrix. For numerical results comparing the accuracy of Algorithm 6.1 (with and without iterative refinement) and the LINPACK method see Eldén and Waldén [1988].

6.6 Accessing a Computed Factor

It is important to note that even a backward stable method for updating or downdating will not compute an accurate factor $\bar{R}$ if $\bar{R}$ is ill-conditioned. Indeed, Stewart [1977] has proved the following result. Let X and $X + E$

be of full column rank and have the R factors R and $R+U$, respectively. Then, neglecting perturbations of second order, we have

$$\frac{\|U\|_2}{\|R\|_2} \leq (3\kappa(R)+1)\frac{\|E\|_2}{\|X\|_2}. \tag{6.17}$$

A similar result holds for the perturbations in Q. Hence, the accuracy of the R factor itself is not essential for the accuracy of the solution to the least squares solution!

Let $\tilde{R}$ be a computed (up/downdated) factor and $\bar{R}$ the corresponding exact result. By the above perturbation result it follows that the quantity

$$\eta_1 = \|\tilde{R} - \bar{R}\|_2 \tag{6.18}$$

is not a suitable measure for the goodness of $\tilde{R}$. If η is small, then $\tilde{R}$ must be a good R factor since

$$\|\bar{Q}\begin{pmatrix}\tilde{R}\\0\end{pmatrix} - \|\bar{Q}\begin{pmatrix}\bar{R}\\0\end{pmatrix}\|_2 = \|\tilde{X} - \bar{X}\|_2 = \eta_1.$$

However, from the above it follows that η_1 can be of order κ even for a backward stable algorithm.

On the other hand we have

$$\|\tilde{X} - \bar{X}\|_2 \geq \max_i |\tilde{\sigma}_i - \bar{\sigma}_i| = \eta_2, \tag{6.19}$$

which gives a *lower bound* of the deviation. Ideally, in comparing different algorithms we would like to compute the error measure

$$\eta = \min_{\tilde{Q}} \|\tilde{Q}\begin{pmatrix}\tilde{R}\\0\end{pmatrix} - \bar{X}\|_2. \tag{6.20}$$

This is the Procrustes problem, see Golub and Van Loan [1983, p. 426]. A solution can be expressed in terms of the SVD

$$\bar{R}\tilde{R}^T = U\Sigma V^T.$$

The optimal $\tilde{Q}$ is given by UV^T and

$$\eta = \|\tilde{R}UV^T - \bar{R}\|_2,$$

see Golub and Van Loan [1983, p.426]. Computationally, one should not form the product matrix $\bar{R}\tilde{R}^T$ (if $\tilde{R} = \bar{R}$ this squares the condition number!) but instead use the method by Heath, Laub, Paige and Ward [1986], which works directly with $\bar{R}$ and $\tilde{R}$.

References

[1] Alexander, S.T., Pan, C.-T., and Plemmons, R.J. [1988]: Analysis of a recursive least squares hyperbolic scheme for signal processing, Lin. Alg. Appl. **98**, pp. 3–40.

[2] Arioli, M., Demmel, J.W., and Duff, I.S. [1988]: Solving sparse linear systems with sparse backward error. Report CSS 214, Harwell Laboratory.

[3] Björck, Å. [1987]: Stability analysis of the method of semi-normal-equations for linear least squares problems, Linear Algebra Appl. **88/89**, 31–48.

[4] Björck, Å. [1989]: Least squares methods. in *Handbook of Numerical Analysis, Vol.II: Finite Difference Methods-Solution of Equations in* R^n. Eds. P.G. Ciarlet and J.L. Lions, Elsevier/North Holland.

[5] Bunch, J.R. [1987]: The weak and strong stability of algorithms in numerical linear algebra, Linear Algebra Appl. **88/89**, 49–66.

[6] Chan, T.F. [1987]: Rank revealing QR-factorizations, Linear Algebra Appl. **88/89**, 67–82.

[7] Cybenko, G. [1980]: The numerical stability of the Levinson-Durbin algorithm for Toeplitz systems of equations, SIAM J. Sci. Statist. Comput. **1**, 303–319.

[8] Daniel, J.W., Gragg, W.B., Kaufman, L. and Stewart, G.W. [1976]: Reorthogonalization and stable algorithms for updating the Gram-Schmidt QR factorization. Math. Comput. **30**, pp. 772–795.

[9] Eldén, L. and Waldén, B. [1988]: Downdating QR decompositions with improved stability, Report LiTH-MAT-R-1988, Linköping, Sweden.

[10] Foster, L. [1988]: The probability of large diagonal entries in the QR factorizations, submitted to SIAM J. Sci. Statist. Comput.

[11] Golub, G.H. and Van Loan, C.F. [1983]: *Matrix Computations*, John Hopkins University Press.

[12] Hager, W.W. [1984]: Condition estimators, SIAM J. Sci. Statist. Comput., **5**, 311–316.

[13] Heath, M.T., Laub, A.J., Paige, C.C., and Ward, R.C. [1986]: Computing the SVD of a product of two matrices. SIAM J.Sci. Stat. Comput. **7**,1147–1149.

[14] The reference is de Jong, L.S. [1977]: Towards a formal definition of numerical stability. Numer. Math. Vol. **28**, pp. 211-220.

[15] Reichel, L. and Gragg, W.B. [1988]: FORTRAN subroutines for updating the QR decomposition. ACM Trans. Math. Software, to appear.

[16] Skeel, R.D. [1979]: Scaling for numerical stability in Gaussian elimination. *Journal of the Association for Computing Machinery*, **26**, 494-526.

[17] Stewart, G.W. [1977]: Perturbation bounds for the QR factorization of a matrix, SIAM J. Numer. Anal. **14**, 509–518.

[18] Stewart, G.W. [1979]: The effects of rounding errors on an algorithm for downdating a Cholesky factorization, J. Inst. Maths. Applics. **23**, 203–213.

[19] Wedin, P-Å. [1973]: Perburbation theory for pseudo-inverses. BIT **13**, 217–32.

[20] Wilkinson, J.H., [1965]: *The Algebraic Eigenvalue Problem*, Oxford University Press, London.

[21] Wilkinson, J.H. [1986]: Error analysis revisited. *IMA Bulletin*, **22**, 192-200.

Parallel Algorithms for Toeplitz Systems*

Richard P. Brent

Computer Sciences Laboratory
Australian National University
Canberra, ACT 2601, Australia

Abstract

We describe some parallel algorithms for the solution of Toeplitz linear systems and Toeplitz least squares problems. First we consider the parallel implementation of the Bareiss algorithm (which is based on the classical Schur algorithm). The alternative Levinson algorithm is less suited to parallel implementation because it involves inner products.

The Bareiss algorithm computes the LU factorization of the Toeplitz matrix T without pivoting, so can be unstable. For this reason, and also for the application to least squares problems, it is natural to consider algorithms for the QR factorization of T. The first $O(n^2)$ serial algorithm for this problem was given by Sweet, but Sweet's algorithm seems difficult to implement in parallel. Also, despite the fact that it computes an orthogonal factorization of T, Sweet's algorithm can be numerically unstable.

We describe an algorithm of Bojanczyk, Brent and de Hoog for the QR factorization problem, and show that it is suitable for parallel implementation. This algorithm overcomes some (but not all) of the numerical difficulties of Sweet's algorithm. We briefly compare some other algorithms, such as the "lattice" algorithm of Cybenko and the "generalized Schur" algorithm of Chun, Kailath and Lev-Ari.

*A preliminary version of this paper appeared in [9]

NATO ASI Series, Vol. F 70
Numerical Linear Algebra, Digital Signal Processing
and Parallel Algorithms
Edited by G. H. Golub and P. Van Dooren

1 Introduction

A Toeplitz matrix $A = (a_{i,j})$ is one in which $a_{i,j}$ is a function of $j - i$. It is convenient to write a_{j-i} for $a_{i,j}$ and consider the $m+1$ by $n+1$ matrix $A = (a_{j-i})_{i=0,\ldots,m;j=0,\ldots,n}$. Toeplitz matrices often arise in digital signal processing, e.g. in linear prediction problems [17, 43] as well as in the discretization of certain integral equations and elsewhere in numerical analysis [13]. We are often interested in solving one or more linear systems

$$Ax = b \tag{1.1}$$

where A is square ($m = n$) and nonsingular, often symmetric positive definite; or in solving linear least squares problems

$$\min \|Ax - b\|_2 \tag{1.2}$$

(where generally A is rectangular with $m > n$). We shall assume throughout that A, b etc are real and that A has rank $n+1$. We shall not consider block Toeplitz matrices, although many of the algorithms mentioned here are applicable (with appropriate modifications) to block Toeplitz problems.

Consider first the linear system (1.1). The "classical" Gaussian elimination or Cholesky decomposition algorithms [28, 57] involve $O(n^3)$ arithmetic operations. However, for over forty years "fast" algorithms involving only $O(n^2)$ operations have been known [1, 3, 20, 21, 22, 31, 41, 45, 47, 56, 58, 59].

The Toeplitz structure of A implies that the matrix-vector product Ax reduces to a convolution problem and can be computed via the fast Fourier transform in $O(n \log n)$ arithmetic operations. This observation suggests that $o(n^2)$ algorithms for (1.1) might exist, and indeed such algorithms were found about ten years ago by Brent et al [10], Bitmead & Anderson [4], and Morf [44]. More such algorithms have been found recently by de Hoog [29], Musicus [46], Kumar [36], and Ammar & Gragg [2]. These algorithms all require $O(n(\log n)^2)$ arithmetic operations and are termed "superfast" by Ammar & Gragg [2], although we prefer the description "asymptotically fast" as n may have to be very large before they are actually faster than the $O(n^2)$ algorithms (see Section 4). It is conceivable that (asymptotically) even faster algorithms exist.

In practice operation counts are not the only criterion. The speed of an implementation may depend on how readily the algorithm can be implemented on a pipelined or parallel architecture, and this may depend on whether inner products need to be computed or not. The overall cost depends on the difficulty of programming and debugging, so simple algorithms are generally preferable to complex ones. Numerical stability (or the lack of it) is an important consideration [13, 16, 18, 42].

Algorithms for solving (1.1) split into two main classes –

- (A) those that compute A^{-1} or a factorization of A^{-1}, and
- (B) those that compute a factorization of A.

In class (A) we have for example the Levinson [41], Durbin [21] and Trench [54] algorithms which, in the symmetric case, are closely related to the classical Szegö recursions for polynomials orthogonal on the unit circle [2, 35, 53]. These algorithms generally require inner products and are stable if A is positive definite [13].

In class (B) we have for example the algorithms of Bareiss [3], Brent, Kung & Luk [11, 12], and Kung and Hu [38], which are related to the classical algorithm of Schur [48] for the continued fraction representation of a holomorphic function in the unit disk [2, 32, 35].

The fast algorithms of Bitmead & Anderson [4] and Morf [44] are based on the concept of displacement rank [25, 26, 33, 34] and may be placed in class (A), while those of Brent et al [10], de Hoog [29], Musicus [46], and Ammar & Gragg [2] are related to continued fractions and generalized Schur algorithms, so may be placed in class (B).

Consider now the full rank linear least squares problem (1.2). The solution of (1.2) is

$$x = (A^TA)^{-1}A^Tb,$$

but it is generally undesirable to compute x from this formula, for numerical reasons [28] and because A^TA is not Toeplitz (although it does have low displacement rank). A better approach is to compute an "orthogonal factorization" of A, i.e. a factorization

$$A = QR \tag{1.3}$$

where Q is an $m+1$ by $n+1$ matrix with orthonormal columns and R is an $n+1$ by $n+1$ upper triangular matrix. Once (1.3) is known, it is easy to solve (1.2), using

$$Rx = Q^T b \tag{1.4}$$

or, to avoid explicit computation of Q, the "semi-normal equations"

$$R^T R x = A^T b \tag{1.5}$$

As in the discussion of (1.1), there are two classes of algorithms – some algorithms find the inverse factorization $A\bar{R} = Q$ instead of $A = QR$. The algorithms which compute $\bar{R}$ are related to the "lattice" algorithm [17, 18, 19, 30, 43], while algorithms which compute R include those of Sweet [52], Bojanczyk, Brent & Kung [6, 42] and Chun, Kailath & Lev-Ari [14]. When considering rectangular matrices, "fast" means algorithms requiring $O(mn)$ operations, since the classical algorithms require $\Omega(mn^2)$ operations.

Clearly with exact arithmetic we have

$$\bar{R} = R^{-1},$$

but numerically there is a difference between algorithms which compute $\bar{R}$ and those which compute R. Cybenko [18] has analysed the classical lattice algorithm and he suggests that its good numerical properties may carry over to his algorithm [19] for the solution of (1.2). This is by no means certain, because of the use of non-Euclidean inner products. On the other hand, Sweet's algorithm can be unstable [6, 42], even though it computes the orthogonal decomposition (1.3). The algorithms of Bojanczyk, Brent & de Hoog [6] and Chun, Kailath & Lev-Ari [14] involve the downdating of a Cholesky factorization, and this is known to be a poorly conditioned problem [8, 51]. It is an open problem whether there is a fast ($O(mn)$), unconditionally stable algorithm for the computation of the QR factorization of a Toeplitz matrix. Such algorithms do exist when the Toeplitz matrix has a special structure [5, 18].

Another approach [27, 50] to the solution of (1.2) is to solve the $m+n+2$ by $m+n+2$ system

$$\begin{pmatrix} 0 & A^T \\ A & -I \end{pmatrix} \begin{pmatrix} x \\ r \end{pmatrix} = \begin{pmatrix} 0 \\ b \end{pmatrix} \tag{1.6}$$

to find both the solution vector x and the residual vector $r = Ax - b$. This can be done in $O(m(\log m)^2)$ operations, by generalizations of some of the algorithms mentioned above, for the matrix in (1.6) has low displacement rank. This approach has the advantage that iterative refinement [28, 57] may easily be applied to improve the accuracy of the solution, if this is necessary. Its disadvantage is its obvious inefficiency if $m \gg n$, unless the special structure is used to advantage [50].

In practice space requirements need to be considered. The problem (1.1) can be defined with $O(n)$ parameters, and its solution can be expressed with $O(n)$ parameters, but many algorithms require $\Omega(n^2)$ words of intermediate storage. Clearly this is necessary if the triangular factors of A or A^{-1} are stored explicitly, since these factors are not Toeplitz. However, with care and a little extra computation, it is possible to implement some of the algorithms mentioned above with only "linear", i.e. $O(n)$, storage. Similar remarks apply to algorithms for the solution of (1.2).

In Section 2 we describe a version of the Bareiss algorithm for solving the linear Toeplitz system (1.1), and show how it can be implemented efficiently in parallel, using only linear storage. We briefly compare some other fast algorithms for solving (1.1). In Section 3 we consider algorithms for solving (1.2), and in particular we describe the algorithm of Bojanczyk, Brent & de Hoog and its parallel implementation. Finally, in Section 4 we comment on some of the asymptotically fast algorithms for solving (1.1).

When considering parallel algorithm we shall concentrate on algorithms suitable for systolic arrays [37]. Because systolic arrays are such a restricted class of parallel machines, they can be simulated efficiently on most other classes of parallel machines (e.g. hypercubes or shared memory machines). Thus, our algorithms for systolic arrays are easily converted into efficient algorithms for these other classes of parallel machines (but not conversely).

2 A parallel Bareiss algorithm

Write (1.1) as

$$A^{(0)}x = b^{(0)}.$$

At iteration k of the Bareiss algorithm ($k = 1, \ldots, n$) we have two linear systems

$$A^{(\pm(k-1))}x = b^{(\pm(k-1))}$$

which are both equivalent to (1.1), and these are transformed into two further systems

$$A^{(\pm k)}x = b^{(\pm k)}.$$

The transformations are chosen to eliminate *diagonals*, rather than rows or columns. Because of the Toeplitz nature of A, each transformation requires only $O(n)$ operations. After k iterations, the matrix $A^{(+k)}$ has k zero diagonals above the main diagonal, and the matrix $A^{(-k)}$ has k zero diagonals below the main diagonal. Formally, if

$$Z_k = (\delta_{i-j+k})_{i=0,\ldots,n;j=0,\ldots,n}$$

is a shift matrix (premultiplication by Z_{+k} shifts k rows up with zero fill; premultiplication by Z_{-k} shifts k rows down with zero fill), then the transformations are defined by

$$A^{(-k)} = A^{(-(k-1))} - m_{-k}Z_{-k}A^{(k-1)},$$

$$b^{(-k)} = b^{(-(k-1))} - m_{-k}Z_{-k}b^{(k-1)},$$

$$A^{(+k)} = A^{(k-1)} - m_k Z_k A^{(-k)},$$

$$b^{(+k)} = b^{(k-1)} - m_k Z_k b^{(-k)}$$

for $k = 1, \ldots, n$, where

$$m_{-k} = a_{k,0}^{(-(k-1))}/a_0$$

and

$$m_{+k} = a_{0,k}^{(k-1)}/a_{n,n}^{(-k)}.$$

To ensure that the multipliers $m_{\pm k}$ are well-defined, we must assume that all the leading principal submatrices of A are nonsingular.

After n iterations we obtain an upper triangular system

$$A^{(-n)}x = b^{(-n)} \tag{2.1}$$

which may easily be solved in $O(n^2)$ operations. In fact, we have computed a triangular factorization of A, for it may be shown that $A = LU$, where $a_0 L = (A^{(n)})^{T2}$, $U = A^{(-n)}$, and "$T2$" denotes matrix transpose about the main antidiagonal [52]. Thus, the Bareiss algorithm has essentially the same numerical properties as Gaussian elimination without pivoting – it is numerically stable if A is positive definite or diagonally dominant (as is often the case in applications), but it is unstable in general. Provided real-time response is not required, it is possible to handle this potential instability by computing the residual vector $r = A\bar{x} - b$ (where $\bar{x}$ is the computed, possibly inaccurate solution), then applying iterative refinement [28, 57] or switching to a slower but more stable algorithm if $\|r\|$ is unacceptably large.

Note that L and U are not Toeplitz. The Toeplitz structure gradually disappears during the iterations of the Bareiss algorithm. In fact, all but the top k rows of $A^{(-k)}$ are Toeplitz, and all but the bottom k rows of $A^{(+k)}$ are Toeplitz. This seems to imply that $\Omega(n^2)$ storage space is required. However, it was shown by Brent et al [11, 12] that space $O(n)$ is sufficient to solve (2.1), as we can generate the rows of $A^{(-n)}$ as required for the backsubstitution process by running the Bareiss algorithm backwards (using the stored multipliers $m_{\pm k}$) –

$$A^{(k-1)} = A^{(k)} + m_k Z_k A^{(-k)}$$

and

$$A^{(-(k-1))} = A^{(-k)} + m_{-k} Z_{-k} A^{(k-1)}$$

for $k = n, \ldots, 1$. For details and numerical examples, see Brent & Luk [12].

It is easy to convert the Bareiss algorithm into a "semisystolic" algorithm [40, 55] in which the multipliers $m_{\pm k}$ are broadcast to all processing elements in a linear array of $n+1$ systolic cells. By a standard technique [40, 55] this semisystolic algorithm can be converted into a true systolic algorithm in which no broadcasting is required, only nearest neighbour communication. For details, we again refer to Brent & Luk [12]. Here we merely note that one systolic processor needs to perform divisions (to compute the multipliers); the others need only perform "multiply and accumulate" operations of the form $\alpha \leftarrow \alpha + \beta \cdot \gamma$.

The systolic implementation of the Bareiss algorithm uses $n+1$ systolic cells and solves (1.1) in time $O(n)$ and space $O(n)$, i.e. constant space per systolic cell. On other parallel architectures such as transputer arrays, hypercubes, etc with p processors, we can simulate the systolic array in time $O(n^2/p)$, provided $p = O(n)$. Thus, the speedup is of order p and the efficiency is of order unity.

The algorithm described above does not take advantage of any symmetry. However, a symmetric version of the algorithm exists, and saves some time and communications overhead if A is symmetric [12, 38].

The algorithms in class (A) generally require less arithmetic operations than the Bareiss algorithm (by a moderate constant factor), but they involve a sequence of $O(n)$ inner products which need to be performed sequentially. On a parallel machine with $p = O(n)$ processors, each inner product takes time $\Omega(n/p + \log p)$, so a parallel implementation of these algorithms will take time $\Omega(n^2/p + n \log p)$. In particular, if $p \sim n$, the time is $\Omega(n \log n)$, greater than the time $O(n)$ for the Bareiss algorithm.

3 Fast Toeplitz QR Factorization

Sweet [52] gave the first fast ($O(mn)$) algorithm for the orthogonal factorization (1.3) of a Toeplitz matrix. The classical Givens and Householder algorithms [28, 39, 50, 57] form Q as a product of elementary orthogonal matrices, so the computed matrix Q is guaranteed to be very close to orthogonal. Unfortunately, this is not the case for Sweet's algorithm – it turns out that the computed Q can deviate significantly from orthogonality because of the effect of rounding errors during the computation. (Similar remarks apply to algorithms which compute R or R^{-1} and then find Q from $Q = AR^{-1}$ – this approach is unsatisfactory if R is poorly conditioned.) Luk & Qiao [42] show (for the case $m = n$) that Sweet's algorithm is unstable if the matrix formed by deleting the first row and last column of A is poorly conditioned. We recommend the numerical examples given by

Luk & Qiao to anyone interested in understanding the numerical behaviour of the Sweet, Bareiss and Trench algorithms.

Bojanczyk, Brent & de Hoog [6] (BBH for short) discovered a different Toeplitz QR factorization algorithm while trying to understand the reason for numerical instability in Sweet's algorithm. The key idea is to obtain a recursion for the rows of R (and, optionally, for the columns of Q) from the shift-invariance property of Toeplitz matrices. We partition the Toeplitz matrix A in two ways –

$$A = \begin{pmatrix} a_0 & y^T \\ z & A_{-1} \end{pmatrix} = \begin{pmatrix} A_{-1} & \bar{y} \\ \bar{z}^T & a_{n-m} \end{pmatrix}$$

and partition R in two corresponding ways –

$$R = \begin{pmatrix} r_{0,0} & u^T \\ 0 & R_b \end{pmatrix} = \begin{pmatrix} R_t & \bar{u} \\ 0 & r_{n,n} \end{pmatrix},$$

where u, $\bar{u}$, y, $\bar{y}$, z and $\bar{z}$ are column vectors, A_{-1} is the principal (top left or bottom right – they are identical) m by n submatrix of A, R_t is the top left principal n by n submatrix of R, and R_b is the bottom right principal n by n submatrix of R. Using the relation $R^T R = A^T A$ we obtain

$$r_{0,0} = (a_0^2 + z^T z)^{1/2}, \tag{3.1}$$

$$r_{0,0} u = a_0 y + A_{-1}^T z, \tag{3.2}$$

and

$$R_b^T R_b = R_t^T R_t + y y^T - \bar{z}\bar{z}^T - u u^T. \tag{3.3}$$

Relations (3.1) and (3.2) define the first row of R. Relation (3.3) shows that R_b differs from R_t by a matrix of rank 3. In fact, R_b can be obtained from R_t by a rank-1 update followed by two rank-1 downdates. This allows us to calculate the kth row of R_b from the first k rows of R_t. However, the kth row of R_b defines the $(k+1)$st row of R_t. Thus, we have a recursion for calculating the rows of R, starting from the first. It is possible to obtain a similar recursion for the columns of Q, although this is not necessary if (1.5) is used to solve the least squares problem (1.2).

Despite their similar derivations, the BBH algorithm and Sweet's algorithm are different. The operation count for the BBH algorithm is slightly lower than for Sweet's algorithm [6]. While Sweet's algorithm appears difficult to implement in parallel, the BBH algorithm can be implemented on a systolic array and runs in time $O(mn)$ with $n+1$ processors [7]. Also, because the rows of R can be generated in reverse order by running the algorithm backwards (as for the Bareiss algorithm described in Section 2), the linear least squares problem (1.2) can be solved with space $O(m)$. Finally, the numerical properties of the BBH algorithm and Sweet's algorithm differ – numerical experiments suggest that the BBH algorithm is never much less accurate than Sweet's algorithm, and often much more accurate.

Updating Cholesky factors after a (positive) rank-1 correction can be done in a numerically stable way using plane rotations (i.e. circular transformations) [24]. However, it is known that downdating of Cholesky factors is numerically dangerous [8, 24, 51]. This is to be expected because the (real) Cholesky factorization of $R^T R - vv^T$ only exists if $R^T R - vv^T$ has no negative eigenvalues. Thus, we would anticipate numerical difficulties if $R_b^T R_b$ is poorly conditioned. We conjecture that the BBH algorithm is *weakly stable* in the sense that R is computed accurately provided that $R^T R$ $(= A^T A)$ is well-conditioned. Similarly, we conjecture that the BBH algorithm combined with the seminormal equations (1.5) gives a weakly stable algorithm for the solution of the linear least squares problem (1.2) so long as $\min \|Ax - b\|_2 \ll \|b\|_2$.

Recently Chun, Kailath & Lev-Ari [14] (CKL for short) presented a Toeplitz QR factorization algorithm (or family of algorithms) based on the fact that R is a Cholesky factor of the matrix $A^T A$, and this matrix has low displacement rank. They were able to interpret the BBH algorithm in this context, and show that it is closely related to the CKL algorithm. The algorithms have similar operation counts – we shall not be specific because the operation counts depend on details such as whether "fast" Givens transformations [23] are used. Because the algorithms are so closely related, it is not surprising that the CKL algorithm has a nice parallel implementation similar to that of the BBH algorithm.

It might be supposed that the CKL algorithm would have better numerical properties than the BBH algorithm because the CKL algorithm uses less hyperbolic transformations (and correspondingly more circular transformations) than the BBH algorithm. However, this is not necessarily the case. Equation (3.3) suggests that the critical factor, at least for the BBH algorithm, is the condition of $R_b^T R_b$, which is dependent on the original Toeplitz matrix A but not on the number of hyperbolic transformations performed by the algorithm. There is no need to invoke several hyperbolic transformations as a cause of instability when a single one is sufficient. As an analogy, we mention that the presence of just one large multiplier during Gaussian elimination without pivoting is sufficient to cause instability.

Cybenko [19] has recently suggested a quite different Toeplitz orthogonalization algorithm. His algorithm finds $\bar{R}$ and Q such that $A\bar{R} = Q$, and is based on a generalization of the "lattice" algorithm [17, 30, 43], which solves the same problem when A is restricted to have a special structure. Because Cybenko's algorithm uses orthogonalization with respect to a non-Euclidean inner product, its numerical properties are uncertain.

The comparison of Cybenko's algorithm with the BBH and CKL algorithms is analogous to the comparison of Levinson's algorithm with the Bareiss algorithm. Cybenko's algorithm uses less arithmetic operations and has different numerical properties, but it is less well suited to parallel implementation because it requires the evaluation of inner products.

4 Asymptotically fast algorithms

The asymptotically fast algorithms mentioned in Section 1 require $O(n(\log n)^2)$ operations to solve the Toeplitz linear system (1.1). The theory of these algorithms is fascinating as it exhibits connections between many apparently unrelated topics – Szegö polynomials, Schur's algorithm, Padé approximants, the Euclidean algorithm, etc. Space does not permit a discussion of these topics, so we refer to the literature [2, 10, 32, 35].

Sexton et al [49] evaluated the constant hidden in the "O" notation for an asymptotically fast algorithm based on the concept of displacement rank [4], and found that the constant was so large that the algorithm was impractical. More recent asymptotically fast algorithms, such as that of de Hoog [29], appear to have a smaller constant. Ammar & Gragg [2] have made a thorough attempt to minimize the constant and claim that their generalized Schur algorithm should be faster than Levinson's algorithm for $n > 256$ (approximately) when applied to a positive definite system. The comparison here is with the classical Levinson algorithm rather than with the slightly faster "split Levinson" algorithm.

All the asymptotically fast algorithms involve a "doubling step" in which the problem of size n is split into two subproblems of size approximately $n/2$, the results of which can be combined by some operations equivalent to convolution. Thus the time $T(n)$ required satisfies

$$T(n) = 2T(n/2) + C(n), \tag{4.1}$$

where $C(n)$ is the time required for a convolution of size n. Using the fast Fourier transform (or other fast convolution algorithm) gives $C(n) = O(n \log n)$ and thus (4.1) implies $T(n) = O(n(\log n)^2)$. Of course, the constant factor here depends on the constant for the convolution algorithm.

When we attempt to implement the asymptotically fast algorithms on a parallel machine, (4.1) still holds because the two subproblems are dependent – the second can not be performed until the first is complete. On a linear systolic array [37] we have $C(n) = O(n)$, so (4.1) implies that $T(n) = O(n \log n)$. Even on a more general parallel machine such as a hypercube, with $C(n) = O((\log n)^\alpha)$ for some positive constant α, we only get $T(n) = O(n)$, which is disappointing as the same result can be obtained much more easily by the parallel Bareiss algorithm (Section 2). On a hypercube with enough processors it is certainly possible to obtain $T(n) = O((\log n)^2)$, but we only know how to do this by ignoring the Toeplitz structure and using a fast but very inefficient algorithm such as that of Csanky [15], which requires $O(n^{3.5})$ processors.

5 Concluding remarks and open problems

Toeplitz systems are of interest to mathematicians because they display connections between various apparently unrelated areas of mathematics. They are of interest to numerical analysts and engineers because they arise in many applications and need to be solved quickly and accurately. A large literature exists, and many interesting algorithms have been developed, but there is still room for progress. We mention a few interesting problem areas in which progress on algorithms might be made –

- 1. The development of fast, provably stable algorithms for computing the QR decomposition of a Toeplitz matrix A and solving the linear least squares problem (1.2).
- 2. The discovery of "asymptotically fast" algorithms which are actually fast (compared to $O(n^2)$ algorithms) for moderate values of n.
- 3. The development of good parallel algorithms which solve the Toeplitz system (1.1) in time $O((\log n)^2)$ on a hypercube, using only a moderate number of processors (say $O(n^2)$).
- 4. Generalization of 1-3 above to algorithms with provably good numerical properties for low displacement rank matrices.
- 5. The discovery of good algorithms for other problems involving Toeplitz and low displacement rank matrices, e.g. eigenvalue and inverse eigenvalue problems.

References

[1] H. A. Ahmed, J-M. Delosme and M. Morf, "Highly concurrent computing structures for matrix arithmetic and signal processing", *IEEE Transactions on Computers* 15 (1982), 65-82.

[2] G. S. Ammar and W. B. Gragg, "Superfast solution of real positive definite Toeplitz systems", *SIAM J. Matrix Anal. Appl.* 9 (1988), 61-76.

[3] E. H. Bareiss, "Numerical solution of linear equations with Toeplitz and vector Toeplitz matrices", *Numerische Mathematik* 13 (1969), 404-424.

[4] R. R. Bitmead and B. D. O. Anderson, "Asymptotically fast solution of Toeplitz and related systems of linear equations", *J. Linear Algebra Appl.* 34 (1980), 103-116.

[5] A. W. Bojanczyk and R. P. Brent, "Parallel solution of certain Toeplitz least squares problems", *J. Linear Algebra Appl.* 77 (1986), 43-60.

[6] A. W. Bojanczyk, R. P. Brent and F. R. de Hoog, "QR factorization of Toeplitz matrices", *Numerische Mathematik* 49 (1986), 81-94.

[7] A. W. Bojanczyk, R. P. Brent and F. R. de Hoog, "Linearly connected arrays for Toeplitz least squares problems", Report CMA-R06-85, Centre for Mathematical Analysis, Australian National University, May 1985.

[8] A. W. Bojanczyk, R. P. Brent, P. Van Dooren and F. R. de Hoog, "A note on downdating the Cholesky factorization", *SIAM J. Scientific and Statistical Computing* 8 (1987), 210-221.

[9] R. P. Brent, "Old and new algorithms for Toeplitz systems", *Proceedings SPIE, Volume 975, Advanced Algorithms and Architectures for Signal Processing III* (edited by Franklin T. Luk), Society of Photo-Optical Instrumentation Engineers, Bellingham, Washington, 1989, 2-9.

[10] R. P. Brent, F. G. Gustavson and D. Y. Y. Yun, "Fast solution of Toeplitz systems of equations and computation of Padé approximants", *J. Algorithms* 1 (1980), 259-295.

[11] R. P. Brent, H. T. Kung and F. T. Luk, "Some linear-time algorithms for systolic arrays", in *Information Processing 83* (edited by R.E.A. Mason), North-Holland, Amsterdam, 1983, 865-876.

[12] R. P. Brent and F. T. Luk, "A systolic array for the linear-time solution of Toeplitz systems of equations", *J. of VLSI and Computer Systems* 1 (1983), 1-23.

[13] J. Bunch, "Stability of methods for solving Toeplitz systems of equations", *SIAM J. Scientific and Statistical Computing* 6 (1985), 349-364.

[14] J. Chun, T. Kailath and H. Lev-Ari, "Fast parallel algorithms for QR and triangular factorization", *SIAM J. Scientific and Statistical Computing* 8 (1987), 899-913.

[15] L. Csanky, "Fast parallel matrix inversion algorithms", *SIAM J. on Computing* 5 (1976), 618-623.

[16] G. Cybenko, "The numerical stability of the Levinson-Durbin algorithm for Toeplitz systems of equations", *SIAM J. Scientific and Statistical Computing* 1 (1980), 303-320.

[17] G. Cybenko, "A general orthogonalization method with applications to time series analysis and signal processing", *Mathematics of Computation* 40 (1983), 323-336.

[18] G. Cybenko, "The numerical stability of lattice algorithms for least squares linear prediction problems", *BIT* 24 (1984), 441-455.

[19] G. Cybenko, "Fast Toeplitz orthogonalization using inner products", *SIAM Jnl. Scientific and Statistical Computing* 8 (1987), 734-740.

[20] P. Delsarte, Y. Genin and Y. Kamp, "A polynomial approach to the generalized Levinson algorithm based on the Toeplitz distance", *IEEE Trans. Inform. Theory* IT-29 (1983), 268-278.

[21] J. Durbin, "The fitting of time-series models", *Rev. Inst. Internat. Statist.* 28 (1960), 233-244.

[22] B. Friedlander, M. Morf, T. Kailath and L. Ljung, "New inversion formulas for matrices classified in terms of their distance from Toeplitz matrices", *J. Linear Algebra Appl.* 27 (1979), 31-60.

[23] M. Gentleman, "Least squares computations by Givens transformations without square roots", *J. Inst. Math. Appl.* 12 (1973), 329-336.

[24] P. E. Gill, G. H. Golub, W. Murray and M. A. Saunders, "Methods for modifying matrix factorizations", *Mathematics of Computation* 28 (1974), 505-535.

[25] I. C. Gohberg and I. A. Feldman, *Convolution Equations and Projection Methods for their Solution*, Translations of Mathematical Monographs, Vol. 41, Amer. Math. Soc., Providence, Rhode Island, 1974.

[26] I. C. Gohberg and A. A. Semencul, "On the inversion of finite Toeplitz matrices and their continuous analogs", *Mat. Issled.* 2 (1972), 201-233 (in Russian).

[27] G. H. Golub, "Numerical methods for solving linear least squares problems", *Numerische Mathematik* 7 (1965), 206-216.

[28] G. H. Golub and C. Van Loan, *Matrix Computations*, Johns Hopkins Press, Baltimore, Maryland, 1983.

[29] F. R. de Hoog, "A new algorithm for solving Toeplitz systems of equations", *J. Linear Algebra Appl.* 88/89 (1987), 122-138.

[30] F. Itakura and S. Saito, "Digital filtering techniques for speech analysis and synthesis", *Proc. 7th Internat. Congr. Acoustics*, Budapest (1971), 261-264.

[31] A. K. Jain, "An efficient algorithm for a large Toeplitz set of linear equations", *IEEE Trans. Acoust. Speech Sig. Proc.* 27 (1979), 612-615.

[32] T. Kailath, "A theorem of I. Schur and its impact on modern signal processing", in *Schur methods in Operator Theory and Signal Processing*, edited by I. C. Gohberg, Birkhäuser-Verlag, Basel (1986), 9-30.

[33] T. Kailath, S. Y. Kung and M. Morf, "Displacement ranks of matrices and linear equations", *J. Math. Appl.* 68 (1979), 395-407.

[34] T. Kailath, B. Levy, L. Ljung and M. Morf, "The factorization and representation of operators in the algebra generated by Toeplitz operators", *SIAM J. Appl. Math.* 37 (1979), 467-484.

[35] T. Kailath, A. Viera and M. Morf, "Inverses of Toeplitz operators, innovations, and orthogonal polynomials", *SIAM Review* 20 (1978), 106-119.

[36] R. Kumar, "A fast algorithm for solving a Toeplitz system of equations", *IEEE Trans. Acoust. Speech and Signal Proc.* 33 (1985), 254-267.

[37] H. T. Kung, "Why systolic architectures ?", *IEEE Computer* 15, 1 (1982), 37-46.

[38] S. Y. Kung and Y. H. Hu, "A highly concurrent algorithm and pipelined architecture for solving Toeplitz systems", *IEEE Trans. Acoust. Speech and Signal Proc.* ASSP-31 (1983), 66-76.

[39] C. L. Lawson and R. J. Hanson, *Solving Least Squares Problems*, Prentice-Hall, Englewood Cliffs, New Jersey, 1974.

[40] C. E. Leiserson and J. B. Saxe, "Optimizing synchronous systems", *J. of VLSI and Computer Systems* 1 (1983), 41-67.

[41] N. Levinson, "The Wiener RMS (root mean square) error criterion in filter design and prediction", *J. Math. Phys.* 25 (1947), 261-278.

[42] F. T. Luk and S. Qiao, "A fast but unstable orthogonal triangularization technique for Toeplitz matrices", *J. Linear Algebra Appl.* 88/89 (1987), 495-506.

[43] J. Makhoul, "Linear prediction: a tutorial review", *Proceedings IEEE*, 63 (1975), 561-580.

[44] M. Morf, "Doubling algorithms for Toeplitz and related equations", *Proc. IEEE Internat. Conf. on Acoustics, Speech and Signal Processing*, Denver, Colorado (April 1980), 954-959.

[45] M. Morf and J-M. Delosme, "Matrix decompositions and inversions via elementary signature-orthogonal transformations", *Proc. Internat. Symposium on Mini- and Micro-computers in Control and Measurement*, San Francisco, California (May 1981).

[46] B. R. Musicus, "Levinson and fast Choleski algorithms for Toeplitz and almost Toeplitz matrices", Report, Research Lab. of Electronics, MIT, Cambridge, Massachusetts, 1984.

[47] J. Rissanen, "Solution of linear equations with Hankel and Toeplitz matrices", *Numerische Mathematik* 22 (1974), 361-366.

[48] J. Schur, "Über Potenzreihen, die im Innern des Einheitskreises beschränkt sind", *J. Reine Angew. Math.* 147 (1917), 205-232.

[49] H. Sexton, M. Shensa and J. Speiser, "Remarks on a displacement-rank inversion method for Toeplitz systems", *J. Linear Algebra Appl.* 45 (1982), 127-130.

[50] G. W. Stewart, *Introduction to Matrix Computations*, Academic Press, New York, 1973.

[51] G. W. Stewart, "The effects of rounding error on an algorithm for downdating a Cholesky factorization", *J. Inst. Math. Appl.* 23 (1979), 203-213.

[52] D. R. Sweet, "Fast Toeplitz orthogonalization", *Numer. Math.* 43 (1984), 1-21.

[53] G. Szegö, *Orthogonal Polynomials*, third edition, AMS Colloquium Pub. vol. 23, Amer. Math. Soc. , Providence, Rhode Island (1967).

[54] W. F. Trench, "An algorithm for the inversion of finite Toeplitz matrices", *J. Soc. Indust. Appl. Math.* 12 (1964), 515-522.

[55] J. D. Ullman, *Computational Aspects of VLSI*, Computer Science Press, Maryland, 1984.

[56] G. A. Watson, "An algorithm for the inversion of block matrices of Toeplitz form", *J. ACM* 20 (1973), 409-415.

[57] J. H. Wilkinson, *The Algebraic Eigenvalue Problem*, Clarendon Press, Oxford, 1965.

[58] S. Zohar, "Toeplitz matrix inversion: the algorithm of W. F. Trench", *J. ACM* 16 (1969), 592-601.

[59] S. Zohar, "The solution of a Toeplitz set of linear equations", *J. ACM* 21 (1974), 272-276.

Parallel Algorithms for Digital Signal Processing

Richard P. Brent

Computer Sciences Laboratory
Australian National University
Canberra, ACT 2601, Australia

1 Introduction

This paper provides an introduction to some parallel algorithms relevant to digital signal processing. First we introduce some basic concepts such as *speedup* and *efficiency* of parallel algorithms. We also outline some practical parallel computer architectures – pipelined, SIMD and MIMD machines, hypercubes and systolic arrays.

To illustrate the basic concepts and key issues, we consider the problem of parallel solution of a nonsingular linear system. Gaussian elimination with pivoting is difficult to implement on many parallel architectures, but omission of pivoting leads to numerical instability. A solution is to implement a parallel version of the orthogonal (QR) decomposition instead of the triangular (LU) decomposition obtained by Gaussian elimination. We consider as an application the solution of linear least squares problems.

Many problems in digital signal processing are easy to solve if we can find the singular value decomposition (SVD) of a rectangular matrix, or the eigenvalues and eigenvectors of a symmetric (or Hermitian) matrix. We describe some good parallel algorithms for these problems. Often the parallel algorithms are based on old ideas (such as Jacobi's method for finding the eigenvalues of a symmetric matrix) rather than on a straightforward adaption of the best serial algorithms. The parallel algorithms can

NATO ASI Series, Vol. F 70
Numerical Linear Algebra, Digital Signal Processing and Parallel Algorithms
Edited by G. H. Golub and P. Van Dooren

be implemented efficiently on systolic arrays, but we also mention how they might be implemented on other parallel architectures.

Toeplitz systems often arise in digital signal processing. However, we do not discuss their solution here as they are considered in the companion paper [6].

2 Basic concepts

We assume that a parallel machine with P processors is available. Thus P measures the degree of parallelism; $P = 1$ is just the familiar serial case. When considering the solution of a particular problem, we let T_P denote the time required to solve the problem using (at most) P processors. The *speedup* S_P is defined by

$$S_P = T_1/T_P,$$

and the *efficiency* $E_P = S_P/P$.

When converting a serial algorithm into a parallel algorithm, our aim is usually to attain constant efficiency, i.e.

$$E_P \geq c$$

for some positive constant c independent of P. This may be written as

$$E_P = \Omega(1).$$

Equivalently, we want to attain linear speedup, i.e.

$$S_P \geq cP,$$

which may be written as

$$S_P = \Omega(P).$$

2.1 Amdahl's Law

Suppose a positive fraction f of a computation is "essentially serial", i.e. not amenable to any speedup on a parallel machine. Then we would expect

$$T_P = fT_1 + (1 - f)T_1/P$$

so the overall speedup

$$S_P = \frac{1}{f + (1-f)/P} \leq \frac{1}{f}, \tag{2.1}$$

i.e. the speedup is *bounded*, not linear. The inequality (2.1) is called *Amdahl's Law* and is often used as an argument against parallel computation. However, what it shows is that the speedup is bounded as we increase the number of processors for a *fixed* problem. In practice, it is more likely that we shall want to solve larger problems as the number of processors increases.

Let N be a measure of the problem size. For many problems it is reasonable to assume that

$$f \leq K/N \tag{2.2}$$

for some constant K. For example, in problems involving N by N matrices, we may have $\Omega(N^3)$ arithmetic operations and $O(N^2)$ serial input/output.

Suppose also that N increases at least linearly with P, i.e.

$$N \geq KP. \tag{2.3}$$

(2.2) and (2.3) imply that $fP \leq 1$, so from (2.1) we have

$$S_P = \frac{P}{fP + (1-f)} \geq \frac{P}{2-f} \geq \frac{P}{2}.$$

Thus we get linear speedup, with efficiency $E_P \geq 1/2$.

2.2 Virtual processors

In practice any parallel machine has a fixed maximum number (say P) of processors imposed by hardware constraints. When analysing parallel algorithms it is often convenient to ignore this fact and assume that we have as many (say p) processors as desired. We may think of these p processors as *virtual processors* or *processes*. Each *real processor* can simulate $\lceil p/P \rceil$ virtual processors (provided each real processor has enough memory). Thus, ignoring overheads associated with the simulation, we have

$$S_P \geq S_p / \lceil p/P \rceil. \tag{2.4}$$

If our analysis for p processors gives us a lower bound on S_p, then (2.4) can be used to obtain a lower bound on S_P.

2.3 Parallel architectures

Many varieties of parallel computer architecture have been proposed in recent years. They include –

- Pipelined vector processors such as the Cray 1 or Cyber 205, in which there is a single instruction stream and the parallelism is more or less hidden from the programmer.
- Single-instruction multiple-data (SIMD [16]) machines such as the Illiac IV or ICL DAP, in which a number of simple processing elements (PEs or cells) execute the same instruction on local data and communicate with their nearest neighbours on a square grid or torus. There is usually a general-purpose controller which can broadcast instructions and data to the cells.
- Multiple-instruction multiple-data (MIMD) machines such as transputer systems, C.mmp, CM*, and most hypercube machines.
- Hypercube machines, in which 2^k processors are connected like the vertices of a k-dimensional cube (i.e. if the processors are identified by k-bit binary numbers, they are connected to the processors whose numbers differ by exactly one bit from their own). These include both SIMD machines (e.g. the Connection Machine) and MIMD machines (e.g. the Caltech "Cosmic Cube", the Intel iPSC, and the NCUBE).
- Shared-memory multiprocessors such as the Alliant FX/8, the Encore Multimax, and the Sequent Symmetry.
- Systolic arrays [30], which are 1 or 2-dimensional arrays of simple processors (cells) connected to their nearest neighbours. The cells on the edge of the array are usually connected to a general-purpose machine which acts as a controller. Examples are the Warp and iWarp machines [1, 5] and several machines described in [43]. Variations on the idea of systolic arrays are wavefront arrays [37] and instruction systolic arrays [45].

In view of the diversity of parallel computer architectures, it is difficult to describe practical parallel algorithms in a machine-independent manner. In some cases an algorithm intended for one class of parallel machine can easily be converted for another (more general) class. For example, an algorithm designed for a systolic array can be mapped onto a hypercube without much loss of efficiency, but not conversely (in general). An algorithm designed for a local memory machine can easily be implemented on a shared memory machine, but often the converse is not true.

Since systolic arrays are very restrictive, it is usually possible to map systolic array algorithms onto other parallel machines without a great loss in efficiency. On the other hand, systolic arrays are sufficient and cost-effective for problems arising in digital signal processing [30, 36, 37, 43]. Thus, we shall generally describe parallel algorithms as though they are to be implemented on systolic arrays – the reader should be able to translate to other parallel computer architectures.

3 Systolic algorithms for signal processing

Many compute-bound computations with applications in signal processing have good parallel algorithms which can be implemented on systolic arrays. For example, we mention:

- 1-D convolution, FIR and IIR filtering [11, 28, 29, 30]
- 2-D convolution and correlation [2, 32, 33, 34, 50]
- discrete Fourier transform [28, 29]
- interpolation [32]
- 1-D and 2-D median filtering [15]
- matrix-vector and matrix-matrix multiplication [31, 48]
- solution of triangular linear systems [31]
- LU and QR factorization of square matrices, and solution of full-rank linear systems [4, 19, 31]

- QR factorization of rectangular matrices, and solution of linear least squares problems [19, 38]
- solution of the symmetric and Hermitian eigenvalue problems [3, 7, 8]
- computation of the singular value decomposition (SVD) [9, 10, 39, 40, 44]
- solution of Toeplitz linear systems and least squares problems [6].

In Section 4 we consider a "simple" example – the solution of square, full-rank linear systems. Then, in Section 5, we consider the computation of the SVD and the solution of the symmetric eigenvalue problem.

4 The solution of linear systems

Suppose we want to solve a nonsingular n by n linear system

$$Ax = b$$

on a systolic array or other parallel machine for which a 2-dimensional mesh is a natural interconnection pattern. It is easy to implement Gaussian elimination *without* pivoting, because multipliers can be propagated along rows of the augmented matrix $[A|b]$, and it is not necessary for one row operation to be completed before the next row operation starts. Unfortunately, as is well-known [23, 47, 49], Gaussian elimination without pivoting is numerically unstable unless A has some special property such as diagonal dominance or positive definiteness.

On a serial machine the problem of numerical instability could easily be overcome by using partial (or complete) pivoting. However, pivoting is difficult to implement efficiently on a parallel machine because it destroys the regular data flow associated with Gaussian elimination without pivoting [31]. Just to find the first pivot (i.e. $\max_{1 \le i \le n} |a_{i,1}|$) takes time $\Omega(n)$ on a systolic array ($\Omega(\log n)$ on a hypercube), so the complete elimination takes time $\Omega(n^2)$ on a systolic array ($\Omega(n \log n)$ on a hypercube).

If $T_P = \Omega(n^2)$ we are not justified in using more than $P = O(n)$ processors (else $T_P P \gg n^3$ and the efficiency $E_P \ll 1$).

On a linear systolic array of $P = n+1$ processors, we store one column of the augmented matrix $[A|b]$ in each processor, and we can implement partial pivoting efficiently. If $P \gg n$ it is not so easy to implement partial pivoting efficiently. One solution is to replace the elementary transformations

$$\begin{pmatrix} 1 & 0 \\ -m & 1 \end{pmatrix}$$

of Gaussian elimination by plane rotations (Givens transformations)

$$\begin{pmatrix} c & s \\ -s & c \end{pmatrix},$$

where $c^2 + s^2 = 1$. Thus, we construct an *orthogonal* matrix Q such that $QA = R$ is upper triangular and the linear system $Ax = b$ is reduced to the upper triangular system $Rx = \bar{b}$ (where $\bar{b} = Qb$) in a *numerically stable* manner. The operation count is roughly four times that for Gaussian elimination, but this is an acceptable price to pay for numerical stability on a parallel machine. The factor four could be reduced by the use of "fast" Givens transformations [18]. A closely related alternative, with similar data flow requirements but inferior numerical properties, is to use "pairwise pivoting", i.e. pivoting between adjacent rows.

4.1 Implementation of parallel Givens – the BBK algorithm

To simplify communication on a systolic array or mesh-connected parallel machine it is desirable to restrict attention to plane rotations which modify *adjacent* rows. Since each rotation modifies only two rows, it is possible to do up to $\lfloor n/2 \rfloor$ rotations in parallel. One of many ways to order the rotations necessary to reduce A to upper triangular form was suggested by Bojanczyk, Brent and Kung (BBK) [4] and is illustrated below (where $n = 8$ and the numbers in the illustration indicate the time step at which the corresponding matrix element is zeroed):

$$\begin{pmatrix} \cdot \\ 7 & \cdot \\ 6 & 9 & \cdot \\ 5 & 8 & 11 & \cdot \\ 4 & 7 & 10 & 13 & \cdot \\ 3 & 6 & 9 & 12 & 15 & \cdot \\ 2 & 5 & 8 & 11 & 14 & 17 & \cdot \\ 1 & 4 & 7 & 10 & 13 & 16 & 19 & \cdot \end{pmatrix}$$

The rotation parameters (c, s) propagate right along rows and it is not necessary to wait for one row transformation to be completed before others commence. Details of the implementation on a triangular systolic array may be found in Bojanczyk, Brent and Kung [4].

4.2 Solution of least squares problems – the Gentleman and Kung algorithm

The BBK algorithm was intended for the solution of the n by n linear system $Ax = b$. When applied to the m by n linear least squares problem

$$min\|Ax - b\|_2$$

it is inefficient because it uses a triangular systolic array of $m^2/2 + O(m)$ processors, and usually in such problems $m \gg n$. Gentleman and Kung [19] (and later Luk [38]) transposed the data flow, so the skewed matrix A is fed into the systolic array by columns rather than by rows. This is advantageous because a triangular systolic array of only $n^2/2 + O(n)$ processors is required. In fact, it is not even necessary to know m in advance when using the Gentleman-Kung algorithm.

A good survey of systolic array algorithms for computing the QR decomposition of a (square or) rectangular matrix may be found in Luk [38].

5 The SVD and symmetric eigenvalue problems

A singular value decomposition (SVD) of a real m by n matrix A is its factorization into the product of three matrices:

$$A = U\Sigma V^T, \tag{5.1}$$

where U is an m by n matrix with orthonormal columns, Σ is an n by n nonnegative diagonal matrix, and V is an n by n orthogonal matrix (we assume here that $m \geq n$). The diagonal elements σ_i of Σ are the *singular values* of A. The singular value decomposition is extremely useful in digital signal processing [21, 36].

The SVD is usually computed by a two-sided orthogonalization process, e.g. by two-sided reduction to bidiagonal form (possibly preceded by a one-sided QR reduction [12]) followed by the QR algorithm [20, 22, 49]. On a systolic array it is simpler to avoid bidiagonalization and to use the two-sided orthogonalization method of Kogbetliantz et al [9, 10, 17, 26, 27] rather than the standard Golub-Kahan-Reinsch algorithm [20, 22]. However, it is even simpler to use a one-sided orthogonalization method due to Hestenes [25]. The idea of Hestenes is to iteratively generate an orthogonal matrix V such that AV has orthogonal columns. Normalizing the Euclidean length of each nonnull column to unity, we get

$$AV = \tilde{U}\Sigma \tag{5.2}$$

As a null column of $\tilde{U}$ is always associated with a zero diagonal element of Σ, there is no essential difference between (5.1) and (5.2).

There is clearly a close connection between the Hestenes method for finding the SVD of A and the classical Jacobi method [49] for finding the eigenvalues and eigenvectors of A^TA. This is discussed in Section 5.4.

5.1 Implementation of the Hestenes method

Let $A_1 = A$ and $V_1 = I$. The Hestenes method uses a sequence of plane rotations Q_k chosen to orthogonalize two columns in $A_{k+1} = A_kQ_k$. If the matrix V is required, the plane rotations are accumulated using $V_{k+1} = V_k Q_k$. Under certain conditions (discussed below) $\lim Q_k = I$, $\lim V_k = V$ and $\lim A_k = AV$. The matrix A_{k+1} differs from A_k only in two columns, say columns i and j. In fact

$$\left(a_i^{(k+1)}, a_j^{(k+1)}\right) = \left(a_i^k, a_j^k\right) \begin{pmatrix} \cos\theta & \sin\theta \\ -\sin\theta & \cos\theta \end{pmatrix}$$

where the rotation angle θ is chosen so that the two new columns $a_i^{(k+1)}$ and $a_j^{(k+1)}$ are orthogonal. This can always be done with an angle θ satisfying

$$|\theta| \le \pi/4, \tag{5.3}$$

see for example [8].

It is desirable for a "sweep" of $n(n-1)/2$ rotations to include all pairs (i, j) with $i < j$. On a serial machine a simple strategy is to choose the "cyclic by rows" ordering

$$(1,2), (1,3), \cdots, (1,n), (2,3), \cdots, (n-1,n).$$

Forsythe and Henrici [17] have shown that the cyclic by rows ordering and condition (5.3) ensure convergence of the Jacobi method applied to $A^T A$, and convergence of the cyclic by rows Hestenes method follows.

5.2 The chess tournament analogy

On a parallel machine we would like to orthogonalize several pairs of columns simultaneously. This should be possible so long as no column occurs in more than one pair. The problem is similar to that of organizing a round-robin tournament between n players. A game between players i and j corresponds to orthogonalizing columns i and j, a round of several games played at the same time corresponds to orthogonalizing several pairs of (disjoint) columns, and a tournament where each player plays each other player once corresponds to a sweep in which each pair of columns is orthogonalized. Thus, schemes which are well-known to chess (or other two-person game) players can be used to give orderings amenable to parallel computation. It is usually desirable to minimize the number of parallel steps in a sweep, which corresponds to the number of rounds in the tournament.

On a parallel machine with restricted communication paths, such as a systolic array, there are constraints on the orderings which we can implement efficiently. A useful analogy is a tournament of lazy chess players. After each round the players want to walk only a short distance to the board where they are to play the next round.

Using this analogy, suppose that each chess board corresponds to a systolic processor and each player corresponds to a column of the matrix (initially A but modified as the computation proceeds). A game between two players corresponds to orthogonalization of the corresponding columns. Thus we suppose that each systolic processor has sufficient memory to store and update two columns of the matrix. If the chess boards (processors) are arranged in a linear array with nearest-neighbour communication paths, then the players should have to walk (at most) to an adjacent board between the end of one round and the beginning of the next round, i.e. columns of the matrix should have to be exchanged only between adjacent processors. Several orderings satisfying these conditions have been proposed [7, 8, 41, 44].

Since A has n columns and at most $\lfloor n/2 \rfloor$ pairs can be orthogonalized in parallel, a sweep requires as least $n-1$ parallel steps (n even) or n parallel steps (n odd). The "Brent-Luk" ordering [8] attains this minimum, and convergence can be guaranteed if n is odd [41, 46]. It is an open question whether convergence can be guaranteed for the Brent-Luk ordering (or any other ordering which requires only the minimum number of parallel steps) when n is even – the problem in proving convergence is illustrated in [24]. However, in practice lack of convergence is not a problem, and it is easy to ensure convergence by the use of a "threshold" strategy [49], or by taking one additional parallel step per sweep when n is even [42].

5.3 A heuristic argument regarding the number of sweeps

In practice we observe linear convergence until the columns of the (updated) matrix A become close to orthogonal, i.e. until the off-diagonal elements of $A^T A$ become small. Subsequently convergence is at least quadratic. It is difficult to quantify what "small" means here unless the eigenvalues of $A^T A$

are distinct. If we assume that "small" means $O(1/n^c)$ for some positive constant c, and that the rate of linear convergence is independent of n, then it should take $O(\log n)$ sweeps before quadratic convergence sets in. This heuristic argument is supported by empirical evidence [8] – certainly the average number of sweeps required to give convergence to a fixed tolerance for random matrices A appears to be $O(\log n)$.

5.4 The symmetric eigenvalue problem

As noted above, there is a close connection between the Hestenes method for finding the SVD of a matrix A and the Jacobi method for finding the eigenvalues of a symmetric matrix $B = A^T A$. An important difference is that the formulas defining the rotation angle θ involve elements $b_{i,j}$ of B rather than inner products of columns of A, and transformations must be performed on the left and right instead of just on the right (since $(AV)^T(AV) = V^T BV$). Instead of permuting columns of A as described in Section 5.2, we have to apply the same permutation to both rows and columns of B. This is easy if we use a square systolic array of $n/2$ by $n/2$ processors with nearest-neighbour connections (assuming, for simplicity, that n is even). Details are given in [8].

If less than $n^2/4$ processors are available, we can use the virtual processor concept described in Section 2.2. For example, on a linear systolic array with $P \le n/2$ processors, each processor can simulate $\sim n/(2P)$ columns of $n/2$ virtual processors. Similarly, on a square array of $P \le n^2/4$ processors, each processor can simulate a block of $\sim n^2/(4P)$ virtual processors. In both cases, communication paths between virtual processors map onto paths between real processors or communication internal to a real processor.

5.5 Other SVD and eigenvalue algorithms

In Section 5.2 we showed how the Hestenes method could be used to compute the SVD of an m by n matrix in time $O(mn^2S/P)$ using $P = O(n)$ processors in parallel. Here S is the number of sweeps required (conjectured to be $O(\log n)$). In Section 5.4 we sketched how Jacobi's method could be used to compute the eigen-decomposition of a symmetric n by n matrix in time $O(n^3S/P)$ using $P = O(n^2)$ processors. It is natural to ask if we can use more than $\Omega(n)$ processors efficiently when computing the SVD. The answer is yes – Kogbetliantz [26, 27] and Forsythe & Henrici [17] suggested an analogue of Jacobi's method, and this can be used to compute the SVD of a *square* matrix using a parallel algorithm very similar to the parallel implementation of Jacobi's method. The result is an algorithm which requires time $O(n^3S/P)$ using $P = O(n^2)$ processors. Details and a discussion of several variations on this theme may be found in [9].

In order to find the SVD of a rectangular m by n matrix A using $O(n^2)$ processors, we first compute the QR factorization $QA = R$ (see Section 4), and then compute the SVD of the principal n by n submatrix of R (i.e. discard the $m - n$ zero rows of R). It is possible to gain a factor of two in efficiency by preserving the upper triangular structure of R [39].

The Hestenes/Jacobi/Kogbetliantz methods are not often used on a serial computer, because they are slower than methods based on reduction to bidiagonal or tridiagonal form followed by the QR algorithm [49]. Whether the fast serial algorithms can be implemented efficiently on a parallel machine depends to some extent on the parallel architecture. For example, on a square array of n by n processors it is possible to reduce a symmetric n by n matrix to tridiagonal form in time $O(n \log n)$ [3]. On a serial machine this reduction takes time $O(n^3)$. Thus, a factor $O(\log n)$ is lost in efficiency, which roughly equates to the factor $O(S)$ by which Jacobi's method is slower than the QR algorithm on a serial machine. It is an open question whether the loss in efficiency by a factor $O(\log n)$ can be avoided on a parallel machine with $P = \Omega(n^2)$ processors. When $P = O(n)$, "block" versions of the usual serial algorithms are attractive on certain architectures [13], and may be combined with the "divide and conquer" strategy [14]. Generally, these more complex algorithms are attractive on shared memory MIMD machines with a small number of processors, while the simpler algorithms described in Sections 5.1 to 5.4 are attractive on systolic arrays and SIMD machines.

References

[1] M. Annaratone, E. Arnould, T. Gross, H. T. Kung, M. Lam, O. Men-

zilcioglu and J. A. Webb, "The Warp computer: architecture, implementation and performance", *IEEE Transactions on Computers*, C-36 (1987), 1523-1538.

[2] J. Blackmer, P. Kuekes and G. Frank, "A 200 MOPS systolic processor", Proc. SPIE, Vol. 298, *Real-Time Signal Processing IV*, Society of Photo-Optical Instrumentation Engineers, 1981.

[3] A. W. Bojanczyk and R. P. Brent, "Tridiagonalization of a symmetric matrix on a square array of mesh-connected processors", *J. Parallel and Distributed Computing* 2 (1985), 261-276.

[4] A. W. Bojanczyk, R. P. Brent and H. T. Kung, *Numerically stable solution of dense systems of linear equations using mesh-connected processors*, Tech. Report TR-CS-81-119, Department of Computer Science, Carnegie-Mellon University, Pittsburgh, Pa., May 1981. Also appeared in *SIAM J. Scientific and Statistical Computing* 5 (1984), 95-104.

[5] S. Borkar, R. Cohn, G. Cox, S. Gleason, T. Gross, H. T. Kung, M. Lam, B. Moore, C. Peterson, J. Pieper, L. Rankin, P. S. Tseng, J. Sutton, J. Urbanski and J. Webb, "iWarp: An integrated solution to high-speed parallel computing", *Proc. Supercomputing 1988 Conference*, Orlando, Florida, Nov. 1988, 330-339.

[6] R. P. Brent, "Parallel algorithms for Toeplitz systems", these proceedings.

[7] R. P. Brent, H. T. Kung and F. T. Luk, "Some linear-time algorithms for systolic arrays", in *Information Processing 83* (edited by R.E.A. Mason), North-Holland, Amsterdam, 1983, 865-876.

[8] R. P. Brent and F. T. Luk, "The solution of singular-value and symmetric eigenvalue problems on multiprocessor arrays", *SIAM J. Scientific and Statistical Computing* 6 (1985), 69-84.

[9] R. P. Brent, F. T. Luk and C. F. Van Loan, "Computation of the singular value decomposition using mesh-connected processors", *J. of VLSI and Computer Systems* 1, 3 (1983-1985), 242-270.

[10] R. P. Brent, F. T. Luk and C. F. Van Loan, "Computation of the generalized singular value decomposition using mesh-connected processors", *Proc. SPIE, Volume 431, Real Time Signal Processing VI*, Society of Photo-Optical Instrumentation Engineers, Bellingham, Washington, 1983, 66-71.

[11] P. R. Cappello and K. Steiglitz, "Digital signal processing applications of systolic algorithms", in [35], 245-254.

[12] T. F. Chan, "An improved algorithm for computing the singular value decomposition", *ACM Trans. Math. Software* 8 (1982), 72-83.

[13] J. J. Dongarra, S. J. Hammarling and D. C. Sorensen, "Block reduction of matrices to condensed forms for eigenvalue computations", *LAPACK Working Note #2*, MCS Division, Argonne National Labs., Argonne, Illinois, Sept. 1987.

[14] J. Dongarra and D. Sorensen, "A fully parallel algorithm for the symmetric eigenproblem", *SIAM J. Scientific and Statistical Computing* 8 (1987), 139-154.

[15] A. Fisher, "Systolic algorithms for running order statistics in signal and image processing", in [35], 265-272.

[16] M. J. Flynn, "Some computer organizations and their effectiveness", *IEEE Transactions on Computers*, C-21 (1972), 702-706.

[17] G. E. Forsythe and P. Henrici, "The cyclic Jacobi method for computing the principal values of a complex matrix", *Trans. Amer. Math. Soc.* 94 (1960), 1-23.

[18] W. M. Gentleman, "Least squares computations by Givens transformations without square roots", *J. Inst. Math. Appl.* 12 (1973), 329-336.

[19] W. M. Gentleman and H. T. Kung, "Matrix triangularization by systolic arrays", *Proc. SPIE, Volume 298, Real-Time Signal Processing IV*, Society of Photo-Optical Instrumentation Engineers, 1981.

[20] G. H. Golub and W. Kahan, "Calculating the singular values and pseudo-inverse of a matrix", *J. SIAM Ser. B: Numer. Anal.* 2 (1965), 205-224.

[21] G. H. Golub and F. T. Luk, "Singular value decomposition: applications and computations", ARO Report 77-1, *Trans. 22nd Conference of Army Mathematicians*, 1977, 577-605.

[22] G. H. Golub and C. Reinsch, "Singular value decomposition and least squares solutions", *Numer. Math.* 14 (1970), 403-420.

[23] G. H. Golub and C. Van Loan, *Matrix Computations*, Johns Hopkins Press, Baltimore, Maryland, 1983.

[24] E. R. Hansen, "On cyclic Jacobi methods", *J. SIAM* 11 (1963), 448-459.

[25] M. R. Hestenes, "Inversion of matrices by biorthogonalization and related results", *J. SIAM* 6 (1958), 51-90.

[26] E. Kogbetliantz, "Diagonalization of general complex matrices as a new method for solution of linear equations", *Proc. Internat. Congress of Mathematicians*, Amsterdam, Vol. 2, 1954, 356-357.

[27] E. Kogbetliantz, "Solution of linear equations by diagonalization of coefficient matrices", *Quart. Appl. Math.* 13 (1955), 123-132.

[28] H. T. Kung, "Let's design algorithms for VLSI systems", *Proc. Conf. on Very Large Scale Integration: Architecture, Design, Fabrication*, California Institute of Technology, Pasadena, Jan. 1979, 65-90.

[29] H. T. Kung, "Special-purpose devices for signal and image processing: an opportunity in VLSI", *Proc. SPIE, Volume 241, Real-Time Signal Processing III*, Society of Photo-Optical Instrumentation Engineers, July 1980, 76-84.

[30] H. T. Kung, "Why systolic architectures ?", *IEEE Computer* 15, 1 (1982), 37-46.

[31] H. T. Kung and C. E. Leiserson, "Systolic arrays (for VLSI)", *Sparse Matrix Proc. 1978*, SIAM, Philadelphia, 1979, 256-282.

[32] H. T. Kung and R. L. Picard, "Hardware pipelines for multi-dimensional convolution and resampling", *Proc. Workshop on Computer Architecture for Pattern Analysis and Image Database Management*, IEEE, Nov. 1981, 273-278.

[33] H. T. Kung, L. M. Ruane and D. W. L. Yen, "A two-level pipelined systolic array for convolutions", in [35], 255-264.

[34] H. T. Kung and S. W. Song, *A systolic 2-D convolution chip*, Technical Report CMU-CS-81-110, Carnegie-Mellon University, Pittsburgh, March 1981.

[35] H. T. Kung, R. F. Sproull and G. L. Steele, Jr. (eds.), *VLSI Systems and Computations*, Computer Science Press, 1981.

[36] S. Y. Kung (editor), *VLSI and Modern Signal Processing*, Proceedings of a Workshop held at Univ. of Southern California, Nov. 1982.

[37] S. Y. Kung, *VLSI Array Processors*, Prentice-Hall International, 1988.

[38] F. T. Luk, "A rotation method for computing the QR-decomposition", *SIAM J. Scientific and Statistical Computing* 7 (1986), 452-459.

[39] F. T. Luk, "A triangular processor array for computing singular values", *J. of Linear Algebra and its Applications* 77 (1986), 259-273.

[40] F. T. Luk, "Architectures for computing eigenvalues and SVDs", *Proc. SPIE, Vol. 614, Highly Parallel Signal Processing Architectures*, 1986, 24-33.

[41] F. T. Luk and H. Park, "On parallel Jacobi orderings", *SIAM J. Scientific and Statistical Computing* 10 (1989), 18-26.

[42] F. T. Luk and H. Park, "A proof of convergence for two parallel Jacobi SVD algorithms", *IEEE Transactions on Computers* 30 (1989), 806-811.

[43] W. Moore, A. McCabe and R. Urquhart (editors), *Systolic Arrays*, Adam Hilger, Bristol, 1987.

[44] D. E. Schimmel and F. T. Luk, "A new systolic array for the singular value decomposition", *Proc. Fourth MIT Conference on Advanced Research in VLSI*, 1986, 205-217.

[45] H. Schröder, "The instruction systolic array – A tradeoff between flexibility and speed", *Computer Systems Science and Engineering* 3 (1988), 83-90.

[46] G. Shroff and R. Schreiber, *On the convergence of the cyclic Jacobi method for parallel block orderings*, Report 88-11, Department of Computer Science, Rensselaer Polytechnic Institute, Troy, New York, May 1988.

[47] G. W. Stewart, *Introduction to Matrix Computations*, Ac. Press, New York, 1973.

[48] U. Weiser and A. Davis, "A wavefront notation tool for VLSI array design", in [35], 226-234.

[49] J. H. Wilkinson, *The Algebraic Eigenvalue Problem*, Clarendon, Oxford, 1965.

[50] D. W. L. Yen and A. V. Kulkarni, "The ESL systolic processor for signal and image processing", *Proc. 1981 IEEE Computer Society Workshop on Computer Architecture for Pattern Analysis and Image Database Management*, Nov. 1981, 265-272.

An Introduction to the Class of Split Levinson Algorithms

P. Delsarte and Y. Genin

Philips Research Laboratory Brussels
Av. Van Becelaere 2
B-1170 Brussels, Belgium

Abstract

The split Levinson algorithms constitute a new class of efficient procedures to solve a linear system of equations exhibiting the positive definite Toeplitz structure. They can be derived from the classical Levinson algorithm by a kind of splitting operation, which results in a more efficient algorithm processing some well-defined symmetric polynomials (with complex coefficients). The algorithm thus obtained is based on a simple two-step recurrence relation satisfied by these polynomials. The paper provides a self-contained introduction to the whole class of split Levinson algorithms, including a detailed technical derivation. This class is shown to depend on two unit modulus parameters, which can be chosen at will by the user.

1 Introduction

The celebrated Levinson algorithm is an efficient method to solve systems of linear equations that exhibit the positive definite Toeplitz structure. Such an algebraic structure is met very often in digital signal processing applications (among other areas), where positive definite Toeplitz matrices occur either as covariance matrices of stationary stochastic processes or as autocorrelation matrices of sampled signal records. This explains the fact

NATO ASI Series, Vol. F 70
Numerical Linear Algebra, Digital Signal Processing and Parallel Algorithms
Edited by G. H. Golub and P. Van Dooren

that the Levinson algorithm is used extensively to deal with various types of important digital signal processing problems.

Recently, the Levinson algorithm has been shown to be redundant in computational complexity and in memory requirement. Roughly speaking, it can be split into two parts, either of which needs only to be processed. The resulting algorithm, first derived for real data, has been called the split Levinson algorithm [6], [7]. It should be pointed out that a method very similar to one of the versions of this algorithm can be found in an earlier publication [5]; the method in question is proposed as an efficient procedure to solve the discrete Gopinath-Sondhi equation.

There is a close connection between the split Levinson algorithm and a new technique, discovered earlier by Bistritz, for testing whether a given polynomial has all its zeros inside the unit circle [1], [2]. In fact, although the problems dealt with are quite different, both algorithms are based on the same type of recurrence relations. Just as the Bistritz test can be extended to complex polynomials [3], [12], the split Levinson algorithm can be extended to complex Toeplitz matrices [8], [15]. (However, it should be pointed out that the latter extension contains some technical difficulties which are not present in the former.) Furthermore, within a simple modification, the split Levinson method applies to matrices exhibiting a structure slightly more general than the Toeplitz structure [4], [17].

The split Levinson algorithm is interesting not only for its reduced complexity but also for a more fundamental reason. Indeed, it can be viewed as the "algebraic facet" of a new function theoretic approach to some important interpolation and approximation problems in the class of positive (Carathéodory) functions [8]. This explains why the ideas behind the split Levinson algorithm are currently receiving much attention; they have led to novel efficient solution methods in a collection of problems of theoretical and practical interest. An account of these developments is given in a companion paper [9].

As alluded to above, the split Levinson algorithm presented in the literature can take several forms. However, the question of the natural structure of the whole family of these methods is not fully understood, because of the lack of a sufficiently general setting. This paper aims at clarifying that question by providing a comprehensive introduction to the class of split Levinson algorithms for positive definite Hermitian Toeplitz matrices. In particular, this class will be shown to depend on two complex parameters of unit modulus.

Section 2 contains some background material concerning the classical Levinson algorithm for computing the predictor polynomials $a_k(z)$ relative to the Toeplitz sections C_k of a given Toeplitz matrix C_{n-1} of order n. Then, it is suggested how suitable modifications $p_k(z)$ of the "symmetric parts" of these predictors could be computed by a more economical algorithm, producing the desired predictor $a_{n-1}(z)$ in a final stage. The basis of such an algorithm should be a one-parameter two-step recurrence relation of the form $p_{k+1}(z) = (\alpha_k + \overline{\alpha}_k z)p_k(z) - zp_{k-1}(z)$.

A useful duality (associated with inversion) for Toeplitz matrices is described in section 3. The dual predictors $s_k(z)$ admit a sequence of Schur-Szegö parameters (reflection coefficients) equal to the conjugate mirror image of the original sequence, within multiplication by a constant ρ_n of unit modulus. This leads to a remakable identity linking the families of primal and dual predictors (in two independent variables).

Section 4 introduces some ρ_n-symmetric polynomials $\pi_k(z)$ involving, in a simple manner, the classical predictors $a_{k-1}(z)$ and the values $s_{n-k}(\zeta_0)$ for a fixed number ζ_0 of unit modulus. The main properties of the family of polynomials $\pi_k(z)$, with given ρ_n and ζ_0, are derived. They include a recurrence relation that has nearly the required structure.

A further step towards the split Levinson algorithm is made in section 5, where suitable symmetric polynomials $p_k(z)$ are defined as normalized versions, $p_k(z) = g_k\pi_k(z)$, of the preceding polynomials $\pi_k(z)$. The normalizing factors g_k are identified in terms of Jacobi determinants, and some useful formulas are obtained for the ratios $\lambda_k = g_k/g_{k-1}$ (termed Jacobi parameters). Special attention is given to the "initial conditions" $p_0(z)$ and $p_1(z)$, which parametrize the complete sequence of polynomials $p_k(z)$. In addition to ζ_0, there is a second unit modulus parameter involved (in one-to-one correspondence with ρ_n).

Section 6 is devoted to a derivation and a detailed description of the general split Levinson algorithm. The core of the method is a well-defined implementation of the recurrence formula giving $p_{k+1}(z)$ from $p_k(z)$ and $p_{k-1}(z)$. In addition, the algorithm has to compute the last Jacobi parameter λ_n (and, in certain cases, the whole λ_k sequence); this is required to determine the predictor $a_{n-1}(z)$ at the end of the procedure. A distinction has to be made between the "regular case" where $p_k(\zeta_0)$ differs from zero for all k and the "singular case" where $p_k(\zeta_0)$ equals zero for all k. (There are no other possibilities.) In the singular case, the Jacobi parameters have to be computed in a recursive manner. It is briefly explained how this case can in principle be reduced to the regular case.

An outline of the split Levinson algorithms for real Toeplitz matrices is given in section 7 as an illustration of the general theory. In this situation, it is shown that there exist essentially four distinct versions processing only real data.

2 Classical definitions and preliminary steps

Given a real number c_0, and $n-1$ complex numbers $c_1, \ldots, c_{n-1}$, consider the $n \times n$ *Hermitian Toeplitz matrix*

$$C_{n-1} = \begin{bmatrix} c_0 & \overline{c}_1 & \overline{c}_2 & \cdots & \overline{c}_{n-1} \\ c_1 & c_0 & \overline{c}_1 & \cdots & \overline{c}_{n-2} \\ c_2 & c_1 & c_0 & \cdots & \overline{c}_{n-3} \\ \vdots & \vdots & \vdots & & \vdots \\ c_{n-1} & c_{n-2} & c_{n-3} & \cdots & c_0 \end{bmatrix}, \tag{1}$$

where the bar denotes the complex conjugate. Throughout the paper, C_{n-1} is assumed to be *positive definite*. With the nested sequence $C_0, \ldots, C_k, \ldots, C_{n-1}$ of principal Toeplitz submatrices of C_{n-1} we associate the family of systems of linear equations

$$C_k[1, a_{k,1}, \ldots, a_{k,k}]^T = [\sigma_k, 0, \ldots, 0]^T, \tag{2}$$

for $k = 0, 1, \ldots, n-1$. The solution vector $[a_{k,0}, \ldots, a_{k,k}]^T$, with $a_{k,0} = 1$, equals the first column of the inverse of C_k within the positive normalization factor

$$\sigma_k = \det C_k / \det C_{k-1}. \tag{3}$$

From this vector and from its conjugate mirror image let us construct the polynomial $a_k(z)$ and its reciprocal $\hat{a}_k(z)$, given by

$$a_k(z) = \sum_{i=0}^{k} a_{k,i} z^i, \tag{4}$$

$$\hat{a}_k(z) = z^k \, \overline{a}_k(1/\overline{z}) = \sum_{i=0}^{k} \overline{a}_{k,k-i} \, z^i . \tag{5}$$

In the context of linear least-squares estimation theory, the polynomials $a_k(z)$ and $\hat{a}_k(z)$ are usually referred to as the *forward* and *backward predictor polynomials* of order k (relative to the Toeplitz matrix C_{n-1}), and σ_k is called the (squared) *error prediction norm* [16]. The backward predictors $\hat{a}_k(z)$, with $k = 0, 1, \ldots, n-1$, are known to constitute a family of *Szegő polynomials*, i.e., a family of polynomials orthogonal on the unit circle [18].

The *linear prediction problem* requires to compute $a_{n-1}(z)$ and σ_{n-1} from the given Toeplitz matrix C_{n-1}. The classical method to solve this problem is the *Levinson algorithm*, which is specifically tailored for the Toeplitz structure [13], [16]. The Levinson algorithm determines the sequence of predictors $a_k(z)$, together with the error norms σ_k, by recursive use of the formulas

$$\rho_k = -\sigma_{k-1}^{-1} \sum_{i=0}^{k} c_{k-i} \, a_{k-1,i} \, , \tag{6}$$

$$a_k(z) = a_{k-1}(z) + \rho_k \, z \, \hat{a}_{k-1}(z) \, , \tag{7}$$

$$\sigma_k = \sigma_{k-1}(1 - |\rho_k|^2) \, , \tag{8}$$

for $k = 1, 2, \ldots, n-1$, with the initial values $a_0(z) = 1$ and $\sigma_0 = c_0$. The number ρ_k in (6), which is involved in the updating formulas (7) and (8), is termed the kth *Schur-Szegő parameter* or *reflection coefficient.* Note the identity $\rho_k = a_{k,k}$. In view of (3) and (8), the positive definiteness of C_{n-1} amounts to the set of inequalities $|\rho_k| < 1$, for all k, together with $c_0 > 0$. According to the Schur-Cohn criterion, this shows that all predictors $a_k(z)$ are devoid of zeros in the closed unit disc $|z| \leq 1$.

It is seen that the Levinson algorithm takes benefit of the Toeplitz structure (1) of the system matrix C_{n-1} to solve the linear prediction problem in $O(n^2)$ flops, instead of $O(n^3)$ flops as it would be the case for an unstructured matrix. This algorithm has however been recently shown to be redundant in computational complexity; a more economical method, called the split Levinson algorithm, has been proposed to solve the same problem [7]. To explain the basic idea in simple terms, let us first assume the existence of an efficient algorithm for computing, in a recursive manner, not the full predictor polynomials $a_k(z)$ but only their "symmetric parts"

$\phi_k(z)$, defined by

$$\phi_k(z) = a_k(z) + \hat{a}_k(z). \tag{9}$$

Thus, $\phi_k(z)$ equals its reciprocal $\hat{\phi}_k(z)$. In view of the Szegö-Levinson recurrence relation (7), the polynomial $\phi_k(z)$ is equivalently given by

$$\phi_k(z) = (1 + \overline{\rho}_k) a_{k-1}(z) + (1 + \rho_k) z \, \hat{a}_{k-1}(z). \tag{10}$$

It is seen that $a_{k-1}(z)$ and ρ_k are directly available from the pair of symmetric polynomials $\phi_{k-1}(z)$ and $\phi_k(z)$. Indeed, (9) and (10) yield the relations

$$a_{k-1}(z) = \frac{\phi_k(z) - \overline{\phi}_k(0) z \, \phi_{k-1}(z)}{\phi_k(0) - \overline{\phi}_k(0) z}, \qquad \rho_k = 1 - \overline{\phi}_k(0). \tag{11}$$

At this point, the main observations are the following.

1. The sequence of symmetric polynomials $\phi_k(z)$ with $k = 0, 1, \ldots, n$ contains exactly the same information as the sequence of predictors $a_k(z)$ with $k = 0, 1, \ldots, n-1$.

2. Due to the symmetry property $\phi_k(z) = \hat{\phi}_k(z)$, i.e., $\phi_{k,k-i} = \overline{\phi}_{k,i}$ for $0 \le i \le k$, only half the number of coefficients of $\phi_k(z)$ have to be computed and stored.

To take real profit of these attractive properties, one has to devise an appropriate method for computing recursively the symmetric polynomials $\phi_k(z)$. In principle, there is no difficulty, since inserting the expression (11) of $a_{k-1}(z)$ into the Szegö-Levinson formula (7) produces a two-step recurrence relation, involving the three successive polynomials $\phi_{k-1}(z)$, $\phi_k(z)$ and $\phi_{k+1}(z)$. The resulting procedure is however more expensive than the Levinson algorithm itself, to the extent that the new approach seems, at first glance, to have little practical interest.

Fortunately, the situation becomes much better if we introduce symmetric polynomials $p_k(z)$ of a slightly more general form than the $\phi_k(z)$ above. Instead of (10) let us set the definition

$$p_k(z) = w_k \, a_{k-1}(z) + \overline{w}_k \, z \, \hat{a}_{k-1}(z), \tag{12}$$

for $k = 1, \ldots, n$, with suitable complex numbers $w_1, \ldots, w_n$. By construction, the polynomials (12) enjoy the same *symmetry property*,

$$\hat{p}_k(z) = p_k(z)\,, \tag{13}$$

as the polynomials $\phi_k(z)$. Furthermore, there exist appropriate substitutes for the relations (11). Hence observations 1 and 2 above remain valid for the $p_k(z)$ family.

The efficiency of this general setting lies in the following fact, which will be proved in sections 4 and 5. The numbers w_k can be chosen in such a way that the polynomials $p_k(z)$ satisfy a *one-parameter two-step recurrence relation* of the form

$$p_{k+1}(z) = (\alpha_k + \overline{\alpha}_k\, z)p_k(z) - z\, p_{k-1}(z)\,, \tag{14}$$

for well-defined complex numbers $\alpha_1, \alpha_2, \ldots, \alpha_{n-1}$ which can be computed from the data (together with the polynomials $p_k(z)$ themselves).

The class of split Levinson algorithms treated in this paper consists of the numerical procedures that make use of the symmetry and recursivity properties (13) and (14) to solve the linear prediction problem relative to a given Toeplitz matrix at a low computational cost. As explained in the sequel, there is an infinite number of choices for the sequence (w_k); hence there exists an infinity of split Levinson algorithms. In fact, this is a "small infinity"; roughly speaking, the whole class is parametrized by two points belonging to the unit circle.

3 Predictor polynomials relative to the inverse of a Toeplitz matrix

Although the inverse of the Toeplitz matrix C_{n-1} does generally not have the Toeplitz structure, it is "close to Toeplitz" in a well-defined sense [14]. More precisely, it admits a factorization of the form

$$C_{n-1}^{-1} = L_{n-1}\, T_{n-1}\, L_{n-1}^*\,, \tag{15}$$

where L_{n-1} is a lower-triangular Toeplitz matrix with unit diagonal and T_{n-1} is a (positive definite) Hermitian Toeplitz matrix [10], [11]. As explained below, the pair (L_{n-1}, T_{n-1}) is not unique; it depends on an arbitrary complex number of unit modulus.

For any decomposition (15) consider the principal Toeplitz submatrix T_k of order $k+1$ of T_{n-1}, and define $s_k(z)$ to be the forward predictor polynomial relative to T_k, for $k = 0, 1, \ldots, n-1$. Thus, the coefficient vector of $s_k(z)$ obeys a system of linear equations of the form (2), that is

$$T_k[1, s_{k,1}, \ldots, s_{k,k}]^T = [\sigma'_k, 0, \ldots, 0]^T . \tag{16}$$

The predictors $s_k(z)$ satisfy the Szegö-Levinson relation

$$s_k(z) = s_{k-1}(z) + \rho'_k\, z\, \hat{s}_{k-1}(z) , \tag{17}$$

together with $\sigma'_k = \sigma'_{k-1}(1 - |\rho'_k|^2)$, where the numbers ρ'_k are the Schur-Szegö parameters relative to T_{n-1}. It can be shown that ρ'_k is given by the remarkable expression

$$\rho'_k = \rho_n\, \overline{\rho}_{n-k} , \tag{18}$$

for $k = 1, 2, \ldots, n-1$, where ρ_n is a complex number subject to the only restriction $|\rho_n| = 1$ (see [11]). In other words, the sequence of reflection coefficients relative to T_{n-1} equals the conjugate mirror image of that of C_{n-1}, within the "rotation factor" ρ_n. This factor is in one-to-one correspondence with the chosen representation (15) of C_{n-1}^{-1}.

Remark. The reader will note that the polynomials $s_k(z)$ can be defined directly from (17) and (18), without explicit reference to (15) and (16).

Consider now any particular choice of ρ_n (with $|\rho_n| = 1$). Let C_n denote the corresponding *singular Toeplitz extension* of order $n+1$ of C_{n-1}. It is implicitly defined through the identity

$$a_n(z) = a_{n-1} + \rho_n\, z\, \hat{a}_{n-1}(z) , \tag{19}$$

where $a_n(z)$ is the forward predictor relative to C_n. Thus, the coefficient vector of $a_n(z)$ is the solution of the Toeplitz system (2), with $k = n$ and $\sigma_n = 0$. As for the "unknown" entry c_n of C_n, it is determined by the first relation (6), which yields $c_n = -\rho_n\, \sigma_{n-1} - \sum_{i=1}^{n-1} c_{n-i}\, a_{n-1,i}$. In view of (19), the predictor $a_n(z)$ is *ρ_n-symmetric*, in the sense that it satisfies

$$\hat{a}_n(z) = \overline{\rho}_n\, a_n(z) . \tag{20}$$

It is well known that all zeros of $a_n(z)$ are simple and are located on the unit circle $|z| = 1$.

With the help of the duality formula (18) and of both recurrences (7) and (17), we obtain a remarkable two-variable cross relation between the families of predictors $a_k(z)$ and $s_\ell(\zeta)$, namely

$$s_{n-k}(\zeta)a_{k-1}(z)+\rho_n\hat{s}_{n-k}(\zeta)z\,\hat{a}_{k-1}(z)=s_{n-k-1}(\zeta)a_k(z)+\rho_n\,\zeta\,\hat{s}_{n-k-1}(\zeta)\hat{a}_k(z)\,, \tag{21}$$

for $k=1,2,\ldots,n-1$. In particular, it follows from (19) and (21) that, for each k with $1\leq k\leq n$, the predictor $a_n(z)$ can be expressed in the form

$$a_n(z)=s_{n-k}(z)\,a_{k-1}(z)+\rho_n\,z\,\hat{s}_{n-k}(z)\hat{a}_{k-1}(z)\,. \tag{22}$$

Relation (21) will play a major role in our approach to the class of split Levinson algorithms.

4 Family of ρ_n-symmetric predictors

In addition to ρ_n, let us select *any point* ζ_0 *on the unit circle* ($|\zeta_0|=1$). For $k=1,2,\ldots,n$, define the polynomial

$$\pi_k(z)=\zeta_0^{(k-n)/2}[s_{n-k}(\zeta_0)a_{k-1}(z)+\rho_n\,\hat{s}_{n-k}(\zeta_0)z\,\hat{a}_{k-1}(z)]\,. \tag{23}$$

Here and in the sequel, $\zeta_0^{1/2}$ denotes either of the square roots of ζ_0. Since the expression between square brackets in (23) equals the left-hand side of (21) with $\zeta=\zeta_0$, we can given an alternative formula for $\pi_k(z)$, involving $a_k(z)$ instead of $a_{k-1}(z)$. In view of (22), this shows that a natural choice for π_0 is the constant $\pi_0(z)=\zeta_0^{-n/2}a_n(\zeta_0)$.

Let us now mention five interesting properties of the polynomials $\pi_k(z)$. They follow readily from (21) and (23), except the second one which is somewhat less elementary (see [8]).

1. Polynomial $\pi_k(z)$ enjoys the ρ_n-*symmetry* property

$$\hat{\pi}_k(z)=\overline{\rho}_n\,\pi_k(z)\,. \tag{24}$$

2. The zeros of $\pi_k(z)$ are simple and are located on the unit circle; they interlace the zeros of $\pi_{k-1}(z)$.

3. The given predictor $a_n(z)$ is *embedded* in the $\pi_k(z)$ family, in the sense that we have $\pi_n(z) = a_n(z)$.

4. The value $\zeta_0^{-k/2}\pi_k(\zeta_0)$ does not depend on k; we have the *invariance* property
$$\pi_k(\zeta_0) = \zeta_0^{(k-n)/2} a_n(\zeta_0)\,, \qquad \text{all } k\,. \tag{25}$$

5. The initial polynomials are given by
$$\pi_0(z) = \zeta_0^{-n/2} a_n(\zeta_0)\,, \quad \pi_1(z) = \zeta_0^{(1-n)/2}[s_{n-1}(\zeta_0) + \rho_n\, \hat{s}_{n-1}(\zeta_0)z]\,. \tag{26}$$

Note that the constant $\pi_0(z)$ vanishes if and only if the point ζ_0 is a zero of $a_n(z)$; in this case, $\pi_k(\zeta_0)$ vanishes for *all* k. On the other hand, it is obvious from (23) that we have $\pi_k(0) \neq 0$ for all $k \geq 1$; hence $\pi_k(z)$ has exact degree k. In view of their meaning in the context of linear prediction theory, the polynomials $\pi_k(z)$ will be termed the *ρ_n-symmetric predictors* relative to the Toeplitz matrix C_{n-1}.

Let us now examine how the classical predictors $a_k(z)$ and the corresponding Schur-Szegö parameters $\rho_k = a_{k,k}$ can be retrieved from the pair of ρ_n-symmetric predictors $\pi_k(z)$ and $\pi_{k+1}(z)$, for $1 \leq k \leq n-1$. Using (21) and (23) we obtain, after straightforward computation, the formula
$$\pi_{k+1}(z) - \zeta_0^{-1/2}\, z\, \pi_k(z) = \pi_{k+1}(0)(1 - \zeta_0^{-1}\, z) a_k(z)\,, \tag{27}$$
from which we can deduce $a_k(z)$ by a simple division. (This should be compared with (11).) Then, equating the leading coefficients in both sides of (27) and making use of ρ_n-symmetry, we obtain ρ_k in the form
$$\rho_k = \zeta_0 \rho_n[\zeta_0^{-1/2}\, \overline{\pi}_k(0) - \overline{\pi}_{k+1}(0)]\pi_{k+1}^{-1}(0)\,. \tag{28}$$

By inserting the expression of $a_k(z)$ given by (27) into the Szegö-Levinson formula (7) we derive a *two-step recurrence relation* for the ρ_n-symmetric predictors. It can be written as
$$\pi_{k+1}(z) = (\beta_k + \overline{\beta}_k\, z)\pi_k(z) - \sigma_k \sigma_{k-1}^{-1}\, |\beta_k|^2 z\, \pi_{k-1}(z)\,, \tag{29}$$
for $k = 1, 2, \ldots, n-1$, where β_k is given by
$$\beta_k = \frac{\pi_{k+1}(0)}{\pi_k(0)} = \frac{\zeta_0^{1/2}\, s_{n-k-1}(\zeta_0)}{s_{n-k}(\zeta_0)}\,. \tag{30}$$

It should be pointed out that the family of ρ_n-symmetric predictors $\pi_k(z)$, with $k = 0, 1, \ldots, n$, is determined completely from the parameters $\beta_1, \beta_2, \ldots, \beta_{n-1}$ and the initial polynomials $\pi_0(z)$, $\pi_1(z)$. Indeed, it follows from (17) and (30) that the coefficients involved in (29) are bound by the relation

$$\zeta_0^{-1/2}\beta_k + \zeta_0^{1/2}\overline{\beta}_k - 1 = \sigma_k\sigma_{k-1}^{-1}\,|\beta_k|^2\,. \tag{31}$$

5 Normalized symmetric predictors

The polynomials $\pi_k(z)$ constructed in the preceding section have the desired form (12), within a constant phase factor. To achieve our objective we have still to determine the normalization that transforms the recurrence (29) into the simple "one-parameter" form (14). Let us introduce the *normalized polynomials*

$$p_k(z) = g_k\,\pi_k(z)\,, \tag{32}$$

with suitable nonzero complex numbers g_k, for $k = 0, 1, \ldots, n$. Then it is seen that (29) reduces to the desired form (14) under the necessary and sufficient conditions

$$\alpha_k g_k = \beta_k\, g_{k+1}\,, \qquad \overline{\alpha}_k g_k = \overline{\beta}_k\, g_{k+1}\,,$$

$$g_{k-1} = g_{k+1}\sigma_k\,\sigma_{k-1}^{-1}\,|\beta_k|^2\,, \tag{33}$$

with $1 \leq k \leq n-1$. Define the ratios

$$\lambda_k = g_k/g_{k-1}\,, \tag{34}$$

which will play an important rôle in the sequel. When expressed in terms of the numbers (34), the constraints (33) on the normalizing factors g_k can be written in the form

$$\lambda_k = \overline{\lambda}_k\,, \qquad \sigma_k\sigma_{k-1}^{-1}\,|\beta_k|^2\,\lambda_k\,\lambda_{k+1} = 1\,. \tag{35}$$

In view of (31) and (35), the sequence $(\lambda_1, \lambda_2, \ldots, \lambda_n)$ is determined uniquely from the parameters $\beta_1, \beta_2, \ldots, \beta_{n-1}$, for any choice of the initial value λ_1 (a nonzero real number). For convenience, we choose $\lambda_1 = 1$, i.e., $g_1 = g_0$. This entails no real loss of generality, since any other choice, say

$\lambda_1' = c$, yields a sequence of real numbers λ_k' which can be expressed in terms of the original sequence by $\lambda_k' = c\,\lambda_k$ for odd k and $\lambda_k' = c^{-1}\lambda_k$ for even k. Note that the coefficients α_k involved in the normalized recurrence (14) are given by $\alpha_k = \lambda_{k+1}\,\beta_k$.

The real numbers $\lambda_1 = 1, \lambda_2, \ldots, \lambda_n$ are clearly *positive*, in view of (35). They will be referred to as the *Jacobi parameters* of the problem, since they can be expressed as ratios of Jacobi determinants in the natural manner explained below. From the coefficients α_j of the recurrence formula (14), define the $k \times k$ Jacobi (tridiagonal symmetric) matrix

$$J_k = \begin{bmatrix} 1 & 1 & & \\ 1 & 2\mathrm{Re}(\zeta_0^{-1/2}\,\alpha_1) & \cdot & \\ & \cdot & \cdot & 1 \\ & & 1 & 2\mathrm{Re}(\zeta_0^{-1/2}\,\alpha_{k-1}) \end{bmatrix}, \tag{36}$$

for $k = 1, 2, \ldots, n$. It follows from (31), (33) and (35) that the Jacobi parameters (34) obey the recurrence relation

$$\lambda_{k+1} = 2\mathrm{Re}(\zeta_0^{-1/2}\,\alpha_k) - \lambda_k^{-1}\,. \tag{37}$$

Applying the Laplace expansion rule to the determinant of (36) and comparing the result with (37) we obtain the expression

$$\lambda_k = \det J_k / \det J_{k-1}\,, \tag{38}$$

with the convention $\det J_0 = 1$. Equivalently, we have $g_k = g_0 \det J_k$, in view of (34). Note that J_n is *positive definite*, as a consequence of the property $\lambda_k > 0$ for all k. Without going into detail let us mention that, in the case $a_n(\zeta_0) \neq 0$, there exists a simple congruence relation between the Toeplitz matrix C_{n-1} and the Jacobi matrix J_n, involving the coefficients of the polynomials $p_k(z)$; see [8].

In the "regular case" where $a_n(\zeta_0)$ is not zero, the Jacobi parameters can be written in terms of the (nonzero) values $p_k(\zeta_0)$ in the simple form

$$\lambda_k = p_k(\zeta_0)/[\zeta_0^{1/2}\,p_{k-1}(\zeta_0)]\,. \tag{39}$$

This is true for $k = 1$ as a consequence of (26), in view of our choice $g_0 = g_1$. The general result (39) is proved by induction, with the help of (14) and (37).

For $k = 1, 2, \dots, n$, let us define the complex number ω_k, of unit modulus, to be the "pseudo reflection coefficient" $p_{k,k}/p_{k,0}$ of $p_k(z)$. In view of (23), it is given by

$$\omega_k = \rho_n \, \hat{s}_{n-k}(\zeta_0)/s_{n-k}(\zeta_0) . \tag{40}$$

At this point we consider only the ω_1 coefficient, and we examine the "initial conditions" in some detail. Denoting by $\omega_1^{1/2}$ either of the square roots of ω_1 we can write $p_0(z) = 2\gamma \mathrm{Re}(\omega_1^{1/2} \, \zeta_0^{1/2})$ and $p_1(z) = \gamma(\omega_1^{-1/2} + \omega_1^{1/2} z)$, for a certain constant γ. This follows from (26), (32) and our assumption $g_0 = g_1$. We can now choose the number g_0 so as to make γ real. As a result, we obtain $\hat{p}_0(z) = p_0(z)$ and $\hat{p}_1(z) = p_1(z)$, whence the desired *symmetry property* (13), that is $\hat{p}_k(z) = p_k(z)$ for all k, by induction based on (14). Without loss of generality we specify g_0 through the requirement $\gamma = 1$. The initial polynomials are now given by

$$p_0(z) = 2\mathrm{Re}(\omega_1^{1/2} \, \zeta_0^{1/2}) , \qquad p_1(z) = \omega_1^{-1/2} + \omega_1^{1/2} \, z . \tag{41}$$

The corresponding value of g_0 (equal to g_1) is

$$g_0 = \omega_1^{-1/2} \, \pi_1^{-1}(0) = \omega_1^{-1/2} \, \zeta_0^{(n-1)/2} \, s_{n-1}^{-1}(\zeta_0) . \tag{42}$$

It is interesting to see how the parameter ω_1 can be used to distinguish between the regular case $a_n(\zeta_0) \neq 0$ and the singular case $a_n(\zeta_0) = 0$. In view of (25) and (41) we have the equivalence

$$(a_n(\zeta_0) = 0) \Leftrightarrow (\omega_1 \zeta_0 = -1) . \tag{43}$$

6 Derivation of the general split Levinson algorithm

From (2), (21), (23) and (32) it follows that the coefficient vector of the symmetric polynomial $p_k(z) = \sum_{j=0}^{k} p_{k,j} \, z^j$ satisfies the Toeplitz system

$$C_k[p_{k,0}, p_{k,1}, \dots, p_{k,k}]^T = [\overline{\tau}_k, 0, \dots, 0, \tau_k]^T , \tag{44}$$

where τ_k (equal to $\sum_{j=0}^{k} c_{k-j}\, p_{k,j}$) is given by

$$\tau_k = \rho_n g_k \sigma_k \zeta_0^{(n-k)/2}\, \overline{s}_{n-k-1}(\zeta_0)\,, \tag{45}$$

for $k = 1, 2, \ldots, n$. Writing the recurrence formula (14) in vector form, multiplying both sides by C_{k+1}, and using (44), we obtain a very useful relation between the parameters, namely

$$\alpha_k = \tau_{k-1}/\tau_k\,. \tag{46}$$

As we have $\alpha_k = p_{k+1}(0)/p_k(0)$, by (14), this implies that the product $\tau_{k-1}p_k(0)$ is independent of k. Computing that product for $k = 2$, by use of (23), (33), (42) and (45), we obtain the value

$$\tau_{k-1}\, p_k(0) = c_0\, \zeta_0^{1/2}\,. \tag{47}$$

When extended to the case $k = 1$, this yields an appropriate definition for τ_0, that is $\tau_0 = c_0\omega_1^{1/2}\zeta_0^{1/2}$ (in agreement with (45)). Hence the substitute for the system (44) when $k = 0$ should read $c_0 p_{0,0} = 2\mathrm{Re}\, \tau_0$.

As explained in section 3, the unit modulus value assigned to the reflection coefficient ρ_n produces a well-defined singular extension C_n of C_{n-1}. The corresponding value of ω_1, which specifies our family of symmetric polynomials $p_k(z)$, is determined from ρ_n through the relations (17), (18) and (40). Conversely, let us verify that any choice of ω_1, with $|\omega_1| = 1$, yields an associated singular extension C_n. Consider the sequence of pseudo reflection coefficients $\omega_1, \omega_2, \ldots, \omega_n$ given by (40). Note the property $\omega_n = \rho_n$. In view of (17) and (18), this sequence satisfies the Schur type recurrence relation

$$\omega_{k+1} = \zeta_0^{-1}(\omega_k - \rho_k)/(1 - \overline{\rho}_k\, \omega_k)\,, \tag{48}$$

for $k = 1, \ldots, n-1$. Starting from ω_1 we can use (48) to determine successively $\omega_2, \ldots, \omega_n$ (in terms of the given parameters $\rho_1, \ldots, \rho_{n-1}$). By induction, we obtain $|\omega_k| = 1$ for all k. The final element, $\rho_n = \omega_n$, provides the required extension C_n.

From an algorithmic viewpoint, the main conclusion is that the family of normalized symmetric polynomials $p_k(z)$ can be computed recursively from the data $c_0, c_1, \ldots, c_{n-1}$, on the basis of (14), (41), (44) and (46). This method yields a substitute for the Levinson algorithm, provided it is supplemented by a procedure to recover $a_{n-1}(z)$ and σ_{n-1} from the polynomials $p_k(z)$. To that end we can make use of the simple formulas

$$a_k(z) = \frac{p_{k+1}(z) - \lambda_{k+1}\, \zeta_0^{-1/2} z\, p_k(z)}{p_{k+1}(0)(1 - \zeta_0^{-1} z)}, \tag{49}$$

$$\sigma_k = c_0\, \lambda_{k+1}\, |p_{k+1}(0)|^{-2}. \tag{50}$$

Expression (49) is a mere translation of (27), making use of (32) and (34). As for (50), it follows easily from (33) and (35), by induction, in view of the identity $p_{k+1}(0) = \alpha_k\, p_k(0)$. Recall that the Jacobi parameter λ_{k+1} involved in (49) and (50) can be determined either in a recursive manner, by use of (37), or in a direct manner, by use of (39). However, the latter method applies only when we have $a_n(\zeta_0) \neq 0$, i.e., $\omega_1\, \zeta_0 \neq -1$.

We are now in a position to describe the *class of split Levinson algorithms* in a precise manner. As the standard Levinson method, each of these algorithms computes the predictor $a_{n-1}(z)$ and the prediction error σ_{n-1} (or/and the reflection coefficients $\rho_1, \ldots, \rho_{n-1}$) relative to a given positive definite Toeplitz matrix C_{n-1}. A typical split Levinson algorithm is determined by a pair of complex numbers $\zeta_0^{1/2}$ and $\omega_1^{1/2}$, each having unit modulus. A simple implementation (valid in the regular case $a_n(\zeta_0) \neq 0$) consists of the following stages.

1. Select two unit modulus numbers $\zeta_0^{1/2}$ and $\omega_1^{1/2}$, with the restriction that their product is neither i nor $-i$.

2. Initialize the algorithm with $\tau_0 = c_0 \omega_1^{1/2} \zeta_0^{1/2}$, $p_0(z) = 2\mathrm{Re}(\omega_1^{1/2}\, \zeta_0^{1/2})$, $p_1(z) = \omega_1^{-1/2} + \omega_1^{1/2}\, z$.

3. For $k = 1, 2, \ldots, n-1$, compute recursively

$$\tau_k = \sum_{j=0}^{k} c_{k-j}\, p_{k,j}, \qquad \alpha_k = \tau_{k-1}/\tau_k\,,$$

$$p_{k+1}(z) = (\alpha_k + \overline{\alpha}_k z) p_k(z) - z\, p_{k-1}(z)\,.$$

4. Determine the solution $(a_{n-1}(z),\ \sigma_{n-1})$ by

$$\lambda_n = p_n(\zeta_0)/[\zeta_0^{1/2}\, p_{n-1}(\zeta_0)]\,,$$

$$a_{n-1}(z) = \frac{p_n(z) - \lambda_n \zeta_0^{-1/2} z\, p_{n-1}(z)}{p_n(0)(1 - \zeta_0^{-1} z)}, \qquad \sigma_{n-1} = c_0 \lambda_n |p_n(0)|^{-2}.$$

If we make the special choice $\omega_1 \zeta_0 = -1$, then the algorithm has to be slightly modified, since the formula given in stage 4 for the Jacobi parameter λ_n cannot be applied. Instead of it, we have to use the recurrence formula (37), which we now insert in stage 3 of the algorithm above, with the initialization $\lambda_1 = 1$ (in stage 2). The latter method is completely general (valid for all ζ_0 and ω_1) and is especially interesting when the algorithm is required to provide the sequence of Schur-Szegö parameters ρ_k relative to C_{n-1}. Indeed, translating (28) in terms of the current variables we obtain the expression

$$\rho_k = \overline{p}_k(0) p_k^{-1}(0)[1 - \zeta_0^{1/2} \lambda_k^{-1} \alpha_k^{-1}]. \tag{51}$$

Let us now explain in some detail why the split Levinson algorithms are more economical than the classical Levinson algorithm. First, due to the symmetry property (13), i.e., $p_{k,j} = \overline{p}_{k,k-j}$, the memory space required to store $p_k(z)$, instead of $a_k(z)$, is reduced by a factor 2. Next, let us compare the scalar product computation of τ_k in stage 3 above with the analogous computation of ρ_k in (6). In the formula $\tau_k = \sum_{j=0}^{k} c_{k-j}\, p_{k,j}$ we examine the sum $\tau_k(j)$ of the j-term and of the $(k-j)$-term, for a given j. Owing to symmetry, $\tau_k(j)$ equals the sum of $(c_{k-j} + c_j)\mathrm{Re}\, p_{k,j}$ and $i(c_{k-j} - c_j)\mathrm{Im}\, p_{k,j}$. As a result, $\tau_k(j)$ involves 6 real additions and 4 real multiplications (instead of 8 multiplications in the nonsymmetric case). As for a comparison between the recursive computations (14) and (7) of $p_k(z)$ and $a_k(z)$, we note that the updated coefficient $p_{k+1,j}$ (given by the sum of $\alpha_k p_{k,j}$, $\overline{\alpha}_k\, p_{k,j-1}$ and $-p_{k-1,j-1}$) has to be calculated only for $j \le (k+1)/2$. In summary, there is an overall saving of approximately 50% in the number of multiplications and no change in the number of additions.

Finally, let us make a short comment concerning the singular case $\omega_1 \zeta_0 = -1$. As pointed out above, this assumption means that all polynomials $p_k(z)$ vanish at the point $z = \zeta_0$. It can be shown that the family of *reduced symmetric polynomials* $\tilde{p}_k(z)$, defined (within a simple normalization) by

$$\tilde{p}_k(z) = p_{k+1}(z)/p_1(z), \tag{52}$$

for $k = 0, 1, \ldots, n-1$, fits exactly into our general theory. The entries of the corresponding Toeplitz matrix $\tilde{C}_{n-2}$ have the form

$$\tilde{c}_k = 2c_k - \zeta_0^{-1} c_{k-1} - \zeta_0 c_{k+1}, \tag{53}$$

with $c_{-1} = \bar{c}_1$. This could lead to an interesting "deflated split Levinson algorithm". We shall not go into detail about that subject.

7 Application to real data

Let us consider the important particular case where the data c_k are *real*. We shall see how the different versions of the "real split Levinson algorithm" proposed in the literature fit nicely into the general theory worked out in the preceding sections. Roughly speaking, these versions involve the families of the symmetric or antisymmetric parts of the classical predictors $a_k(z)$, with a possibility of interleaving them [6], [7].

7.1 Symmetric version $[\omega_k = 1]$

Choose $\zeta_0^{1/2} = \omega_1^{1/2} = 1$. The initialization is $\tau_0 = c_0$, $p_0(z) = 2$, $p_1(z) = 1 + z$. All numbers τ_k, α_k and $p_{k,j}$ are real. We have $p_k(z) = p_k(0)$ $[a_{k-1}(z) + z\,\hat{a}_{k-1}(z)]$ and $\lambda_k = p_k(1)/p_{k-1}(1)$. The recurrence relation (14) reads $p_{k+1}(z) = \alpha_k(1+z)p_k(z) - z\,p_{k-1}(z)$. The reflection coefficients ρ_k can be computed by $\rho_k = 1 - (\lambda_k \alpha_k)^{-1}$ with the help of the recurrence λ_{k+1} $= 2\alpha_k - \lambda_k^{-1}$. Note that this yields $\rho_{k+1} = 1 - [\alpha_k\,\alpha_{k+1}(1+\rho_k)]^{-1}$.

7.2 Antisymmetric version $[\omega_k = -1]$

Choose $\zeta_0^{1/2} = 1$, $\omega_1^{1/2} = i$. The initialization is $\tau_0 = i\,c_0$, $p_0(z) = 0$, $p_1(z) = -i(1-z)$. The recursive part (stage 3) of the split Levinson algorithm can be applied to the modified polynomials $p'_k(z) = i\,p_k(z)$, which are real, and to the corresponding real numbers $\tau'_k = i\,\tau_k$. The initial conditions are now $\tau'_0 = -c_0$, $p'_0(z) = 0$, $p'_1(z) = 1 - z$. Polynomial $p'_k(z)$ has the antisymmetry property $\hat{p}'_k(z) = -p'_k(z)$; it has the form $p'_k(z) = p'_k(0)$ $[a_{k-1}(z) - z\,\hat{a}_{k-1}(z)]$. The recurrence relation for the modified polynomials $p'_k(z)$ is formally the same as in § 7.1, with $\alpha_k = \tau'_{k-1}/\tau'_k$.

All polynomials $p'_k(z)$ have a zero at the point $z = 1$. The Jacobi parameters have to be computed by the recursive formula $\lambda_{k+1} = 2\alpha_k - \lambda_k^{-1}$ (the same as in § 7.1). The reflection coefficients are given by $\rho_k = (\lambda_k \alpha_k)^{-1} - 1$.

7.3 Even alternating version $[\omega_k = (-1)^k]$

Choose $\zeta_0^{1/2} = \omega_1^{1/2} = i$. The initialization is $\tau_0 = -c_0$, $p_0(z) = -2$, $p_1(z) = -i(1-z)$. As in the preceding case, the algorithm can be brought to a real setting by a simple transformation. Here we define $p'_k(z) = i^k p_k(z)$ and $\tau'_k = i^k \tau_k$. This yields the real numbers $\alpha'_k = \tau'_{k-1}/\tau'_k = -i\,\alpha_k$. The polynomials $p'_k(z)$ are real and enjoy the even alternating symmetry property $\hat{p}'_k(z) = (-1)^k p'_k(z)$. We have $p'_k(z) = p'_k(0)\ [a_{k-1}(z) + (-1)^k z\,\hat{a}_{k-1}(z)]$ and $\lambda_k = -p'_k(-1)/p'_{k-1}(-1)$. The recurrence relation becomes $p'_{k+1}(z) = -\alpha'_k(1-z)p'_k(z) + z\,p'_{k-1}(z)$, with the initial conditions $p'_0 = -2$ and $p'_1(z) = 1 - z$. The Jacobi parameters satisfy the recurrence formula $\lambda_{k+1} = 2\alpha'_k - \lambda_k^{-1}$, and the reflection coefficients are given by $\rho_k = (-1)^k\ [1 - (\lambda_k \alpha_k)^{-1}]$.

7.4 Odd alternating version $[\omega_k = (-1)^{k-1}]$

Choose $\zeta_0^{1/2} = i$, $\omega_1^{1/2} = 1$. The initialization is $\tau_0 = i\,c_0$, $p_0(z) = 0$, $p_1(z) = 1 + z$. Here, a suitable transformation is given by $p'_k(z) = i^{k-1} p_k(z)$ and $\tau'_k = i^{k-1}\tau_k$. This yields $\alpha'_k = -i\,\alpha_k$ (as in § 7.3). The polynomials $p'_k(z)$ are real and enjoy the odd alternating symmetry property $\hat{p}'_k(z) = -(-1)^k p'_k(z)$. They have the form $p'_k(z) = p'_k(0)\ [a_{k-1}(z) - (-1)^k z\,\hat{a}_{k-1}(z)]$. The recurrence relation is formally the same as in § 7.3; the initial conditions are $\tau'_0 = c_0$, $p'_0(z) = 0$, $p'_1(z) = 1 + z$.

All polynomials $p'_k(z)$ have a zero at the point $z = -1$. The Jacobi parameters λ_k have to be computed by the recurrence $\lambda_{k+1} = 2\alpha'_k - \lambda_k^{-1}$ (the same as in § 7.3). The reflection coefficients are given by $\rho_k = (-1)^k$ $[(\lambda_k \alpha'_k)^{-1} - 1]$.

Acknowledgment

The authors are grateful to T. Kailath and D. Slock for calling their attention to the paper by Bube and Burridge.

References

[1] Y. Bistritz, "A new unit circle stability criterion", *Proc Internat. Symp. Mathematical Theory of Networks and Systems*, Beer Sheva, 1983, in Lecture Notes in Control and Information Sciences no. 58, P.A. Fuhrmann ed., New York: Springer-Verlag, 1984, pp. 69–87.

[2] Y. Bistritz, "Zero location with respect to the unit circle of discrete-time linear system polynomials", *IEEE Proc.*, **72**, pp. 1131–1142, 1984.

[3] Y. Bistritz, "A circular stability test for general polynomials", *Systems Control Lett.*, **7**, pp. 89–97, 1986.

[4] Y. Bistritz, H. Lev-Ari, and T. Kailath, "Immittance-domain Levinson algorithms, *Proc. IEEE Internat. Conf. Acoustics Speech Signal Processing*, Tokyo, 1986, pp. 253–256.

[5] K.B. Bube and R. Burridge, "The one-dimensional inverse problem of reflection seismology", *SIAM Review*, **25**, pp. 497–559, 1983.

[6] P. Delsarte and Y. Genin, "The split Levinson algorithm", *Philips Res. Lab. Brussels Manuscript* M99, Jan. 1985.

[7] P. Delsarte and Y. Genin, "The split Levinson algorithm", *IEEE Trans. Acoust. Speech Signal Process.*, **ASSP-34**, pp. 470–478, 1986.

[8] P. Delsarte and Y. Genin, "The tridiagonal approach to Szegö's orthogonal polynomials, Toeplitz linear systems, and related interpolation problems", *SIAM J. Math. Anal.*, **19**, pp. 718–735, 1988.

[9] P. Delsarte and Y. Genin, "On the split approach based algorithms for DSP problems", *this volume.*

[10] P. Delsarte, Y. Genin, and Y. Kamp, "On the class of positive definite matrices equivalent to Toeplitz matrices", *Proc. Internat. Symp. Mathematical Theory of Networks and Systems*, Santa Monica, 1981, N. Levan ed., North Hollywood: Western Periodicals, pp. 40–45.

[11] P. Delsarte, Y. Genin, and Y. Kamp, "Equivalence classes of Hermitian matrices and their Schur parametrization", *SIAM J. Alg. Disc. Meth.*, **3**, pp. 279–298, 1983.

[12] P. Delsarte, Y. Genin, and Y. Kamp, "Application of the index theory of pseudo-lossless functions to the Bistritz stability test", *Philips J. Res.*, **39**, pp. 226–241, 1984.

[13] G.H. Golub and C.F. Van Loan, *Matrix Computations*, Oxford: North Academic, 1983.

[14] T. Kailath, S.-Y. Kung, and M. Morf, "Displacement ranks of matrices and linear equations", *J. Math. Anal. Appl.*, **68**, pp. 395–407, 1979.

[15] H. Krishna and S.D. Morgera, "The Levinson recurrence and fast algorithms for solving Toeplitz systems of linear equations", *IEEE Trans. Acoust. Speech Signal Process.*, **ASSP-35**, pp. 839–848, 1987.

[16] J. Makhoul, "Linear prediction: a tutorial review", *IEEE Proc.*, **63**, pp. 561–580, 1975.

[17] S.D. Morgera and H. Krishna, "Generalized Levinson/Szegö complex recurrences for a class of second-order nonstationary stochastic processes", *Proc. Internat. Symp. Circuits and Systems*, Philadelphia, 1987, pp. 84–90.

[18] G. Szegö, Orthogonal Polynomials, New York: American Mathematical Society, 1959.

On the Split Approach Based Algorithms for DSP Problems

P. Delsarte and Y. Genin

Philips Research Laboratory Brussels
Av. Van Becelaere 2
B-1170 Brussels, Belgium

Abstract

This paper shows that the "split approach" produces new efficient algorithms to solve not only the Toeplitz linear prediction problem, as explained in a companion paper, but also a collection of other important DSP problems belonging to the Toeplitz environment. It emphasizes both the algorithmic procedures themselves, and their connections with standard mathematical subjects such as the theory of orthogonal polynomials and the theory of positive functions.

1 Introduction

In a companion paper, a general framework is presented for the class of *split Levinson algorithms*, which are attractive substitutes for the classical Levinson algorithm to solve a Toeplitz system of linear equations [14]. This research subject seems to be especially rich, mainly for the following reason. The ideas behind the split Levinson algorithm open the door to an efficient new approach to a collection of algebraic and function theoretic problems belonging to the environment of positive definite Toeplitz matrices. Many of these problems are quite significant from the application viewpoint, especially in the digital signal processing (DSP) area (see e.g. [24] and [26]). It is therefore not surprising that the "*split approach*" is currently receiving a good deal of attention in DSP research.

NATO ASI Series, Vol. F 70
Numerical Linear Algebra, Digital Signal Processing and Parallel Algorithms
Edited by G. H. Golub and P. Van Dooren

The present contribution aims at giving an overview of the main results obtained in this area. (Further details and additional references can be found in [13].) To emphasize the fundamentals rather than the technicalities of the questions dealt with, we will focus on the case of *real scalar data*, although most methods and results can be extended to more general settings. We will also trade generality for simplicity in considering a single representative of the class of split Levinson algorithms (section 2).

The classical Levinson algorithm for the linear prediction problem has some close companions, namely the *Schur algorithm* [28] and the *lattice algorithm* [23], which are especially attractive to solve the same problem in specific applications. Just as the Levinson algorithm, both of them admit *split counterparts* of reduced complexity [9]. The situation is similar in the more general context of the joint process estimation problem (leading to a Toeplitz system with an arbitrary right-hand side) [15]. These results are explained in section 3.

The Schur-Cohn test to check polynomial stability (in the "discrete time" sense) is known to be intimately related to the Levinson algorithm. In fact, both procedures are based (formally) on the same recurrence formula, with the essential difference that they process this recurrence either for decreasing degree or for increasing degree, respectively. It turns out that there is an analogous formal relation between the *Bistritz stability test* [4] (an efficient substitute for the Schur-Cohn test) and the split Levinson algorithm. This is shown in section 4.

The recurrence formula underlying the Levinson algorithm is identical to that occurring in Szegö's theory of orthogonal polynomials on the unit circle [33]. As explained in section 5, the split Levinson recurrence enjoys an analogous property; within a simple transformation, it coincides with the three-term formula satisfied by a well-defined family of *orthogonal polynomials on the interval* [-1,1] *of the real line* [8], [19].

Positive definite Toeplitz matrices belong to the natural environment of the theory of *positive* (Carathéodory) and *bounded* (Schur) *functions*. The coefficient characterization problems and the associated interpolation problems for these two classes of functions are of direct significance in various questions of engineering interest (such as inverse scattering) [18]. It is briefly indicated in section 6 how these problems can be approached and solved by a new method, originating from the split Levinson recursion, which can be used as a substitute for the methods based on the Carathéodory-Toeplitz theorem and on the Schur theorem [10]. This method is shown to provide the natural function theoretic background of the split Schur algorithm.

Section 7 contains a review of various problems and techniques that belong naturally to the scope of the "split approach". It is mentioned in section 8 how the split theory can be extended to the case of complex data [10], [27], and even to the case of matrix data [11], [12], with some noticeable exceptions; finally, the possibility of generalizing the split Levinson algorithm to non-Toeplitz matrices is alluded to.

2 Split Levinson algorithm

Consider a real symmetric positive definite $n \times n$ Toeplitz matrix $C_{n-1} = [c_{i-j} : 0 \leq i, j \leq n-1]$. There exist four versions of the split Levinson algorithm, processing only *real numbers*, to solve the linear prediction problem relative to such a matrix [14]. Here we single out one of these versions, namely the *symmetric split Levinson algorithm* (see section 7.1 of [14]). It turns out that any choice would lead to similar results.

The algorithm in question processes symmetric polynomials $p_k(z) = \sum_{i=0}^{k} p_{k,i}\, z^i$ (with $p_{k,i} = p_{k,k-i}$) of the form

$$p_k(z) = p_{k,0}\left[a_{k-1}(z) + z\, \hat{a}_{k-1}(z)\right], \tag{1}$$

where $a_{k-1}(z)$ is the classical *predictor polynomial* relative to C_{k-1}. For further reference let us write down the *Levinson recurrence relation*

$$a_k(z) = a_{k-1}(z) + \rho_k\, z\, \hat{a}_{k-1}(z), \tag{2}$$

and $\sigma_k = \sigma_{k-1}(1 - \rho_k^2)$, with $\rho_k = -\sigma_{k-1}^{-1} \sum_{i=0}^{k-1} c_{k-i}\, a_{k-1,i}$ denoting the kth *reflection coefficient.* Our version of the split Levinson algorithm consists of the following steps.

1. Set the initial values $\tau_0 = c_0$, $p_0(z) = 2$, $p_1(z) = 1 + z$.

2. For $k = 1, 2, \ldots, n-1$, compute recursively

$$\tau_k = \sum_{i=0}^{k} c_i\, p_{k,i}\,, \qquad \alpha_k = \tau_{k-1}/\tau_k\,, \tag{3}$$

$$p_{k+1}(z) = \alpha_k(1+z)\,p_k(z) - z\,p_{k-1}(z). \tag{4}$$

3. Compute the solution $a_{n-1}(z)$, σ_{n-1} by

$$\lambda_n = p_n(1)/p_{n-1}(1), \tag{5}$$

$$a_{n-1}(z) = \frac{p_n(z) - \lambda_n z\,p_{n-1}(z)}{p_{n,0}(1-z)}, \quad \sigma_{n-1} = \frac{c_0\lambda_n}{p_{n,0}^2}. \tag{6}$$

In DSP applications one is often more interested in determining the sequence of reflection coefficients $\rho_1, \ldots, \rho_{n-1}$ rather than the predictor $a_{n-1}(z)$ itself. It can be obtained by inserting the relations

$$\lambda_{k+1} = 2\alpha_k - \lambda_k^{-1}, \qquad \rho_k = 1 - (\lambda_k\,\alpha_k)^{-1} \tag{7}$$

in the recursive part (step 2) of the split Levinson algorithm; the initialization is given by $\lambda_1 = 1$. The numbers λ_k are termed *Jacobi parameters*. Recall that they can be defined by $\lambda_k = p_k(1)/p_{k-1}(1)$.

3 Companion split algorithms

The reflection coefficients ρ_k relative to C_{n-1} are classically computed by means of the Levinson algorithm as explained above or, sometimes preferably, by means of the *Schur algorithm* [28]. The latter method processes the variables u and v defined by

$$u_{k-1,j} = \sum_{i=0}^{k-1} c_{k+j-i}\,a_{k-1,i}, \quad v_{k-1,j} = \sum_{i=0}^{k-1} c_{j+i}\,a_{k-1,i}. \tag{8}$$

It is based on the formulas

$$\rho_k = -u_{k-1,0}/v_{k-1,0}, \qquad \sigma_{k-1} = v_{k-1,0}, \tag{9}$$

$$u_{k,j} = u_{k-1,j+1} + \rho_k\,v_{k-1,j+1}, \quad v_{k,j} = \rho_k\,u_{k-1,j} + v_{k-1,j}, \tag{10}$$

with $k = 1, \ldots, n-2$ and $j = 0, \ldots, n-k-2$; the initial values are $u_{0,j} = c_{j+1}$ and $v_{0,j} = c_j$ for $j = 0, \ldots, n-2$. These results can be proved by use of relation (2).

By combining the Schur variables (8) in agreement with (1) we define the "split Schur variables"

$$w_{k,j} = p_{k,0}(u_{k-1,j} + v_{k-1,j}) = \sum_{i=0}^{k} c_{j+i}\, p_{k,i}\,. \tag{11}$$

Translating the split Levinson formula (4) in terms of these new variables, we obtain the recurrence relation

$$w_{k+1,j} = \alpha_k(w_{k,j} + w_{k,j+1}) - w_{k-1,j+1}\,, \tag{12}$$

for $k = 1, \ldots, n-1$ and $j = 0, \ldots, n-k-2$, with

$$\alpha_k = w_{k-1,0}/w_{k,0}\,, \tag{13}$$

in view of (3) and (11). The *split Schur algorithm* [9] is an efficient method to compute the reflection coefficients by use of (7), (12) and (13). The initial values are $w_{0,0} = c_0$, $w_{0,j} = 2c_j$ for $j = 1, \ldots, n-1$, and $w_{1,j} = c_j + c_{j+1}$ for $j = 0, \ldots, n-2$. This algorithm exhibits a complexity reduction (with respect to its classical counterpart) comparable to that of the split Levinson algorithm. Note that neither the Schur algorithm nor the split Schur algorithm involve any scalar product computation.

The *lattice algorithm* [7], [23] is another interesting method to compute reflection coefficients; it is closely related to the Levinson and Schur algorithms. It can be applied in those situations where the entries c_k of C_{n-1} are the *autocorrelation lags*,

$$c_k = \sum_{t=0}^{N-1} s(t)\, s(t+k)\,, \tag{14}$$

of a given finite sequence $s(0), s(1), \ldots, s(N-1)$ of *signal samples* $s(t)$, under the windowing assumption $s(t) = 0$ for $t \geq N$ and $s(t) = 0$ for $t \leq -1$. The lattice algorithm processes the signal variables f and b, termed forward and backward prediction errors, defined by

$$f_{k-1}(t) = \sum_{i=0}^{k-1} a_{k-1,i}\, s(t-i), \quad b_{k-1}(t) = \sum_{i=0}^{k-1} a_{k-1,k-1-i}\, s(t-i)\,, \tag{15}$$

for $t = 0, \ldots, N+k-2$. They satisfy the recurrence relations

$$f_k(t) = f_{k-1}(t) + \rho_k\, b_{k-1}(t-1)\,, \quad b_k(t) = \rho_k\, f_{k-1}(t) + b_{k-1}(t-1)\,, \tag{16}$$

which follow readily from (2). In this context, the reflection coefficients can be determined from the identities

$$\sigma_{k-1} = \sum_{t=0}^{N+k-2} f_{k-1}^2(t) = \sum_{t=0}^{N+k-2} b_{k-1}^2(t),\ \sigma_{k-1}\,\rho_k = -\sum_{t=1}^{N+k-2} f_{k-1}(t)\, b_{k-1}(t-1)\,. \tag{17}$$

The lattice algorithm computes the sequence of reflection coefficients ρ_k by use of (16) and (17). The initial conditions are given $f_0(t) = b_0(t) = s(t)$ for $t = 0, \ldots, N-1$.

A split version of this algorithm can be obtained in a way very similar to that by which the split Levinson and split Schur algorithms are derived from their classical counterparts [9]. It is clear from (1) and (15) that an appropriate definition for the "split lattice variables" is

$$x_k(t) = p_{k,0}[f_{k-1}(t) + b_{k-1}(t-1)] = \sum_{i=0}^{k} p_{k,i}\, s(t-i)\,. \tag{18}$$

Translating the split Levinson formula (4) in terms of these variables produces the recurrence relation

$$x_{k+1}(t) = \alpha_k[x_k(t) + x_k(t-1)] - x_{k-1}(t-1)\,, \tag{19}$$

for $k = 1, \ldots, n-1$ and $t = 0, \ldots, N-k$. Moreover, we have

$$\alpha_k = \frac{c_0}{2}\left[\sum_{k=0}^{N+k-1} x_k^2(t)\right]^{-1}\,. \tag{20}$$

To prove (20) we use the identity $2\tau_k p_{k,0} = \sum x_k^2(t)$, resulting from (3), (14) and (18), together with the identity $\tau_{k-1} p_{k,0} = c_0$, given in [14]. The *split lattice algorithm* [9] computes the sequence of reflection coefficients with the help of (7), (19) and (20); the initialization is provided by $c_0 = \sum s^2(t)$, $x_0(t) = s(t)$ and $x_1(t) = s(t) + s(t-1)$. The complexity of this algorithm, measured as the number of multiplications involved, is approximately two thirds of that of the classical lattice algorithm.

The last part of this section is concerned with the linear estimation problem for a pair of jointly stationary stochastic processes. It is a natural extension of the linear prediction problem; it underlies many important DSP adaptive filtering techniques [22]. This *joint process estimation problem* amounts to a Toeplitz system of equations of the form

$$[b_{n-1,0}, b_{n-1,1}, \ldots, b_{n-1,n-1}]C_{n-1} = [d_0, d_1, \ldots, d_{n-1}], \tag{21}$$

with an arbitrary right-hand side (consisting of real numbers d_i). The solution is classically computed by means of a simple *extension of the Levinson algorithm* [21], which is obtained by inserting the additional formulas

$$\gamma_k = \sum_{i=0}^{k-1} c_{k-i}\, b_{k-1,i}, \qquad \nu_k = \sigma_k^{-1}(d_k - \gamma_k), \tag{22}$$

$$b_k(z) = b_{k-1}(z) + \nu_k\, \hat{a}_k(z), \tag{23}$$

in the recursive loop (2) of the "linear prediction version"; the extra initial condition is $b_0(z) = c_0^{-1}d_0$. Indeed, it can be shown that the coefficient vector of the polynomial $b_k(z) = \sum_{i=0}^{k} b_{k,i}\, z^i$ generated by this algorithm is the solution of the Toeplitz system $[b_{k,0}, \ldots, b_{k,k}]C_k = [d_0, \ldots, d_k]$, for $k = 0, 1, \ldots, n-1$.

There exists a more economical "*split counterpart*" of the generalized Levinson algorithm just described [15]. It processes the symmetric polynomials $p_k(z)$ together with the symmetric polynomials $q_k(z)$ defined by

$$q_k(z) = b_k(z) + \hat{b}_k(z). \tag{24}$$

These polynomials satisfy the identity

$$q_k(z) = (1+z)q_{k-1}(z) - z\, q_{k-2}(z) + \eta_k p_k(z), \tag{25}$$

where the coefficient η_k is obtainable by means of

$$\xi_k = \sum_{i=0}^{k-1} c_{i+1}\, q_{k-1,i}, \qquad \eta_k = \tau_k^{-1}(d_k - d_{k-1} - \xi_k + \xi_{k-1}). \tag{26}$$

The solution $b_{n-1}(z)$ of the problem (21) can be determined from $q_{n-2}(z)$, $q_{n-1}(z)$ and $p_n(z)$ by use of

$$\mu_{n-1} = \frac{q_{n-1}(1) - q_{n-2}(1)}{p_n(1)}, \quad b_{n-1}(z) = \frac{q_{n-1}(z) - z\, q_{n-2}(z) - \mu_{n-1}p_n(z)}{1-z}. \tag{27}$$

These results are proved in [15]. Supplementing steps 2 and 3 of the split Levinson algorithm above with the relations (25) (26) and (27), respectively, we obtain an efficient solution method, more economical than the classical procedure based on (2) and (23), for the joint process estimation problem (21). The extra initial conditions (in step 1) are $q_0(z) = 2c_0^{-1}d_0$ and $q_1(z) = (c_0 + c_1)^{-1}\,(d_0 + d_1)(1 + z)$.

Without going into detail let us mention that there exist Schur type and lattice type algorithms for the joint process estimation problem, and that both of them also admit economical split counterparts [15].

4 Polynomial stability testing

Consider a real polynomial $a_{n-1}(z)$ of degree $n - 1$, with the normalization $a_{n-1}(0) = 1$. The problem of verifying whether $a_{n-1}(z)$ is *stable*, i.e., devoid of zeros in the closed unit disc $|z| \leq 1$, is solved classically by means of the *Schur-Cohn stability test* [29]. This method generates a sequence of polynomials $a_k(z)$, of decreasing degree k, by exploiting the Levinson recursion (2) in reverse order, i.e., in the form

$$a_{k-1}(z) = (1 - \rho_k^2)^{-1}[a_k(z) - \rho_k\hat{a}_k(z)]\,, \quad \text{with } \rho_k = a_{k,k}\,, \tag{28}$$

for $k = n, n-1, \ldots, 1$. The Schur-Cohn criterion states that $a_{n-1}(z)$ is stable if and only if the leading coefficients $\rho_k = a_{k,k}$ satisfy $|\rho_k| < 1$ for $k = 1, \ldots, n-1$.

A more economical solution method for the same problem has been discovered recently by Bistritz [4]. It is based on a two-step recurrence relation that is formally identical to the reverse version of the split Levinson formula (4). Thus, the Bistritz method generates a sequence of symmetric polynomials $p_k(z)$, of decreasing degree k, by means of the formula

$$p_{k-1}(z) = z^{-1}[\alpha_k(1 + z)p_k(z) - p_{k+1}(z)]\,, \quad \text{with } \alpha_k = p_{k+1}(0)/p_k(0)\,, \tag{29}$$

for $k = n-1, n-2, \ldots, 1$. A simple initialization (different from that proposed originally) is given by

$$p_n(z) = a_{n-1}(z) + z\,\hat{a}_{n-1}(z)\,, \quad p_{n-1}(z) = a_{n-1}(z) + \hat{a}_{n-1}(z)\,. \tag{30}$$

The *Bistritz criterion* states that $a_{n-1}(z)$ is stable if and only if all numbers $p_k(1)$ are different from zero and have the *same sign*, for $k = 0, \ldots, n-1$. This result can be proved in a simple manner by noting that the ratios $\lambda_k = p_k(1)/p_{k-1}(1)$ are related to the Schur-Cohn parameters $\rho_k = a_{k,k}$ by the identity $\lambda_k\lambda_{k+1} = (1-\rho_k)/(1+\rho_k)$, for $k = 1, \ldots, n-1$ (see [13]).

5 Orthogonal polynomials

The polynomial recurrence relation (2) underlying the Levinson algorithm has been discovered first in the framework of Szegö's theory of orthogonal polynomials on the unit circle [33]. A connection between both settings is obtained by interpreting the entries c_k of the positive definite Toeplitz matrix C_{n-1} as the first n *trigonometric moments* (or Fourier coefficients)

$$c_k = \int_0^{2\pi} e^{-ik\theta}\, d\omega(\theta)\,, \qquad k = 0, 1, \ldots, n-1\,, \tag{31}$$

of a suitable symmetric positive measure $d\omega$. It is then easy to show that the backward predictors $\hat{a}_k(z)$ constitute a family of *orthogonal polynomials on the unit circle*, in the sense that they satisfy

$$\int_0^{2\pi} \bar{\hat{a}}_k(e^{i\theta})\hat{a}_\ell(e^{i\theta})d\omega(\theta) = \sigma_k\delta_{k,\ell}, \qquad \text{for } 0 \le k, \ell \le n-1\,. \tag{32}$$

It turns out that the symmetric polynomials $p_k(z)$ processed by the split Levinson algorithm enjoy also some interesting orthogonality properties (in the real case). To explain this let us transform $p_k(z)$ into the function $f_k(x)$ defined by means of the simple formula

$$\phi_k(x) = z^{-k/2}\, p_k(z)\,, \qquad \text{with } x = \frac{1}{2}(z^{1/2} + z^{-1/2})\,. \tag{33}$$

The symmetry property $\hat{p}_k(z) = p_k(z)$ implies that $\phi_k(x)$ is a polynomial of degree k in the variable x having the even/odd property $\phi_k(-x) = (-1)^k\phi_k(x)$. It can be deduced from (32) that the polynomials $\phi_k(x)$ constitute a family of *orthogonal polynomials on the real interval* [-1,1], in the sense that they satisfy

$$\int_{-1}^{1} \phi_k(x)\,\phi_\ell(x)\,d\mu(x) = 2c_0\alpha_k^{-1}\delta_{k,\ell}\,, \qquad \text{for } 0 \le k, \ell \le n\,, \tag{34}$$

with $\alpha_0 = 1/2$, the same $\alpha_1, \ldots, \alpha_{n-1}$ as in (4), and a suitable α_n; the appropriate measure $d\mu$ is given simply by $d\mu(x) = d\omega(\theta)$ for $x = \cos\theta/2$, in agreement with (33). Note that this result induces a one-to-one correspondence between the families of real polynomials orthogonal on the unit circle and the families of even/odd polynomials orthogonal on the interval [-1,1].

Translating the split Levinson relation (4) by means of (33) we obtain the *three-term recurrence formula* for the family of orthogonal polynomials $\phi_k(x)$ in the form

$$\phi_{k+1}(x) = 2\alpha_k\, x\, \phi_k(x) - \phi_{k-1}(x)\,. \tag{35}$$

The initialization is $\phi_{-1}(x) = 0$ and $\phi_0(x) = 2$; it yields $\phi_1(x) = 2x$.

The result (35) provides an interesting numerical method to *compute the zeros ζ_k of the polynomial $p_n(z)$*. These zeros are known to be simple and to be located on the unit circle; they appear in $\lfloor n/2 \rfloor$ distinct conjugate pairs $\zeta_k = \exp(\pm i\theta_k)$, together with the point $\zeta_n = -1$ when n is odd. In view of (6) it is clear that computing these numbers ζ_k is equivalent to computing the zeros ξ_k of the polynomial $\phi_n(x)$; the connection is given by $\xi_k = \cos\theta_k/2$. By a standard argument in the theory of orthogonal polynomials (see e.g. [30]), it follows from (35) that *the zeros ξ_k of $\phi_n(x)$ are the eigenvalues of the tridiagonal matrix*

$$T_n = \begin{bmatrix} \beta_1 & 0 & \cdot & \\ & \cdot & \cdot & \beta_{n-1} \\ & & \beta_{n-1} & 0 \end{bmatrix}, \qquad \text{with } \beta_k = \frac{1}{2}(\alpha_{k-1}\alpha_k)^{-1/2}\,. \tag{36}$$

This is a useful result since there exist efficient algorithms to solve the eigenvalue problem for T_n or, equivalently, to solve the singular value problem for the $\lceil n/2 \rceil \times \lfloor n/2 \rfloor$ bidiagonal matrix associated with T_n; see [20]. Further details on that subject can be found in [8] and [13].

6 Positive real functions

The concepts of positive real function and of bounded real function are pervasive in applied mathematics; they are introduced in a natural way from "physical constraints" in various areas such as classical circuit theory, inverse scattering theory, wave digital filter theory, and stochastic process theory [18].

A function $f(z)$, with real Maclaurin coefficients, is said to be *positive real* if it is analytic and satisfied Re $f(z) \geq 0$ in the unit disc $|z| < 1$. A function $s(z)$, with real Maclaurin coefficients, is said to be *bounded real* if it is analytic and satisfies $|s(z)| \leq 1$ in the unit disc. Positive real functions and bounded real functions are set in one-to-one correspondence through the self-inverse bilinear transform $s(z) = [1 - f(z)]/[1 + f(z)]$.

Characterizing positive real functions and bounded real functions in terms of their Maclaurin coefficients are very important questions, both in theory and in applications. Answers to these questions are provided by two celebrated theorems, due to Carathéodory and Toeplitz, and to Schur, respectively [1]. Let us now briefly explain how a suitable function theoretic interpretation of the split Levinson recurrence relation leads to a novel solution method for the characterization problem of positive real functions [10].

To start with, consider the sequence of rational functions

$$f_k(z) = \frac{p_{n-k}(z)}{(1-z)p_{n-k-1}(z)}, \qquad \text{for } k = 0, 1, \ldots, n-1, \tag{37}$$

defined from the symmetric polynomials $p_\ell(z)$ processed by the split Levinson algorithm. It can be shown that $f_k(z)$ is a positive real function of the *lossless* type, i.e., satisfying Re $f(e^{i\theta}) = 0$ almost everywhere. It follows from (4) that the functions $f_k(z)$ are related by the recurrence formula

$$f_k(z) = \alpha_k' \, \frac{1+z}{1-z} - \frac{z}{(1-z)^2 f_{k+1}(z)}, \tag{38}$$

with $\alpha_k' = \alpha_{n-k-1} = f_k(0)$. Note that the Jacobi parameter λ_{n-k} is the limit of $(1-z)f_k(z)$ when z tends to 1. The property of $f_0(z)$ being a positive real function can be characterized by the set of inequalities $\lambda_{n-k} > 0$ for $k = 0, 1, \ldots, n-1$.

This result can be generalized considerably as we now explain. If $f_0(z)$ is a positive real function (not necessarily lossless, or rational), then so are all functions $f_k(z)$ generated by the recurrence relation (38), used in "ascending form", with $\alpha_k' = f_k(0)$, for $k = 0, 1, 2$, etc. From these numbers

α'_k let us construct the sequence of "Jacobi parameters" λ'_k by means of the formula $\lambda'_{k+1} = 2\alpha'_k - \lambda'^{-1}_k$, with $\lambda'_1 = 2\alpha'_0$. It can be shown that $f_0(z)$ is *a positive real function under the necessary and sufficient condition* $\lambda'_k \geq 0$ *for all* $k \geq 1$. (If $\lambda'_m = 0$ for a certain m, then the Jacobi sequence has finite length m; this case corresponds exactly to rational lossless functions.)

The method assumes a simpler form when it is expressed in terms of the functions $w_k(z)$ defined through the identity

$$f_k(z) = \frac{w_{k-1}(z)}{(1-z)w_k(z)}, \qquad \text{with } w_{-1}(z) = 1 - z, \quad w_0(z) = f_0(z). \tag{39}$$

Indeed, the recurrence relation (38) can be rewritten as

$$w_{k+1}(z) = z^{-1}[\alpha'_k(1+z)w_k(z) - w_{k-1}(z)], \tag{40}$$

with $\alpha'_k = w_{k-1}(0)/w_k(0)$. Note that the split Schur recursion (12) can be viewed as a direct implementation of (40), where $w_0(z)$ is any positive real function of the form $w_0(z) = c_0 + 2\sum_{i=1}^{n-1} c_i z^i + O(z^n)$.

The new approach to the theory of positive real functions based on (38) can be applied successfully to the Carathéodory-Fejér and Nevanlinna-Pick interpolation problems, which occur as important modelling problems in DSP applications (among other areas). A detailed treatment of this subject can be found in [10].

7 Miscellaneous results

In this section it is shown how some interesting results published recently in the DSP literature (for most of them) fit nicely into the framework of the split Levinson algorithm.

The *LSP* ("*line spectral pairs*") technique is an efficient method of signal analysis and synthesis; it is especially useful in the case of speech signals [32]. It involves symmetric or antisymmetric polynomials of the same type as those processed by the split Levinson algorithm. Therefore, this algorithm yields an economical implementation of the LSP technique. In the speech processing application, the zeros of the successive polynomials generated by the algorithm are known to be closely related to the formants of a short time speech record (whose autocorrelation matrix is the appropriate

Toeplitz matrix of the theory). The split Levinson approach has been shown to yield a new formant extraction algorithm outperforming other methods especially at the transition zone between voiced and unvoiced speech, where the formants are somewhat blurred and hence difficult to compute [34].

Linear phase filters are of common use in DSP. In the context of linear estimation, the transfer function of such a filter is a symmetric polynomial of the form (24). As a result, the split approach is seen to be specifically suited to deal with that type of filters (see [3]).

The "split approach" to Toeplitz systems also yields an efficient algorithm to compute the sinusoidal components of the *Pisarenko model*, which is widely used in DSP applications. Roughly speaking, this model is a stationary stochastic process resulting from the parallel connection of a white noise generator and of a certain number of sinusoidal wave generators [31]. The frequencies of the model waves are given by the zeros of a symmetric polynomial $p_n(z)$ relative to a well-defined Toeplitz matrix C_{n-1}. Therefore, as explained at the end of section 5, the Pisarenko frequencies can be obtained by solving an eigenvalue problem for a tridiagonal matrix with zero diagonal. (It turns out that the wave amplitudes can be determined from the corresponding eigenvectors.)

Closely related to the same question is the *spectral decomposition problem* for *orthogonal Hessenberg matrices*. An interesting solution method based on techniques very similar to those mentioned above has been obtained recently [2], [13].

Let us finally mention that the split approach ideas are currently receiving a great deal of attention in some important application areas that are clearly beyond the scope of this paper. In particular, they have been applied to adaptive filtering problems [3], and extended (in some sense) to three-dimensional random field estimation problems [35].

8 Generalizations and open questions

The split Levinson algorithm and its relatives can be generalized so as to accommodate *complex* (scalar) *data* [10], [27], and even *matrix data* [11], [12]. As for the Bistritz stability test, it can be extended to complex polynomials (but not to matrix polynomials); more precisely, it can be generalized to an efficient algorithm for *computing the number of zeros of a complex polynomial inside the unit circle* [5], [16].

Let us stress that the duality between orthogonal polynomials on the unit circle and orthogonal polynomials on the interval [-1,1], treated in section 5, is restricted strictly to the case of *real* (scalar) data. On the other hand, the new approach to the class of positive real functions, outlined in section 6, can be extended to the case of positive functions (or Carathéodory functions) with complex Maclaurin coefficients [10] and even with matrix Maclaurin coefficients [11].

In DSP problems one has often to deal with positive definite matrices that are not Toeplitz but have a certain (hidden) "Toeplitz-like structure", which can be characterized by their *displacement rank* [25]. It is well known that the linear prediction problem relative to a given matrix can be solved by a "generalized Levinson algorithm" at a computational cost which is a linear function of the displacement rank of this matrix. The question arises then whether the split Levinson algorithm admits a similar generalization. In that respect, let us first mention that the split Levinson algorithm can be applied to a class of matrices which have the same displacement rank (two) as Toeplitz matrices but are slightly more general than them [6]. There exist theoretical reasons to believe that the split Levinson algorithm actually admits a generalization (of the kind alluded to above) to "arbitrary matrices". First, the generalized (classical) Levinson algorithm is obtainable from the block Levinson algorithm through a simple embedding of the given matrix into a well-defined block Toeplitz matrix (see e.g. [17]). Next, as mentioned in the beginning of this section, there exists an extension of the split Levinson algorithm that processes block Toeplitz matrices [11], [12]. These two arguments tend to establish the existence of a certain "generalized split Levinson algorithm based on the displacement rank concept". However, disclosing the detailed structure and assessing the efficiency of such an algorithm are questions open for research.

References

[1] N.I. Akhiezer, The Classical Moment Problem, London: Oliver and Boyd, 1965.

[2] G.S. Ammar, W.B. Gragg, and L. Reichel, "On the eigenvalue problem for orthogonal matrices", *Proc. 25th IEEE Conf. Decision and Control*, Athens, 1986.

[3] J.R. Bellegarda and D.C. Farden, "A new structure for adaptive linear-phase filtering", *IEEE Trans. Circuits and Systems*, **CAS-34**, pp. 712–720, 1987.

[4] Y. Bistritz, "Zero location with respect to the unit circle of discrete-time linear system polynomials", *IEEE Proc.*, **72**, pp. 1131–1142, 1984.

[5] Y. Bistritz, "A circular stability test for general polynomials", *Systems Control Lett.*, **7**, pp. 89–97, 1986.

[6] Y. Bistritz, H. Lev-Ari, and T. Kailath, "Immittance-domain Levinson algorithms, *Proc. IEEE Internat. Conf. Acoustics Speech Signal Processing*, Tokyo, 1986, pp. 253–256.

[7] G. Cybenko, "A general orthogonalization technique with applications to time series analysis and signal processing", *Math. Comp.*, **40**, pp. 323–336, 1983.

[8] P. Delsarte and Y. Genin, "The split Levinson algorithm", *IEEE Trans. Acoust. Speech Signal Process.*, **ASSP-34**, pp. 470–478, 1986.

[9] P. Delsarte and Y. Genin, "On the splitting of classical algorithms in linear prediction theory", *IEEE Trans. Acoust. Speech Signal Process.*, **ASSP-35**, pp. 645–653, 1987.

[10] P. Delsarte and Y. Genin, "The tridiagonal approach to Szegö's orthogonal polynomials, Toeplitz linear systems, and related interpolation problems", *SIAM J. Math. Anal.*, **19**, pp. 718–735, 1988.

[11] P. Delsarte and Y. Genin, "Multichannel singular predictor polynomials", *IEEE Trans. Circuits and Systems*, **CAS-35**, pp. 190–200, 1988.

[12] P. Delsarte and Y. Genin, "The multichannel split Levinson algorithm", in Linear Circuits, Systems and Signal Processing: Theory and Applications, C.I. Byrnes, C.F. Martin, and R.E. Saeks eds, Amsterdam: North-Holland, 1988, pp. 183–190.

[13] P. Delsarte and Y. Genin, "A survey of the split approach based techniques in digital signal processing applications", *Philips J. Res.*, **43**, pp. 346–374, 1988.

[14] P. Delsarte and Y. Genin, "An introduction to the class of split Levinson algorithms", *this volume.*

[15] P. Delsarte and Y. Genin, "An extension of the split Levinson algorithm and its relatives to the joint process estimation problem", *IEEE Trans. Inform. Theory*, to appear.

[16] P. Delsarte, Y. Genin, and Y. Kamp, "Application of the index theory of pseudo-lossless functions to the Bistritz stability test", *Philips J. Res.*, **39**, pp. 226–241, 1984.

[17] P. Delsarte, Y. Genin, and Y. Kamp, "Positive-definite Toeplitz embedding based on the cyclic extension of a data matrix", *IEEE Trans. Acoust. Speech Signal Process.*, **ASSP-33**, pp. 393–400, 1985.

[18] Y. Genin, "An introduction to the modern theory of positive functions and some of its today applications to signal processing, circuits and systems problems, *Proc. Europ. Conf. Circuit Theory and Design*, Paris, 1987, pp. 195–234.

[19] Y. Genin, "On a duality relation in the theory of orthogonal polynomials and its application in signal processing, *Proc. 1st Internat. Conf. Industrial and Applied Mathematics*, Paris, 1987, pp. 102–113.

[20] G.H. Golub, "Least squares, singular values and matrix approximations", *Apl. Mat.*, **13**, pp. 44–51, 1968.

[21] G.H. Golub and C.F. Van Loan, *Matrix Computations*, Oxford: North Academic, 1983.

[22] M.L. Honig and D.G. Messerschmitt, Adaptive Filters, Structures, Algorithms, and Applications, Boston: Kluwer Academic, 1984.

[23] F. Itakura and S. Saito, "Digital Filtering techniques for speech analysis and synthesis", *Proc. 7th Internat. Congr. Acoustics*, Budapest, 1971, pp. 261–264.

[24] T. Kailath, "A view of three decades of linear filtering theory", *IEEE Trans. Inform. Theory*, **IT-20**, pp. 145–181, 1974.

[25] T. Kailath, S.-Y. Kung, and M. Morf, "Displacement ranks of matrices and linear equations", *J. Math. Anal. Appl.*, **68**, pp. 395–407, 1979.

[26] S.M. Kay and S.L. Marple, Jr., "Spectrum analysis — a modern perspective", *IEEE Proc.*, **69**, pp. 1380–1419, 1983.

[27] H. Krishna and S.D. Morgera, "The Levinson recurrence and fast algorithms for solving Toeplitz systems of linear equations", *IEEE Trans. Acoust. Speech Signal Process.*, **ASSP-35**, pp. 839–848, 1987.

[28] J. Le Roux and C. Gueguen, "A fixed point computation of partial correlation coefficients", *IEEE Trans. Acoust. Speech Signal Process.*, **ASSP-25**, pp. 257–259, 1977.

[29] M. Marden, Geometry of Polynomials, Providence: American Mathematical Society, 1966.

[30] B.N. Parlett, The Symmetric Eigenvalue Problem, Englewood Cliffs: Prentice-Hall, 1980.

[31] V.P. Pisarenko, "The retrieval of harmonics from a covariance function", *Geophys. J.R. Astr. Soc.*, **33**, pp. 347–366, 1973.

[32] N. Sugamura and F. Itakura, "Speech analysis and synthesis methods developed at ECL in NTT — from LPC to LSP", *Speech Communication*, **5**, pp. 199–215, 1986.

[33] G. Szegö, Orthogonal Polynomials, New York: American Mathematical Society, 1959.

[34] L.F. Willems, "Robust formant analysis for speech synthesis applications", *Proc. Europ. Conf. on Speech Technology*, Edinburgh, 1987, pp. 250–253.

[35] A.E. Yagle, "Generalized split Levinson, Schur, and lattice algorithms for three-dimensional random field estimation problems", preprint.

Updating and Downdating of Orthogonal Polynomials with Data Fitting Applications

Sylvan Elhay

Computer Science Department
University of Adelaide
Adelaide, South Australia 5000, Australia

Gene H. Golub

Computer Science Department
Stanford University
Stanford, CA 94305, USA

Jaroslav Kautsky

School of Mathematical Sciences
Flinders University
Bedford Park, South Australia 5042, Australia

Abstract

We derive and assess new methods for updating and downdating least squares polynomial fits to discrete data using polynomials orthogonal on *all* the data points being used. Rather than fixing on one basis throughout, the methods adaptively update and downdate both the least squares fit and the polynomial basis. This is achieved by performing similarity transformations on the tridiagonal Jacobi matrices representing the basis. Although downdating is potentially unstable,

NATO ASI Series, Vol. F 70
Numerical Linear Algebra, Digital Signal Processing
and Parallel Algorithms
Edited by G. H. Golub and P. Van Dooren

experimental results show that the methods give satisfactory results for low degree fits. The most economical of the algorithms implementing the methods needs $14n+O(1)$ flops and $2n$ square roots to update a fit of order n.

Keywords: updating, downdating, least squares, polynomial fits, discrete inner products, orthogonal polynomials, Jacobi matrices.

1 Introduction

Updating and downdating are commonly applied to the decomposition of an $m \times n$ matrix $X = QR$, $m > n$ into orthogonal factor Q and upper triangular factor R, when new rows of X are added or old ones removed. While the original factoring requires $O(mn^2)$ operations, the updates need only $O(n^2)$ operations. In some applications the $m \times m$ factor Q need not be stored but if it is needed it can be saved in product form using $O(mn)$ locations; Q and R therefore represent a powerful data compression device. The up/downdating thus leads to savings in both computer operations and storage.

Updating is usually done by perfectly stable orthogonal rotations. Downdating a row of X can be achieved by rotations if we know the corresponding row of Q (see [6]). Downdating can also be done by hyperbolic rotations ([5]) the condition of which depends on the data. Hyperbolic rotations (called *hyper-rotations* here) and the stability of downdating have attracted some attention in the literature ([1], [12], [2] and [10]).

In the context of polynomial least squares data fitting the techniques mentioned above apply if we use a fixed polynomial basis to represent the required fits. The matrix X is then a Vandermonde-like matrix of the values of the basis polynomials at the data points.

The choice of the polynomial basis strongly influences the stability of the computation. Thus a basis chosen to suit some given input data may be inappropriate once more data points are added or existing data points are removed. For every data set there is, however, a natural polynomial basis - the polynomials orthogonal on that data. This then leads to the problem of adaptively up/downdating the orthogonal polynomial basis together with the coefficients of the least squares fit. The solution of this problem, rather than the up/downdating of the QR factoring, is the central task of our paper which is organised as follows [1].

In §2 we define the problem in detail and in §3 we derive some well known properties of orthogonal polynomials expressed in matrix notation. The presentation is essentially self-contained: we find it easier - and possibly instructive - to prove the properties in question rather than to quote the facts and to translate them. The inspiration for this approach stems from the works of Wilf [13] and Golub-Welsch [7] and has been already used by the authors extensively (see e.g. [9] for further references). Some preliminary work in this area was also done by Parker in 1982 (private communication).

§4 deals with updating methods. We begin with the details of the updating and downdating problems in terms of the chosen representation.

Theorem 4.1 in §4.2 gives a constructive characterization of the solution based on orthogonal similarity. This is a special case of the technique, widely used in numerical linear algebra, which transforms a symmetric matrix into a similar tridiagonal matrix. In this instance the transformation is achieved more economically by rotations than by elementary Householder matrices because of the structure of the problem. It would appear that the square-root free version of such a method [8] does not allow the updating of the least squares fit polynomial.

In §4.3 we present methods which bear the same relation to those in §4.2 as the LR factoring does to the QR factoring, i.e. the orthogonal similarity is replaced by a triangular similarity. We call the methods *Lanczos-type methods* because the construction of the similarity follows the same lines as the Lanczos method. Two special Lanczos-type methods are derived in §4.4 and §4.5. The first, based on determinantal relations of Rutishauser [11], is the most computationally economical algorithm for updating the basis but is not well suited to updating the least squares fit. The second, based on the special structure of the similarity matrix observed in [4], leads to the most economical of all the algorithms here.

[1]This work appears in a more fully documented form, including detailed algorithms for the methods developed and their numerical testing, in S. Elhay, G.H. Golub & J. Kautsky, *Updating and downdating of orthogonal polynomials with data fitting applications*, Numerical Analysis Project Manuscript NA-89-04, Computer Science Department, Stanford University.

All the methods mentioned so far produce a representation of the least squares polynomials of all possible degrees, including the maximal degree polynomial fit that interpolates the given data. Limiting the calculation to produce all fits up to some prescribed lower degree introduces new aspects of the problem which are dealt with in §4.6. For example, updating by a data point which is new to the full data set may be equivalent to updating by a point which already belongs to the (different) data set associated with the currently held representation. Effectively this means *changing* the weight of an existing data point. Theorem 4.2 in §4.7 characterizes the existence of the solution to the updating problem for this situation.

§5.1 defines the downdating problem and §5.2 introduces downdating methods which are closely related to the updating methods of §4. In addition we give in Theorem 5.1 of §5.3 a downdating which is based on the reversal of the orthogonal rotations method. This leads naturally in Theorem 5.2 of §5.4 to conditions for the existence of the downdate. Essentially, downdating is possible if the weight involved is sufficiently small; for larger weights it may be necessary to reduce the degree of the polynomial fit.

Limited testing indicates that

(a) although accuracy deteriorates as the degree of the fit increases, the methods are satisfactory for low degree polynomials,

(b) computational complexity and convenience rather than other considerations are the main determinants in choosing between the four methods.

2 The problem

Problem 1 (weighted least square fitting): Given the discrete data

$$Y_N = \{x_j, w_j, y_j\}_{j=1}^N$$

find polynomial q of degree n such that

$$\sum_{j=1}^{N} w_j^2 (y_j - q(x_j))^2 \tag{2.1}$$

is minimised.

Suppose Y_{N+1} is obtained from Y_N by augmenting a new triplet of data. Solving Problem 1 for Y_{N+1} *assuming* the knowledge of its solution q for Y_N is called *updating* the least squares fit. Conversely, solving Problem 1 for Y_N assuming the knowledge of its solution for Y_{N+1} is called *downdating*. In other words, updating and downdating means modifying the solutions of Problem 1 when one data set (triple) is added to or removed from the data.

The importance of updating and downdating is three-fold:

- Problem 1 can be solved by updating, starting from the trivial solution for $N = 1$ (taking in account obvious restrictions on distinctness of x_j's and relations between n and N).

- Using downdating and updating we can calculate "sliding" least square fits - include new measurements and discard old ones or even change the weights of selected data sets.

- Using suitable representation of the solutions we do not have to store - or refer to - data already used.

The keyword here is representation. We choose to represent the least squares fit q by the coefficients of its expansion in terms of the polynomials orthonormal with respect to the discrete inner product

$$[f, g]_{W_N} = \sum_{j=1}^{N} w_j^2 f(x_j) g(x_j) \tag{2.2}$$

where we have denoted $W_N = \{x_j, w_j\}_{j=1}^N$. We need therefore to consider also the following problem:

Problem 2 (discrete orthonormal polynomials): Given the discrete data W_N find polynomials p_k of exact degree k, $k = 0, 1, \ldots, n$ such that

$$[p_i, p_k]_{W_N} = \delta_{ik} \quad , i, k = 0, 1, \ldots, n \tag{2.3}$$

where $i \neq k$ implies $\delta_{ik} = 0$ and $\delta_{ii} = 1$.

Here we choose to represent the orthonormal polynomials by the coefficients of the three-term relation which they satisfy. This representation has several advantages, say in comparison with standard powers expansions:

- It allows for stable evaluation of the values of the polynomials and other manipulations with them.

- It is concise - for Problem 2 we have $2n$ values representing $\frac{1}{2}n^2$ power coefficients of $p_0, p_1, \ldots, p_n$. For Problem 1 it appears generous to use $3n$ values to represent one polynomial but these $3n$ values give *all* least square fits up to degree n !

- Using matrix notation the Problems 1 and 2 can be re-formulated, in the terms of the chosen representation, as problems in numerical linear algebra. This gives a new insight into the existing methods and leads to simple derivation of both old and new algorithms.

We note that the comments on the importance of updating and downdating made above with reference to Problem 1 apply equally to the Problem 2.

3 Matrix notation and basic relations

For any vector of functions $\mathbf{u} = (u_1, \ldots, u_n)^T$, $\mathbf{v} = (v_1, \ldots, v_m)^T$ and scalar product $[.,.]$, discrete as in (2.2) or a continuous one, we denote by $[\mathbf{u}, \mathbf{v}^T]$ the constant $n \times m$ matrix with elements $[u_i, v_j]$. Some obvious rules apply - for example if A and B are constant matrices of suitable sizes then $[A\mathbf{u}, \mathbf{v}^T B] = A[\mathbf{u}, \mathbf{v}^T]B$ - which we shall use freely.

For any polynomials p_j of exact degree j, $j = 0, 1, \ldots, n$ there exists a unique lower Hessenberg matrix J and scalar β_n such that

$$t\mathbf{p}_n = J\mathbf{p}_n + \beta_n p_n \mathbf{e}_n \tag{3.1}$$

where we have denoted $\mathbf{p}_n = (p_0, p_1, \ldots, p_{n-1})^T$, t the identity function and $\mathbf{e}_n$ the n-th column of the identity matrix. The matrix J is unreduced, i.e. its superdiagonal element $\beta_j = \mathbf{e}_{j+1}^T J \mathbf{e}_j$, $j = 1, \ldots, n-1$, together with β_n, do not vanish.

Given an unreduced lower Hessenberg matrix J and non-zero scalars p_0 and β_n, the polynomials $p_1, p_2, \ldots, p_n$ can be determined recursively from (3.1). We call J the *recurrence matrix* for polynomials $\mathbf{p}_n$, p_n; because of the one-to-one correspondence it can be used to represent the polynomials.

The polynomials $\mathbf{p}_n$, p_n will be orthonormal with respect to a scalar product $[.,.]$ (for example, the scalar product $[.,.]_{W_N}$ corresponding to the data W_N in (2.2)) if

$$[\mathbf{p}_n, \mathbf{p}_n^T] = I_n \ , \ [\mathbf{p}_n, p_n] = \mathbf{0} \ , \ [p_n, p_n] = 1. \tag{3.2}$$

Combining (3.2) and (3.1) we now have $[t\mathbf{p}_n, \mathbf{p}_n^T] = [J\mathbf{p}_n, \mathbf{p}_n^T] + \beta_n \mathbf{e}_n [p_n, \mathbf{p}_n^T] = J$ from which it follows that, the matrix on the left being symmetric, the recurrence matrix J for orthonormal polynomials must also be symmetric and thus three-diagonal.

The *Jacobi* (symmetric, three-diagonal, unreduced) *matrix* J_n, with diagonal elements $\alpha_1, \alpha_2, \ldots, \alpha_n$ and sub-diagonal elements $\beta_1, \beta_2, \ldots, \beta_{n-1}$, has real distinct eigenvalues, say $\lambda_1, \ldots, \lambda_n$. From (3.1) it is immediate that all eigenvalues λ_j are roots of p_n and $\mathbf{p}_n(\lambda_j)$ are the corresponding eigenvectors. So denoting $P = (\mathbf{p}_n(\lambda_1), \ldots, \mathbf{p}_n(\lambda_n))$ and $\Lambda = \text{diag}\ (\lambda_1, \ldots, \lambda_n)$, we have $JP = P\Lambda$. As J is symmetric, P has orthogonal columns,i.e.

$$P^T P = diag(\nu_1^{-2}, \ldots, \nu_n^{-2}) = D^{-2} \tag{3.3}$$

where $\nu_j = (\mathbf{p}_n(\lambda_j)^T \mathbf{p}_n(\lambda_j))^{-\frac{1}{2}}$, $j = 1, 2, \ldots, n$. By (3.3) PD is orthogonal so that $PD^2P^T = I$. This shows that polynomials $\mathbf{p}_n$ orthonormal with respect to some scalar product are also orthonormal with respect to the discrete scalar product $[.,.]_{\tilde{W}_n}$ given by the data $\tilde{W}_n = \{\lambda_j, \nu_j\}_{j=1}^n$. Finally we note that $\mathbf{e}_1^T PD = (\nu_1, \ldots, \nu_n)p_0$ which allows, up to the sign (not relevant), to recover the scaled weights ν_j from any normalised eigenvectors of J.

Turning to the least square fit problem, let us denote temporarily $\mathbf{p}_N$, J_N, P_N, D_N the above solution for our data Y_N (i.e. $\tilde{W}_n = W_N$) and $\mathbf{y}_N = (y_1, \ldots, y_N)^T$. For any polynomial $q = \mathbf{p}_N^T \mathbf{d}$ with coefficients $\mathbf{d} = (d_0, \ldots, d_{N-1})^T$ the quantity (2.1) to be minimised is (we note that for $\mathbf{x}_N = (x_1, \ldots, x_N)^T$ we have $q(\mathbf{x}_N) = P_N^T \mathbf{d}$)

$$\begin{aligned} [y - q, y - q]_{W_N} &= \|D_N(\mathbf{y}_N - q(\mathbf{x}_N))\|_2^2 \\ &= \|D_N(\mathbf{y}_N - P_N^T \mathbf{d})\|_2^2 \\ &= \|P_N D_N^2 \mathbf{y}_N - \mathbf{d}\|_2^2, \end{aligned} \tag{3.4}$$

the last equality being the consequence of the above observation that $P_N D_N$ is orthogonal. Now choosing $\mathbf{d} = P_N D_N^2 \mathbf{y}_N$ gives the solution to the *interpolating* problem for data Y_N as the expansion in the polynomials orthonormal on W_N (the weights are not of importance in that case). However, from (3.4) it follows that for any $n < N$, $\mathbf{d}_n = (d_0, \ldots, d_{n-1},\ 0, 0, \ldots, 0)^T$ gives the solution to the least square fit problem by polynomials of degree less than n. The shortened vector $\mathbf{d}_n = (d_0, \ldots, d_{n-1})^T$ is now given by

$$\mathbf{d}_n = P_{n,N} D_N^2 \mathbf{y}_N \tag{3.5}$$

where $P_{n,N}$ is the matrix comprising the first n rows of P_N.

The solution to the Problem 2 is now given by the first n diagonal elements $\alpha_1, \ldots, \alpha_n$ and $n-1$ off-diagonal elements $\beta_1, \ldots, \beta_{n-1}$ of the matrix J_N (we denote the submatrix J_n) which together with the zero-th moment $\mu_0 = \sum_{j=1}^N w_j^2 = [1,1]_{W_N}$ represent the first n orthonormal polynomials $\mathbf{p}_n$. The solution to the Problem 1 is then given by μ_0, J_n and $\mathbf{d}_n$ for which $q = \mathbf{d}_n^T \mathbf{p}_n$.

The evaluation of $q(t)$ can conveniently be done in two ways. Denote by

$$K(t)\mathbf{p}_n(t) = p_0(t)\mathbf{e}_1 \tag{3.6}$$

the system of equations defined by the first n rows of the $n+1$ equations in

$$\begin{pmatrix} \mathbf{e}_1^T \\ (J_n - tI) \end{pmatrix} \mathbf{p}_n(t) = \begin{pmatrix} p_0(t) \\ -\beta_n p_n(t)\mathbf{e}_n \end{pmatrix} .$$

Given the vector $\mathbf{d}_n$, the approximating polynomial $q(t)$ can be evaluated at a point by a forward-substitution on (3.6) for $\mathbf{p}_n$ and then an inner product with $\mathbf{d}_n$.

A second more direct way is the Clenshaw recursion in which we have $q = p_0 \mathbf{e}_1^T \mathbf{y}$ where $\mathbf{y}$ is the solution obtained by a back-substitution on the system $K^T \mathbf{y} = \mathbf{d}_n$.

4 Updating methods

4.1 A reformulation of the updating problem

The problem of updating is now, with respect to the representation of the solutions introduced in §3, as follows:

Given $\mu_0 > 0, J_n$ a Jacobi matrix of size n and vector $\mathbf{d}_n$ representing the solution of Problem 1 for some data set Y_N, and given also an integer $\tilde{n} = n$ or $n+1$ and triplet $\{x, w, y\}$ find $\tilde{\mu}_0, \tilde{\mathbf{d}}_{\tilde{n}}$ and $\tilde{J}_{\tilde{n}}$ representing the solution for the data set $Y_N \cup \{x, w, y\}$.

We have formulated the updating problem only for Problem 1 - the corresponding version for Problem 2 is obtained by omitting the coefficients $\mathbf{d}$ and the function values.

As we shall see, the data set Y_N does not enter into the calculations; it is of course important for the formulation of the problem and for the following brief discussion of the existence and meaningfulness of the solutions.

The solution of the updating problem always exists for $\tilde{n} = n$. Requiring the increase of size $\tilde{n} = n + 1$ implies that $N = n$, that is that the points x_j and weights w_j of Y_N are the eigenvalues λ_j and coefficients ν_j of J_n discussed in §3. The increase is then possible iff x is not one of the λ_j's.

Finally, as the solution for the one-point set $y_1 = \{x, w, y\}$ is trivially $\mu_0 = w^2$, $J_1 = (x)$, $d_0 = y|w|$, we can indeed use updating (with $\tilde{n} = n+1$) to build up the solution to Problems 1 and 2 of any required combinations of values n and N without storing the data set Y_N - as long as we increase n only at the beginning of the process.

4.2 Rotation method

To describe the solution of the updating problem, we will temporarily deal with the case $n = N$ and $\tilde{n} = N + 1$. We denote here $\mathbf{w}_N = (w_1, \ldots, w_N)^T$ the vector of weights and $\Lambda_N = diag(x_1, \ldots, x_N)$ the diagonal matrix of the points of our data set. We also introduce $\sigma_N = \sqrt{\mu_0^{(N)}} = (w_1^2 + \cdots + w_N^2)^{\frac{1}{2}} = \|\mathbf{w}_N\|$.

The following theorem gives the solution of the updating problem:

Theorem 4.1 *Given σ_N, J_N and $\mathbf{d}_N$ for the data Y_N and also $\{x, w, y\}$ the solution for $Y_{N+1} = Y_N \cup \{x, w, y\}$ is*

$$\tilde{J}_{N+1} = Q \begin{pmatrix} J_N & \mathbf{0} \\ \mathbf{0}^T & x \end{pmatrix} Q^T, \tag{4.1}$$

$$\tilde{\sigma}_{N+1} = (\sigma_N^2 + w^2)^{\frac{1}{2}}, \tag{4.2}$$

$$\tilde{\mathbf{d}}_{N+1} = Q\begin{pmatrix} \mathbf{d}_N \\ wy \end{pmatrix} \tag{4.3}$$

where the orthogonal similarity matrix Q is uniquely determined by requiring $\tilde{J}_{N+1}$ to be tridiagonal and Q to satisfy

$$Q(\sigma_N \mathbf{e}_1 + w\mathbf{e}_{N+1}) = \tilde{\sigma}_{N+1}\mathbf{e}_1. \tag{4.4}$$

The orthogonal similarity Q transforming the given matrix in (4.1) into an upper Hessenberg matrix $\tilde{J}_{N+1}$ is fully determined by its first row $Q^T\mathbf{e}_1 = \frac{1}{\tilde{\sigma}_{N+1}}(\sigma_N\mathbf{e}_1 + w\mathbf{e}_{N+1})$, as specified in (4.4).

For algorithmic implementation Q is constructed as

$$Q = R_N R_{N-1} \dots R_1 \tag{4.5}$$

where R_j is a rotation between the j-th and $(N+1)$-st rows, $j = 1, 2, \dots, N$. These rotations are determined in such a way that R_1 achieves (4.4) which, by definition, will not be effected by the other rotations determined so that the intermediate matrices

$$K_n = R_n R_{n-1} \cdots R_2 R_1 \begin{pmatrix} J_N & \mathbf{0} \\ \mathbf{0}^T & 1 \end{pmatrix} R_1^T R_2^T \cdots R_{n-1}^T R_n^T$$

the first $n-1$ elements of the last (i.e. $N+1$ st) row vanish. An inspection shows that K_n is tridiagonal up to the last row and column which has non-zero elements only in nth, $n+1$st and $N+1$st positions (so that $J_{N+1} = K_N$ is indeed tridiagonal). Furthermore, the step $K_{n+1} = R_{n+1}K_nR_{n+1}^T$ involves, besides the last row and column, only

- the n-th off diagonal element of K_n to determine the rotation R_{n+1} and its new value in K_{n+1},
- the $n+1$-st diagonal and off-diagonal elements to determine their new values.

These observations lead to the following

Corollary 4.1 *The leading size n submatrix $\tilde{J}_n$ of $\tilde{J}_{N+1}$ is the leading submatrix of K_n and only the leading submatrix J_n of J_N (and $\{x, w, y\}$) is needed to determine rotations $R_1, \dots, R_n$ and $\tilde{J}_n$. Similarly, only the first n elements of $\mathbf{d}_N$ and the rotations $R_1, \dots, R_n$ are sufficient to obtain the first n elements of $\mathbf{d}_{N+1}$.*

4.3 Lanczos-type methods

The method we wish to describe now is a modification of the Lanczos method where the orthogonal matrix Q is essentially replaced by a lower triangular matrix (it has been described and called LTL - the lower triangular Lanczos method - in [9]). We shall describe it here again for two reasons. Firstly, unlike the case in [9] we deal here with polynomials orthogonal with respect to a *finite* discrete scalar product. Secondly, we will use it as a starting point from which to derive two new methods.

Let us assume, as before, that the polynomials $\mathbf{p}_N, p_N$, orthonormal with respect to $[.,.]_{W_N}$, satisfy the three term relation

$$t\mathbf{p}_N = J_N\mathbf{p}_N + \beta_N\mathbf{e}_N p_N. \tag{4.6}$$

We note that p_N cannot be normalized by the scalar product ($x_1,\ldots,x_N$ are the eigenvalues of J_N and thus roots of p_N) as $[p_N, f]_{W_N} = 0$ for any f. The β_N can be chosen in such a way that p_N is, say, monic. This is not necessarily a good choice for numerical calculation (which is in fact independent of β_N) but is convenient for the present discussion. We are now seeking the Jacobi matrix $\tilde{J}_{N+1}$ for the polynomials $\tilde{\mathbf{p}}_{N+1}, \tilde{p}_{N+1}$ orthonormal with respect to $[.,.]_{W_{N+1}}$. They must satisfy

$$t\tilde{\mathbf{p}}_{N+1} = \tilde{J}_{N+1}\tilde{\mathbf{p}}_{N+1} + \tilde{\beta}_{N+1}\mathbf{e}_{N+1}\tilde{p}_{N+1}. \tag{4.7}$$

Obviously ($\tilde{\beta}_{N+1}$ being again chosen to make $\tilde{p}_{N+1}$ monic), $\tilde{p}_{N+1}(t) = (t - x_{N+1})p_N(t)$, so that by denoting $\mathbf{p}_{N+1}^T = (\mathbf{p}_N^T,\ p_N)$ we obtain

$$t\mathbf{p}_{N+1} = J_{N+1}\mathbf{p}_{N+1} + \mathbf{e}_{N+1}\tilde{p}_{N+1} \tag{4.8}$$

where $J_{N+1} = \begin{pmatrix} J_N & \beta_N\mathbf{e}_N \\ \mathbf{0}^T & x_{N+1} \end{pmatrix}$. There exists a constant nonsingular lower triangular matrix L such that

$$\mathbf{p}_{N+1} = L\tilde{\mathbf{p}}_{N+1}. \tag{4.9}$$

By substituting (4.9) into (4.8) and comparing it with (4.7), premultiplied by L, we obtain immediately

$$\begin{aligned} J_{N+1}L &= L\tilde{J}_{N+1} \\ \tilde{\beta}_{N+1}\mathbf{e}_{N+1}^T L\mathbf{e}_{N+1} &= 1. \end{aligned} \tag{4.10}$$

Denoting the elements of $\tilde{J}_{N+1}$ similarly as in J_N and $\mathbf{v}_j = L\mathbf{e}_j$ the columns of L, we have from (4.10) the recurrence

$$(1-\delta_{j,N+1})\tilde{\beta}_j\mathbf{v}_{j+1}+\tilde{\alpha}_j\mathbf{v}_j+\tilde{\beta}_{j-1}\mathbf{v}_{j-1}(1-\delta_{j1}) = J_{N+1}\mathbf{v}_j \quad j=1,2,\ldots,N+1. \tag{4.11}$$

As discussed in detail in [9], $\tilde{J}_{N+1}$ can be explicitly constructed by evaluating alternatively its elements and the columns of L using the fact that $\mathbf{e}_k^T\mathbf{v}_j = 0$ for $k < j$ (clearly analogous to the Lanczos method) *and* from the knowledge of $\mathbf{v}_1$.

It remains to show how to find $\mathbf{v}_1 = L\mathbf{e}_1$ for this particular case. We have

$$M_{N+1} := [\mathbf{p}_{N+1}, \mathbf{p}_{N+1}^T]_{W_{N+1}} = LL^T \tag{4.12}$$

so that, with $\rho_1 = \mathbf{e}_1^T L\mathbf{e}_1 = \mathbf{e}_1^T\mathbf{v}_1$, we have $\rho_1\mathbf{v}_1 = LL^T\mathbf{e}_1 = [\mathbf{p}_{N+1}, p_0]_{W_{N+1}}$ $= \mathbf{e}_1 + p_0 w_{N+1}^2 \mathbf{p}_{N+1}(x_{N+1})$ which determines $\mathbf{v}_1$ explicitly.

We remark that this LTL method, described and devised here in matrix notation, is equivalent to the modified Chebyshev's algorithm discussed in [3] (p. 295) and suffers the instabilities mentioned there, particularly when the knots x_j are uniformly spaced. As the rotation method provides stable answers, it is not likely, however, that the underlying map itself is ill-conditioned (as suggested in [3]) but the LTL method only is then unsuitable. We note that the elements σ_{kl} defined in (2.12) in [3] are indeed those of L^T as $L^T = [\tilde{\mathbf{p}}_{N+1}, \tilde{\mathbf{p}}_{N+1}^T]_{W_{N+1}} L^T = [\tilde{\mathbf{p}}_{N+1}, \mathbf{p}_{N+1}^T]_{W_{N+1}}$.

We will now show that we can proceed with the recurrence constructing $\tilde{J}_{N+1}$ using the diagonal and subdiagonal elements of L only.

Denoting $\mathbf{v}_j = (\underbrace{0,\ldots,0}_{j-1}, \rho_j, \tau_j, \ldots)^T$ we obtain, for tridiagonal $\tilde{J}_{N+1}$, from (4.11) taken element-wise,

$$\tilde{\beta}_{j-1}\rho_{j-1} = \mathbf{e}_{j-1}^T \tilde{J}_{N+1}\mathbf{v}_j = \beta_{j-1}\rho_j \quad j = 2,3,\ldots,N+1 \tag{4.13}$$

$$\tilde{\alpha}_j\rho_j + \tilde{\beta}_{j-1}\tau_{j-1} = \mathbf{e}_j^T J_{N+1}\mathbf{v}_j = \begin{cases} \alpha_j\rho_j + \beta_j\tau_j & 1 \le j \le N \\ x\rho_{N+1} & j = N+1 \end{cases}. \tag{4.14}$$

$$\tilde{\beta}_{N+1} = 1/\rho_{N+1}$$

the last equation coming from (4.10). These relations lead to the following algorithmic recurrences for updating.

$\alpha_1 = \alpha_1 + \beta_1 \frac{\tau_1}{\rho_1}$
for $j = 2, 3, \ldots, N$
$\quad \alpha_j = \alpha_j + \beta_j \frac{\tau_j}{\rho_j} - \beta_{j-1} \frac{\tau_{j-1}}{\rho_{j-1}}$
$\quad \beta_{j-1} = \beta_{j-1} \rho_j / \rho_{j-1}$
endfor j:
$\alpha_{N+1} = x_{N+1} - \beta_N \frac{\tau_N}{\rho_N}$
$\beta_N = \beta_N \rho_{N+1} / \rho_N$

Notice that for this calculation the term β_N is needed at start to advance the algorithm. However, $\tilde{J}_{N+1}$ is independent of the input value of β_N and we can set $\beta_N = 1$ at start rather than normalizing the root polynomial by setting $\beta_{N+1} = 1/\rho_{N+1}$ at finish.

Unfortunately, we cannot generate the ρ_j and τ_j using (4.11) without calculating the other elements of L, too. In the next two sections we present two other ways to determine ρ_j, τ_j without using (4.11).

Turning now to the updating of the Fourier coefficients $\mathbf{d}$, we have

$$L\tilde{\mathbf{d}}_{N+1} = \begin{pmatrix} \mathbf{d}_N + w^2 y \mathbf{p}_N(x) \\ w^2 y p_N(x) \end{pmatrix} \tag{4.15}$$

with Q_{N+1} replaced by PD where P and D are the matrices of §3. The updated coefficients can be found by a forward substitution.

4.4 Solution by determinants

In this section we derive formulae (see [11]) for the ρ_j and τ_j from subdeterminants of the Gram matrix of modified moments M_{N+1} of (4.12).

Let us denote

$$M_k = [\mathbf{p}_k, \mathbf{p}_k^T]_{W_{N+1}} = \begin{pmatrix} M_{k-1} & \mathbf{m}_k \\ \mathbf{m}_k^T & \mu_k \end{pmatrix} = L_k L_k^T$$

and $L_k = \begin{pmatrix} L_{k-1} & \mathbf{0} \\ \mathbf{r}_k^T & \rho_k \end{pmatrix}$. We have, generally, $det(M_k) = det(L_k)^2 = \rho_1^2 \rho_2^2 \cdots \rho_k^2$ so that $\rho_1 = M_1^{\frac{1}{2}}$ and

$$\rho_k = (det(M_k)/det(M_{k-1}))^{\frac{1}{2}}, k = 2, 3, \ldots \tag{4.16}$$

We now derive an explicit expression for τ_k. Defining two $k \times k+1$ selection matrices

$$S = (I_k\ 0),\ \tilde{S} = \begin{pmatrix} I_{k-1} & 0 & 0 \\ 0^T & 0 & 1 \end{pmatrix} \text{ we have } \tilde{M}_k = \tilde{S} M_{k+1} S^T = \begin{pmatrix} M_{k-1} & \mathbf{m}_k \\ & \mathbf{m}_{k+1}^T \end{pmatrix}$$

and

$$\tilde{S} L_{k+1} L_{k+1}^T S^T = \begin{pmatrix} L_{k-1} & 0 & 0 \\ \mathbf{r}_{k+1} & & \rho_{k+1} \end{pmatrix} \begin{pmatrix} L_{k-1} & 0 & 0 \\ \mathbf{r}_k^T & \rho_k & 0 \end{pmatrix}^T$$

in which, the last columns not contributing, the right hand side is a product of two triangular matrices. Taking determinants, we have $det(\tilde{M}_k) = \rho_1^2 \rho_2^2 \cdots \rho_{k-1}^2 \rho_k \tau_k$ from which

$$\tau_k = det(\tilde{M}_k)(det(M_k)det(M_{k-1}))^{-\frac{1}{2}}. \tag{4.17}$$

So far we worked with general M_k, that is the Gram matrix of the "old" polynomials $\mathbf{p}$ with respect to the inner product determining the "new" orthogonal polynomials. In this sense, the above formulae could be used to improve the LTL method of [9] whenever the above determinants are obtainable. This is the case in the present situation as

$$[f,g]_{W_{N+1}} = [f,g]_{W_N} + w_{N+1}^2 f(x_{N+1}) g(x_{N+1})$$

so that

$$M_k = I_k + \mathbf{u}_k \mathbf{u}_k^T \tag{4.18}$$

where $\mathbf{u}_k = w_{N+1} \mathbf{p}_k(x_{N+1})$. Denoting $\psi_j = w_{N+1} p_{j-1}(x_{N+1})$, the j-th element of $\mathbf{u}_k, j \leq k$ and independent of k, we have also $\mathbf{m}_k = \psi_k \mathbf{u}_{k-1}$, $\mu_k = 1 + \psi_k^2$ and

$$\tilde{M}_k = \begin{pmatrix} I + \mathbf{u}_{k-1}\mathbf{u}_{k-1}^T & \psi_k \mathbf{u}_{k-1} \\ & \psi_{k+1} \mathbf{u}_k^T \end{pmatrix} = \begin{pmatrix} I + \mathbf{u}_{k-1}\mathbf{u}_{k-1}^T & \psi_k \mathbf{u}_{k-1} \\ \psi_{k+1} \mathbf{u}_{k-1}^T & \psi_{k+1}\psi_k \end{pmatrix}.$$

Using the relation $det\begin{pmatrix} A & B \\ C & D \end{pmatrix} = det(D)det(A - BD^{-1}C)$ we have

$$det(\tilde{M}_k) = \psi_k \psi_{k+1} det(I + \mathbf{u}_{k-1}\mathbf{u}_{k-1}^T - \psi_k \mathbf{u}_{k-1} \psi_{k+1} \mathbf{u}_{k-1}^T / \psi_k \psi_{k+1}) = \psi_k \psi_{k+1}. \tag{4.19}$$

The other determinant follows easily from (4.18):

$$det(M_k) = 1 + \mathbf{u}_k^T \mathbf{u}_k = 1 + \psi_1^2 + \psi_2^2 + \ldots + \psi_k^2. \tag{4.20}$$

To update the Fourier coefficients we need to solve (4.15). The right hand side is known because $w^2 y \mathbf{p}_{N+1}(x) = wy\mathbf{u}_{N+1}$. By transposing (4.10), one can derive a recurrence for the rows of L which is similar to that in (4.11) for the columns of L but with J and $\tilde{J}$ interchanged. The forward substitution can thus be performed as the elements of $\tilde{J}$ are obtained.

4.5 Solution using the special form of L

Although the calculation of the new Jacobi matrix by the method in the previous section is convenient, the computation of the rows of the L matrix requires $O(n^2)$ operations while other methods need $O(n)$. As pointed out in [4] the matrix L, being the Cholesky factor of a rank-one update to a diagonal matrix, has a special form. We exploit this property to calculate the updated Fourier coefficients more efficiently. In fact, the special form of L leads to a new method which we now describe.

From (4.12) and from (4.18), we have

$$LL^T = M_{N+1} = \text{diag}\{I_N, 0\} + \mathbf{u}_{N+1}\mathbf{u}_{N+1}^T. \tag{4.21}$$

The form referred to in [4] is special in that the sub-diagonal elements of the factor L are the elements of the strictly lower triangular part of a rank-one matrix. Comparing the elements of (4.21) we have for $j < k$ $\mathbf{e}_k^T L \mathbf{e}_j = \psi_k q_j$ where the diagonal elements ρ_j of L and q_j must satisfy

$$\rho_j^2 = 1 - \delta_{j,N+1} + \psi_j^2(1 - (q_1^2 + q_2^2 + \ldots + q_{j-1}^2)) \tag{4.22}$$

$$\rho_j q_j = \psi_j(1 - (q_1^2 + q_2^2 + \ldots + q_{j-1}^2)) \,. \tag{4.23}$$

We can therefore evaluate ρ_j and τ_j alternatively and this is sufficient to proceed with the recurrence in (4.13) and(4.14) for which we need ρ_j and the ratio

$$\frac{\tau_j}{\rho_j} = \frac{\psi_{j+1}\psi_j}{\psi_j^2 + 1/(1 - (q_1^2 + q_2^2 + \ldots + q_{j-1}^2))} \,.$$

The forward substitution for the Fourier coefficients now simplifies significantly to

$$\tilde{d}_j = (d_j + wy\psi_j - \psi_j \sum_{k=1}^{j-1} \tilde{d}_k q_k)/\rho_j$$

where the sum is accumulated through the process.

4.6 Updating a partial matrix

The four methods in §4.2, 4.3, 4.4, 4.5 were described for the case when $n = N$ (see the formulation in §4.1). For practical purposes we are mostly interested in the situation where $n << N$. We have already pointed out in Corollary 4.1 (§4.2) that when updating by the Rotations method it is sufficient to store and use the submatrix J_n of J_N (corresponding to W_N) of size $n < N$ to calculate the submatrix $\tilde{J}_n$ of $\tilde{J}_{N+1}$ (corresponding to $W_{N+1} = W_N \cup \{x_{N+1}, w_{N+1}\}$). The same observation applies to the other three methods of §4.3, 4.4, 4.5 because of their forward recurrence character.

It is worth noting that, once a restriction to some size n has been adopted, further increase is not meaningfully possible i.e. to represent the full data set Y_N. In fact the partial solution of size $n < N$ may be visualized as representing a class of data sets, the minimal size of which is the inner product generated by the Gauss quadrature corresponding to the matrix J_n (its eigenvalues as knots, etc.) and the function values of the least squares fit polynomial at the quadrature knots. Any increase in the size of J_n (i.e. any increase in the degree of the least squares polynomial fit) then corresponds to augmenting the new data points to this minimal data set.

4.7 Changing the weight of an existing data point

Theorem 4.1 was derived under the assumption that the knots x_j in Y_{N+1} were distinct. We now want to discuss what happens when x is equal to one of the knots in Y_N, say $x = x_N$. Let us consider (as in Theorem 4.1) the size N solution $\tilde{J}_N$, $\tilde{\mathbf{d}}_N$ for the data $Y_{N+1} = \{x_j, w_j, y_j\}_{j=1}^{N+1}$ where $\{x_{N+1}, w_{N+1}, y_{N+1}\} \to \{x_N, w, y\}$ for some w and y. The limit of such a change will correspond to the solution for the data set $\tilde{Y}_N = \{x_j, \tilde{w}_j, \tilde{y}_j\}_{j=1}^{N}$ where

$$\tilde{w}_j = w_j, \quad \tilde{y}_j = y_j, \quad j = 1, \ldots, N-1$$

$$\tilde{w}_N = \sqrt{w_N^2 + w^2} \tag{4.24}$$

$$\tilde{y}_N = (w_N^2 y_N + w^2 y)/(w_N^2 + w^2). \tag{4.25}$$

Here (4.24) is obvious from the definition of $\tilde{\sigma}_N$ and implies that the limiting $\tilde{J}_N$ corresponds to $\tilde{W}_N = \{x_j, \tilde{w}_j\}_{j=1}^{N}$ independently of $\tilde{y}_j$. Denoting (as in

(3.5) for Y_{N+1})

$$P_{N,N+1} = (\, P \quad \mathbf{p}_N(x_N) \quad \mathbf{p}_N(x_{N+1})\,) \ ,$$

we have
$\mathbf{d}_N = P_{N,N+1} D_{N+1}^2 \mathbf{y}_{N+1} = P D_{N-1}^2 \mathbf{y}_{N-1} + w_N^2 y_N \mathbf{p}_N(x_N) + w_{N+1}^2 y_{N+1} \mathbf{p}_N(x_{N+1})$ so that in the limit $\tilde{\mathbf{d}}_N = P D_{N-1}^2 \mathbf{y}_{N-1} + (w_N^2 y_N + w^2 y)\mathbf{p}_N(x_N)$ from which (4.25) follows directly. We thus have a situation similar to the case discussed in §4.2.

Theorem 4.2 *Given σ_N, J_N and $\mathbf{d}_N$ for the data Y_N and also $\{x, w, y\}$, where $x = x_N$, the solution for $\tilde{Y}_N$ is $\tilde{J}_N = Q J_N Q^T$, $\tilde{\sigma}_N = (\sigma_N^2 + w^2)^{\frac{1}{2}}$, $\tilde{\mathbf{d}}_N = Q(\mathbf{d}_N + (\tilde{w}_N \tilde{y}_N - w_N y_N)\mathbf{q}_N)$ where the orthogonal similarity matrix Q is uniquely determined by requiring $\tilde{J}_N$ to be tridiagonal and $Q(\sigma_N \mathbf{e}_1 + (\tilde{w}_N - w_N)\mathbf{q}_N) = \tilde{\sigma}_N \mathbf{e}_1$. Here $\mathbf{q}_N$ is the eigenvector of J_N corresponding to x_N, scaled to have norm 1 and a positive first element.*

This theorem shows that it is possible to modify the weight of a chosen knot by orthogonal similarity as in §4.2. However, the eigenvector $\mathbf{q}_N$ would then have to be evaluated. On the other hand, the four updating methods discussed so far can be applied to the present situation - the only difference is that the resulting matrix $\tilde{J}_{N+1}$ cannot then be unreduced because it has two equal eigenvalues. It is then a consequence of Theorem 4.2 that $\tilde{\beta}_N = 0$.

We note that the proof of Theorem 4.1 depends on the matrix $\tilde{J}_{N+1}$ having uniquely determined set of eigenvectors. Here this is not the case as the two equal eigenvalues have a two dimensional invariant subspace. In fact, condition (4.4) determining the first row of Q is based on one choice of eigenvectors in this subspace while the required modification of the weight in the resulting matrix corresponds to another choice of these eigenvectors.

5 Downdating methods

5.1 A reformulation of the downdating problem

The problem of downdating is now, with respect to the representation of the solutions introduced in §3, as follows:

Given $\mu_0 > 0, J_n$ a Jacobi matrix of size n and vector $\mathbf{d}_n$ representing the solution of Problem 1 for some data set Y_{N+1}, and given a triple $\{x, w, y\} \in Y_{N+1}$ find $\tilde{\mu}_0, \tilde{\mathbf{d}}_n$ and $\tilde{J}_n$ representing the solution for the data set $\tilde{Y}_N = Y_{N+1} \backslash \{x, w, y\}$.

As before, we have formulated the downdating problem only for the Problem 1 - the corresponding version for Problem 2 is obtained by omitting the coefficients $\mathbf{d}$ and values y.

The problem as formulated here always has a solution. In practical cases the set Y_{N+1} may have been discarded and so one would not know if the given triple belonged to the set.

The rest of this section is therefore set out as follows. We first derive downdating methods closely related to the four updating methods of §4 for the case $n = N + 1$, i.e. when the solution (of size $n = N - 1$) is known to exist. Next we derive another downdating method based on reversing the rotations method. We then turn to the problem of downdating a partial matrix ($n < N + 1$) and characterize the existence of the solution in that case - the downdating may, in general, decrease the size $\tilde{n}$ of the solution or the solution may not exist at all. This leads to stopping criteria for all the methods discussed.

5.2 Methods based on a complex weight

As formulated, Theorem 4.2 describes only an increase of the weight of the data triple $\{x_N, w_N, y_N\}$, i.e. $w^2 > 0$ implies $\tilde{w}_N > w_N$. However, replacing w^2 in Theorem 4.2 by $-w^2$ (note that only w^2 is used there, not w) leads to a solution, $\tilde{J}_N$, (with $\tilde{w}_N < w_N$) which remains real and unreduced of size N as long as $w^2 < w_N^2$. In the limit, when $\tilde{w}_N = 0$, we will have $\tilde{\beta}_{N-1} = 0$, $\tilde{\alpha}_N = x_N$ and the size $N - 1$ submatrix of $\tilde{J}_N$ will be the result of the downdating problem of §5.1 (although for $n = N$ rather than $N + 1$).

We have pointed out in §4.7 that the modification of a weight can be achieved by any of the three methods for updating. For downdating this simply means replacing w^2 by $-w^2$ and thus, if necessary, w by iw, $i^2 = -1$.

It turns out that, throughout the calculation, the $N \times N$ submatrix which is required remains real and those elements of the calculations which are complex quantities remain pure imaginary. Thus the whole calculation can be done very conveniently in real arithmetic. In the case of the rotation method the updating similarity matrices are products of (orthogonal) plane rotations: a $\begin{pmatrix} c & s \\ -s & c \end{pmatrix}$, $c^2 + s^2 = 1$ matrix imbedded in an identity. The similarity matrices here are products of elementary matrices which are of the form $\begin{pmatrix} c & is \\ -is & c \end{pmatrix}$, $c^2 - s^2 = 1$ imbedded in an identity. These matrices, which may be called hyper-rotations because the sines and cosines are replaced by hyperbolic sines and cosines, are (complex) orthogonal but not unitary. While plane rotations always have condition unity the condition of a hyper-rotation is $(|c| + |s|)/(|c| - |s|)$ which can be arbitrarily large. The inherent instability of downdating is manifest here in the possible ill-conditioning of these hyper-rotation matrices.

5.3 An eigenvector method for downdating

The downdating problem in §5.1 is a converse of the updating problem as formulated §4.1 (up to exchanging J and $\tilde{J}$, etc.).We can therefore attempt to solve it by reversing one of the methods for updating - this is particularly suitable for the rotation method. The analogue of Theorem 4.1 now is:

Theorem 5.1 *Given σ_{N+1}, J_{N+1} and $\mathbf{d}_{N+1}$ for the data $Y_{N+1} = Y_N \cup \{x, w, y\}$ the solution for Y_N is*

$$\begin{aligned} \begin{pmatrix} J_N & \mathbf{0} \\ \mathbf{0}^T & x \end{pmatrix} &= QJ_{N+1}Q^T, \\ \tilde{\sigma}_N &= (\sigma_N^2 - w^2)^{\frac{1}{2}}, \\ \begin{pmatrix} \tilde{\mathbf{d}}_N \\ wy \end{pmatrix} &= Q\mathbf{d}_{N+1} \end{aligned} \tag{5.1}$$

where the orthogonal similarity matrix Q is uniquely determined by requiring $\tilde{J}_N$ to be tridiagonal and $Q(\sigma_{N+1}\mathbf{e}_1 - w\mathbf{q}) = \tilde{\sigma}_N\mathbf{e}_1$ where $\mathbf{q}$ is the eigenvector of J_{N+1} corresponding to x, scaled to have norm 1 and a positive first element.

The similarity Q is now the product of the same rotations as in (4.5): $Q^T = R_N R_{N-1} \ldots R_1$. In fact these can again be calculated implicitly after finding R_N first and applying it to J_{N+1}. For R_N, only the ratio of the last two elements of the eigenvector $\mathbf{q}$ is required so $\mathbf{q}$ need not even be scaled. These unscaled last two elements could, in theory, be obtained by one step of the back-substitution on $J_{N+1} - xI$; however, as we shall see in the next section, this is not practical.

5.4 Downdating a partial matrix and the existence of the solution

When we have available only a submatrix J_n of the whole matrix J_{N+1}, $n < N+1$ and the triple to be downdated $\{x, w, y\}$, the question of existence of the solution is equivalent to the existence of some matrix J_{N+1}, $N+1 > n$ with the properties

- the given J_n is a submatrix of J_{N+1}
- x is an eigenvalue of J_{N+1} with the weight w scaled to the current zero-th moment σ^2_{N+1}.

The smallest ($n = N$) matrix with these properties will have the form $\begin{pmatrix} J_n & \beta\mathbf{e}_n \\ \beta\mathbf{e}_n^T & \alpha \end{pmatrix}$ where α and β are to be chosen so that J_{N+1} has eigenvalue x with weight w. Denoting the corresponding normalized eigenvector $\begin{pmatrix} \mathbf{q} \\ \eta \end{pmatrix}$ this leads to equations

$$\begin{aligned} \mathbf{e}_1^T\mathbf{q} &= w/\sigma_{N+1} \\ (J_n - xI)\mathbf{q} &= -\beta\eta\mathbf{e}_n \\ \beta\mathbf{e}_n^T\mathbf{q} + \alpha\eta &= x\eta. \end{aligned} \tag{5.2}$$

The first n of these $n+2$ equations form a lower triangular system for $\mathbf{q}$ (independent of α and β) which can thus be determined uniquely. The solution $\mathbf{q}$ is proportional to w, so for w not too large we have $\|\mathbf{q}\| < 1$ and a real $\eta = \sqrt{1 - \mathbf{q}^T\mathbf{q}}$ exists. Otherwise, for larger w, we might decrease the size n to $\tilde{n}$ so that the norm of the shorter vector $\mathbf{q}$ is less than 1 - this will imply the existence of the solution of the downdating problem for that size submatrix $J_{\tilde{n}}$. The unknown α and β are now easily obtained from the last two equations of (5.2); the sign of η may be chosen so that β is positive. The solution of the downdating problem can now be obtained as discussed in §5.3 from the last two elements of the eigenvector $(\mathbf{q}^T, \eta)^T$. We note here that, contrary to the case discussed in §5.3, the complete forward substitution is indeed necessary as we do not know the whole matrix J_{N+1} needed in order to do only one step of back substitution as suggested there.

We have thus established the following result.

Theorem 5.2 *Given σ_{N+1} and J_n, $n \leq N$, the solution for some Y_{N+1} and the downdating pair $\{x, w\}$, denote $\mathbf{q}$ the solution of just the first $n-1$ equations of $(J_n - xI)\mathbf{q} = \mathbf{0}$ satisfying $\mathbf{e}_1^T\mathbf{q} = w/\sigma_{N+1}$. Let $\tilde{n}$ be the largest integer such that $\tilde{\mathbf{q}}^T\tilde{\mathbf{q}} < 1$ where $\tilde{\mathbf{q}}$ is the vector of the first $\tilde{n}$ elements of $\mathbf{q}$. Then $\tilde{n}$ is the largest size of the leading submatrix $J_{\tilde{n}}$ of J_n from which it is possible to downdate by the pair $\{x, w\}$.*

Obviously, if $w \geq \sigma_{N+1}$ then $\tilde{n} = 0$ and no downdating is possible. Furthermore, all three downdating methods in §5.2 must produce a solution of size $\tilde{n}$; suitable criteria for recognizing this critical size correspond to identifying the depletion of the total *mass*, i.e. the current zero-th moment, during the process.

The constructive nature of Theorem 5.2, for downdating a Jacobi matrix, means that a practical method can be based on it. The vector, $\mathbf{d}_{\tilde{n}}$, of given Fourier coefficients has also to be extended by an unknown element, say $d_{\tilde{n}+1}$, so that after the transformation (5.1) the last element of the left hand side is wy. We achieve this by the following method. Note first that elements $\mathbf{e}_j^T\mathbf{d}_{\tilde{n}+1}$ are known for the range $j = 1, 2, \ldots, \tilde{n}$ and wy is known. We may rewrite (5.1) as

$$\begin{pmatrix} \tilde{\mathbf{d}}_{\tilde{n}} \\ 0 \end{pmatrix} + wy\mathbf{e}_{\tilde{n}+1} = Q \begin{pmatrix} \hat{\mathbf{d}}_{\tilde{n}} \\ 0 \end{pmatrix} + \gamma Q\mathbf{e}_{\tilde{n}+1}, \qquad (5.3)$$

for some scalar γ. As the $R_{\tilde{n}}, R_{\tilde{n}-1}, \ldots, R_1$ become available we apply them to the known $\hat{\mathbf{d}}_{\tilde{n}}$ and $\mathbf{e}_{\tilde{n}+1}$. After all rotations have been applied we use the last row of (5.3) to solve for γ. The right hand side of (5.3) then gives the solution.

The elementary matrices used are real (orthogonal) rotations. So the stability of the process depends on the accuracy with which the first rotation used, R_n, is determined.

6 Concluding remarks

We refer to the methods discussed as follows:

RHR - rotations update (§4.2), and a downdate based on (§5.2),

REV - rotations eigenvector downdate (§5.3 and 5.4),

TLD - triangular Lanczos method using determinants: (§4.4) update and a downdate based on (§5.2),

TLS - triangular Lanczos method using the special form of L: (§4.5) update and a downdate based on (§5.2).

The table below shows the complexity of algorithms, based on the methods of the previous sections, as the term with the highest power of n in the number of *flops* (essentially the number of $\times$ and $\div$ operations) per up/downdate. The updating and downdating algorithms for the RHR, TLD and TLS methods essentially differ only in that certain signs in the update version are reversed in the downdate version. Thus the numbers of flops shown apply to both up and downdating.

Complexity

Method	Up/downdate J_n $\surd$	Up/downdate J_n flops	Up/downdate $\mathbf{d}_n$ $\surd$	Up/downdate $\mathbf{d}_n$ flops
RHR	n	$15n$	-	$4n$
REV	n	$23n$	-	$11n$
TLD	n	$10n$	n	$\frac{7}{2}n^2$
TLS	n	$11n$	n	$3n$

A feature common to the TLD and TLS algorithms is that they require the calculation of the elements $\psi_j = wp_{j-1}(x)$, $j = 1, 2, \ldots$ of an eigenvector $\mathbf{u}$ by the forward substitution $\beta_j\psi_{j+1} = (x - \alpha_j)\psi_j - \beta_{j-1}\psi_{j-1}$. When x lies inside the interval containing the eigenvalues of J_n, the values of the ψ_{j-1} are bounded by a reasonably small constant and may even vanish. For x outside this interval however, the ψ_{j-1} will never vanish and indeed will grow rapidly with j, possibly causing overflow. To avoid this possibility a scaled version of the algorithms which uses the ratios ψ_j/ψ_{j-1} should be used in this case.

In our experience the updating methods not based on rotations and *all* downdating methods may suffer from instability if the degree of the fitted polynomial approaches the degree of the interpolating polynomial. This has also been observed elsewhere [3], [1]. Nevertheless, our preliminary testing (computing a moving least squares polynomial fit to noisy data) indicates that the methods will be very satisfactory for polynomial least squares fits of low degree, such as occur in many real situations. Moreover, on this quite severe preliminary test, the methods are all equally accurate. Thus complexity considerations are likely to be the main factor determining which method will be chosen.

7 Acknowledgements

The work of the second-mentioned author (GHG) was in part supported by the US Army (DAAL03-87-K-0095), the John Simon Guggenheim Memorial Foundation and the Flinders University of South Australia Research Grant. Some preliminary work was also done while he was a guest worker at the Bell Telephone Laboratories (summer, 1987).

8 References

[1] S.T. Alexander, C.T. Pan & R.J. Plemmons, "Analysis of a Recursive Least Squares Hyperbolic rotation algorithm for signal processing", *Linear Algebra Appl.,* 1988, v. **98**, pp. 3-40.
[2] A.W. Bojanczyk, R.P. Brent, P. Van Dooren & F.R. de Hoog, "A note on downdating the Choleski factorization", *SIAM J. Sci. Stat. Comput.,* 1987, v. **8**, pp. 210-221.
[3] W. Gautschi, "On Generating Orthogonal Polynomials", *SIAM J. Sci. Statist. Comput.,* 1982, v. **3**, pp. 289-317.
[4] P.E. Gill, G.H. Golub, W. Murray & M.A. Saunders, "Methods for modifying matrix factorizations", *Math. Comp.,* 1974, v. **28**, pp. 505-535.

[5] G.H. Golub & M.A. Saunders, "Linear least squares and quadratic programming", *Integer and non-linear programming, J. Abadie (Ed), North-Holland Publishing Co.,* 1970, pp. 229-256.

[6] G.H. Golub & C.F. Van Loan, "Matrix computations", *North Oxford Academic Publishing,* 1983,

[7] G.H. Golub & J.H. Welsch, "Calculation of Gauss quadrature rules", *Math. Comp.,* 1969, v. **23,** pp. 221-230.

[8] W.B. Gragg & W.J. Harrod, "The numerically stable reconstruction of Jacobi matrices from spectral data", *Numer. Math.,* 1984, v. **44,** pp. 317-335.

[9] J. Kautsky & G.H. Golub, "On the calculation of Jacobi matrices", *Linear Algebra Appl.,* 1983, v. **52/53,** pp. 439-455.

[10] C. Rader & A.O. Steinhardt, "Hyperbolic Householder transformations", *IEEE Trans. Acoust. Speech Signal Process.,* 1986, pp. 1589-1602.

[11] H. Rutishauser, "Further contributions to the solution of simultaneous linear equations and the determination of eigenvalues", *US Dept of Commerce, NBS, Applied Maths Series 49,* 1958,

[12] G.W. Stewart, "The effects of rounding error on an algorithm for downdating a Choleski factorization", *J. Inst. Math. Appl.,* v. **23,** 1979, pp. 203-213.

[13] H.S. Wilf, "Mathematics for the physical sciences", 1962, *Wiley, New York,* pp. Chapter 2.

Parallel Algorithms for Singular Value Problems

Sven Hammarling

Numerical Algorithms Group Ltd.
Wilkinson House
Jordan Hill Road
Oxford, OX2 8DR, UK

1 Introduction

We look at algorithms for computing the singular value decomposition (SVD) and related problems, that are suitable for parallel environments.

High performance and parallel computing is a fast developing field and while we are striving to find algorithms that are portable across a wide range of environments, that has certainly not been reached in the field of the SVD. This paper gives a brief survey of the available methods, as well as plenty of references to the available literature.

2 Factorizations and Algorithm Tools

The singular value decomposition of an m by $n, m \geq n$ real matrix is given by

$$A = U \begin{pmatrix} \Sigma \\ 0 \end{pmatrix} V^T,$$

where U is an m by m orthogonal matrix, V is an n by n orthogonal matrix and Σ is diagonal. The first n columns of U are the left singular vectors of

NATO ASI Series, Vol. F 70
Numerical Linear Algebra, Digital Signal Processing
and Parallel Algorithms
Edited by G. H. Golub and P. Van Dooren

A, the columns of V are the right singular vectors of A and the diagonal elements of Σ are the singular values of A, which may be ordered so that

$$\sigma_1 \geq \sigma_1 \geq ... \geq \sigma_n \geq 0.$$

Here we just concentrate on the real case with $m \geq n$. The generalizations to $m < n$ and to the complex case are straightforward [36].

An important factorization that is often used in conjunction with the SVD is the QR factorization given by

$$A = Q\begin{pmatrix} R \\ 0 \end{pmatrix},$$

where Q is orthogonal and R is upper triangular. If we now perform the SVD of R as

$$R = \tilde{U}\Sigma V^T,$$

then

$$A = Q\begin{pmatrix} \tilde{U} & 0 \\ 0 & I \end{pmatrix}\begin{pmatrix} \Sigma \\ 0 \end{pmatrix}V^T = U\begin{pmatrix} \Sigma \\ 0 \end{pmatrix}V^T,$$

where

$$U = Q\begin{pmatrix} \tilde{U} & 0 \\ 0 & I \end{pmatrix},$$

so that A and R have the same singular values and right singular vectors. It is usual, particularly when $m \gg n$, to perform a QR factorization first and then to find the SVD of R, thus regarding the QR factorization as a preprocessor for the SVD. We could for example exploit sparsity in the QR step, even if we cannot subsequently exploit sparsity in the (much smaller) SVD part. We can also use the technique of updating the QR factorization if the data is coming in real-time, or A has too many rows to fit in main memory.

These ideas are particularly important for problems such as least squares where we may wish to stop after the QR factorization if R is well-conditioned, and proceed to the SVD only if we wish to get a reliable determination of the rank of A (see for example [38]).

One of the important transformations used in algorithms for finding the above factorizations is the plane rotation. The two by two plane rotation matrix is given by

$$J = \begin{pmatrix} c & s \\ -s & c \end{pmatrix}, \quad c^2 + s^2 = 1$$

and a rotation for the (i,j) plane is obtained simply by embedding J in positions (i,i), (i,j), (j,i) and (j,j) of the unit matrix. Plane rotations are used in Givens transformations and Jacobi-type transformations [36]. Another important transformation matrix is the elementary reflector given by

$$P = I - \mu u u^T, \quad \mu u^T u = 2,$$

which is used in Householder transformations. The vector u is often called the Householder vector. Both the above matrices are orthogonal and hence, so long as they are computed properly, give stable transformations [64].

These transformations are at the heart of two types of algorithm: direct reduction to canonical form and iterative reduction to canonical form. The direct methods give canonical forms, such as the QR factorization, for which the subsequent computation of the SVD is, in some sense, easier. The iterative reductions are necessary at some point since the computation of the singular values is, of course, a nonlinear problem. QR type algorithms are usually based upon an initial direct reduction to bidiagonal form, whereas Jacobi-type algorithms iterate directly on either A or R [36].

Methods which do not directly utilize the above transformations, but which are important in a parallel setting include Lanczos, bisection and multisection, and inverse iteration [36,45].

3 Paradigms to Aid Parallelism

Important aims in devising algorithms for parallel machines are to look for independent operations at the correct level of granularity for the target architecture and to reduce data movement, in terms of storage or communication.

On a systolic array one may be thinking of operations at the level of setting up and applying plane rotations, while on a multiprocessor Cray the level of granularity needs to be much larger. Many modern machines have a hierarchy of memory such as global memory, high-speed cache or local memory and vector registers and we need to aim for full reuse of the data held in the cache or local memory. Similarly, on distributed memory systems we need to utilise the processors and avoid slowing down the processors with the need to communicate data.

Useful techniques to help realise those aims include restructuring, block algorithms and divide and conquer algorithms [52,19,43,1]. As examples, restructuring might involve rearranging the algorithm so as to expose the vector operations, by describing the algorithm clearly in terms of matrix-vector operations and inserting calls to the Level 2 BLAS [18,20], or reordering a sequence of plane rotations in order to perform them in parallel [49,27]. Block algorithms are considered when the level of granularity needs to be raised to avoid excessive data movement, and so that parallelism can be exploited, either by performing operations on distinct blocks in parallel, or by performing scalar and vector operations within blocks in parallel. This might, for example, involve deriving a block version of a point algorithm, describing the algorithm in terms of matrix-matrix operations and inserting calls to the Level 3 BLAS [17,13,7,21], with the aim of obtaining the so called surface to volume effect. With a divide and conquer approach one divides the problem up into independent subproblems, which then have to be successively combined together until one finally conquers the original problem [11,43,1].

Whatever technique we use, we must remember to ensure that numerical stability is not compromised.

4 Algorithms for the QR Factorization

If J_{ij} and J_{kl} are plane rotation matrices for the (i, j) and (k, l) planes respectively, with $i \neq k, l$ and $j \neq k, l$, then J_{ij} and J_{kl} are called disjoint rotations [31]. Since

$$J_{ij}J_{kl} = J_{kl}J_{ij}$$

transformations by these matrices can be performed independently and this is the basis for a number of methods for the QR factorization based upon Givens transformations (see for example [49] and the references given there). This approach has principally been used on distributed memory systems [23]. Givens transformations are also commonly used for updating the

QR factorization and, in a systolic setting, we can readily define construction and application processors to set up plane rotations and to transform elements in parallel, the best known example being the Gentleman-Kung triangular array [32].

Givens transformations are also important in methods for obtaining the QR factorization of a sparse matrix [41].

For coarse grained parallel machines and for shared memory machines, block methods based upon Householder transformations [7,59] may be preferable, because of their higher level of granularity (see [5] and the references given there). This is the approach being taken by the LAPACK project [13,6]. In essence this approach uses the fact that a product of r Householder matrices may be represented in the form

$$Q = I - USU^T,$$

where U is an m by r matrix, whose kth column is the kth Householder vector, and S is an r by r upper triangular matrix. We now have the potential for matrix-matrix operations, rather than matrix-vector operations.

5 Algorithms for the SVD

Algorithms for computing the SVD of general dense problems fall roughly into two categories. In either case we may start off with an initial QR factorization of A in order to obtain an upper triangular matrix R. Firstly are the methods which perform a reduction to bidiagonal form, followed by an iterative method on the bidiagonal form to find the singular values, such as the QR algorithm to further reduce the bidiagonal form to diagonal form, and secondly are the Jacobi-type methods which iterate directly on either A or R. In a parallel setting the first category has not received anything like the attention that the Jacobi-type methods have attracted, but it does not yet seem clear that this is justified. The Jacobi-type algorithms generally require considerably more floating-point operations than the standard QR algorithm, but they do possess more inherent parallelism.

For the first category, the standard approach is to perform a Householder reduction to bidiagonal form, followed by the Golub-Kahan-Reinsch version of the QR algorithm to obtain the diagonal form [33,35,61,16].

A block reduction to bidiagonal form has recently been developed [21], whereby at least a large part of the computation can be performed as matrix-matrix operations. If one, or both, of the orthogonal transformation matrices is required, in order that the singular vectors may be obtained, then these can be found by means of the product form representation of the Householder transformation matrices. Demmel and Kahan have developed a version of the QR algorithm that guarantees high accuracy in the singular values of the bidiagonal matrix [15], and which increases the potential for pipelining. In any case, when the singular vectors are required, the corresponding Givens rotations that constitute a sweep can be saved and then applied, as a sequence, thus greatly increasing the granularity and enhancing the prospect for parallelism, a technique that can be used in many situations where plane rotations are being applied, [22, Part 1, Section 7 for example].

A promising alternative to the QR algorithm is the divide and conquer approach, whereby the bidiagonal matrix is split into a number of smaller bidiagonal matrices, each of these sub-problems is solved and then glued together as rank-one or rank-two modifications [43,1]. It may be possible to interleave the divide and conquer part with the block reduction to bidiagonal form [21]. For the future we might hope for the development of block QR algorithms, or at least multishift QR algorithms along the lines of that proposed for the unsymmetric eigenproblem in LAPACK [3]. The current open question for block QR algorithms is how to choose the shifts?

Bisection and multisection are alternative inherently parallel algorithms that can be used on the bidiagonal matrix to find singular values, with inverse iteration an obvious contender for obtaining the corresponding singular vectors. Some of the relevant issues are raised in [14].

Turning now to the second category, interest in Jacobi-type algorithms has, to a large extent, been driven by considerations of fine grain parallelism, particularly systolic arrays for real-time signal processing applications and SIMD architectures such as the DAP and the Connection Machine [50,62,24], although the methods have been used with success on a number of other architectures. The classical Jacobi algorithm solves the symmetric eigenvalue problem by means of (serial) Jacobi rotations, where each transformation is chosen to solve a two-by-two symmetric eigenvalue problem so that

$$\begin{pmatrix} c & s \\ -s & c \end{pmatrix} \begin{pmatrix} x & y \\ y & z \end{pmatrix} \begin{pmatrix} c & -s \\ s & c \end{pmatrix} = \begin{pmatrix} x' & 0 \\ 0 & z' \end{pmatrix},$$

the aim being to reduce the symmetric matrix to diagonal form [36], this aim being motivated by the fact that

$$(x')^2 + (y')^2 \geq x^2 + y^2.$$

The more recent work has been concerned with alternative orderings to enable rotations to be performed in parallel, together with analyses of the convergence of the resulting algorithms, see for example [8,58,49,27,25,48] and the further references given in those papers. There are basically two Jacobi-type algorithms for the SVD, the first a direct generalization due to Kogbetliantz [44,30], where each transformation solves the two-by-two SVD problem, so that

$$\begin{pmatrix} c_1 & s_1 \\ -s_1 & c_1 \end{pmatrix} \begin{pmatrix} x & y \\ w & z \end{pmatrix} \begin{pmatrix} c_2 & -s_2 \\ s_2 & c_2 \end{pmatrix} = \begin{pmatrix} x' & 0 \\ 0 & z' \end{pmatrix},$$

and once again

$$(x')^2 + (y')^2 \geq x^2 + y^2.$$

A number of architectures have been proposed for the Kogbetliantz method, mostly with systolic implementations in mind [8,9,10,60]. For the case where a QR factorization has been performed, a triangular array that can be used to find the SVD of R has been proposed by Luk [47], and an attempt to propose a unified architecture for the QR factorization, eigenvalue and singular value problems is given in [29].

The second type of approach uses a one-sided algorithm due to Hestenes [42] which aims to reduce A (or R) to a matrix with either orthogonal columns, or orthogonal rows. Thus in the version where A is transformed to a matrix $\hat{V}$ with orthogonal rows, an orthogonal matrix U, made up as a sequence of plane rotations, is found so that

$$U^T A = \hat{V},$$

and if we now scale $\hat{V}$ to be orthonormal, so that the scaled matrix is orthogonal then, if necessary with suitable row permutations,

$$A = U \begin{pmatrix} \Sigma \\ 0 \end{pmatrix} V^T,$$

gives the SVD of A. The plane rotations are chosen to orthogonalize two rows of A and correspond to Jacobi rotations applied to AA^T [51]. Since we are applying one-sided transformations to the columns of A, this method is particularly appropriate for machines with vector processors [12]. Once again we can precede the one-sided method by a QR factorization [39,4].

To raise the level of granularity of Jacobi-type methods a number of people have considered block versions of the methods [9,63,26]

If A is sparse then the method of Lanczos can be used to obtain a bidiagonal matrix having the required singular values [53,36]. The computationally intense part of the Lanczos algorithm involves computing Ax and A^Ty for given vectors x and y at each iteration, and it is here where sparsity and parallelism can be exploited. If A has some special structure, such as a Toeplitz matrix, then once again this structure could be exploited in the matrix vector products of the Lanczos algorithm. A block Lanczos method for the SVD is given in [34] and a Lanczos method for solving least squares problems directly is given in [56].

6 Related Factorizations

A number of the methods for finding the SVD can be adapted to related factorizations such as the the generalized SVD (GSVD), [57,36] and the product induced SVD (ΠSVD), [28]. In particular, a number of implicit Kogbetliantz algorithms and architectures have been investigated for these factorizations [54,40,2].

Just as the QR factorization can be used as a preprocessor for the SVD, so the generalized QR factorization [37,55] can be used as a preprocessor to the GSVD and the ΠSVD, which can then be solved using an extension of the Luk triangular array [54,46,28].

References

[1] P. Arbenz. Divide-and-conquer algorithms for the computation of the SVD of bidiagonal matrices. In J. J. Dongarra, I. S. Duff, P. Gaffney, and S. McKee, editors, *Vector and Parallel Computing*, pages 1–10, Ellis Horwood, Chichester, 1989.

[2] Z. Bai. *The Direct GSVD Algorithm and its Parallel Implementations.* Technical Report CS-TR-1901, University of Maryland, Department of Computer Science, College Park, Maryland 20742, 1987. Ph.D. Thesis.

[3] Z. Bai and J. Demmel. On a block implementation of Hessenberg multishift QR iteration. *Int. J. High Speed Comput.*, 1:97–112, 1989.

[4] M. Berry and A. Sameh. A multiprocessor scheme for the singular value decomposition. In G. Rodrigue, editor, *Parallel Processing for Scientific Computing*, pages 67–71, SIAM, Philadelphia, 1989.

[5] C. H. Bischof. *QR Factorization Algorithms for Coarse-Grained Distributed Systems.* Technical Report TR 88-939, Cornell University, Department of Computer Science, Ithaca, New York 14853-7501, 1988. Ph.D. Thesis.

[6] C. H. Bischof, J. Demmel, J. J. Dongarra, J. Du Croz, A. Greenbaum, S. Hammarling, and D. C. Sorensen. *LAPACK Working Note No.5 - Provisional Contents.* Technical Report ANL-88-38, Argonne National Laboratory, Mathematics and Computer Science Division, Argonne, Illinois 60439, 1988.

[7] C. H. Bischof and C. F. Van Loan. The WY representation of products of Householder matrices. *SIAM J. Sci. Stat. Comput.*, 8:s2–s13, 1987.

[8] R. P. Brent and F. T. Luk. The solution of singular-value and symmetric eigenvalue problems on multiprocessor arrays. *SIAM J. Sci. Stat. Comput.*, 6:69–84, 1985.

[9] R. P. Brent, F. T. Luk, and C. F. Van Loan. Computation of the singular value decomposition using mesh-connected processors. *J. VLSI Comput. Syst.*, 1:242–270, 1985.

[10] J. P. Charlier, M. Vanbegin, and P. Van Dooren. On efficient implementations of Kogbetliantz's algorithm for computing the singular value decomposition. *Numer. Math.*, 52:279–300, 1988.

[11] J. J. M. Cuppen. A divide and conquer method for the symmetric tridiagonal eigenproblem. *Numer. Math.*, 36:177–195, 1981.

[12] P. P. M. De Rijk. A one-sided Jacobi algorithm for computing the singular value decomposition on a vector computer. *SIAM J. Sci. Stat. Comput.*, 10:359–371, 1988.

[13] J. Demmel, J. J. Dongarra, J. Du Croz, A. Greenbaum, S. Hammarling, and D. C. Sorensen. A project for developing a linear algebra library for high-performance computers. In M. Wright, editor, *Aspects of Computation on Asynchronous Parallel Processors*, pages 87–92, North-Holland, Amsterdam, 1989.

[14] J. Demmel, J. Du Croz, S. Hammarling, and D. C. Sorensen. *LAPACK Working Note No.4 - Guidelines for the Design of Symmetric Eigenroutines, SVD, and Iterative Refinement and Condition Estimation for Linear Systems.* Technical Memorandum 111, Argonne National Laboratory, Mathematics and Computer Science Division, Argonne, Illinois 60439, 1988.

[15] J. Demmel and W. Kahan. *LAPACK Working Note No.3 - Computing Small Singular Values of Bidiagonal Matrices with Guaranteed High Relative Accuracy.* Technical Memorandum 110, Argonne National Laboratory, Mathematics and Computer Science Division, Argonne, Illinois 60439, 1988.

[16] J. J. Dongarra, J. R. Bunch, C. B. Moler, and G. W. Stewart. *LINPACK Users' Guide.* SIAM, Philadelphia, 1978.

[17] J. J. Dongarra, J. Du Croz, I. S. Duff, and S. Hammarling. A set of Level 3 Basic Linear Algebra Subprograms. *ACM Trans. Math. Software*, 1990. To appear.

[18] J. J. Dongarra, J. Du Croz, S. Hammarling, and R. J. Hanson. An extended set of FORTRAN Basic Linear Algebra Subprograms. *ACM Trans. Math. Software*, 14:1–32, 1988.

[19] J. J. Dongarra and S. Hammarling. Evolution of numerical software for dense linear algebra. In M. G. Cox and S. Hammarling, editors, *Reliable Numerical Computation*, pages 297–327, Oxford University Press, Oxford, 1990.

[20] J. J. Dongarra, L. Kaufman, and S. Hammarling. Squeezing the most out of eigenvalue solvers on high-performance computers. *Linear Algebra Appl.*, 77:113–136, 1986.

[21] J. J. Dongarra, D. C. Sorensen, and S. Hammarling. Block reduction of matrices to condensed forms for eigenvalue computations. *JCAM*, 27:215–227, 1989.

[22] J. Du Croz and P. Mayes. *NAG Fortran Library Vectorization Review.* Technical Report TR6/89, Numerical Algorithms Group, Wilkinson House, Jordan Hill Road, Oxford OX2 8DR, 1989.

[23] L. Eldén. *A Parallel QR Decomposition Algorithm.* Technical Report Lith-MAT-R-1988-02, Linköping University, Department of Mathematics, S-581 83 Linköping, 1988.

[24] L. M. Ewerbring and F. T. Luk. Computing the singular value decomposition on the connection machine. In Ed. F. Deprettere, editor, *SVD and Signal Processing: Algorithms, Applications and Architectures*, pages 407–424, North-Holland, Amsterdam, 1988.

[25] K. V. Fernando. Linear convergence of the row cyclic Jacobi and Kogbetliantz methods. *Numer. Math.*, 56:73–91, 1989.

[26] K. V. Fernando and S. Hammarling. On block Kogbetliantz methods for computation of the SVD. In Ed. F. Deprettere, editor, *SVD and Signal Processing: Algorithms, Applications and Architectures*, pages 349–356, North-Holland, Amsterdam, 1988.

[27] K. V. Fernando and S. Hammarling. Parallel eigenvalue and singular value algorithms for signal processing. In M. Wright, editor, *Aspects of Computation on Asynchronous Parallel Processors*, pages 13–22, North-Holland, Amsterdam, 1989.

[28] K. V. Fernando and S. Hammarling. A product induced singular value decomposition (ΠSVD) for two matrices and balanced realization. In B. N. Datta, C. R. Johnson, M. A. Kaashoek, R. J. Plemmons, and E. D. Sontag, editors, *Linear Algebra in Signals, Systems, and Control*, pages 128–140, SIAM, Philadelphia, 1988.

[29] K. V. Fernando and S. Hammarling. *Unified Mesh-Connected Architecture for Eigenvalue, Singular Value and QR-Decompositions.* Technical Report TR2/88, Numerical Algorithms Group, Wilkinson House, Jordan Hill Road, Oxford OX2 8DR, 1988.

[30] G. E. Forsythe and P. Henrici. The cyclic Jacobi method for computing the principal values of a complex matrix. *Trans. Amer. Math. Soc.*, 94:1–23, 1960.

[31] W. M. Gentleman. Error analysis of QR decomposition by Givens transformations. *Linear Algebra Appl.*, 10:189–197, 1975.

[32] W. M. Gentleman and H. T. Kung. Matrix triangularization by systolic arrays. In *Real-Time Signal Processing IV, SPIE Proceedings 298*, pages 19–26, The Society of Photo-Optical Engineers, Bellingham, Washington, 1981.

[33] G. H. Golub and W. Kahan. Calculating the singular values and pseudo-inverse of a matrix. *SIAM J. Num. Anal., Ser. B*, 2:205–224, 1965.

[34] G. H. Golub, F. T. Luk, and M. Overton. A block Lanczos method for computing the singular values and corresponding singular vectors of a matrix. *ACM Trans. Math. Software*, 7:149–169, 1981.

[35] G. H. Golub and C. Reinsch. Singular value decomposition and least squares solutions. In J. H. Wilkinson and C. Reinsch, editors, *Handbook for Automatic Computation, Vol.2, Linear Algebra*, pages 134–151, Springer-Verlag, Berlin, 1971.

[36] G. H. Golub and C. F. Van Loan. *Matrix Computations.* The Johns Hopkins University Press, Baltimore, 2nd edition, 1989.

[37] S. Hammarling. The numerical solution of the general Gauss-Markov linear model. In T. S. Durrani, J. B. Abbiss, J. E. Hudson, R. N. Madan, J. G. McWhirter, and T. A. Moore, editors, *Mathematics in Signal Processing*, pages 441–456, Oxford University Press, Oxford, 1987.

[38] S. Hammarling. The singular value decomposition in multivariate statistics. *ACM Signum Newsletter*, 20:2–25, 1985.

[39] P. C. Hansen. Reducing the number of sweeps in Hestenes' method. In Ed. F. Deprettere, editor, *SVD and Signal Processing: Algorithms, Applications and Architectures*, pages 357–368, North-Holland, Amsterdam, 1988.

[40] M. T. Heath, A. J. Laub, C. C. Paige, and R. Ward. Computing the SVD of a product of two matrices. *SIAM J. Sci. Stat. Comput.*, 7:1147–1159, 1986.

[41] M. T. Heath and D. C. Sorensen. A pipelined Givens method for computing the QR factorization of a sparse matrix. *Linear Algebra Appl.*, 77:189–203, 1986.

[42] M. R. Hestenes. Inversion of matrices by biorthogonalization and related results. *J. SIAM*, 6:51–90, 1958.

[43] E. R. Jessup and D. C. Sorensen. A divide and conquer algorithm for computing the singular value decomposition. In G. Rodrigue, editor, *Parallel Processing for Scientific Computing*, pages 61–66, SIAM, Philadelphia, 1989.

[44] E. Kogbetliantz. Solution of linear equations by diagonalization of coefficients matrix. *Quart. Appl. Math.*, 13:123–132, 1955.

[45] S. Lo, B. Philippe, and A. Sameh. A multiprocessor algorithm for the symmetric tridiagonal eigenvalue problem. *SIAM J. Sci. Stat. Comput.*, 8:s155–s165, 1987.

[46] F. T. Luk. A parallel method for computing the generalized singular value decomposition. *J. Parallel Distrib. Comput.*, 2:250–260, 1985.

[47] F. T. Luk. A triangular processor array for computing singular values. *Linear Algebra Appl.*, 77:259–274, 1986.

[48] F. T. Luk and H. Park. On parallel Jacobi orderings. *SIAM J. Sci. Stat. Comput.*, 10:18–26, 1989.

[49] J. J. Modi. *Parallel Algorithms and Matrix Computation.* Oxford University Press, Oxford, 1988.

[50] J. J. Modi and J. D. Pryce. Efficient implementation of Jacobi's diagonalization method on the DAP. *Numer. Math.*, 46:443–454, 1985.

[51] J. C. Nash. *Compact Numerical Methods for Computers.* Adam Hilger, Bristol, 1979.

[52] J. M. Ortega and R. G. Voigt. Solution of partial differential equations on vector and parallel computers. *SIAM Review*, 27:149–240, 1985.

[53] C. C. Paige. Bidiagonalization of matrices and solution of linear equations. *SIAM J. Num. Anal.*, 11:197–209, 1974.

[54] C. C. Paige. Computing the generalized singular value decomposition. *SIAM J. Sci. Stat. Comput.*, 7:1126–1146, 1986.

[55] C. C. Paige. Some aspects of generalized QR factorizations. In M. G. Cox and S. Hammarling, editors, *Reliable Numerical Computation*, pages 73–91, Oxford University Press, Oxford, 1990.

[56] C. C. Paige and M. A. Saunders. LSQR: an algorithm for sparse linear equations and sparse least squares. *ACM Trans. Math. Software*, 8:43–71,195–209, 1982.

[57] C. C. Paige and M. A. Saunders. Towards a generalized singular value decomposition. *SIAM J. Num. Anal.*, 18:398–405, 1981.

[58] C. C. Paige and P. Van Dooren. On the quadratic convergence of Kogbetliantz's algorithm for computing the singular value decomposition. *Linear Algebra Appl.*, 77:301–313, 1986.

[59] R. Schreiber and C. F. Van Loan. A storage-efficient WY representation for products of Householder transformations. *SIAM J. Sci. Stat. Comput.*, 10:53–57, 1989.

[60] W. D. Shoaff and C. D. Chan. A parallel algorithm for the singular value decomposition of rectangular matrices. In G. Rodrigue, editor, *Parallel Processing for Scientific Computing*, pages 72–76, SIAM, Philadelphia, 1989.

[61] B. T. Smith, J. M. Boyle, Y. Ikebe, V. C. Klema, and C. B. Moler. *Matrix Eigensystem Routines: EISPACK Guide.* Springer-Verlag, Berlin, 2nd edition, 1970.

[62] J. M. Speiser. Signal processing computational needs. In J. M. Speiser, editor, *Advanced Algorithms and Architectures for Signal Processing, SPIE Proceedings 696*, pages 2–6, The Society of Photo-Optical Engineers, Bellingham, Washington, 1987.

[63] C. F. Van Loan. The block Jacobi method for computing the singular value decomposition. In C. I. Byrnes and A. Lindquist, editors, *Computational and Combinatorial Methods in Systems Theory*, pages 245–255, North-Holland, Amsterdam, 1986.

[64] J. H. Wilkinson. *The Algebraic Eigenvalue Problem.* Oxford University Press, London, 1965.

Some Remarks on the Generalised Bareiss and Levinson Algorithms

Ilse Ipsen*

Department of Computer Science
Yale University
51 Prospect Street
New Haven, CT 06511, USA

Abstract

The Bareiss (or Schur) and Levinson algorithms are the most popular algorithms for solving linear systems with dense $n \times n$ Toeplitz coefficient matrix in $O(n^2)$ arithmetic operations. Both algorithms have been generalised to solve linear systems whose $n \times n$ coefficient matrices A are not necessarily Toeplitz (in $O(n^3)$ operations). We show in this paper that the generalised Levinson algorithm is a direct consequence of the generalised Bareiss algorithm, thereby considerably simplifying its presentation in comparison to previous work.

1 Introduction

The solution of linear systems of equations $Ax = b$ with real dense, nonsingular coefficient matrix A is usually accomplished by first determining a triangular factorisation of A [9]. There are two types of triangular factorisations: lower-upper and upper-lower. A lower-upper factorisation is given by $A = L\tilde{U}$, where L is a unit lower triangular and $\tilde{U}$ an upper triangular

*The work presented in this paper was supported by DARPA under contract N00014-88-K-0573.

NATO ASI Series, Vol. F 70
Numerical Linear Algebra, Digital Signal Processing and Parallel Algorithms
Edited by G. H. Golub and P. Van Dooren

matrix, while an upper-lower factorisation provides $A = U\tilde{L}$, where U is unit upper triangular and $\tilde{L}$ lower triangular (a unit triangular matrix is a triangular matrix with ones on the main diagonal). For non-singular A the factorisations are unique. The number of arithmetic operations required to factor a $n \times n$ matrix is $O(n^3)$.

In Section 2 we give a simple, structural view of these triangular factorisations, which is intended to lead to and facilitate the explanation of the generalised Bareiss algorithm in Section 3. The generalised Bareiss algorithm [5] computes a matrix Q and the two triangular matrices $\tilde{U}$ and $\tilde{L}$ such that

$$Q\begin{pmatrix} A \\ A \end{pmatrix} = \begin{pmatrix} \tilde{U} \\ \tilde{L} \end{pmatrix}. \qquad (*)$$

As a product of 2×2 elementary transformations, Q is of the form

$$Q = \begin{pmatrix} L_1 & L_2 \\ U_1 & U_2 \end{pmatrix},$$

where L_1+L_2 is unit lower triangular and U_1+U_2 unit upper triangular. The uniqueness of the factorisations implies $L^{-1} = L_1 + L_2$ and $U_1 + U_2 = U^{-1}$. Thus, the generalised Bareiss algorithm computes explicitly the triangular factors $\tilde{U}$ and $\tilde{L}$ of A. The unit triangular factors L^{-1} and U^{-1} of A^{-1} are obtained implicitly: in factored form as the product Q of 2×2 elementary transformations.

In Section 4 the generalised Bareiss algorithm is scaled to yield a symmetric factorisation for symmetric positive-definite matrices A, the so-called Hyperbolic Cholesky algorithm [4].

Upon multiplication of the components in equation (*) by A^{-1} one obtains

$$Q\begin{pmatrix} I \\ I \end{pmatrix} = \begin{pmatrix} L^{-1} \\ U^{-1} \end{pmatrix},$$

where I is the identity matrix. This is the generalised Levinson algorithm [6] of Section 5; it explicitly computes the matrix Q and the unit triangular factors L^{-1} and U^{-1} of A^{-1}. The two triangular factors of A are available implicitly, they may be determined from $\tilde{U} = L^{-1}A$ and $\tilde{L} = U^{-1}A$.

We also discuss how the matrix Q can be used to accomplish forward elimination and backsubstitution in order to determine the solution x to the linear system $Ax = b$, when A is symmetric positive-definite.

Section 6 concludes the paper with a combination of the Bareiss and Levinson algorithms for the fast parallel solution of linear systems with persymmetric coefficient matrices. For the special case of Toeplitz matrices, it seems to be identical to an algorithm from [8], but we provide a much simpler description.

In each section, we discuss how the algorithms take advantage of the situation where the coefficient matrix A is a Toeplitz matrix. A $n \times n$ matrix T is a *Toeplitz matrix* if its elements are constant along the diagonals:

$$T = \begin{pmatrix} t_0 & t_1 & \dots & t_{n-1} \\ t_{-1} & t_0 & \dots & t_{n-2} \\ \vdots & \vdots & \ddots & \vdots \\ t_{-n+1} & t_{-n+2} & \dots & t_0 \end{pmatrix}$$

Linear systems of equations $Tx = b$ with Toeplitz coefficient matrix T arise, for instance, from time series analysis, Padé approximations, and discretisations of partial differential and integral equations [3]. Because a $n \times n$ Toeplitz matrix contains at most $2n - 1$ *different* elements, Toeplitz linear systems can be solved with $O(n^2)$ or fewer arithmetic operations. In this case, the generalised Bareiss and Levinson algorithms reduce to the classical Schur [2,13] and Levinson [12] algorithms, which perform the solution of $n \times n$ Toeplitz systems in $O(n^2)$ operations.

The purpose of this paper is to provide simple algebraic explanations for the algorithms. We hope that our strategy of first presenting the methods for non-Toeplitz matrices followed by the subsequent restriction to Toeplitz matrices gives a much clearer idea of the underlying structure of the algorithms. The numerical issues associated with the methods, which we ignored in this paper, are discussed in [3,9]. We only consider here the case of real matrices, but the methods generalise readily to the complex case. Reference [9] contains all the relevant facts about matrix theory.

2 Solution of Linear Systems based on Gaussian Elimination

Let A be a real non-singular coefficient matrix and b the right-hand side vector of the system of simultaneous linear equations $Ax = b$. When A is a dense matrix such systems are usually solved by computing a triangular factorisation of A.

2.1 General Matrices

There are essentially two types of triangular factorisations. If all leading principal submatrices of A are non-singular then the factorisation $A = L\tilde{U}$ exists and is unique, where L is a unit lower triangular matrix (i.e. a triangular matrix with ones on the diagonal) and $\tilde{U}$ is an upper triangular matrix with non-zero diagonal elements. The linear system $Ax = b$ can then be solved in three steps

1. Factorisation $A = L\tilde{U}$

2. Forward Elimination $Lc = b$

3. Backsubstitution $\tilde{U}x = c$.

Alternatively, if all trailing principal submatrices of A are non-singular, then the factorisation $A = U\tilde{L}$ exists and is unique, where U is a unit upper triangular matrix and $\tilde{L}$ is a lower triangular matrix with non-zero diagonal elements, so $Ax = b$ may be solved by computing

$$A = U\tilde{L}, \quad Ud = b, \quad \tilde{L}x = d.$$

Applying Gaussian Elimination (without pivoting) to factor the $n \times n$ matrix A requires $\frac{1}{3}n^3 + O(n^2)$ arithmetic operations, and performing the two subsequent steps by triangular system solution requires another $O(n^2)$ operations.

In many applications the matrix A satisfies a stronger condition than the two above: it is symmetric positive-definite (and so are all its contiguous principal submatrices). In this case, $\tilde{U} = D_L L^T$ and $\tilde{L} = D_U U^T$ and the two symmetric factorisations $A = LD_L L^T$ and $A = UD_U U^T$ exist, where D_L and D_U are diagonal matrices with positive elements on the main diagonal. Symmetry saves half of the work in the factorisation step, so only $\frac{1}{6}n^3 + O(n^2)$ arithmetic operations are necessary.

2.2 Toeplitz Matrices

The Levinson algorithm exploits the *persymmetry* of Toeplitz matrices T: $T^T = JTJ$, where J is the permutation matrix, also called 'exchange matrix', with ones on the anti-diagonal [9]. Obviously, if T is persymmetric, so is its inverse.

Persymmetry implies that the triangular factors from the two types of factorisations are permutations of each other. In particular, let $\tilde{U} = D_L U_L$, where D_L is a diagonal matrix and U_L a unit upper triangular matrix, and similarly $\tilde{L} = D_U L_U$, where D_U is diagonal and L_U unit lower triangular. Then

$$T = LD_LU_L = UD_UL_U, \qquad \text{and} \qquad T = JT^TJ = (JL_U^TJ)(JD_UJ)(JU^TJ),$$

implying that $JL_U^TJ = L$, $JD_UJ = D_L$ and $JU^TJ = U_L$.

If T is also symmetric, i.e. $t_i = t_{-i}$ for $1 \leq i \leq n-1$, then

$$JTJ = T = LD_LL^T = UD_UU^T,$$

so $L = JUJ$. In particular, the last row of T is the reverse of its first row, and its last column the reverse of its first column.

The solution methods outlined above are not able to exploit any Toeplitz structure. That is, even though the $n \times n$ matrix T is described by at most $2n-1$ parameters, the operation count for the linear system solution is still $O(n^3)$. In order to explain algorithms that solve Toeplitz systems in $O(n^2)$ arithmetic operations, it is helpful to first consider how Gaussian elimination changes the zero structure of a general matrix during the factorisation process.

2.3 Structural View of the Factorisations

Gaussian elimination (without pivoting) accomplishes the factorisation $A = L\tilde{U}$ by computing successive columns of the unit lower triangular matrix L and successive rows of the upper triangular matrix $\tilde{U}$ by zeroing out successive columns in A. The zero-structure of the matrices occuring in the evolution from the original matrix A to the final upper triangular $\tilde{U}$ are depicted in the 4×4 example of Figure 1. There, x represents an element which is generally non-zero, u an element of the final matrix $\tilde{U}$, a blank

$$A = \begin{bmatrix} u & u & u & u \\ \otimes & x & x & x \\ \otimes & x & x & x \\ \otimes & x & x & x \end{bmatrix} \rightarrow \begin{bmatrix} u & u & u & u \\ \hline & u & u & u \\ & \otimes & x & x \\ & \otimes & x & x \end{bmatrix} \rightarrow \begin{bmatrix} u & u & u & u \\ & u & u & u \\ \hline & & u & u \\ & & \otimes & x \end{bmatrix} \rightarrow \begin{bmatrix} u & u & u & u \\ & u & u & u \\ & & u & u \\ & & & u \end{bmatrix} = \tilde{U}$$

Figure 1: Evolving Non-Zero Structure of a 4×4 Matrix During the Computation of $A = L\tilde{U}$.

a zero element and $\otimes$ an element doomed for elimination in the current step. The partitioning distinguishes those elements in the upper part of the matrix that have ceased to participate in computation. In each step, an appropriate multiple α of the latest row of $\tilde{U}$, which is the one just underneath the horizontal partitioning line, is subtracted from each row beneath it in order to remove the elements in the next column. So, all operations are of the form

$$\text{lower row} = \text{lower row} - \alpha * \text{latest row of } \tilde{U}.$$

In this process no previously introduced zeros are destroyed, because the latest row of $\tilde{U}$ has the same zero structure as all the lower rows to which it is added. Note that in each step all the rows can be modified simultaneously and independently.

The computation of the factorisation $A = U\tilde{L}$ proceeds in a similar manner.

3 The Generalised Bareiss Algorithm

In [2] Bareiss proposed an algorithm for solving square, non-symmetric Toeplitz systems. The process of factoring the matrix in Bareiss algorithm is identical to the Schur algorithm, which is based on [13]; in addition, Bareiss realised that forward elimination could also be accomplished by recursions similar to those used for the factorisation. In [5] it was shown how to generalise this algorithm to non-Toeplitz matrices. In contrast to Gaussian elimination, which computes either $A = L\tilde{U}$ or $A = U\tilde{L}$, the generalised Bareiss algorithm computes factors of both factorisations $A =$

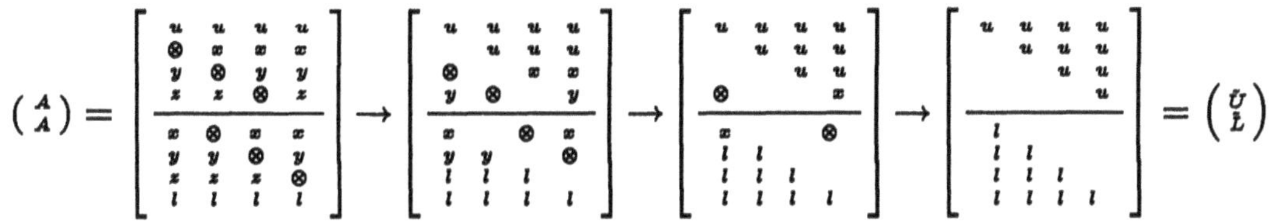

Figure 2: Evolving Non-Zero Structure of a 4 × 4 Matrix During the Generalised Bareiss Algorithm.

$L\tilde{U}$ and $A = U\tilde{L}$ by working with two copies of the matrix A:

$$Q\begin{pmatrix} A \\ A \end{pmatrix} = \begin{pmatrix} \tilde{U} \\ \tilde{L} \end{pmatrix} = \begin{bmatrix} \cdot & \cdots & \cdot \\ & \ddots & \vdots \\ & & \cdot \\ \cdot & & \\ \vdots & \ddots & \\ \cdot & \cdots & \cdot \end{bmatrix}.$$

The next section explains how to reduce two copies of the original matrix A to an upper triangular matrix and a lower triangular matrix, and the one following argues that these triangular matrices are (in exact arithmetic) identical to the ones obtained from Gaussian elimination.

3.1 Structural View of the Factorisation

In contrast to Gaussian elimination, which eliminates columns, the generalised Bareiss algorithm eliminates diagonals, which turns out to be crucial for the exploitation of Toeplitz properties. This is shown in the 4 × 4 example of Figure 2; there an element of the final matrix $\tilde{L}$ is denoted by l, and the letters x, y, z denote elements that are generally non-zero. The upper copy of A is transformed to upper triangular form and the lower copy to lower triangular form. This is accomplished by 'rotating' in step k rows $i+k$ in the upper matrix with rows i in the lower matrix (in Figure 2 those rows are made up of identical letters x, y or z) so as to eliminate one element in each of them, namely $(i+k,i)$ and $(i,i+k)$:

$$\begin{pmatrix} \text{row } i+k \text{ in upper matrix} \\ \text{row } i \text{ in lower matrix} \end{pmatrix} = \begin{pmatrix} 1 & -\alpha_{i,i+k} \\ -\beta_{i+k,i} & 1 \end{pmatrix} \begin{pmatrix} \text{row } i+k \text{ in upper matrix} \\ \text{row } i \text{ in lower matrix} \end{pmatrix}$$

The two rows are selected to have the same zero structure, so already introduced zeros are preserved. Note that in each step all pairs of rows can be modified simultaneously and independently. The transformation matrix Q is thus made up as a product of these 2×2 'rotations'.

We can now express the generalised Bareiss algorithm formally. As a matter of simplicity, we just refer to entire rows of the matrix rather than individual elements and do not show how one can save operations by taking advantage of the increasing number of zero entries in the matrix. Below, the vectors $a_i^{(0)}$ are initialised to be the ith row of the matrix A. From now on, quantities carrying a positive superscript will be associated with the upper matrix, and those with a negative superscript will be associated with the lower matrix. In the kth step, the kth subdiagonal of the upper matrix and the kth superdiagonal of the lower matrix are removed; that is, elements $(i+k,i)$ in the upper matrix and elements $(i,i+k)$ in the lower matrix are removed by appropriately combining the $(i+k)$th row $a_{i+k}^{(k-1)}$ of the upper matrix with the ith row $a_i^{(-k+1)}$ of the lower matrix.

Factorisation in the Generalised Bareiss Algorithm

$$\begin{array}{ll} 1 \le i \le n, & a_i^{(0)} = a_i \\ 1 \le k \le n-1, & 1 \le i \le n-k, \\ & \alpha_{i,i+k} = a_{i+k,i}^{(k-1)} / a_{i,i}^{(-k+1)}, \quad \beta_{i+k,i} = a_{i,i+k}^{(-k+1)} / a_{i+k,i+k}^{(k-1)} \\ & \begin{pmatrix} a_{i+k}^{(k)} \\ a_i^{(-k)} \end{pmatrix} = \begin{pmatrix} 1 & -\alpha_{i,i+k} \\ -\beta_{i+k,i} & 1 \end{pmatrix} \begin{pmatrix} a_{i+k}^{(k-1)} \\ a_i^{(-k+1)} \end{pmatrix}. \end{array}$$

Now the kth rows of $\tilde{U}$ and $\tilde{L}$ are respectively given by $a_k^{(k-1)}$ and $a_k^{(-n+k)}$.

3.2 Correctness of the Factorisation

It remains to explain why indeed the generalised Bareiss algorithm computes triangular factorisations of the matrix A.

As illustrated in the previous section, the transformation matrix Q that reduces the two copies of a A to triangular matrices consists of products of 2×2 'rotations', and 'rotations' belonging to the same step are disjoint. The matrix Q for the 4×4 example in Figure 2 is of the form

$$Q = \left[\begin{array}{cccc|cccc} 1 &&&&&&& \\ & 1 &&&&&& \\ && 1 &&&&& \\ &&& 1 & x &&& \\ \hline &&& x & 1 &&& \\ &&&&& 1 && \\ &&&&&& 1 & \\ &&&&&&& 1 \end{array}\right] \left[\begin{array}{cccc|cccc} 1 &&&&&&& \\ & 1 &&&&&& \\ && 1 && x &&& \\ &&& 1 && x && \\ \hline && x && 1 &&& \\ &&& x && 1 && \\ &&&&&& 1 & \\ &&&&&&& 1 \end{array}\right] \left[\begin{array}{cccc|cccc} 1 &&&&&&& \\ & 1 &&& x &&& \\ && 1 &&& x && \\ &&& 1 &&& x & \\ \hline & x &&& 1 &&& \\ && x &&& 1 && \\ &&& x &&& 1 & \\ &&&&&&& 1 \end{array}\right]$$

The right-most of the three matrices eliminates the first pair of off-diagonals. The x's in the first subdiagonal of its top right part represent the multipliers for removing the first subdiagonal in the upper matrix; similarly the x's in the first superdiagonal of the bottom left part are the multipliers for removing the first superdiagonal in the lower matrix.

Because subdiagonals are removed in the upper matrix and superdiagonals are removed in the lower matrix, the transformation matrices applied at each step consist of two lower triangular matrices in the top half and two upper triangular matrices in the bottom half of the matrix. Due to the particular zero structure of these triangular matrices the product Q of these transformation matrices is shown in [5] to have the form

$$Q = \left[\begin{array}{cccc|cccc} 1 &&&&&&& \\ \vdots & \ddots &&& \cdot &&& \\ \vdots && \ddots && \vdots & \ddots && \\ \cdot & \cdots & \cdots & 1 & \cdot & \cdots & \cdot & \\ \hline & \cdot & \cdots & \cdot & 1 & \cdots & \cdots & \cdot \\ && \ddots & \vdots && \ddots && \vdots \\ &&& \cdot &&& \ddots & \vdots \\ &&&&&&& 1 \end{array}\right] = \begin{pmatrix} L_1 & L_2 \\ U_1 & U_2 \end{pmatrix}.$$

If A is a $n \times n$ matrix, then the $2n \times 2n$ matrix Q consists of four $n \times n$ triangular matrices, whereby L_1 is unit lower triangular, L_2 is strictly lower triangular, U_1 is strictly upper triangular, and U_2 is unit upper triangular. But $L_1 + L_2$ is a unit lower triangular matrix, and so is its inverse; similarly

$U_1 + U_2$ and its inverse are unit upper triangular matrices. Moreover, in

$$Q\begin{pmatrix} A \\ A \end{pmatrix} = \begin{pmatrix} (L_1 + L_2)A \\ (U_1 + U_2)A \end{pmatrix}$$

$(L_1 + L_2)A$ is upper triangular, and $(U_1 + U_2)A$ is lower triangular. Due to the uniqueness of the triangular factorisations we must then have $L = (L_1 + L_2)^{-1}$, $U = (U_1 + U_2)^{-1}$, $(L_1 + L_2)A = \tilde{U}$ and $(U_1 + U_2)A = \tilde{L}$. Consequently,

$$Q\begin{pmatrix} A \\ A \end{pmatrix} = \begin{pmatrix} L^{-1}A \\ U^{-1}A \end{pmatrix} = \begin{pmatrix} \tilde{U} \\ \tilde{L} \end{pmatrix}.$$

Thus, the generalised Bareiss determines explicitly the triangular factors $\tilde{U}$ and $\tilde{L}$ of A and implicitly (as the product of the 'rotations' making up Q) the unit triangular factors L^{-1} and U^{-1} of A^{-1}.

In [5] it is proved that the generalised Bareiss algorithm does not break down if all *contiguous* principal submatrices of A are non-singular. This is a more stringent condition than needed for the independent computations of $A = L\tilde{U}$ and $A = U\tilde{L}$ by Gaussian elimination, which only require all *leading* or all *trailing* principal submatrices of A to be non-singular.

3.3 Forward Elimination and Backsubstitution

Instead of solving a triangular system the transformation matrix Q may be used to perform forward elimination, as Bareiss [2] already realised in the Toeplitz case. From the last equation in the previous section we see that the application of Q amounts to a premultiplication by L^{-1} and U^{-1}, thus Q can also be applied to the right-hand side b in order to effect forward elimination

$$Q\begin{pmatrix} b \\ b \end{pmatrix} = \begin{pmatrix} L^{-1}b \\ U^{-1}b \end{pmatrix} = \begin{pmatrix} c \\ d \end{pmatrix}.$$

Backsubstitution can now be performed by solving either $\tilde{U}x = c$ or $\tilde{L}x = d$ in $\frac{n^2}{2} + O(n)$ arithmetic operations. Employing a trick from Bareiss [2], one could alternatively determine the upper half of x from $\tilde{U}x = c$ and the lower half from $\tilde{L}x = d$, which would require only $\frac{n^2}{4} + O(n)$ operations.

If A is symmetric positive-definite, the application of Q^T can replace the triangular system solution for backsubstitution. To this end, let I denote the identity matrix of the same size as A, and

$$\begin{aligned} y &\equiv (I \quad I)\,Q^T \begin{pmatrix} D_L^{-1} & \\ & D_U^{-1} \end{pmatrix} \begin{pmatrix} c \\ d \end{pmatrix} \\ &= (L_1 + L_2)^T D_L^{-1} c + (U_1 + U_2)^T D_U^{-1} d \\ &= L^{-T} D_L^{-1} c + U^{-T} D_U^{-1} d \end{aligned}$$

because $c = L^{-1}b$ and $d = U^{-1}b$, so

$$\begin{aligned} y &= L^{-T} D_L^{-1} L^{-1} b + U^{-T} D_U^{-1} U^{-1} b \\ &= A^{-1}b + A^{-1}b = 2A^{-1}b = 2x. \end{aligned}$$

In order to obtain x, rather than $2x$, observe that c and d each give rise to one x, and applying the transformations instead to $(\,\lambda c \quad (1-\lambda)d\,)^T$, where λ is any real number, yields

$$\begin{aligned} y &= (I \quad I)\,Q^T \begin{pmatrix} D_L^{-1} & \\ & D_U^{-1} \end{pmatrix} \begin{pmatrix} \lambda c \\ (1-\lambda)d \end{pmatrix} \\ &= \lambda A^{-1}b + (1-\lambda)A^{-1}b = \lambda x + (1-\lambda)x = x. \end{aligned}$$

In particular, λ may be set to either zero or one so as to require only one of the right-hand sides c or d as input.

For symmetric positive-definite matrices A the generalised Bareiss algorithm permits the solution of the linear system $Ax = b$ with the following three steps:

1. Determine Q such that $Q\begin{pmatrix} A \\ A \end{pmatrix} = \begin{pmatrix} \tilde{U} \\ \tilde{L} \end{pmatrix}$

2. Determine $\begin{pmatrix} c \\ d \end{pmatrix} = Q\begin{pmatrix} b \\ b \end{pmatrix}$

3. Determine $x = (I \quad I)\,Q^T \begin{pmatrix} D_L^{-1} & \\ & D_U^{-1} \end{pmatrix} \begin{pmatrix} \lambda c \\ (1-\lambda)d \end{pmatrix}.$

Because the generalised Bareiss algorithm determines a non-symmetric factorisation $\alpha_{i,i+k} \neq \beta_{i+k,i}$, and it can *generally* not take advantage of the symmetry of a matrix in order to reduce the number of arithmetic operations – except if the matrix is a symmetric Toeplitz matrix: then the operation count decreases by fifty percent compared to the non-symmetric Toeplitz case, as shown in the next section.

$$
\left[\begin{array}{cccc}
t_0 & t_1 & t_2 & t_3 \\
t_{-1} & t_0 & t_1 & t_2 \\
t_{-2} & t_{-1} & t_0 & t_1 \\
t_{-3} & t_{-2} & t_{-1} & t_0 \\
\hline
s_0 & s_1 & s_2 & s_3 \\
s_{-1} & s_0 & s_1 & s_2 \\
s_{-2} & s_{-1} & s_0 & s_1 \\
s_{-3} & s_{-2} & s_{-1} & s_0
\end{array}\right]
\rightarrow
\left[\begin{array}{cccc}
t_0 & t_1 & t_2 & t_3 \\
 & t'_0 & t'_1 & t'_2 \\
t'_{-2} & & t'_0 & t'_1 \\
t'_{-3} & t'_{-2} & & t'_0 \\
\hline
s'_0 & & s'_2 & s'_3 \\
s'_{-1} & s'_0 & & s'_2 \\
s'_{-2} & s'_{-1} & s'_0 & \\
s_{-3} & s_{-2} & s_{-1} & s_0
\end{array}\right]
\rightarrow
\left[\begin{array}{cccc}
t_0 & t_1 & t_2 & t_3 \\
 & t'_0 & t'_1 & t'_2 \\
 & & t''_0 & t''_1 \\
t''_{-3} & & & t''_0 \\
\hline
s''_0 & & & s''_3 \\
s''_{-1} & s''_0 & & \\
s'_{-2} & s'_{-1} & s'_0 & \\
s_{-3} & s_{-2} & s_{-1} & s_0
\end{array}\right]
$$

$$
\rightarrow
\left[\begin{array}{cccc}
t_0 & t_1 & t_2 & t_3 \\
 & t'_0 & t'_1 & t'_2 \\
 & & t''_0 & t''_1 \\
 & & & t'''_0 \\
\hline
s'''_0 & & & \\
s''_{-1} & s''_0 & & \\
s'_{-2} & s'_{-1} & s'_0 & \\
s_{-3} & s_{-2} & s_{-1} & s_0
\end{array}\right]
$$

Figure 3: Evolving Structure of a 4×4 Toeplitz Matrix During Bareiss Algorithm; Initially $s_i = t_i$.

3.4 Specialisation to Toeplitz Matrices

If the matrix T to be factored is a Toeplitz matrix, then the generalised Bareiss algorithm is identical to the original Bareiss algorithm [2], which exploits the structure of a Toeplitz matrix. Figure 3 displays the evolution of the structure of the 4×4 Toeplitz matrix

$$
T = \begin{pmatrix}
t_0 & t_1 & t_2 & t_3 \\
t_{-1} & t_0 & t_1 & t_2 \\
t_{-2} & t_{-1} & t_0 & t_1 \\
t_{-3} & t_{-2} & t_{-1} & t_0
\end{pmatrix}
$$

during the course of Bareiss algorithm. Although the triangular factors of a Toeplitz matrix are generally not Toeplitz, the 'working part' of the matrix in Bareiss algorithm is Toeplitz as shown in Figure 3 and below, because identical elements along each diagonal result in identical $\alpha_{i+k,i}$ and $\beta_{i,i+k}$ for each step; this makes it possible to exploit the Toeplitz structure.

In particular, denote the elements of T by $t_{ij} \equiv t_{j-i}$, $1 \leq i, j \leq n$. Rows a_1 and a_n constitute the respective first row of $\tilde{U}$ and last row of $\tilde{L}$. Since T is Toeplitz, at the beginning of the first step all the elements necessary for further computation are contained in rows a_2 and a_n of the upper matrix and rows a_1 and a_{n-1} of the lower matrix. The multipliers are

$$\alpha_{i,i+1} = \frac{t_{i+1,i}}{t_{i,i}} = \frac{t_1}{t_0} = \alpha_1, \qquad \beta_{i+1,i} = \frac{t_{i,i+1}}{t_{i+1,i+1}} = \frac{t_{-1}}{t_0} = \beta_1, \qquad 1 \le i \le n-1.$$

These multipliers are used in the rotations, as above, to construct rows $a_{i+1}^{(1)}$ of the upper matrix and rows $a_i^{(-1)}$ of the lower matrix, $1 \le i \le n-1$. Hence rows $3, \ldots, n$ of the upper matrix and rows $1, \ldots, n-2$ of the lower matrix retain their Toeplitz structure, and rows $a_3^{(1)}$, $a_n^{(1)}$, $a_{n-2}^{(-1)}$ and $a_1^{(-1)}$ encompass all necessary information for further computation, see Figure 3.

In general at the beginning of step k, rows $k+1, \ldots, n$ of the upper matrix as well as rows $1, \ldots, n-k$ of the lower matrix have Toeplitz structure. If we represent the upper and lower triangular parts of the two matrices by separate vectors

$$\begin{aligned}
b_{k+1}^{(k-1)} &= (0 \;\ldots\; 0 \;\; a_{k+1,k+1}^{(k-1)} \;\ldots\; a_{k+1,n}^{(k-1)}), \\
b_n^{(k-1)} &= (a_{n,1}^{(k-1)} \;\ldots\; a_{n,n-k}^{(k-1)} \;\; 0 \;\ldots\; 0), \\
b_1^{(-k+1)} &= (0 \;\ldots\; 0 \;\; a_{1,k+1}^{(-k+1)} \;\ldots\; a_{1,n}^{(-k+1)}), \\
b_{n-k}^{(-k+1)} &= (a_{n-k,1}^{(-k+1)} \;\ldots\; a_{n-k,n-k}^{(-k+1)} \;\; 0 \;\ldots\; 0),
\end{aligned}$$

then $b_{k+1}^{(k-1)}$ and $b_n^{(k-1)}$ contain all the distinct elements in the respective upper and lower triangular parts of the upper matrix, while $b_{n-k}^{(-k+1)}$ and $b_1^{(-k+1)}$ contain all distinct elements in the respective lower and upper triangular parts of the lower matrix, see Figure 3. Thus, the following two 'rotations' embody all the distinct computations on the Toeplitz matrix during step k:

$$\begin{pmatrix} b_{k+1}^{(k)} \\ b_1^{(-k)} \end{pmatrix} = \begin{pmatrix} 1 & -\alpha_k \\ -\beta_k & 1 \end{pmatrix} \begin{pmatrix} b_{k+1}^{(k-1)} \\ b_1^{(-k+1)} \end{pmatrix}, \quad \begin{pmatrix} b_n^{(k)} \\ b_{n-k}^{(-k)} \end{pmatrix} = \begin{pmatrix} 1 & -\alpha_k \\ -\beta_k & 1 \end{pmatrix} \begin{pmatrix} b_n^{(k-1)} \\ b_{n-k}^{(-k+1)} \end{pmatrix}.$$

The transition to the next step is accomplished by considering the next lower row in the upper matrix and the next higher row in the lower matrix:

$$b_{k+2}^{(k)} = b_{k+1}^{(k)} Z, \quad b_{n-k+1}^{(k)} = b_{n-k}^{(k)} Z^T, \qquad \text{where} \quad Z = \begin{pmatrix} 0 & 1 & & \\ & \ddots & \ddots & \\ & & 0 & 1 \\ & & & 0 \end{pmatrix}.$$

This leads to the following form of the factorisation process in Bareiss Algorithm, which is also called 'Schur algorithm' [13]. Below, 0_k denotes a row vector consisting of k zeros, and the rows $b_i^{(j)}$ are represented by their individual elements.

Schur Algorithm: Factorisation Part of Bareiss Algorithm

$$(\,t_0^{(0)} \;\; t_1^{(0)} \;\; \ldots \;\; t_{n-1}^{(0)}\,) = (\,t_0 \;\; t_1 \;\; \ldots \;\; t_{n-1}\,)$$

$$(\,t_{-n+1}^{(0)} \;\; \ldots \;\; t_{-1}^{(0)} \;\; t_0^{(0)}\,) = (\,t_{-n+1} \;\; \ldots \;\; t_{-1} \;\; t_0\,)$$

$$1 \le k \le n-1, \qquad \alpha_k = t_{-k}^{(k-1)}/t_0^{(-k+1)}, \quad \beta_k = t_k^{(-k+1)}/t_0^{(k-1)}$$

$$\begin{pmatrix} 0_k & t_0^{(k)} & t_1^{(k)} & \ldots & t_{n-k-1}^{(k)} \\ 0_k & 0 & t_{k+1}^{(-k)} & \ldots & t_{n-1}^{(-k)} \end{pmatrix} = \begin{pmatrix} 1 & -\alpha_k \\ -\beta_k & 1 \end{pmatrix} \begin{pmatrix} 0_k & t_0^{(k-1)} & t_1^{(k-1)} & \ldots & t_{n-k-1}^{(k-1)} \\ 0_k & t_k^{(-k+1)} & t_{k+1}^{(-k+1)} & \ldots & t_{n-1}^{(-k+1)} \end{pmatrix}$$

$$\begin{pmatrix} t_{-n+1}^{(k)} & \ldots & t_{-k-1}^{(k)} & 0 & 0_k \\ t_{-n+k+1}^{(-k)} & \ldots & t_{-1}^{(-k)} & t_0^{(-k)} & 0_k \end{pmatrix} = \begin{pmatrix} 1 & -\alpha_k \\ -\beta_k & 1 \end{pmatrix} \begin{pmatrix} t_{-n+1}^{(k-1)} & \ldots & t_{-k-1}^{(k-1)} & t_{-k}^{(k-1)} & 0_k \\ t_{-n+k+1}^{(-k+1)} & \ldots & t_{-1}^{(-k+1)} & t_0^{(-k+1)} & 0_k \end{pmatrix}.$$

Now the kth rows of $\tilde{U}$ and $\tilde{L}$ are respectively given by

$$(\,0_k \;\; t_0^{(k)} \;\; t_1^{(k)} \;\; \ldots \;\; t_{n-k-1}^{(k)}\,) \text{ and } (\,t_{-n+k+1}^{(-k)} \;\; \ldots \;\; t_{-1}^{(-k)} \;\; t_0^{(-k)} \;\; 0_k\,).$$

The recursions for forward elimination and backsubstitution are similar to those in the factorisation; see also [4].

If the Toeplitz matrix T is symmetric positive-definite, then $\alpha_1 = \beta_1$ and, as explained in Section 2, the last row of the Toeplitz matrix is the reverse of the first one, so its computation may be omitted, hence reducing the operation count by fifty percent. The Schur algorithm for this case follows:

Schur Algorithm for Symmetric Positive-Definite Matrices

$$\begin{pmatrix} t_0^{(0)} & t_1^{(0)} & \dots & t_{n-1}^{(0)} \end{pmatrix} = \begin{pmatrix} t_0 & t_1 & \dots & t_{n-1} \end{pmatrix}$$

$$1 \le k \le n-1, \qquad \rho_k = t_k^{(-k+1)}/t_0^{(k-1)},$$

$$\begin{pmatrix} t_0^{(k)} & t_1^{(k)} & \dots & t_{n-k-1}^{(k)} \\ 0 & t_{k+1}^{(-k)} & \dots & t_{n-1}^{(-k)} \end{pmatrix} = \begin{pmatrix} 1 & -\rho_k \\ -\rho_k & 1 \end{pmatrix} \begin{pmatrix} t_0^{(k-1)} & t_1^{(k-1)} & \dots & t_{n-k-1}^{(k-1)} \\ t_k^{(-k+1)} & t_{k+1}^{(-k+1)} & \dots & t_{n-1}^{(-k+1)} \end{pmatrix}$$

4 The Hyperbolic Cholesky Algorithm

The hyperbolic Cholesky algorithm [4] (see also [1,7]) computes both Cholesky factorisations $A = \mathcal{U}\mathcal{U}^T$ and $A = \mathcal{L}\mathcal{L}^T$, where $\mathcal{U}^T = D_U^{1/2} U^T$ and $\mathcal{L}^T = D_L^{1/2} L^T$, of the symmetric positive-definite matrix A. If D is a diagonal matrix whose diagonal equals the diagonal of A, then

$$Q_H \begin{pmatrix} D^{-1/2} & \\ & D^{-1/2} \end{pmatrix} \begin{pmatrix} A \\ A \end{pmatrix} = \begin{pmatrix} \mathcal{L}^T \\ \mathcal{U}^T \end{pmatrix} = \begin{pmatrix} D_L^{1/2} L^T \\ D_U^{1/2} U^T \end{pmatrix}.$$

4.1 Connection to the Generalised Bareiss Algorithm

Premultiplying both sides of the equation by $\begin{pmatrix} D_L^{1/2} & \\ & D_U^{1/2} \end{pmatrix}$ gives, due to the uniqueness of the factorisation, the following relations between the quantities computed by the Bareiss and Hyperbolic Cholesky algorithms:

$$\begin{pmatrix} D_L^{1/2} & \\ & D_U^{1/2} \end{pmatrix} Q_H \begin{pmatrix} D^{-1/2} & \\ & D^{-1/2} \end{pmatrix} \begin{pmatrix} A \\ A \end{pmatrix} = \begin{pmatrix} D_L L^T \\ D_U U^T \end{pmatrix} = \begin{pmatrix} \tilde{U} \\ \tilde{L} \end{pmatrix}$$

and

$$\begin{pmatrix} D_L^{1/2} & \\ & D_U^{1/2} \end{pmatrix} Q_H \begin{pmatrix} D^{-1/2} & \\ & D^{-1/2} \end{pmatrix} = Q.$$

Thus, as also shown in [5], the Hyperbolic Cholesky algorithm is a 'scaled' version of the generalised Bareiss algorithm.

Instead of the 'rotations' of the generalised Bareiss algorithm, the Hyperbolic Cholesky algorithm uses hyperbolic rotations to eliminate a pair of elements. In particular, if the generalised Bareiss algorithm removes element $(i+k,i)$ in the upper matrix and element $(i,i+k)$ in the lower matrix by applying

$$\begin{pmatrix} 1 & -\alpha_{i,i+k} \\ -\beta_{i+k,i} & 1 \end{pmatrix}$$

to rows $i+k$ and i, then the Hyperbolic Cholesky algorithm removes elements in the same positions via the hyperbolic rotation

$$\frac{1}{\sqrt{1-\rho_{i+k,i}^2}} \begin{pmatrix} 1 & -\rho_{i+k,i} \\ -\rho_{i+k,i} & 1 \end{pmatrix}$$

where $\rho_{i+k,i} = \sqrt{\alpha_{i,i+k}\beta_{i+k,i}}$. This can easily be seen in a 2×2 example, where

$$A = \begin{pmatrix} a & b \\ b & c \end{pmatrix}.$$

The generalised Bareiss algorithm computes

$$\left(\begin{array}{cc|cc} 1 & & & \\ & 1 & -\alpha & \\ \hline & -\beta & 1 & \\ & & & 1 \end{array}\right) \left(\begin{array}{cc} a & b \\ b & c \\ \hline a & b \\ b & c \end{array}\right) = \left(\begin{array}{cc} a & b \\ 0 & c - b^2/a \\ \hline a - b^2/c & 0 \\ b & c \end{array}\right),$$

where $\alpha = b/a$ and $\beta = b/c$. The Hyperbolic Cholesky algorithm computes

$$\frac{1}{\sqrt{1-\rho^2}} \left(\begin{array}{cc|cc} 1 & & & \\ & 1 & -\rho & \\ \hline & -\rho & 1 & \\ & & & 1 \end{array}\right) \left(\begin{array}{cc} \sqrt{a} & b/\sqrt{a} \\ b/\sqrt{c} & \sqrt{c} \\ \hline \sqrt{a} & b/\sqrt{a} \\ b/\sqrt{c} & \sqrt{c} \end{array}\right) = \left(\begin{array}{cc} \sqrt{a} & b/\sqrt{a} \\ 0 & \sqrt{c - b^2/a} \\ \hline \sqrt{a - b^2/c} & 0 \\ b/\sqrt{c} & \sqrt{c} \end{array}\right),$$

where $\rho = b/\sqrt{ac}$. Thus, $\rho = \sqrt{\alpha\beta}$.

The process of forward elimination and backsubstitution proceeds as in the generalised Bareiss but with Q replaced by

$$Q_H \begin{pmatrix} D^{-1/2} & \\ & D^{-1/2} \end{pmatrix}.$$

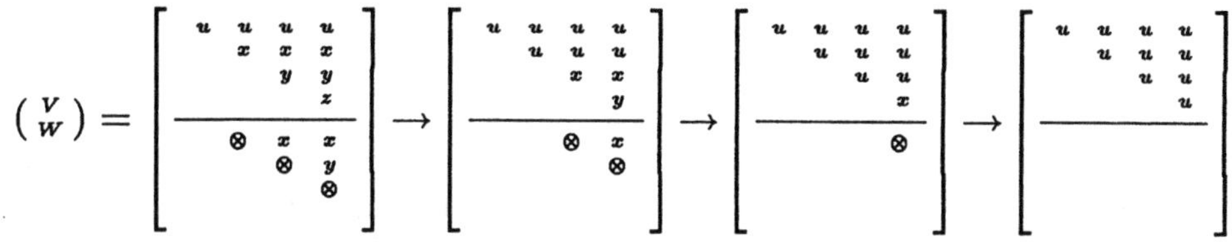

(a) Removing Successive Diagonals.

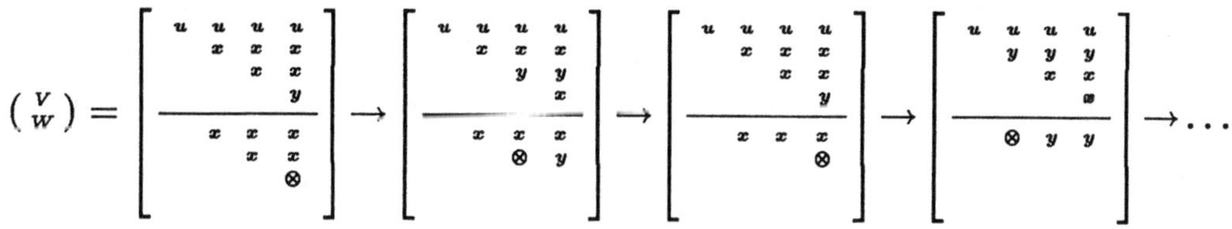

(b) Removing Successive Rows.

Figure 4: Two Different Orders of Rotations in the Hyperbolic Cholesky Algorithm Applied to a 4×4 Matrix.

As a consequence, the Hyperbolic Cholesky algorithm gives rise to a scaled Schur algorithm based on rotations

$$\frac{1}{\sqrt{1-\rho_k^2}} \begin{pmatrix} 1 & -\rho_k \\ -\rho_k & 1 \end{pmatrix}.$$

4.2 View of Hyperbolic Cholesky Factorisation as a Downdating Process

Unlike the generalised Bareiss algorithm the Hyperbolic Cholesky algorithm can take advantage of the symmetry of non-Toeplitz matrices. If only one Cholesky factor is desired, it suffices to compute

$$Q_H \begin{pmatrix} D^{-1/2} & \\ & D^{-1/2} \end{pmatrix} \begin{pmatrix} A_+ \\ A_\# \end{pmatrix} = \begin{pmatrix} \mathcal{L}^T \\ 0 \end{pmatrix},$$

where A_+ consists of the upper triangle of A including the diagonal, and $A_\#$ consists of the strict upper triangle of A excluding the diagonal. As a product of hyperbolic rotations, the matrix Q_H is pseudo-orthogonal, that is,

$$Q_H^T \begin{pmatrix} I & \\ & -I \end{pmatrix} Q_H = \begin{pmatrix} I & \\ & -I \end{pmatrix}.$$

Multiplying the previous equation by its transpose and exploiting the pseudo-orthogonality of Q_H yields $\mathcal{L}\mathcal{L}^T = VV^T - WW^T$, where $V = D^{-1/2}A_+$ is an upper triangular matrix with positive diagonal elements, and $W = D^{-1/2}A_\#$ is a strictly upper triangular matrix. Denote by l_i the columns of W, so $\mathcal{L}\mathcal{L}^T = VV^T - \sum_i l_i l_i^T$; and $V^{(0)} \equiv V$. We can view the Hyperbolic Cholesky algorithm as computing a sequence of Cholesky downdating problems: each problem consists of premultiplying $\begin{pmatrix} V^{(i)} \\ l_i^T \end{pmatrix}$ by hyperbolic rotations resulting in $\begin{pmatrix} V^{(i+1)} \\ 0 \end{pmatrix}$ and

$$V^{(i+1)}V^{(i+1)^T} = V^{(i)}V^{(i)^T} - l_i l_i^T$$

such that $V^{(i+1)}$ is an upper triangular matrix with positive diagonal elements, see for example [11]. The Hyperbolic Cholesky algorithm computes the rotations in the same order as the generalised Bareiss algorithm. This is also the order that a downdating process would choose, as Figure 4 illustrates. Instead of removing successive diagonals, as shown in Figure 4(a), one can also pipeline the rotations so as to remove successive rows; see Figure 4(b).

Reducing the Hyperbolic Cholesky algorithm to a sequence of downdating problems may facilitate its round-off error analysis due to the availability of round-off error analyses for downdating problems.

5 The Generalised Levinson Algorithm

The first explicit algorithm for solving $n \times n$ Toeplitz systems with $O(n^2)$ operations was introduced by Levinson in 1947 [12]. More recently, Levinson's algorithm was extended to non-Toeplitz matrices by Delsarte, Genin and Kamp [6], just as Bareiss algorithm [2] was extended to the generalised Bareiss algorithm [5]. In previous work [10] we have shown that the Schur algorithm can be derived as the result of trying to increase the degree of parallelism in the Levinson algorithm (i.e. by getting rid of the inner products). Conversely, we will now show that the generalised Levinson algorithm is a simple consequence of the generalised Bareiss algorithm. Our version of the generalised Levinson algorithm is simpler and more intuitive than the one in [6] as it does without the exchange matrix J, and it establishes a direct connection to the other factorisation methods.

5.1 From the Generalised Bareiss to the Generalised Levinson Algorithm

The generalised Bareiss algorithm computes

$$Q\begin{pmatrix} A \\ A \end{pmatrix} = \begin{pmatrix} \tilde{U} \\ \tilde{L} \end{pmatrix},$$

where $A = L\tilde{U}$ and $A = U\tilde{L}$ are the triangular factorisations of a non-singular matrix A.

Postmultiplying both sides of the equation by $\begin{pmatrix} A^{-1} & \\ & A^{-1} \end{pmatrix}$ gives

$$Q\begin{pmatrix} I \\ I \end{pmatrix} = \begin{pmatrix} \tilde{U}A^{-1} \\ \tilde{L}A^{-1} \end{pmatrix} = \begin{pmatrix} L^{-1} \\ U^{-1} \end{pmatrix}$$

since $A^{-1} = \tilde{U}^{-1}L^{-1}$ and $A^{-1} = \tilde{L}^{-1}U^{-1}$. Thus, by applying the matrix Q to the identity matrix the generalised Levinson algorithm determines explicitly the unit triangular factors L^{-1} and U^{-1} of A^{-1} and implicitly (by forming $\tilde{U} = L^{-1}A$ and $\tilde{L} = U^{-1}A$) the non-unit triangular factors of A.

Now we consider how the elements of Q are actually determined by the generalised Levinson algorithm. If initially $a_i^{(0)} = a_i$, $1 \leq i \leq n$, where a_i is the ith row of A then the kth step of the generalised Bareiss algorithm combines the $(i+k)$th row $a_{i+k}^{(k-1)}$ of the upper matrix with the ith row $a_i^{(-k+1)}$ of the lower matrix:

$$\begin{pmatrix} a_{i+k}^{(k)} \\ a_i^{(-k)} \end{pmatrix} = \begin{pmatrix} 1 & -\alpha_{i,i+k} \\ -\beta_{i+k,i} & 1 \end{pmatrix} \begin{pmatrix} a_{i+k}^{(k-1)} \\ a_i^{(-k+1)} \end{pmatrix}, \qquad 1 \leq i \leq n-k,$$

where

$$\alpha_{i,i+k} = a^{(k-1)}_{i+k,i} / a^{(-k+1)}_{i,i}, \qquad \beta_{i+k,i} = a^{(-k+1)}_{i,i+k} / a^{(k-1)}_{i+k,i+k}$$

to remove element $(i+k,i)$ in the upper matrix and element $(i,i+k)$ in the lower matrix. Eventually, $a_k^{(k-1)}$ is the kth row of $\tilde{U}$ and $a_k^{(-n+k)}$ is the kth row of $\tilde{L}$.

This implies that the kth row of $\tilde{U}A^{-1} = L^{-1}$ is $\psi_k^{(k-1)} \equiv a_k^{(k-1)}A^{-1}$ and the kth row of $\tilde{L}A^{-1} = U^{-1}$ is $\psi_k^{(-n+k)} \equiv a_k^{(-n+k)}A^{-1}$. In particular, $\psi_1^{(0)} = a_1A^{-1} = e_1^T$ and $\psi_n^{(0)} = a_nA^{-1} = e_n^T$, where e_i are the $n \times 1$ canonical vectors with a one in position i and zeros everywhere else. This motivates

$$\psi_i^{(0)} \equiv a_iA^{-1} = e_i^T, \qquad 1 \le i \le n.$$

If

$$\psi_{i+k}^{(k)} = a_{i+k}^{(k)}A^{-1}, \qquad \psi_i^{(-k)} = a_i^{(-k)}A^{-1}, \qquad 1 \le i \le n-k,$$

then it follows by induction that

$$\alpha_{i,i+k} = \psi_{i+k}^{(k-1)}Ae_i / \psi_i^{(-k+1)}Ae_i, \qquad \beta_{i+k,i} = \psi_i^{(-k+1)}Ae_{i+k} / \psi_{i+k}^{(k-1)}Ae_{i+k},$$

and

$$\begin{pmatrix} \psi_{i+k}^{(k)} \\ \psi_i^{(-k)} \end{pmatrix} = \begin{pmatrix} 1 & -\alpha_{i,i+k} \\ -\beta_{i+k,i} & 1 \end{pmatrix} \begin{pmatrix} \psi_{i+k}^{(k-1)} \\ \psi_i^{(-k+1)} \end{pmatrix}, \qquad 1 \le i \le n-k.$$

Note that only elements $i \ldots i+k$ of $\psi_{i+k}^{(k)}$ and $\psi_i^{(-k)}$ may be non-zero. To keep the presentation simple, we do not take advantage of this fact in the description of the generalised Levinson algorithm below.

The above derivation shows that, in contrast to the presentation of the generalised Levinson algorithm in [6], there is no need for the exchange matrix J. Its occurrence in the symmetric Levinson algorithm is an artefact of the symmetric Toeplitz structure, as explained in the next section.

We further observe that the inner products in the denominators of $\alpha_{i,i+k}$ and $\beta_{i+k,i}$ can be replaced by recursive equations, thus saving about $O(2k)$ operations in step k of the generalised Levinson algorithm. Define the denominators of $\alpha_{i,i+k}$ and $\beta_{i+k,i}$ by

$$d_i^{(-k+1)} \equiv \psi_i^{(-k+1)}Ae_i, \quad d_{i+k}^{(k-1)} \equiv \psi_{i+k}^{(k-1)}Ae_{i+k},$$

so that

$$\alpha_{i,i+k} = \psi_{i+k}^{(k-1)} A e_i / d_i^{(-k+1)}, \quad \beta_{i+k,i} = \psi_i^{(-k+1)} A e_{i+k} / d_{i+k}^{(k-1)}.$$

We can now derive recursive equations for $d_i^{(-k)}$ and $d_{i+k+1}^{(k)}$ in terms of $d_i^{(-k+1)}$ and $d_{i+k+1}^{(k-1)}$: initially

$$d_i^{(0)} = \psi_i^{(0)} A e_i = e_i^T A e_i = a_{ii}, \qquad 1 \le i \le n.$$

From the above definition and the computation of $\psi_i^{(-k)}$ it follows that

$$\begin{aligned} d_i^{(-k)} &= \psi_i^{(-k)} A e_i = \psi_i^{(-k+1)} A e_i - \beta_{i+k,i} \psi_{i+k}^{(k-1)} A e_i \\ &= d_i^{(-k+1)} - \alpha_{i,i+k} \beta_{i+k,i} d_i^{(-k+1)} \\ &= (1 - \alpha_{i,i+k} \beta_{i+k,i}) d_i^{(-k+1)}. \end{aligned}$$

Similary, one shows that

$$d_{i+k}^{(k)} = (1 - \alpha_{i,i+k} \beta_{i+k,i}) d_{i+k}^{(k-1)}.$$

Our formulation of the generalised Levinson algorithm is summarised below.

Factorisation in the Generalised Levinson Algorithm

$$1 \le i \le n, \qquad \psi_i^{(0)} = e_i^T, \quad d_i^{(0)} = a_{ii}$$

$$1 \le k \le n-1, \quad 1 \le i \le n-k,$$

$$\alpha_{i,i+k} = \psi_{i+k}^{(k-1)} A e_i / d_i^{(-k+1)},$$

$$\beta_{i+k,i} = \psi_i^{(-k+1)} A e_{i+k} / d_{i+k}^{(k-1)},$$

$$d_{i+k}^{(k)} = (1 - \alpha_{i,i+k} \beta_{i+k,i}) d_{i+k}^{(k-1)},$$

$$d_i^{(-k)} = (1 - \alpha_{i,i+k} \beta_{i+k,i}) d_i^{(-k+1)}$$

$$\begin{pmatrix} \psi_{i+k}^{(k)} \\ \psi_i^{(-k)} \end{pmatrix} = \begin{pmatrix} 1 & -\alpha_{i,i+k} \\ -\beta_{i+k,i} & 1 \end{pmatrix} \begin{pmatrix} \psi_{i+k}^{(k-1)} \\ \psi_i^{(-k+1)} \end{pmatrix}.$$

The kth rows of L^{-1} and U^{-1} are respectively given by $\psi_k^{(k-1)}$ and $\psi_k^{(-n+k)}$, and the kth diagonal elements of $\tilde{U}$ and $\tilde{L}$ by $d_k^{(k-1)}$ and $d_k^{(-n+k)}$.

Because it computes the same quantities $\alpha_{i,i+k}$ and $\beta_{i+k,i}$, hence the same matrix Q, as the generalised Bareiss algorithm, the generalised Levinson algorithm does not break down as long as all contiguous principal submatrices are non-singular. Moreover, forward elimination and backsubstitution can be done in the same manner as for the generalised Bareiss algorithm.

For a symmetric positive-definite matrix A, the generalised Levinson algorithm determines L^{-1}, U^{-1}, D_L and D_U. Alternatively, the Hyperbolic Cholesky algorithm can be used to derive a symmetric version of the generalised Levinson algorithm that computes the Cholesky factorisations $A^{-1} = \mathcal{U}^{-T}\mathcal{U}^{-1} = \mathcal{L}^{-T}\mathcal{L}^{-1}$ of the inverse.

5.2 Specialisation to Toeplitz Matrices

If the matrix T at hand is a Toeplitz matrix then the generalised Levinson algorithm reduces to the Levinson algorithm [12]. The Levinson algorithm is derived by applying to the Bareiss algorithm the relations

$$\psi_{i+k}^{(k)} = a_{i+k}^{(k)} A^{-1}, \qquad \psi_i^{(-k)} = a_i^{(-k)} A^{-1}, \qquad 1 \le i \le n-k.$$

Hence, only the vectors $\psi_{k+1}^{(k-1)}$, $\psi_n^{(k-1)}$, $\psi_1^{(-k+1)}$ and $\psi_{n-k}^{(-k+1)}$ are needed; see Section 3.4. Since the generalised Levinson algorithm computes the same multipliers $\alpha_{i,i+k}$ and $\beta_{i+k,i}$ as the generalised Bareiss algorithm, the Toeplitz case simplifies to

$$\alpha_k \equiv \alpha_{i,i+k} = \psi_{k+1}^{(k-1)} T e_1 / \psi_1^{(-k+1)} T e_1$$
$$\beta_k \equiv \beta_{i+k,i} = \psi_1^{(-k+1)} T e_{k+1} / \psi_{k+1}^{(k-1)} T e_{k+1}$$

In contrast to the Bareiss algorithm, which requires the three different elements $a_{k+1,1}^{(k-1)}$, $a_{11}^{(-k+1)}$ and $a_{1,k+1}^{(-k+1)}$, only one set of quantities, $\psi_{k+1}^{(k-1)}$ and $\psi_1^{(-k+1)}$ for instance, suffices to compute *both* multipliers for the Levinson algorithm. Hence, about half of the arithmetic operations in the Levinson algorithm can be saved when only one of the factorisations is desired (that is, the computation of the vectors with subscript n in the algorithm below can be omitted). For simplicity, we write $\phi_1^{(k-1)} \equiv \psi_{k+1}^{(k-1)}$ and $\phi_n^{(-k+1)} = \psi_{n-k}^{(-k+1)}$, and we also exploit the fact that only elements $i, \ldots, i+k$ of $\psi_{i+k}^{(k)}$ and $\psi_i^{(-k)}$ may be non-zero.

Factorisation in the Levinson Algorithm

$$
\begin{aligned}
1 \le i \le n, \qquad & \psi_1^{(0)} = \phi_1^{(0)} = \psi_n^{(0)} = \phi_n^{(0)} = 1, \quad d_1 = t_0 \\
1 \le k \le n-1, \quad & \alpha_k = \phi_1^{(k-1)} (\, t_{-1} \quad \ldots \quad t_{-k} \,)^T / d_k, \\
& \beta_k = \psi_1^{(-k+1)} (\, t_1 \quad \ldots \quad t_k \,)^T / d_k \\
& d_{k+1} = (1 - \alpha_k \beta_k) d_k
\end{aligned}
$$

$$
\begin{pmatrix} \phi_1^{(k)} \\ \psi_1^{(-k)} \end{pmatrix} = \begin{pmatrix} 1 & -\alpha_k \\ -\beta_k & 1 \end{pmatrix} \begin{pmatrix} 0 & \phi_1^{(k-1)} \\ \psi_1^{(-k+1)} & 0 \end{pmatrix}
$$

$$
\begin{pmatrix} \psi_n^{(k)} \\ \phi_n^{(-k)} \end{pmatrix} = \begin{pmatrix} 1 & -\alpha_k \\ -\beta_k & 1 \end{pmatrix} \begin{pmatrix} 0 & \psi_n^{(k-1)} \\ \phi_n^{(-k+1)} & 0 \end{pmatrix}.
$$

The kth rows of L^{-1} and U^{-1} are respectively given by

$$
(\, \phi_1^{(k-1)} \quad 0_{n-k} \,) \qquad \text{and} \qquad (\, 0_{k-1} \quad \phi_n^{(-n+k)} \,),
$$

and the kth diagonal elements of $\tilde{U}$ and $\tilde{L}$ by d_k and d_{n-k+1}.

If the matrix T is symmetric positive-definite then $\psi_1^{(-k)} = \phi_1^{(k)} J$ and $\psi_n^{(k)} = \phi_n^{(-k)} J$, and because of L=JUJ also $\phi_n^{(-k)} = \phi_1^{(k)} J$. Hence the algorithm simplifies as follows.

Factorisation in the Levinson Algorithm for Symmetric Positive-Definite Matrices

$$
\begin{aligned}
1 \le i \le n, \qquad & \phi_1^{(0)} = 1, \quad d_1 = t_0 \\
1 \le k \le n-1, \quad & \rho_k = \phi_1^{(k-1)} (\, t_1 \quad \ldots \quad t_k \,)^T / d_k, \qquad d_{k+1} = (1 - \rho_k^2) d_k
\end{aligned}
$$

$$
\phi_1^{(k)} = (\, 1 \quad -\rho_k \,) \begin{pmatrix} 0 & \phi_1^{(k-1)} \\ \phi_1^{(k-1)} J & 0 \end{pmatrix}.
$$

6 Parallel Algorithm for Persymmetric Systems

During the usual process of solving a linear system $Ax = b$ as described in Section 2.1, the factorisation and forward elimination part can be performed simultaneously. The backsubstitution part, however, cannot be overlapped with the other two because the first step in the solution of the upper triangular system requires the last element obtained from the solution of the lower triangular system in the forward elimination.

We will now show how one can combine the Bareiss and Levinson algorithms in order to perform all three steps of solving a linear system $Ax = b$ in parallel, where A is a persymmetric matrix. It appears that this algorithm when applied to Toeplitz matrices is identical to one given in [8], but we give a much simpler description here.

Suppose that there are enough processors, i.e. $O(n^2)$, available; and that a division, or a multiplication followed by an addition constitutes one arithmetic operation. The computation

$$Q\begin{pmatrix} T & I & b \\ T & I & b \end{pmatrix} = \begin{pmatrix} \tilde{U} & L^{-1} & c \\ \tilde{L} & U^{-1} & d \end{pmatrix}, \qquad \begin{pmatrix} c \\ d \end{pmatrix} = \begin{pmatrix} L^{-1}b \\ U^{-1}b \end{pmatrix},$$

where the multipliers α_k and β_k are computed as in the Bareiss algorithm, performs the factorisation and forward elimination. It requires $2(n-1)$ parallel operations because the computation of the multipliers cannot be overlapped with their application.

Because $T = UD_UL_U = LD_LU_L$ and because of persymmetry we actually have

$$\begin{pmatrix} \tilde{U} & L^{-1} & c \\ \tilde{L} & U^{-1} & d \end{pmatrix} = \begin{pmatrix} D_LU_L & L^{-1} & c \\ D_UL_U & U^{-1} & d \end{pmatrix} = \begin{pmatrix} D_LU_L & L^{-1} & c \\ JD_LL^TJ & JU_L^{-T}J & d \end{pmatrix},$$

and $x = \tilde{U}^{-1}c = U_L^{-1}D_L^{-1}c$. Because the Levinson algorithm delivers the elements of D_L, the divisions $y = D_L^{-1}c$ can be performed concurrently with the forward elimination (possibly with a lag of $O(1)$ steps). Note that the elements of the vectors c and y are computed in the order $1, \ldots, n$, and the rows of $JU_L^{-T}J$ in the order $n, \ldots, 1$. Consequently, the rows of $U_L^{-T}J = (JU_L^{-1})^T$ are available in the order $1, \ldots, n$, so the columns of

JU_L^{-1} are available in the order $1, \ldots, n$ (the premultiplying matrix J just permutes the rows and has no effect on the columns). Thus, the ith column of U_L^{-1} is available by the time the ith element of y has been computed so that the linear combination $x = U_L^{-1}y$ of the columns of U_L^{-1} can be performed concurrently with the factorisation and forward elimination in $2n + O(1)$ parallel operations.

The algorithm for Toeplitz matrices can be implemented on $O(n)$ processors [8] in $2n$ parallel operations, but seems to require a considerable amount of broadcasting. If a truly systolic implementation without broadcasting is desired then we believe that this algorithm has the same time complexity as the one in [4], which uses the matrix Q to perform backsubstitution, see Section 3.3. At last note that the asymptotic time complexity is the same for persymmetric and for Toeplitz systems.

Acknowledgements

I would like to thank Jean-Marc Delosme, Liz Jessup and Jin-Hong Ma for their careful reading of the paper.

References

[1] Ahmed H.M., Delosme J.-M. and Morf M., "Highly Concurrent Computing Structures for Matrix Arithmetic and Signal Processing", IEEE Computer, vol. 15, 1982, pp. 65-82.

[2] Bareiss E.H., "Numerical Solution of Linear Equations with Toeplitz and Vector Toeplitz Matrices, Numer. Math., 13, 1969, pp. 404-24

[3] Bunch J.R., "Stability of Methods for Solving Toeplitz Systems of Equations", SIAM J. Sci. Stat. Comput., vol. 6, pp. 349-64, 1985.

[4] Delosme J.-M. and Ipsen I.C.F., "Parallel Solution of Symmetric Positive Definite Systems with Hyperbolic Rotations", Linear Algebra and its Applications, vol. 77, pp. 75-111, 1986.

[5] Delosme J.-M. and Ipsen I.C.F., "From Bareiss' Algorithm to the Stable Computation of Partial Correlations", Journal of Computational and Applied Mathematics, To appear, 1989.

[6] Delsarte P., Genin Y. and Kamp Y., "A Method of Matrix Inverse Triangular Decomposition, Based on Contiguous Principal Submatrices", Linear Algebra and its Applications, vol. 31, 1980, pp. 199-212.

[7] Delsarte P., Genin Y. and Kamp Y., "Generalized Schur Positivity Test and Levinson Recursion", Proc. Europ. Conf. Circuit Theory and Design, pp. 321-3, 1983.

[8] Gohberg I., Kailath T., Koltracht I. and Lancaster P., "Linear Complexity Parallel Algorithms for Linear Systems of Equations with Recursive Structure", Linear Algebra and its Applications, vol. 88/89, 1987, pp. 271-315.

[9] Golub G.H. and van Loan C.F., "Matrix Computations", The Johns Hopkins Press, 1983.

[10] Ipsen I.C.F., "Systolic Algorithms for the Parallel Solution of Dense Symmetric Positive-Definite Toeplitz Systems", Numerical Algorithms for Modern Parallel Computer Architectures", pp. 85-108, Schultz M.H., Springer Verlag, 1988.

[11] Lawson C.L. and Hanson R.J., "Solving Least Squares Problems", Prentice Hall, 1974.

[12] Levinson N., "The Wiener RMS (Root-Mean-Square) Error Criterion in Filter Design and Prediction", J. Math. Phys., vol. 25, 1947, pp. 261-78.

[13] Schur I., "Ueber Potenzreihen die im Innern des Einheitskreises Beschraenkt Sind", J. Reine Angewandte Mathematik, vol. 147, 1917, pp. 205-32.

Generalized Displacement Structure for Block-Toeplitz, Toeplitz-block, and Toeplitz-derived Matrices

J. Chun and T. Kailath*

Information Systems Laboratory
Stanford University
Stanford, CA 94305, USA

Abstract

The concept of displacement structure has been used to solve several problems connected with Toeplitz matrices and with matrices obtained in some way from Toeplitz matrices (e.g. by combinations of multiplication, inversion and factorization). Matrices of the latter type will be called Toeplitz-derived (or Toeplitz-like). In this paper we shall introduce a generalized definition of displacement for block-Toeplitz and Toeplitz-block matrices. It will turn out that Toeplitz-derived matrices are perhaps best regarded as particular Schur complements obtained from suitably defined block matrices. The displacement structure will be used to obtain a generalized Schur algorithm for the fast triangular and orthogonal factorizations of all such matrices, and well structured fast solutions of the corresponding exact and overdetermined systems of linear equations.

*This work was supported by the U.S. Army Research Office, under Contract DAAL03-86-K-0045, and by the SDIO/IST, managed by the Army Research Office under Contract DAAL03-87-K-0033. This manuscript is submitted for publication with the understanding that the U.S. Government is authorized to reproduce and distribute reprints for Government purpose notwithstanding any copyright notation therein.

NATO ASI Series, Vol. F 70
Numerical Linear Algebra, Digital Signal Processing and Parallel Algorithms
Edited by G. H. Golub and P. Van Dooren

1 Introduction

In multichannel signal processing, system identification and image processing applications, one encounters various forms of structured matrices. One interesting family consists of matrices having a block-Toeplitz form

$$A_1 = \begin{bmatrix} B_0 & B_{-1} & \cdot & B_{-N+1} \\ B_1 & B_0 & \cdot & B_{-N+2} \\ \cdot & \cdot & \cdot & \cdot \\ B_{M-1} & B_{M-2} & \cdot & B_{-N+M} \end{bmatrix}, \quad B_i : \text{ rectangular matrices}, \quad (1.a)$$

or a Toeplitz-block form

$$A_2 = \begin{bmatrix} T_{1,1} & T_{1,2} & \cdot & T_{1,N} \\ T_{2,1} & T_{2,2} & \cdot & T_{2,N} \\ \cdot & \cdot & \cdot & \cdot \\ T_{M,1} & T_{M,2} & \cdot & T_{M,N} \end{bmatrix}, \quad T_{i,j} : \text{ rectangular Toeplitz matrices} \quad (1.b)$$

or often as Schur complements with respect to various entries in A_1 or A_2 (see the examples in Sec 4). Often we call the matrix A_1 an $M \times N$ *block-Toeplitz array*, and the matrix A_2 an $M \times N$ *Toeplitz-block array*; the matrices obtained as Schur complements are often not Toeplitz at all, but have been called near-Toeplitz, or close-to-Toeplitz, or Toeplitz-like or Toeplitz-derived matrices.

For such A, we shall show how to obtain fast triangular factorization $A = LU$, and fast QR factorization, $A = QR$, which among other things will also give us fast nicely structured methods for solving exactly-determined systems of equations,

$$A\mathbf{x} = \mathbf{b}, \quad A \in \mathbf{R}^{n \times n}, \quad A \text{ is strongly nonsingular} \quad (2)$$

and also over-determined systems of equations,

$$A\mathbf{x} = \mathbf{b}, \quad A \in \mathbf{R}^{m \times n}, \quad m \geq n, \ A \text{ has full column rank}. \quad (3)$$

Our results will be based on a generalization of the concept of displacement structure used in earlier work (see e.g., [15]–[18]). Besides enabling us to solve several new problems, this generalized concept will also provide a new and simpler approach to many of the problems studied in [15]–[18] and [5]. However, first we briefly review earlier approaches and results.

For a square block-Toeplitz matrix $A_1 \in \mathbf{R}^{n\times n}$, with square blocks $B_1 \in \mathbf{R}^{r\times r}$, there exist several fast triangular factorization algorithms such as the Bareiss algorithm [1], the multichannel Levinson algorithm [19] and the Schur algorithm [14], [15], [25], all of which require matrix (of the block size, $r \times r$) operations. Our approach will treat block-Toeplitz matrices in essentially the same way as *scalar* Toeplitz matrices, and in particular will use only elementary scalar operations; the absence of matrix operations such as inversion will simplify the design of dedicated hardware implementations. For a square Toeplitz-block matrix $A_2 \in \mathbf{R}^{n\times n}$ with, $T_{ij} \in \mathbf{R}^{m_i\times n_j}$, the previous approaches were first to transform A_2 into a block-Toeplitz matrix by pre- and post-multiplication with permutation matrices, and then apply an algorithm for square block-Toeplitz matrix to get a row- and column- permuted triangular factorization of A_2; there is clearly a difficulty with this approach when $m_i \neq m_j$. Also the permuted matrix might not be strongly nonsingular. Our approach will not have this problem because it directly factorizes A_2 without permutations. Finally, for matrices obtained via Schur complementation, the concept of displacement structure (see e.g. [5], [15]–[18]) has been used to obtain a number of fast algorithms; in particular, several algorithms have recently appeared [2], [5], [9], [22] for the orthogonalization of scalar Toeplitz matrices; our new approach also provides a generalized unification of these algorithms.

Several illustrations and applications of our approach will be given in Sec 4; however there are several others (see [4]) that space does not allow us to describe. Our choice here was made in part to relate to examples and problems that are studied, generally in different-ways, in other chapters of this book.

Our generalized definition of displacement structure is presented in Sec 2. A correspondingly generalized Schur algorithm for matrix factorization is derived in Sec 3. As just mentioned, Sec 4 contains various applications. Finally some computational aspects are elaborated in Sec 5; in particular we may note the introduction of *spinors*, which include as special cases the circular (Givens) and hyperbolic rotations as well as the well-known elementary (or elimination) matrices.

2 Displacement of Matrices

Let $A \in \mathbf{R}^{m\times n}$ be a given matrix, and let F^f and F^b be *strictly* lower triangular matrices. The matrix

$$\nabla_{(F^f,F^b)}A \equiv A - F^f\,A\,F^{bT} \tag{4}$$

will be called the *displacement* of A with respect to the *displacement operators* $\{F^f,\ F^b\}$. Assume that

$$\operatorname{rank}\nabla_{(F^f,F^b)}\,A = \alpha\,.$$

Any matrix pair $\{X,\ Y\}$ such that

$$\nabla_{(F^f,F^b)}\,A = XY^T\,,\quad X \equiv [\mathbf{x}_1, \mathbf{x}_2, \ldots, \mathbf{x}_\alpha]\,,\quad Y \equiv [\mathbf{y}_1, \mathbf{y}_2, \ldots, \mathbf{y}_\alpha]$$

will be called a *generator* of A (with respect to $\{F^f, F^b\}$). The number α will be called the *length* of the generator (with respect to $\{F^f, F^b\}$). A generator of A with the minimal possible length will be called a *minimal* generator. The length of the minimal generator of A (i.e., $\operatorname{rank}(\nabla_{(F^f,F^b)}\,A)$) will be called the *displacement rank* of A (with respect to $\{F^f, F^b\}$), and denoted as $\alpha_{(F^f,F^b)}(A)$.

If $\{X, Y\}$ is a generator of $A \in \mathbf{R}^{n\times n}$ with respect to $\{F^f, F^b\}$), then for any nonsingular matrix $S \in R^{\alpha\times\alpha}$, the matrix pair $\{XS,\ YS^{-T}\}$ is also a generator of A because

$$\nabla_{(F^f,F^b)}A = XY^T = XSS^{-1}\,Y^T\,. \tag{5}$$

Let $\{X, Y\}$ be a generator of a matrix with respect to strictly lower triangular displacement operators $\{F^f, F^b\}$. We say that a generator is *proper (with respect to the column j)* if, for a certain i, all the elements in the ith row of X and above, except for the element $[X]_{i,j}$, are zero, and all elements in the ith row of Y and above, except the element $[Y]_{i,j}$, are zero; for example,

$$X = \begin{bmatrix} 0 & \cdot & 0 & * & 0 & \cdot & 0 \\ * & \cdot & * & * & * & \cdot & * \\ * & \cdot & * & * & * & \cdot & * \end{bmatrix},\quad Y = \begin{bmatrix} 0 & \cdot & 0 & * & 0 & \cdot & 0 \\ * & \cdot & * & * & * & \cdot & * \\ * & \cdot & * & * & * & \cdot & * \end{bmatrix}.$$
$$\qquad\qquad\uparrow_{\,j} \qquad\qquad\qquad\qquad\qquad \uparrow_{\,j}$$

Often we shall denote a proper generator as $\{X_p, Y_p\}$. If $\{X, Y\}$ is not proper, then by choosing appropriate S, we can obtain a proper generator $\{XS, YS^{-T}\}$ under certain conditions on the matrix A (see Sec 5). Note that the displacement of a symmetric matrix A can be written as $\nabla_{(F^f,F^b)}A = X\Sigma X^T$ where Σ is a diagonal matrix with 1 or -1 along the main diagonal; we shall say that A has a *symmetric generator*, $\{X, X\Sigma\}$.

As an example, for a square Toeplitz matrix $T = (t_{i-j}) \in \mathbf{R}^{n\times n}$

$$\nabla_{(Z_n,Z_n)}T = \begin{bmatrix} t_0 & t_{-1} & \cdot & t_{1-n} \\ t_1 & & & \\ \cdot & & & \\ t_{n-1} & & & \end{bmatrix} = X\Sigma X^T,\ X = \begin{bmatrix} t_0 & 0 \\ t_1 & t_1 \\ \cdot & \cdot \\ t_{n-1} & t_{n-1} \end{bmatrix} / t_0^{1/2},$$

$$\Sigma = \begin{bmatrix} 1 & 0 \\ 0 & -1 \end{bmatrix}.$$

Choice of Displacement Operators

Let $\{X, Y\}$ be a generator of length α of A with respect to F^f and F^b. If the matrix-vector multiplications $F^f\mathbf{v}$ and $F^b\mathbf{v}$ takes $f(n)$ operations, then our algorithm in Sec 3 will need $O(\alpha n f(n))$ operations. Therefore, our objective is to choose the "simplest" or sparse (to make $f(n)$ small) strictly lower triangular matrices F^f and F^b that also make α as small as possible. For a scalar $n\times n$ Toeplitz matrix, a natural choice of displacement operator is the simple $n \times n$ *shift matrix*, Z_n, with 1's along the first sub-diagonal, and 0's elsewhere.

For an $M \times N$ block-Toeplitz array with $r \times s$ blocks, the following choice of displacement operators gives the smallest α,

$$F^f = Z^M_{Mr}, \quad F^b = Z^N_{Ns},$$

where Z^k_{kr} is a *block shift matrix*, i.e., a $k \times k$ array with $r \times r$ identity matrices on the 1st block subdiagonal and zeros elsewhere.

For block-Toeplitz or Toeplitz-block matrices, it is straight forward to obtain generators from the displacements by inspection (see [4] for closed forms), as we shall illustrate in several examples.

Example 2.1. *For the following block Toeplitz matrix A, called the "Hurwitz matrix", we can choose $F^f = Z^2$ and $F^b = Z$ to get a rank-2 displacement $\nabla_{(F^f,F^b)}A$,*

$$A = \begin{bmatrix} a_1 & a_3 & a_5 & a_7 & \cdot \\ a_0 & a_2 & a_4 & a_6 & \cdot \\ 0 & a_1 & a_3 & a_5 & \cdot \\ 0 & a_0 & a_2 & a_4 & \cdot \\ \cdot & \cdot & \cdot & \cdot & \cdot \end{bmatrix}, \quad \nabla_{(F^f,F^b)}A = \begin{bmatrix} a_1 & a_3 & a_5 & a_7 & \cdot \\ a_0 & a_2 & a_4 & a_6 & \cdot \\ & & & & \\ & & & & \\ & & & & \end{bmatrix}.$$

Note that $\nabla_{(F^f,F^b)}A = XY^T$, where

$$X = \begin{bmatrix} 1 & 0 & 0 & \cdot & \cdot \\ 0 & 1 & 0 & \cdot & \cdot \end{bmatrix}^T, \quad Y = \begin{bmatrix} a_1 & a_3 & a_5 & \cdot & \cdot \\ a_0 & a_2 & a_4 & \cdot & \cdot \end{bmatrix}^T.$$

□

For an $M \times N$ Toeplitz-block array with $m_i \times n_j$ Toeplitz matrices, we shall use the displacement operators,

$$F^f = \oplus_{i=1}^{M} Z_{m_i} \in R^{m \times m}, \quad F^b = \oplus_{i=1}^{N} Z_{n_i} \in R^{n \times n},$$

where $\oplus_i^M$ denotes the concatenated direct sum, i.e., $A \oplus B = \begin{bmatrix} A & O \\ O & B \end{bmatrix}$.

Example 2.2 *For the following Toeplitz-block matrix A, which arises in ARMA system identification problems [10], we can choose $F^f = Z_4$ and $F^b = Z_2 \oplus Z_3$ to obtain a rank-3 displacement,*

$$A = \begin{bmatrix} \beta_0 & \beta_{-1} & \gamma_0 & \gamma_{-1} & \gamma_{-2} \\ \beta_1 & \beta_0 & \gamma_1 & \gamma_0 & \gamma_1 \\ \beta_2 & \beta_1 & \gamma_2 & \gamma_1 & \gamma_0 \\ \beta_3 & \beta_2 & \gamma_3 & \gamma_2 & \gamma_1 \end{bmatrix}, \quad \nabla_{(F^f,F^b)}A = \begin{bmatrix} \beta_0 & \beta_{-1} & \gamma_0 & \gamma_{-1} & \gamma_{-2} \\ \beta_1 & & \gamma_1 & & \\ \beta_2 & & \gamma_2 & & \\ \beta_3 & & \gamma_3 & & \end{bmatrix}$$

Example 2.3. *Consider the matrix M,*

$$M = \begin{bmatrix} I & A & O \\ A^T & O & I \\ O & I & O \end{bmatrix}.$$

If A is an $M \times N$ block-Toeplitz array with $r \times s$ blocks, we could choose

$$F^f = F^b = Z_{Mr}^r \oplus Z_{Ns}^s \oplus Z_{Nr} .$$

If A is a Toeplitz-block array, for example, $A = \begin{bmatrix} T_1 \\ T_2 \end{bmatrix}$, where $T_1 \in \mathbf{R}^{m_1 \times n}$ and $T_2 \in \mathbf{R}^{m_2 \times n}$ then we could choose

$$F^f = F^b = Z_{m_1} \oplus Z_{m_2} \oplus Z_n \oplus Z_n .$$

Remark 2.1. One might check that the displacement operators for a block-Toeplitz matrix and a Toeplitz-block matrix are related by

$$Z_{kr}^r = P \cdot [\oplus_{i=1}^r Z_k] \cdot P ,$$

where P is the permutation matrix that transforms the block-Toeplitz matrix into the Toeplitz-block matrix by pre and post-multiplication, vice versa.

3 Generalized Schur Algorithm and Partial Triangularization

A fundamental method for triangular matrix factorization is the so-called *Schur reduction* process, which computes *Schur complements* of leading submatrices iteratively. Lev-Ari and Kailath [17], [18] realized that the classical Schur algorithm amounts to Schur reduction, and gave several important generalizations including one for Hankel matrices. In the rest of this paper, we shall further elaborate the ideas in [18].

Fast Schur Reduction using Displacement Structure.

Our fast algorithms will be based on the following theorem.

Theorem 3.1 *Let $\{X^{(1)}, Y^{(1)}\}$ be a generator of a rectangular matrix $A^{(1)} \in R^{m \times n}$ with respect to $\{F^f, F^b\}$. Also assume that $\{X^{(1)}, Y^{(1)}\}$ is proper with respect to a particular (pivoting) column, which we shall index as "pvt". If we denote the columns of $X^{(1)}$ and $Y^{(1)}$ by*

$$X_p^{(1)} = [\mathbf{x}_1^{(1)}, \ldots, \mathbf{x}_{pvt}^{(1)}, \ldots, \mathbf{x}_\alpha^{(1)}],\ Y_p^{(1)} = [\mathbf{y}_1^{(1)}, \ldots, \mathbf{y}_{pvt}^{(1)}, \ldots, \mathbf{y}_\alpha^{(1)}],$$

then the matrix $A^{(2)}$ *defined by*

$$A^{(2)} \equiv A^{(1)} - \mathbf{x}_{pvt}^{(1)} \mathbf{y}_{pvt}^{(1)T}$$

has null first column and row, and has a generator $\{X^{(2)}, Y^{(2)}\}$, *with respect to* $\{F^f, F^b\}$ *of the form*

$$X^{(2)} = [\mathbf{x}_1^{(1)}, \ldots, F^f \mathbf{x}_{pvt}^{(1)}, \ldots, \mathbf{x}_\alpha], \; Y^{(2)} = [\mathbf{y}_1^{(1)}, \ldots, F^b \mathbf{y}_{pvt}^{(1)}, \ldots, \mathbf{y}_\alpha^{(1)}].$$

Proof

$$\begin{aligned} A^{(2)} - F^f A^{(2)} F^{bT} &= [A^{(1)} - \mathbf{x}_{pvt}^{(1)} \mathbf{y}_{pvt}^{(1)T}] - F^f [A^{(1)} - \mathbf{x}_{pvt}^{(1)} \mathbf{y}_{pvt}^{(1)T}] F^{bT} \\ &= A^{(1)} - F^f A^{(1)} F^{bT} - \mathbf{x}_{pvt}^{(1)} \mathbf{y}_{pvt}^{(1)T} + F \mathbf{x}_{pvt}^{(1)} \mathbf{y}_{pvt}^{(1)T} F^{bT} \\ &= X^{(1)} Y^{(1)T} - \mathbf{x}_{pvt}^{(1)} \mathbf{y}_{pvt}^{(1)T} + F \mathbf{x}_{pvt}^{(1)} \mathbf{y}_{pvt}^{(1)T} F^{bT} \\ &= X^{(2)} Y^{(2)T} \end{aligned}$$

The first column and row of $A^{(2)}$ are null because

$$A^{(2)} \mathbf{e}_1 = [X^{(2)} Y^{(2)T}] \mathbf{e}_1 = 0, \; e_1^T A^{(2)} = \mathbf{e}_1^T [X^{(2)} Y^{(2)T}] = 0,$$

where we have used the facts that F^f and F^b are strictly lower triangular and $\{X^{(1)}, Y^{(1)}\}$ is proper. □

Before presenting the generalized Schur algorithm, we assume that we have a procedure called *MakeProper* that can convert a given generator $\{X, Y\}$ of $A \in \mathbf{R}^{m \times n}$ into a proper generator $\{X_p, Y_p\}$ (whenever A has a proper generator); this can be done with $O(\alpha m)$ operations as will be shown in Sec 5.

By applying the previous theorem using such a proper generator we can obtain a (possibly non-proper) generator of $A^{(2)}$. By repeating this process, r times, we shall generate the matrices

$$\begin{aligned} A^{(r+1)} &= A^{(r)} - \mathbf{x}_{pvt}^{(r)} \mathbf{y}_{pvt}^{(r)T} \\ A^{(r)} &= A^{(r-1)} - \mathbf{x}_{pvt_{r-1}}^{(r-1)} \mathbf{y}_{pvt_{r-1}}^{(r-1)T} \\ &\cdots \\ A^{(2)} &= A^{(1)} - \mathbf{x}_{pvt_1}^{(1)} \mathbf{y}_{pvt_1}^{(1)T}. \end{aligned}$$

It turns out that this process gives a *partial triangular factorization* of $A^{(1)}$; this follows by noting that

$$A^{(1)} = \Sigma_{i=1}^{r} \mathbf{x}_{pvt_i}^{(i)} \mathbf{y}_{pvt_i}^{(i)T} + A^{(r+1)}$$

$$= \begin{bmatrix} \mathbf{x}_{pvt_1}^{(1)} & \mathbf{x}_{pvt_r}^{(r)} \end{bmatrix} \begin{bmatrix} \mathbf{y}_{pvt_1}^{(1)T} \\ \cdot \\ \mathbf{y}_{pvt_r}^{(r)T} \end{bmatrix} + A^{(r+1)}, \; A^{(r)} \equiv \begin{bmatrix} & \\ & S^{(r+1)} \end{bmatrix}.$$

Remark 3.1. If we define $A^{(1)} = \begin{bmatrix} B & D \\ C & E \end{bmatrix}$ then it is easy to check that $S^{(r+1)} = E - CB^{-1}D$. The matrix $S^{(r+1)}$ is called the *Schur complement* of B in $A^{(1)}$. Notice that the above process also gives a generator of $S^{(r+1)}$.

Remark 3.2. The above r step partial triangularization breaks down if and only if there is a singular leading principal submatrix of order less than or equal to r; we shall assume that this is not so.

The above procedure can be summarized in the following algorithm, which we shall call a *generalized Schur algorithm.*

Algorithm (Generalized Schur Algorithm)

Input: A generator $\{X, Y\}$ of $A \in \mathbf{R}^{m\times n}$ with respect to $\{F^f, F^b\}$.

Output: (i) Partial triangular factors $L \in \mathbf{R}^{m\times r}$ and $U \in \mathbf{R}^{r\times n}$ of A.

(ii) A generator $\{X, Y\}$ of the Schur complement of the $r\times r$ leading principal submatrix of A,

Procedure GeneralizedSchur
begin
    for $k := 1$ **to** r **do begin**
        MakeProper;
        The kth column of $L := \mathbf{x}_{pvt}$;   The kth row of $U := \mathbf{y}_{pvt}^T$;
        Replace $\mathbf{x}_{pvt}$ with $F^f\mathbf{x}_{pvt}$;   Replace $\mathbf{y}_{pvt}$ with $F^b\mathbf{y}_{pvt}$;

end ;
return $(L, U, \{X, Y\})$;
end.

Note that the above procedure needs $O(\alpha mr)$ operations, where α is the length of the given generator, assuming that MakeProper takes $O(\alpha m)$ operations (see Sec 5).

Example 3.1 Triangularization of block-Toeplitz or Toeplitz-block matrices.

As trivial examples we can triangularize block-Toeplitz matrices or Toeplitz-block matrices simply by completing the above generalized Schur algorithm. Note that the multiplications $F^f \mathbf{x}_{pvt}$ *and* $F^b \mathbf{y}_{pvt}$ *amount to shifting down "segments" of* $\mathbf{x}_{pvt}$ *and* $\mathbf{y}_{pvt}$.

Example 3.2 Simultaneous Factorization of a Toeplitz matrix and its Inverse [5].

Consider the matrix

$$A = \begin{bmatrix} T & I \\ I & O \end{bmatrix}, \quad T = (t_{i-j}) \in R^{n \times n}, \quad t_0 = 1 \tag{6}$$

which has the following symmetric generator,

$$X = \begin{bmatrix} 1 & t_1 & \cdot & t_{n-1} & 1 & 0 & \cdot & 0 \\ 1 & t_1 & \cdot & t_{n-1} & 1 & 0 & \cdot & 0 \end{bmatrix}^T, \quad \Sigma = \begin{bmatrix} 1 & 0 \\ 0 & -1 \end{bmatrix}.$$

After performing n steps of partial triangular factorization using the generalized Schur algorithm, we shall have

$$A = \begin{bmatrix} L \\ U \end{bmatrix} [L^T, U^T] + \begin{bmatrix} O & O \\ O & S \end{bmatrix}. \tag{7}$$

Now, one can check by comparing the entries of A in (6) and (7), that

$$T = LL^T, \ T^{-1} = UU^T.$$

Remark 3.3 Recall that the classical Schur algorithm [21] gives only the factorization, $T = LL^T$, whereas the Levinson algorithm gives the factorization, $T^{-1} = UU^T$. If one only needs the factorization of T^{-1} the above method is slower than the Levinson algorithm [19]; a derivation of the Levinson algorithm from the generalized Schur algorithm can be found in [4].

Example 3.3 Orthogonalization of a fully windowed Toeplitz matrix.

Let $T = (t_{i-j}) \in \mathbf{R}^{m\times n}$, $m > n$ be a fully windowed Toeplitz matrix, i.e.,

$$t_{i-j} = 0, \quad \text{if } j > i, \quad \text{or } i > m - n + j\,.$$

Then it is easy to check that $C \equiv T^T T$ is also an (unwindowed) Toeplitz matrix. Now consider the following matrix

$$A = \begin{bmatrix} T^T T & T^T \\ T & O \end{bmatrix}, \tag{8}$$

for which it can be checked that a generator of A is

$$X = \begin{bmatrix} c_0 & c_1 & \cdot & c_{n-1} & t_0 & t_1 & \cdot & t_{m-n} & 0 & \cdot & 0 \\ 0 & c_1 & \cdot & c_{n-1} & t_0 & t_1 & \cdot & t_{m-n} & 0 & \cdot & 0 \end{bmatrix}^T \in \mathbf{R}^{(m+n)\times 2}, \Sigma = \begin{bmatrix} 1 & 0 \\ 0 & -1 \end{bmatrix}$$

After performing n steps of partial triangular factorization using the generalized Schur algorithm, we shall have

$$A = \begin{bmatrix} R^T \\ Q \end{bmatrix} [R,\, Q^T] + \begin{bmatrix} O & O \\ O & S \end{bmatrix}. \tag{9}$$

From (9), one can easily see that

$$T^T T = R^T R\,,\; T = QR$$

so that Q is orthogonal because $R^T Q^T Q R = R^T R$.

Remark 3.4 Recall that the (fixed AR) lattice filter operates on a (data) sequence $\{t_i\}$ and performs orthogonalization to get the prediction errors without a knowledge of the covariance matrix (i.e., $T^T T$) of the data sequence. The lattice filter is again a Levinson-like version of the above method (see [4] for details).

4 Applications to non-Toeplitz matrices

Now we shall consider various *extended* matrices M. By applying the generalized Schur algorithm in Sec 3 to judiciously chosen extended matrices, we can obtain interesting results including QR factorizations of block-Toeplitz or Toeplitz-block matrices.

Generators of extended matrices in this section can be also easily found by inspection. For closed forms of various generators, see [4].

A. QR factorization

Let $A \in \mathbf{R}^{m\times n}$ be a block-Toeplitz or a Toeplitz-block matrix, and let us define the block matrix,

$$M \equiv \begin{bmatrix} -I & A & O \\ A^T & O & A^T \\ O & A & I \end{bmatrix}. \tag{10}$$

If we apply the generalized Schur algorithm to (10) then after the mth step we shall have a generator [1] of

$$\begin{bmatrix} A^TA & A^T \\ A & I \end{bmatrix}. \tag{11}$$

After another n-steps of partial triangularization, we shall have

$$\begin{bmatrix} A^TA & A^T \\ A & I \end{bmatrix} = \begin{bmatrix} R^T \\ Q \end{bmatrix} \cdot [R\,Q^T] + \begin{bmatrix} O & O \\ O & S \end{bmatrix}. \tag{12}$$

Now, one can check that the matrices Q and R in (13) satisfy

$$A = QR, \qquad Q^TQ = I,$$

i.e., we obtained the QR factorization of A. This procedure will need $O(mn)$ flops.

[1]One can start with a generator of the extended matrix (11), as in Example 3.3. A closed-form expression for a generator of (11) for block-Toeplitz and Toeplitz-block matrix A can be found in [4].

If one wish to compute R^{-1} directly, then one can perform the $(m+n)$ steps of partial triangularization with the matrix,

$$M = \begin{bmatrix} -I & A & O \\ A^T & O & I \\ O & I & O \end{bmatrix}.$$

Note that

$$\begin{bmatrix} A^TA & I \\ I & O \end{bmatrix} = \begin{bmatrix} R^T \\ U \end{bmatrix} \cdot [R\,U^T] + \begin{bmatrix} O & O \\ O & S \end{bmatrix},$$

and therefore, $U = R^{-1}$ because $UR = I$.

B. Removing Forward Elimination in Square Systems

If one's primary interest in the factorization is in solving a square system of equations,

$$A\mathbf{x} = \mathbf{b}, \tag{13}$$

then one might want to obtain the transformed right-side vector $\mathbf{y} \equiv L^{-1}\mathbf{b}$, during the course of the factorization process. This can be done by performing the following partial triangular factorization of the matrix M,

$$M \equiv \begin{bmatrix} A \\ -\mathbf{b}^T \end{bmatrix} = \begin{bmatrix} L \\ \mathbf{y}^T \end{bmatrix} \cdot L^T$$

whence the solution to (13) can be obtained by solving the triangular system of equations,

$$L^T\mathbf{x} = \mathbf{y}. \tag{14}$$

C. Removing Back-Substitution in Square Systems

From a hardware implementation point of view, the back-substitution step in (14) can still be quite cumbersome [7]. This back-substitution process can also be eliminated by performing the partial factorization of the matrix,

$$M = \begin{bmatrix} A & -\mathbf{b} \\ I & 0 \end{bmatrix}.$$

Notice that the solution $\mathbf{x} = A^{-1}\mathbf{b}$ is the Schur complement of A in M. Therefore, after n steps of partial triangularization, we shall have a generator of the solution, >From which we can read out the solution; see [7] for details.

D. Solving Least Squares Problems without Back-substitution

To solve the weighted least squares problem of minimizing

$$\| A_2(A_1\mathbf{x} - \mathbf{b}) \|_2 ,$$

where A_1 and A_2 are block-Toeplitz or Toeplitz-block matrices, we form the matrix

$$M = \begin{bmatrix} -A_2 & A_1 & -\mathbf{b} \\ A_1^T & O & 0 \\ O & I & 0 \end{bmatrix} .$$

Now notice that the least squares solution,

$$\mathbf{x} = (A_1^T A_2^{-1} A_1)^{-1} A_2^T \mathbf{b} \tag{15}$$

is the Schur complement of the submatrix

$$\begin{bmatrix} -A_2 & A_1 \\ A_1^T & O \end{bmatrix} ,$$

Therefore, after $m + n$ steps of the generalized Schur algorithm, we shall have a generator of the solution (15), >From which the solution can be read out [7].

E. Regularization

If the given Toeplitz least squares system is particularly ill-conditioned, it is meaningless to compute the exact (least squares) solution, since small perturbations of the matrix can cause very large perturbations in the solution. In such cases, we may solve the following regularized system [11], [20]

$$\begin{bmatrix} A \\ \eta I \end{bmatrix} \mathbf{x} = \begin{bmatrix} \mathbf{b} \\ 0 \end{bmatrix}$$

by partial triangularization of the matrix

$$M = \begin{bmatrix} & & A & \mathbf{b} \\ & & \eta I & \mathbf{0} \\ A^T & \eta I & O & \mathbf{0} \\ & & I & \mathbf{0} \end{bmatrix}$$

After $m + 2n$ steps of the generalized Schur algorithm, we shall have the solution. We may remark that this technique of regularization is known as the leakage method (adding white noise with variance η^2 to the data sample) in the signal processing literature (see e.g. [3]).

5 Construction of Proper Generators

We shall present a method for constructing a proper generator using *spinors*; for a method using Householder matrices, see [4]. A spinor $S_{(j|i)} \in \mathbf{R}^{\alpha\times\alpha}$ is defined as the identity matrix except for the following 4 entries,

$$[S_{(j|i)}]_{i,i} = c, \; [S_{(j|i)}]_{i,j} = s_2, \; [S_{(j|i)}]_{j,i} = -s_1, \; [S_{(j|i)}]_{j,j} = c,$$

where $[A]_{i,j}$ denotes the (i,j)th element of the matrix A, and $c^2 + s_1 s_2 = 1$. The parameters $\{c, s_1, s_2\}$ will be called *Schur parameters.* Notice that the inverse of a spinor is also a spinor, viz., $S_{(j|i)}^{-1}$ is the identity matrix except for the following 4 entries,

$$[S_{(j|i)}^{-1}]_{i,i} = c, \; [S_{(j|i)}^{-1}]_{i,j} = -s_2, \; [S_{(j|i)}^{-1}]_{j,i} = s_1, \; [S_{(j|i)}^{-1}]_{j,j} = c,$$

Let $\mathbf{x}^T \in \mathbf{R}^{1\times\alpha}$ and $\mathbf{y}^T \in \mathbf{R}^{1\times\alpha}$ be row vectors. Let c, s_1 and s_2 be chosen as

$$c = \left[\frac{x_i\, y_i}{x_i\, y_i + x_j\, y_j}\right]^{1/2}, \; s_2 = -c \cdot \frac{x_j}{x_i}, \; s_1 = -c \cdot \frac{y_j}{y_i},$$

and define x' and y' by

$$\mathbf{x}'^T \equiv \mathbf{x}^T S_{(j|i)}, \quad \mathbf{y}'^T \equiv \mathbf{y}^T S_{(j|i)}^{-T}.$$

Then it is easy to check that $x'_j = y'_j = 0$ and $\mathbf{x}'^T\mathbf{y}' = \mathbf{x}^T\mathbf{y}$. We shall call the elements x_i and y_i *pivoting elements.* Therefore, by repeating this process we can *annihilate* all elements of $\mathbf{x}$ and $\mathbf{y}$ except the pivoting elements, resulting in

$$[0, \ldots 0, x_i, 0, \ldots, 0] = \mathbf{x}^T \prod_{j \neq i}^{\alpha} S_{(j|i)}, \quad [0, \ldots 0, y_i, 0, \ldots, 0] = \mathbf{y}^T \prod_{j \neq i}^{\alpha} S_{(j|i)}^{-T}.$$

An arbitrary choice of pivoting element or an arbitrary *ordering* of annihilation, might result in $[1 + \frac{x_j y_j}{x_i y_i}] \leq 0$, for which real spinors do not exist. The following function returns an index "pvt", and two sets of indices, FIRST and NEXT; if we annihilate the elements in $\mathbf{x}^T$ and $\mathbf{y}^T$ whose indices are given in the set FIRST, pivoting with the element x_{pvt} and y_{pvt}, before the annihilation of the elements given in the set NEXT, then it is not hard to see that $[1 + \frac{x_j y_j}{x_{pvt} y_{pvt}}] \leq 0$.

Procedure FindOrdering
  begin
    Compute $\gamma_i = x_i y_i$ **for all** $1 \leq i \leq \alpha$
    $s := \Sigma\, \gamma_i$;
    Pset := $\{i | y_i > 0\}$; Nset := $\{i | \gamma_i < 0\}$; Zset := $\{i | \gamma_i = 0\}$;
    if $s > 0$ **then**
      pvt := any $i \in$ Pset;
      FIRST := Pset; NEXT := Nset;
    else if $s < 0$ **then**
      pvt := any $i \in$ Nset;
      FIRST := Nset; NEXT := Pset;
    else /* Cannot rotate */
      return (s);
    Add Zset either to FIRST or NEXT;
    return (pvt, FIRST, NEXT)
  end

With FindOrdering, we can summarize the procedure for constructing proper generators.

Procedure MakeProper
  begin
    $\mathbf{x}^T$:= first non-zero row of X; $\mathbf{y}^T$:= first non-zero row of Y;
    FindOrdering;
    if $s=0$ **then**
      return ("A has a singular minor");
    for each $j \in$ FIRST, **and then for each** $j \in$ NEXT
      Determine $S_{(j|pvt)}$ to annihilate x_j and y_j;
      $X := XS_{(j|pvt)}$; $Y := YS^{-T}_{(j|pvt)}$
    end;
    return $(\{X, Y\})$
  end

Remark 5.1 The quantity $s = \Sigma\gamma_i$ that is used to find the annihilation ordering is the product of the diagonal elements $l_{k,k} \cdot \mathbf{u}_{k,k}$ of the partial triangular matrices L and U obtained by the generalized Schur algorithm. Therefore, $s > 0$ for positive-definite symmetric matrices, and $s < 0$ for negative-definite symmetric matrices. Hence, for these matrices we can choose a single column as a pivoting column throughout the triangularization process.

Some Special Cases

If we are given a symmetric generator of a symmetric matrix A, i.e., if $Y = X\Sigma$, then the updating of Y in the above procedure is redundant, because the updated $\{X', Y'\}$ after annihilating a row is still symmetric. To see this, let

$$\mathbf{x}^T = \mathbf{y}^T = [x_{pvt}, y_j].$$

Then the spinor that annihilates x_j will reduce to a *Givens rotation*,

$$G_{(j|pvt)} = \begin{bmatrix} c & -s \\ s & c \end{bmatrix}, \quad c^2 + s^2 = 1$$

On the other hand, if

$$\mathbf{u}^T = [u_{pvt}, u_j], \quad \mathbf{v}^T = [u_{pvt}, -u_j]$$

the spinor will become a *hyperbolic rotation*,

$$H_{(j|pvt)} = \begin{bmatrix} ch & -sh \\ -sh & ch \end{bmatrix}, \quad ch^2 - sh^2 = 1$$

Notice that Givens and hyperbolic rotations preserve the symmetry of the updated generator, i.e.,

$$YS^{-T} = Y' = X'\Sigma, \; X' = XS, \; S \; : \text{ a Givens or hyperbolic rotation}.$$

As another special case of spinors, consider the two row vectors

$$\mathbf{u}^T = [u_{pvt}, u_j], \quad \mathbf{v}^T = [v_{pvt}, 0].$$

For this case, the spinor that annihilates u_j will reduce to the usual *elimination* matrix,

$$E_{(j|pvt)} = \begin{bmatrix} 1 & -\kappa_E \\ 0 & 1 \end{bmatrix}, \quad \kappa_E = \frac{u_j}{u_{pvt}}. \tag{16}$$

Example 5.1 *The celebrated "Routh procedure" [12] for stability testing is just a triangularization using our generalized Schur algorithm of the Hurwitz matrix of Example 2.1.*

Example 5.2 *Consider the following Toeplitz-block matrix called the "Sylvester matrix" [12],*

$$S = \begin{bmatrix} a_0 & & & & b_0 & & & \\ a_1 & a_0 & & & b_1 & b_0 & & \\ \cdot & a_1 & \cdot & & \cdot & b_1 & \cdot & \\ a_n & \cdot & \cdot & a_0 & b_m & \cdot & \cdot & b_0 \\ & a_n & \cdot & a_1 & & b_m & \cdot & b_1 \\ & & \cdot & \cdot & & & \cdot & \cdot \\ & & & a_n & & & & b_m \end{bmatrix} \in \mathbf{R}^{(m+n)\times(m+n)} \tag{17}$$

A nonsingular Sylvester matrix is always strongly nonsingular, and therefore, we can check whether a Sylvester matrix is singular or not by using the generalized Schur algorithm. The matrix S in (17) has a generator, with respect to $\{Z_{m+n}, Z_m \oplus Z_n\}$,

$$X = \begin{bmatrix} a_0 & a_1 & \cdot & a_n & 0 & 0 & \cdot & 0 \\ b_0 & b_1 & \cdot & \cdot & b_m & 0 & \cdot & 0 \end{bmatrix}^T, Y = \begin{bmatrix} 1 & 0 & \cdot & 0 & 0 & & \cdot & 0 \\ 0 & & \cdot & 0 & 1 & 0 & \cdot & 0 \end{bmatrix}^T \tag{18}$$

When we apply the reduction technique to X and Y in (18), the spinors during the first n steps will be just the elimination matrices of (16).

Remark 5.2 For the triangular factorization of a block-Toeplitz matrix, one may use the *block-spinor*,

$$S = \begin{bmatrix} I & K_1 \\ -K_2^T & I \end{bmatrix} U^{-1}, \; S^{-T} = \begin{bmatrix} I & K_2 \\ -K_1^T & I \end{bmatrix} L^{-T},$$

$$LU = \begin{bmatrix} I + K_1 K_2^T & O \\ O & I + K_2^T K_1 \end{bmatrix}$$

to annihilate $X_{1,2}$, pivoting with $X_{1,1}$, by choosing $K = X_{1,1}^{-1} X_{2,1}$. However, the use of matrix inversion in the block rotation makes the control flow in most hardware implementations complicated, and therefore, the use of block rotations is often discouraged; our recommendation is to use the generalized Schur algorithm, which only operates on selected columns of scalars.

6 Concluding Remarks

We have shown how to obtain triangular factorization and QR factorization of Toeplitz-block or block-Toeplitz matrices in $O(mn)$ flops. Our method is based on the displacement structure properties of matrices. We also presented some other applications of our algorithm.

We have generalized earlier definitions (see e.g. [15]–[18]) of the displacement for Toeplitz-like matrices and presented a correspondingly generalized Schur algorithm for their factorization. The extended definition allows us to handle block-Toeplitz and Toeplitz-block matrices and Schur complements with respect to the leading (block) entries of such matrices. Composite matrices obtained as products and inverses of Toeplitz matrices can be nicely handled by formulating them as Schur complements of entries in a suitably defined block-Toeplitz matrix. Some interesting examples were given in Sec 4; Several of them will be considered in other ways in other chapters in this book.

We also mention that displacement structure can also be introduced for Hankel and Hankel-like matrices (see e.g. [18]) and also Vandermonde-like matrices; analogs of the (generalized) definitions, algorithms and applications in this paper have also been obtained for these matrices (see [4] and [6]).

References

[1] E. Bareiss, *Numerical solution of linear equations with Toeplitz and vector Toeplitz matrices*, Numer. Math., 13: (1969), 404-424.

[2] A. Bojanczyk, R. Brent and F. de Hoog, *QR factorization of Toeplitz matrices*, Numer. Math., 49 (1986), pp. 81-94.

[3] J. Cioffi, *Limited precision effects in adaptive filtering*, IEEE Trans, on Circuit and Systems, 1988.

[4] J. Chun, *Fast algorithms for structured matrix equations*, Ph.D. Thesis, Stanford University, 1989.

[5] J. Chun, T. Kailath and H. Lev-Ari, *Fast parallel algorithms for QR and triangular factorization*, SIAM J. Sci. Stat. Comput., Nov., vol. 8, No. 6, (1987), pp. 899-913.

[6] J. Chun and T. Kailath, *Displacement structure for Hankel- and Vandermonde-like matrices*, Preprint, 1988.

[7] J. Chun, V. Roychowdury and T. Kailath, *Systolic Array for Solving Toeplitz Systems of Equations*, SPIE's 32nd Conference on Advanced Algorithms and Architecture for Signal Processing III, San Diego, Aug., 1988.

[8] G. Cybenko, *A generalized orthogonalization technique with applications to time series analysis and signal processing*, Math. Comp. vol 40, No. 161, (1983), pp. 323-336.

[9] G. Cybenko, *Fast Toeplitz orthogonalization using inner products*, SIAM J. Sci. Stat. Comput., (1987).

[10] P. Eykhoff, *System identification*, John Wiley and Sons, New York, 1974

[11] L. Elden, *An algorithm for the regularization of ill-conditioned least squares problems*, SIAM J. Sci. Stat. Comput., vol 5, No. 1., (1984), pp. 237-254.

[12] F. Gantmacher *The theory of matrices*, vol. 2, Chelsea Publishing Comp., New York, 1960.

[13] T. Kailath, *Linear Systems*, Prentice-Hall, Englewood Cliffs, New Jersey, 1980.

[14] T. Kailath, *Signal processing in the VLSI era*, VLSI and Modern Signal Processing, S. Kung, H. Whitehouse and T. Kailath, eds., Prentice-Hall, Englewood Cliffs, New Jersey, 1985.

[15] T. Kailath, *Signal processing applications of some moment problems*, Proceedings of Symposia in Applied Mathematics, vol. 37, 1987 pp. 71-109.

[16] T. Kailath, S. Kung and M. Morf, *Displacement ranks of matrices and linear equations*, J. Math. Anal. Appl., 68 (1979) pp. 395-407. See also Bull. Amer. Math. Soc., 1 (1979), pp. 769-773.

[17] H. Lev-Ari and T. Kailath, *Lattice filter parameterizations and modeling of nonstationary process*, IEEE Trans. Inform. Theory, IT-30 (1984), pp. 2-16.

[18] H. Lev-Ari and T. Kailath, *Triangular factorization of structured Hermitian matrices*, Operator Theory, Advances and Applications, vol. 18, Birkhäuser, Boston, pp. 301-324, (1986).

[19] N. Levinson, *The Wiener RMS error criterion in filter design and prediction*, J. of Math. Phys. 25 (1947), 261-278.

[20] H. Rutishauser, *Once again: The least squares problem*, Linear Alg. and its Appli., 1. (1968), pp. 471-478.

[21] I. Schur, *Über Potenzreihen, die im Innern des Einheitskreises beschrankt sind*, J. für die Reine und Angewandte Mathematik, 147 (1917), pp. 205-232. English translation appears in Operator Theory, Advances and Applications, vol. 18, Birkhäuser, Boston, pp. 31-60, (1986).

[22] D. Sweet, *Fast Toeplitz orthogonalization*, Numer. Math., 43 (1984), pp. 1-21.

Fault Tolerant Recursive Least Squares Minimization

Franklin T. Luk

School of Electrical Engineering,
Cornell University
Ithaca, NY 14853, USA

Abstract

Existing fault tolerance schemes have often been ignored by systolic array designers because they are too costly and unwieldy to implement. With this in mind, we have developed a new technique specially tailored for recursive least squares minimization that emphasizes simplicity. We propose a new decoding scheme that allows for error detection while wasting no precious processor cycles and preserving the basic structure of the systolic array. We will show that errors can be detected by examining a single scalar. The technique can be implemented with negligible algorithmic modification and little additional hardware. The simplicity of our method invites its use in future systolic arrays.

1 Introduction

The compatibility of systolic arrays and algorithms with both matrix computations and today's VLSI and wafer-scale technology guarantees their future use as key components in any signal processing system. An especially important systolic algorithm is the QR decomposition for least squares minimization, a crucial cog in most adaptive antenna processing algorithms. The importance of these problems is evidenced by the two major systolic array projects; one at MIT's Lincoln Laboratory [1] and the other

NATO ASI Series, Vol. F 70
Numerical Linear Algebra, Digital Signal Processing and Parallel Algorithms
Edited by G. H. Golub and P. Van Dooren

at the United Kingdom's Royal Signals and Radar Establishment (RSRE) [2]. However, traditional fault tolerance techniques such as modular redundancy have been regarded as too costly and unwieldy to implement on these systolic arrays. In this paper, we present a simple fault tolerance scheme for the QR decomposition and show how it can be easily incorporated into the RSRE systolic arrays for recursive least squares minimization.

Our method is a new implementation of the idea of algorithm-based fault tolerance [3],[4]. To review, algorithm-based fault tolerance employs three steps; encode the input data, execute the algorithm on the encoded input to produce encoded output, and decode the output to detect and perhaps correct errors. The beauty of the approach is that the algorithm need not be altered to operate on the encoded input, so that the bulk of the implementational complexity comes from the encoding and decoding stages. Both checksum and weighted checksum encoding schemes have been developed by Abraham et al., who showed that a variety of matrix operations preserves the checksum property. We can therefore detect and correct errors by explicitly summing the rows (or columns) in the resultant matrix. However, this is often expensive; while explicitly summing the rows is a simple operation, the procedure is costly to integrate with most systolic algorithms. We propose a new decoding scheme that allows concurrent error detection, wastes no precious processor cycles, and maintains the basic structure of the systolic array. The scheme presented in this paper does not rely on regulating the sums themselves, but instead we only have to monitor a scalar check quantity which is easily computed with little additional cost. We will show that in most cases we can detect errors by examining this single quantity.

The paper is organized as follows. In Sections 2 and 3 we describe algorithm-based fault tolerance and the QR decomposition. We discuss the error model in Section 4, and review McWhirter's recursive least squares algorithm [5] in the following section. In Section 6 we discuss the detection problem, and give conditions under which we will be able to detect errors. In the last two sections we discuss how the data can correct itself due to the exponential forget factor, and conclude our paper.

2 Checksum Scheme

Let us consider the checksum scheme of Abraham et al. Given an $n \times p$ matrix A and two $p \times 1$ vectors

$$e = (1, 1, \ldots, 1)^T$$

and

$$w = (w_1, w_2, \ldots, w_p)^T,$$

where $w_i > 0$ and $w_i \neq w_j$ for $i \neq j$, we define an $n \times (p+2)$ weighted row checksum matrix A_{rw} by

$$A_{rw} = (\, A \quad Ae \quad Aw\,). \tag{2.1}$$

For the weight vector w, Jou and Abraham [4] proposed the choice $w_j = 2^{j-1}$, for $j = 1, 2, \ldots, p$. Such big weights may cause numerical difficulties. Let $p = 101$ and

$$a_{i1} = \ldots = a_{i,100} = 0, a_{i,101} = 1.$$

Then $w_{101} = 2^{100} > 10^{30}$, and on a 10-decimal-digit machine any perturbation in a_{i1} of size less than 10^{20} would not affect the value of the weighted checksum. We may alleviate the problem by choosing $w_j = j$ [6]. Suppose there is an error of size γ in the (i, j) position of A, i.e., we get the erroneous matrix

$$\tilde{A} = A - \gamma e_i e_j^T, \tag{2.2}$$

where $\gamma \neq 0$, and e_i and e_j denote the ith and jth unit coordinate vectors, respectively. Let us define the erroneous weighted checksum matrix $\tilde{A}_{rw}$ by

$$\tilde{A}_{rw} = (\, \tilde{A} \quad Ae \quad Aw\,). \tag{2.3}$$

We do not consider the case where the error occurs in the checksums; interested reader may consult [3],[4]. Now, we first determine that the ith row of $\tilde{A}$ is inconsistent from

$$\sum_{k=1}^{n} \tilde{a}_{ik} - (Ae)_i = -\gamma.$$

Second, we use the weighted checksum matrix $\tilde{A}_{rw}$ to determine that the jth element of the row causes the inconsistency:

$$\sum_{k=1}^{n} \tilde{a}_{ik} w_k - (Aw)_i = -w_j \gamma.$$

Correction of the error is straightforward via the use of checksums.

3 QR Decomposition

Assume that $n \geq p+2$ and consider the QR factorization of an $n \times p$ matrix A:

$$A = Q \begin{pmatrix} R \\ 0 \end{pmatrix}, \tag{3.1}$$

where the matrix Q is $n \times n$ orthogonal, and R is $p \times p$ upper triangular. The corresponding decomposition of the $n \times (p+2)$ matrix A_{rw} is

$$A_{rw} = Q \begin{pmatrix} R & Re & Rw \\ 0 & 0 & 0 \end{pmatrix}.$$

A systolic array implementing the QR decomposition was presented by Gentleman and Kung [7]. In their array, the annihilations for an 10×8 matrix proceed as follows:

$$\begin{pmatrix} \times & \times & \times & \times & \times & \times & \times & \times \\ 1 & \times & \times & \times & \times & \times & \times & \times \\ 2 & 4 & \times & \times & \times & \times & \times & \times \\ 3 & 5 & 7 & \times & \times & \times & \times & \times \\ 4 & 6 & 8 & 10 & \times & \times & \times & \times \\ 5 & 7 & 9 & 11 & 13 & \times & \times & \times \\ 6 & 8 & 10 & 12 & 14 & 16 & \times & \times \\ 7 & 9 & 11 & 13 & 15 & 17 & 19 & \times \\ 8 & 10 & 12 & 14 & 16 & 18 & 20 & 22 \\ 9 & 11 & 13 & 15 & 17 & 19 & 21 & 23 \end{pmatrix}.$$

The QR decompositions of the erroneous matrices $\tilde{A}$ and $\tilde{A}_{rw}$ are given by

$$\tilde{A} = \tilde{Q}\begin{pmatrix} \tilde{R} \\ 0 \end{pmatrix} \tag{3.2}$$

and

$$\tilde{A}_{rw} = \tilde{Q}\begin{pmatrix} \tilde{R} & \tilde{Q}^T QRe & \tilde{Q}^T QRw \\ 0 & \xi & \eta \\ 0 & 0 & 0 \end{pmatrix}, \tag{3.3}$$

where ξ and η are scalars. Here we must assume that the transient error has not changed the property of the transformations, i.e., $\tilde{Q}$ should be orthogonal. A way to enforce orthogonality is to check that the computed cosine and sine parameters, say $\tilde{c}$ and $\tilde{s}$, of each plane rotation satisfy

$$\tilde{c}^2 + \tilde{s}^2 = 1.$$

If the relation is violated, then we set

$$\tilde{c} = 1 \quad \text{and} \quad \tilde{s} = 0.$$

Note that the two factors $\tilde{Q}$ and $\tilde{R}$ can be completely wrong if a_{11} is erroneous [8], and that the simple error model (2.2) is no longer adequate to describe the transient error that may occur at any step of the decomposition.

4 Error Model

We need to derive an error model for a computational error that arises during a QR decomposition. Let

$$A^{(0)} = A$$

and

$$A^{(l+1)} = Q^{(l)} A^{(l)},$$

where $Q^{(l)}$ denotes the appropriate transformation at the lth time step, for $l = 0, 1, \ldots$. Assume that a transient error occurs at the kth step of the computation:

$$\tilde{A}^{(k)} = A^{(k)} - \gamma e_i e_j^T. \tag{4.1}$$

Since

$$\tilde{A}^{(k)} = Q^{(k-1)} \dots Q^{(0)} A^{(0)} - \gamma e_i e_j^T,$$

we may define an initial error matrix $\tilde{A}$ by

$$\tilde{A} = A - h e_j^T, \tag{4.2}$$

where

$$h = \gamma (Q^{(k-1)} \dots Q^{(0)})^T e_i.$$

Our technique in (4.2) of throwing the error back to the initial matrix is reminiscent of the idea of backward error analysis [9]. The major advantage is that we need not worry about the time when the transient error occurred. We get from (3.3) and (4.2) that

$$\tilde{A}_{rw} = \tilde{Q} \begin{pmatrix} \tilde{R} & \tilde{R}e + \tilde{Q}^T h & \tilde{R}w + w_j \tilde{Q}^T h \\ 0 & \xi & w_j \xi \\ 0 & 0 & 0 \end{pmatrix}. \tag{4.3}$$

The important observations are that the jth column of $\tilde{R}$ is the sole cause for inconsistent checksums, and that the index j can be determined from the ratio η/ξ.

5 Recursive Least Squares

We now review McWhirter's algorithm [5], which uses a sequence of planar rotations to compute the least squares residual without performing a back substitution. We are given an $n \times p$ full rank data matrix $X(n)$, and an $n \times 1$ data vector $y(n)$. Note that the index n can be confusing. It denotes the number of data rows, and so its value increases by one with each new time step due to the recursive nature of the algorithm. For further details see also [10]. The $n \times 1$ least-squares residual vector $r(n)$ is given by

$$r(n) = X(n)w(n) + y(n),$$

where $w(n)$ is the $p \times 1$ vector of weights that minimizes $\|B(n)r(n)\|$, and $B(n)$ is an $n \times n$ diagonal matrix

$$B(n) = diag(\beta^{n-1}, \beta^{n-2}, \ldots, \beta, 1),$$

for $0 < \beta \leq 1$. We use $\|\cdot\|$ to denote the Frobenius norm. Premultiplication by $B(n)$ applies an exponential forget factor to each row of $X(n)$ in favor of the nth row whose weight factor is unity. Let $Z(n)$ be the following $n \times (p+1)$ matrix

$$Z(n) = (\, X(n) \quad y(n)\,), \tag{5.1}$$

and let

$$q = p + 1.$$

The QR decomposition of $B(n)Z(n)$ is an important step in the algorithm, i.e.,

$$B(n)Z(n) = Q(n)R(n),$$

where $Q(n)$ is an $n \times q$ matrix with orthonormal columns and $R(n)$ is a $q \times q$ upper triangular matrix. We simplify our analysis by considering the orthogonal triangularization as a simple series of row operations. On the lth stage of the process, an input row z_l^T to the array will have elements $z_{l,i}$, for $i = 1, 2, \ldots, q$, successively rotated to zero as follows. Element $z_{l,i}$ will be zeroed out by a Givens rotation against the element residing in the (i, i) processor. The rotation parameters will be sent to the east in the array, successively updating rows i and l. It has been shown in [5] that r_n, the residual element produced during stage n, can be determined by computing a product of the cosines generated by the rotations with a scalar.

For fault tolerant computations we will encode the input data as follows. Let d be an $q \times 1$ vector with nonzero entries, e.g., all ones. Define the $n \times (q+1)$ matrix $Z_r(n)$ as

$$Z_r(n) = (\, Z(n) \quad Z(n)d\,). \tag{5.2}$$

Let the QR decomposition of $B(n)Z_r(n)$ be given by

$$B(n)Z_r(n) = P(n)U(n),$$

where $P(n)$ is the $n \times (q+1)$ matrix with orthonormal columns; $P(n) = (\, Q(n) \quad z(n)\,)$ and $z(n)$ is orthogonal to all the columns of $Q(n)$. The matrix $U(n)$ is $(q+1) \times (q+1)$ upper triangular:

$$U(n) = \begin{pmatrix} R(n) & R(n)d \\ 0^T & 0 \end{pmatrix}.$$

As before, we use the "$\tilde{\ }$" symbol to denote a possibly erroneous quantity. The matrix $\tilde{U}(n)$ is given by

$$\tilde{U}(n) = \begin{pmatrix} \tilde{R}(n) & f(n) \\ 0^T & \xi(n) \end{pmatrix}. \tag{5.3}$$

Here $f(n)$ is the possibly erroneous version of $R(n)d$, and $\xi(n)$ may or may not be zero. It should be noted that the algorithm will work on the encoded data exactly the same as it would on non-encoded data.

We do not use weighted checksums because, in a recursive scheme, new data are added continually and the model of a single transient error for the whole computation is too restrictive. So, we relax our assumption to *at most one error per input row*, and locating only one error for the QR decomposition no longer makes sense.

6 Error Detection

We now analyze the fault coverage of our scheme. We show that with high probability, our technique will detect any error quickly. Recall our assumption that any error does not affect the orthogonality property. Suppose also that the $(q+1, q+1)$ processor is fault-free. We will show that we can achieve a high level of fault tolerant protection through merely monitoring the scalar $\xi(n)$ of (5.3). We will give a definition of what we mean by detection and derive some important results.

Definition 6.1. An error is *detected* if $\xi(n) \neq 0$.

Definition 6.2. A *step* in the orthogonal triangularization process is the transformation of the kth row with the ith row to zero out the ith element of the kth row, where $k > i$.

Definition 6.3. An *erroneous step* is a step in which one of the processors errs for a single cycle.

Definition 6.4. A *stage* in the orthogonal triangularization process is the transformation of the kth row into a row with q leading zero elements, where $k > q$.

Definition 6.5. Consider a row of $Z_r(n)$, say $z^T = (\tilde{z}^T, \rho)$, where ρ denotes a scalar. We say that the row is *consistent* if the inner product of the two vectors $\tilde{z}$ and d equals ρ, and that the row is *inconsistent* otherwise.

We will say that the kth input row is inconsistent if an error occurs some time during the kth stage. We consider the matrix $\tilde{U}(k)$ to be inconsistent if an error occurs affecting one of its elements during or before the kth stage.

Lemma 6.1. An erroneous step will transform at least one of the two processed rows into a inconsistent row.

Proof. See Chen and Abraham [11].

Lemma 6.2. For $\beta = 1$, $\xi(k)$ is a non-negative non-decreasing quantity. For $0 < \beta < 1$, $\xi(k)$ is a non-negative quantity.

Proof. Suppose on the $(k+1)$st stage we input row z_{k+1}^T. After performing steps 1 through q in the array, let the $(k, k+1)$ element have value $\delta(k)$. By applying the formula for Givens rotations, we get

$$\xi(k+1) = \sqrt{\beta^2\xi^2(k) + \delta^2(k)}. \tag{6.1}$$

It is obvious that for $\beta = 1$, $\xi(k)$ is non-negative and non-decreasing. Furthermore, for $0 < \beta < 1$, $\xi(k)$ is non-negative.

Proposition 6.1. If, after computing the QR decomposition of $Z_r(n)$, the check quantity $\xi(n)$ is nonzero, then an error has occurred during the computations.

Proof. The only way that $\xi(n)$ can be nonzero is if it has been rotated against a nonzero row; i.e., if it has been rotated against an inconsistent row. Therefore, an error has occurred.

It should be noted that we cannot go the other way. That is, we cannot say that if $\xi(n) = 0$ an error has not occurred. We will examine the cases where this can occur. However, we will show why it is an extremely unlikely event that after an error has occurred the scalar $\xi(n)$ will remain zero for all values of n. In most cases errors will be detected quickly merely through monitoring $\xi(n)$.

Proposition 6.2. Suppose $\tilde{U}(n-1)$ is consistent and $\xi(n-1)=0$. If z_n^T is inconsistent then $\xi(n) \neq 0$.

Proof. Let $\tilde{z}_n^T$ be the erroneous version of z_n^T and let $\sigma = \sum_{j=1}^{q} \tilde{z}_{n,j}$. Since $\tilde{z}_n^T$ is inconsistent, we have that $\tilde{z}_{n,q+1} \neq \sigma$. After q steps, the array has zeroed out elements 1 through q of $\tilde{z}_n^T$. However, $\tilde{z}_{n,q+1}$ will not be zero after the updates since it is not equal to σ. When we rotate the updated value of $\tilde{z}_{n,q+1}$ against the element in the $(q+1,q+1)$ processor, i.e., $\xi(n-1)=0$, the two quantities are switched. Hence, $\xi(n)$ will equal the updated value of $\tilde{z}_{n,q+1}$, which is nonzero.

Proposition 6.3. Suppose $\tilde{U}(n-1)$ is inconsistent with $\xi(n-1)=0$. Then $\xi(n)$ may not be zero; that is, the error may not be detected.

Proof. We will show why it is highly improbable for successive values of n that $\xi(n)$ will remain zero. Suppose that a diagonal element becomes erroneous; say $u(k,k)$, where $u(i,j)$ denotes the (i,j) entry of the matrix $\tilde{U}(n-1)$. Let the next input row z_n^T to the array be consistent. The error in the (k,k) position will remain undetected if, after zeroing out the first $(k-1)$ elements of z_n^T, we have that $z_{n,k}=0$. The error will continue to remain undetected if this holds true for every new input row. This is a very unlikely occurrence. Essentially, in order for an error to remain undetected, identity rotations must be applied whenever the erroneous element is involved in a computation.

Suppose an off-diagonal element of $\tilde{U}(n)$ becomes erroneous and the next input row is consistent. Let $u(i,k)$, where $k>i$, be the erroneous element. Again, if this element does not interact with any new input row, then the error will not be detected by $\xi(n)$ for all values of n. That is, the identity rotation must be applied to the input rows z_n^T for all successive values of n. Again, this seems to be a very unlikely occurrence.

Suppose that the next input row z_n^T is inconsistent, i.e., an error occurs at some time during the processing of this row. Recall that we have assumed one error per input row. If the $(q,q+1)$ processor erroneously sets element

$\tilde{z}_{n,q+1}$ to zero, then $\xi(n) = 0$. This is even in the event that the error in $\tilde{U}(n-1)$ affects row z_n^T. So, we cannot guarantee that $\xi(n)$ is non-zero in this case. In order for the error to remain undetected, the processor would have to permanently set the output to 0. Off-line testing will detect this. The remaining cases involve errors cancelling each other. This seems to be a highly unlikely circumstance, especially when putting into consideration the effects of finite precision arithmetic and rounding error.

Proposition 6.4. Suppose $\tilde{U}(n-1)$ is inconsistent and $\xi(n-1) \neq 0$. Then $\xi(n) \neq 0$.

Proof. The proof follows from (6.1).

7 Error Correction

We now show how we can perform error correction by simply continuing to run the algorithm. The obvious question is; how can the data correct itself? The key to the answer is that the array employs an exponential forget factor β to de-emphasize old data, by performing the QR decomposition of the $n \times q$ matrix $B(n)Z_r(n)$. The result is that, if an error of magnitude Δz occurs while processing the ith row, then k input rows later the magnitude of the error has been reduced to $\beta^k \Delta z$. Clearly, if $\beta < 1$ and the error is transient, the error will eventually disappear. However, this is of no use if we cannot determine when the error has disappeared. It is important that we do not prematurely assert that the error has been eliminated; on the other hand, we do not want to waste thousands of good cycles incorrectly thinking the residual is erroneous. It turns out that monitoring the check value $\xi(n)$ provides an excellent test for determining when the error has corrected itself. We now examine how this may be accomplished.

Definition 7.1. The $n \times 1$ vector $c(n)$ is defined by

$$c(n) = \begin{pmatrix} \tilde{R}(n)d - f(n) \\ -\xi(n) \end{pmatrix}.$$

Proposition 7.1. Assume that the $(n+1)$st stage is error-free. Then for $0 < \beta \leq 1$,

$$\|c(n+1)\| = \beta\|c(n)\|. \tag{7.1}$$

Proof. When $\beta = 1$, a sequence of Givens rotations is applied to the vector $c(n)$. Since we have assumed that the orthogonality property is not violated in the event of a fault, the result follows. Note we have made use of the property that the Frobenius norm is invariant under orthogonal transformations. When $0 < \beta < 1$, we multiply every row of the matrix in the array by a factor of β before we rotate it with the new input row. So, we are in effect, multiplying the vector $c(n)$ by β and then applying a series of orthogonal transformations to it. Thus, we see that (7.1) holds.

Corollary 7.1. If no errors occur after stage k, for some k, and $0 < \beta < 1$, then $\lim_{n\to\infty} \|c(n)\| = 0$.

It should be noted that $|\xi(n)| \leq \|c(n)\|$. Thus we see from the Corollary, that in the event of a single transient error and $0 < \beta < 1$, the $\xi(n)$ check value should quickly approach zero. When $\beta = 1$, $\xi(n)$ is bounded above by $\|c(n)\|$ and should increase to that upper limit. Hence, a heuristic for correction in the case when $0 < \beta < 1$ is that the $\xi(n)$ value can be considered as approximately equal to the value of $\|c(n)\|$. So, when $\xi(n)$ becomes small we will assume that $c(n)$ is small, i.e., the error is correcting itself. Empirically, we have observed this behavior in numerical simulations. In the case of a permanent error, $\xi(n)$ can increase without bound. This behavior gives us a heuristic for determining the difference between transient and permanent errors.

8 Concluding Remarks

In this paper, we introduce a new algorithm-based fault tolerance technique specifically designed for use in recursive least squares minimization. Through monitoring a single scalar $\xi(n)$, we get error protection. The same quantity can also be used as an indicator for correction. This technique applies in equally effective fashion to the other RSRE arrays [2], as it provides the same properties for the QR decomposition phase of the algorithms. Our scheme does not require the summation process during the decoding stage; it only requires the monitoring of $\xi(n)$. Furthermore, due to the flow of data into the array, which is in a wavefront pattern, $\xi(n)$ is the only reasonable quantity to examine continually. The chief attribute of our technique is its remarkable simplicity.

Acknowledgements

This work was supported in part by the U.S. Army Research Office under contract DAAL 03-86-K-0109. The author would like to thank Eric Torng and Cynthia Anfinson for valuable discussions and contributions.

References

[1] C.M. Rader, "Wafer-scale systolic array for adaptive antenna processing," *Proc. IEEE ICASSP, New York, NY*, pp. 2069-2071, April 1988.

[2] J.V. McCanny and J.G. McWhirter, "Some systolic array developments in the United Kingdom," *IEEE Computer, Vol. 20, No. 7*, pp. 51-63, July 1987.

[3] K.H. Huang and J.A. Abraham, "Algorithm-based fault tolerance for matrix operations," *IEEE Transactions on Computers, Vol. C-33, No. 6*, pp. 518-528, June 1984.

[4] J.Y. Jou and J.A. Abraham, "Fault-tolerant matrix arithmetic and signal processing on highly concurrent computing structures," *Proc. IEEE, Vol. 74, No. 5*, pp. 732-741, May 1986.

[5] J.G. McWhirter, "Recursive least-squares minimization using a systolic array," *Proc. SPIE, Vol. 431, Real Time Signal Processing VI*, pp. 105-112, 1983.

[6] F.T. Luk, "Algorithm-based fault tolerance for parallel matrix equation solvers," *Proc. SPIE, Vol. 564, Real Time Signal Processing VIII*, pp. 49-53, Aug. 1985.

[7] W.M. Gentleman and H.T. Kung, "Matrix triangularization by systolic arrays," *Proc. SPIE, Vol. 298, Real Time Signal Processing IV*, pp. 19-26, 1981.

[8] F.T. Luk and H. Park, "Fault-tolerant matrix triangulations on systolic arrays," *IEEE Trans. Comput., Vol. C-37, No. 11*, pp. 1434-1438, Nov. 1988.

[9] J.H. Wilkinson, The Algebraic Eigenvalue Problem, Oxford University Press, London, 1965.

[10] F.T. Luk and S. Qiao, "Analysis of a recursive least squares signal processing algorithm," *SIAM J. Sci. Statist. Comput., Vol. 10*, to appear, 1989.

[11] C.Y. Chen and J.A. Abraham, "Fault-tolerant systems for the computation of eigenvalues and singular values," *Proc. SPIE, Vol. 696, Advanced Algorithms and Architectures for Signal Processing I*, pp. 228-237, 1986.

A Systolic Array for Recursive Least Squares Minimisation and Adaptive Beamforming

J.G. McWhirter

Royal Signals and Radar Establishment
St. Andrew's Road
Malvern, Worcs. WR14 4PS, UK

Abstract

A systolic/wavefront array for linearly constrained recursive least squares minimisation is described in the context of adaptive antenna array beamforming. The architectural and numerical advantages of this signal processing technique are discussed.

1 Introduction

This paper is based on the two lectures which I gave on "Recursive Least Squares" at the NATO ASI in Leuven. Since it was not possible to report both lectures fully in the space available here, I have tried to avoid overlapping with the other lecturers as far as possible and concentrate on those aspects which proved to be of greatest interest to participants at the meeting. In particular, I will describe the application of a systolic array processor to the problem of adaptive beamforming as applied to antenna arrays. The title of my paper has been changed to reflect this emphasis. Since the paper concerns the application of a state of the art numerical algorithm to a demanding signal processing problem by means of parallel processing hardware, it is directly relevant to each of the three technical areas represented at the meeting.

NATO ASI Series, Vol. F 70
Numerical Linear Algebra, Digital Signal Processing and Parallel Algorithms
Edited by G. H. Golub and P. Van Dooren

Section 2 provides a brief introduction to the concept of adaptive antenna arrays and the use of adaptive beamforming to null out sources of interference or deliberate jamming.

Section 3 is devoted to showing how the adaptive beamforming problem may be formulated as a classical least squares minimisation process and discussing some of the disadvantages associated with the conventional algorithm and associated processor architecture. An efficient recursive algorithm based on the numerically stable method of QR decomposition by Givens rotations is then described.

Section 4 describes how a triangular systolic array may be used to implement the QR decomposition algorithm based on either square-root-free or conventional Givens rotations.

The purpose of Section 5 is to show how the least squares residual may be obtained directly from the triangular systolic array without using an additional back substitution processor to derive the weight vector explicitly. It is then shown how the weight vector may be "frozen" and "flushed" out of the array in a very simple manner if required.

In Section 6 it is shown how the triangular systolic array may also be used to solve the least squares minimisation problem subject to one or more linear constraints. These are incorporated using a very simple preprocessor which may be embedded within the main triangular array.

Finally, section 7 is devoted to discussing some properties of the numerical algorithm, presenting the results of a typical computer simulation and describing some recent hardware developments.

2 Adaptive Antenna Arrays

The objective of an adaptive antenna is to select a set of amplitude and phase weights with which to combine the outputs from the elements in an antenna array so as to produce a far-field beam pattern which, in some sense, optimises the reception of a desired signal. The substantial improvements in system performance offered by this form of array processing has meant that it is now becoming an essential requirement for many radar, communications and navigation systems.

The key components of an adaptive antenna system are illustrated in figure 1. The amplitude and phase weights are selected by a beam-pattern controller which continuously updates them in response to the element outputs. In some systems the output from the beamformer is also monitored to provide a feedback control. In all cases the resulting array beam pattern is determined by continuously adjusting the weight vector so as to minimise the output power from the array. The minimisation is carried out subject to a suitable constraint which ensures that the desired signal is protected while the interference and jamming sources are being cancelled.

The most commonly employed technique for deriving the adaptive weight vector uses a closed-loop gradient descent algorithm where the weight updates are derived from estimates of the correlation between the signal in each channel and the summed output of the array. This process can be implemented in an analogue fashion using correlation loops [1] or digitally in the form of the Widrow LMS algorithm [2]. The value of this approach should not be underestimated. Gradient descent algorithms are very cost-effective and extremely robust but unfortunately they are not suitable for all applications. The major problem with an adaptive beamformer based on a gradient descent process is one of poor convergence for a broad dynamic range signal environment. This constitutes a fundamental limitation for many modern systems where features such as improved antenna platform dynamics (in the tactical aircraft environment, for example), sophisticated jamming threats, and agile waveform structures (as produced by frequency hopped, spread spectrum formats) produce a requirement for adaptive systems having rapid convergence and high cancellation performance.

In recent years, there has been considerable interest in the application of direct solution or "open-loop" techniques to adaptive antenna processing in order to accommodate these increasing demands [3]. In the context of adaptive antenna processing, these algorithms have the advantage of requiring only minimal input data to describe accurately the external environment and provide an antenna pattern capable of suppressing a wide dynamic range of jamming signals. Direct solution algorithms may be explained most concisely by expressing the adaptive process as a least squares minimisation problem and, in effect, the least squares algorithm defines the optimal path of adaptation.

The main problem with direct solution techniques is the large amount of computation required in comparison with gradient descent algorithms.

Typically, a direct solution algorithm will require $O(p^2)$ floating point operations per sample time (where p is the number of antenna elements) as opposed to $O(p)$ mostly fixed point operations for the digital implementation of a simple gradient descent algorithm. Until recently, it was impossible or, at best, impractical to perform the typical direct computation in real time for most practical applications. However, as a result of recent advances in VLSI circuit technology and the development of dedicated parallel processing architectures such as systolic and wavefront arrays, the use of open-loop techniques is now quite feasible even for fairly high bandwidth applications.

2.1 Canonical Configuration

Consider the form of adaptive combiner which is illustrated in figure 2. The inputs to the combiner take the form of a primary signal $y(t)$ and set of $p-1$ (complex) auxiliary signals denoted here by the vector $\mathbf{x}(t)$. The weight $\mathbf{w}$ is adjusted to minimise the power of the combined output signal which is given at time t by

$$e(t) = \mathbf{x}^T(t)\,\mathbf{w} + y(t)\,, \tag{2.1}$$

where the superscript T denotes matrix transposition. This type of adaptive linear combiner may be used in a wide range of adaptive antenna applications. It is well known, for example, how it may be applied to adaptive sidelobe cancellation. In this case the primary signal constitutes the output from a main (high gain) antenna while the auxiliary signals are obtained from an array of $p-1$ auxiliary antennae. The adaptive combiner serves to modify the beam pattern of the overall antenna system by directing deep nulls towards jamming waveforms received via the sidelobes of the main antenna.

The type of adaptive combiner illustrated in figure 2 may also be used in the so-called "power inversion" mode which has particular application to communications. In this case the p antenna elements are assumed to be omni-directional and of comparable gain. The received signals are fed into the combiner, one of them going to the primary channel and thus having its weight coefficient constrained to unity. The other $p-1$ signals enter the auxiliary channels with their adaptive weights initialised to zero and so, prior to adaptation, the overall beam pattern is determined solely by the (omni-directional) response of the primary element. This "end-element-clamped" configuration provides no inherent mechanism to inhibit the adaptive process from nulling the desired signal. However, the system is only allowed to adapt when the desired signal is known to be absent. When it is present, the weight vector is frozen, thus allowing signal reception. This is referred to as the "power inversion" mode of operation because the differential interference powers received by the antenna elements are inverted by the combiner.

A particularly important application of adaptive antenna arrays requires the power of a p-element combined signal

$$e(t) = \mathbf{x}^T(t)\,\mathbf{w} \tag{2.2}$$

to be minimised subject to one or more linear beam constraints of the form

$$\mathbf{c}^T\mathbf{w} = \mu\,. \tag{2.3}$$

Each constraint ensures that the gain of the antenna array maintains a constant value μ in a given look direction specified by the vector $\mathbf{c}$. It is worth pointing out that the end-element-clamped configuration described above constitutes a particularly simple form of linearly constrained process in which the constraint vector is given by

$$\mathbf{c}^T = [0,0,\ldots,0,1]\,. \tag{2.4}$$

The incorporation of a general linear constraint is not so straightforward. A number of techniques have been proposed in the literature [4] but in all cases the resulting implementation is extremely cumbersome. However, in section 6 of this paper it will be shown how one or more linear constraints may be imposed using a simple pre-processor in conjunction with an end-element-clamped beamformer of the type illustrated in figure 2. Clearly the end-element-clamped combiner, has a wide range of applications and, since it may be applied to the general problem of linearly constrained adaptive beamforming, as well as to the more obvious sidelobe cancellation problem, it will henceforth be referred to as the "canonical" configuration. In the following section, we shall concentrate on the development of a novel algorithm and corresponding processor architecture which applies specifically to the canonical adaptive combiner.

3 Least Squares Formulation

The function of the adaptive combiner in figure 2 will now be formulated in terms of least squares minimisation. We denote the combined array output at the ith discrete sample time t_i by

$$e(t_i) = \mathbf{x}^T(t_i)\mathbf{w} + y(t_i) \tag{3.1}$$

where $\mathbf{x}(t_i)$ is the vector of (complex) auxiliary signals at time t_i and $y(t_i)$ is the corresponding sample of the (complex) primary signal. The residual signal power at time tn is estimated by the quantity $E^2(n)$ where

$$E(n) = [|e(t_n)|^2 + \beta^2|e(t_{n-1})|^2 + \cdots + \beta^{2(n-1)}|e(t_1)|^2]^{\frac{1}{2}} . \tag{3.2}$$

For the sake of generality this unnormalised estimator includes a simple "forget factor" β $(0 < \beta \leq 1)$ which generates an exponential time window and localizes the averaging procedure. Introducing a more compact matrix notation, the estimator defined in equation (3.2) may be expressed in the form

$$E(n) = \| \mathbf{e}(n) \| \tag{3.3}$$

where

$$\mathbf{e}(n) = \mathbf{B}(n) \begin{bmatrix} e(t_1) \\ e(t_2) \\ \vdots \\ e(t_n) \end{bmatrix} \tag{3.4}$$

and

$$\mathbf{B}(n) = diag[\beta^{n-1}, \beta^{n-2}, \ldots, 1] . \tag{3.5}$$

Now from equation (3.1) it follows that the vector of residuals may be written in the form

$$\mathbf{e}(n) = \mathbf{X}(n)\mathbf{w} + \mathbf{y}(n) \tag{3.6}$$

where

$$\mathbf{X}(n) = \mathbf{B}(n) \begin{bmatrix} \mathbf{x}^T(t_1) \\ \mathbf{x}^T(t_2) \\ \vdots \\ \mathbf{x}^T(t_n) \end{bmatrix} \tag{3.7}$$

and

$$\mathbf{y}(n) = \mathbf{B}(n) \begin{bmatrix} y(t_1) \\ y(t_2) \\ \vdots \\ y(t_n) \end{bmatrix} \tag{3.8}$$

$\mathbf{X}(n)$ is simply the matrix of all data samples received by the antenna array up to time t_n, and $\mathbf{y}(n)$ is the corresponding vector of data in the primary or reference channel. The matrix $\mathbf{B}(n)$ takes account of the exponential time window and, for convenience, it has simply been absorbed into the definition of $\mathbf{e}(n)$, $\mathbf{y}(n)$ and $\mathbf{X}(n)$.

Determination of the weight vector $\mathbf{w}(n)$ which minimises $E^2(n)$ is referred to as least squares estimation [5]. The conventional approach to this problem is to derive an analytical expression for the complex gradient of the quantity $E^2(n)$ and determine the weight vector $\mathbf{w}(n)$ for which it vanishes. This leads to the well known Wiener-Hopf equation

$$\mathbf{M}(n)\,\mathbf{w}(n) + \rho(n) = \mathbf{0} \tag{3.9}$$

where

$$\mathbf{M}(n) = \mathbf{X}^H(n)\,\mathbf{X}(n) \tag{3.10}$$

and

$$\rho(n) = \mathbf{X}^H(n)\,\mathbf{y}(n) \tag{3.11}$$

$\mathbf{M}(n)$ is simply the (estimated) covariance matrix and $\rho(n)$ is the corresponding cross-correlation vector.

In their classic paper, Reed, Mallet and Brennan [3] suggested that the weight vector be obtained by solving equation (3.9) directly and showed that the problems of poor convergence associated with closed-loop algorithms may be avoided in this way. This approach, often referred to as Sample Matrix Inversion, leads directly to the type of signal processing architecture which is illustrated schematically in figure 3. It comprises a number of distinct components - one to form and store the covariance matrix estimate, one to compute the solution of equation (3.9), and one to apply the resulting weight vector to the received signal data. These data must be stored in a suitable memory while the weight vector is being computed. The system also requires a number of high-speed data communication buses and a sophisticated control unit to deliver the appropriate sequence of instructions to each component. This type of architecture is obviously complicated, extremely difficult to design, and not very suitable for VLSI.

Not only does the analytic solution given in equation (3.9) lead to a complicated circuit architecture, it is also very poor from the numerical point of view. The effect of forming the covariance matrix in equation (3.10) is to square the condition number of the problem and, for a given finite wordlength, this leads to a considerable loss in performance which should be avoided if possible.

3.1 QR Decomposition

An alternative approach to the least-squares estimation problem which is particularly good in the numerical sense is that of orthogonal triangularisation [6]. This is typified by the method known as QR decomposition, which we generalise here to the case of complex data. An $(n \times n)$ unitary matrix $\mathbf{Q}(n)$ is generated such that

$$\mathbf{Q}(n)\,\mathbf{X}(n) = \begin{bmatrix} \mathbf{R}(n) \\ \mathbf{0} \end{bmatrix} \tag{3.12}$$

where $R(n)$ is a $(p-1)\times(p-1)$ upper triangular matrix. Then, since $\mathbf{Q}(n)$ is unitary, we have

$$E(n) = \| \mathbf{e}(n) \| = \| \mathbf{Q}(n)\mathbf{e}(n) \| = \left\| \begin{bmatrix} \mathbf{R}(n) \\ \mathbf{0} \end{bmatrix} \mathbf{w}(n) + \begin{bmatrix} \mathbf{u}(n) \\ \mathbf{v}(n) \end{bmatrix} \right\| \tag{3.13}$$

where

$$\mathbf{Q}(n)\,\mathbf{y}(n) = \begin{bmatrix} \mathbf{u}(n) \\ \mathbf{v}(n) \end{bmatrix} . \tag{3.14}$$

It follows that the least squares weight vector $\mathbf{w}(n)$ must satisfy the equation

$$\mathbf{R}(n)\,\mathbf{w}(n) + \mathbf{u}(n) = \mathbf{0}\,, \tag{3.15}$$

and hence

$$E(n) = \| \mathbf{v}(n) \| \ . \tag{3.16}$$

Since the matrix $\mathbf{R}(n)$ is upper triangular, equation (3.15) is much easier to solve than the Wiener-Hopf equation described earlier. The weight vector $\mathbf{w}(n)$ may be derived quite simply by a process of back-substitution. Equation (3.15) is also much better conditioned since the condition number of $\mathbf{R}(n)$ is identical to that of the original data matrix $\mathbf{X}(n)$, the two being related by a unitary transformation.

3.2 Givens Rotations

The triangularisation process may be carried out using either Householder transformations [6] or Givens rotations [7,8,9]. However, the Givens rotation method is particularly suitable for the adaptive antenna application since it leads to a very efficient algorithm whereby the triangularisation process is recursively updated as each new row of data enters the problem. A sequence of Givens rotations may be used to triangularise the matrix $\mathbf{X}(n)$ in the following recursive manner. Assume that the matrix $\mathbf{X}(n-1)$ has already been reduced to triangular form by the unitary transformation

$$\mathbf{Q}(n-1)\,\mathbf{X}(n-1) = \begin{bmatrix} \mathbf{R}(n-1) \\ \mathbf{0} \end{bmatrix}.$$

Now define the unitary matrix

$$\overline{\mathbf{Q}}(n-1) = \begin{bmatrix} \mathbf{Q}(n-1) & \mathbf{0} \\ \mathbf{0} & 1 \end{bmatrix}. \tag{3.18}$$

Clearly,

$$\overline{\mathbf{Q}}(n-1)\,\mathbf{X}(n) = \overline{\mathbf{Q}}(n-1) \begin{bmatrix} \beta\mathbf{X}(n-1) \\ \\ \mathbf{x}^T(t_n) \end{bmatrix} = \begin{bmatrix} \beta\mathbf{R}(n-1) \\ \mathbf{0} \\ \mathbf{x}^T(t_n) \end{bmatrix} \tag{3.19}$$

and so the triangularisation process may be completed by the following sequence of operations. Rotate the p-1 element vector $\mathbf{x}^T(t_n)$ with the first row of $\beta\mathbf{R}(n-1)$ so that the leading element of $\mathbf{x}^T(t_n)$ is eliminated, producing a reduced vector $\mathbf{x}'^T(t_n)$. The first row of $\mathbf{R}(n-1)$ will, of course, be modified in the process. Then rotate the $(p-2)$-element reduced vector $\mathbf{x}'^T(t_n)$ with the second row of $\beta\mathbf{R}(n-1)$ so that the leading element of $\mathbf{x}'^T(t_n)$ is eliminated, and so on until every element associated with the data vector has been rotated away. The resulting triangular matrix $\mathbf{R}(n)$ then corresponds to a complete triangularisation of the matrix $\mathbf{X}(n)$ as defined in equation (3.12). The corresponding unitary matrix $\mathbf{Q}(n)$ is simply given by the recursive expression

$$\mathbf{Q}(n) = \hat{\mathbf{Q}}(n)\,\overline{\mathbf{Q}}(n-1) \tag{3.20}$$

where $\hat{\mathbf{Q}}(n)$ is a unitary matrix representing the sequence of Givens rotation operations described above, ie

$$\hat{\mathbf{Q}}(n)\begin{bmatrix}\beta\mathbf{R}(n-1)\\ \mathbf{0}\\ \mathbf{x}^T(t_n)\end{bmatrix} = \begin{bmatrix}\mathbf{R}(n)\\ \mathbf{0}\\ \mathbf{0}\end{bmatrix}. \tag{3.21}$$

It is not difficult to deduce in addition that

$$\hat{\mathbf{Q}}(n)\begin{bmatrix}\beta\mathbf{u}(n-1)\\ \beta\mathbf{v}(n-1)\\ y(t_n)\end{bmatrix} = \begin{bmatrix}\mathbf{u}(n)\\ \beta\mathbf{v}(n-1)\\ \alpha(n)\end{bmatrix} = \begin{bmatrix}\mathbf{u}(n)\\ \mathbf{v}(n)\end{bmatrix}, \tag{3.22}$$

and this shows how the vector $\mathbf{u}(n)$ can be updated recursively using the same sequence of Givens rotations. The least squares weight vector $\mathbf{w}(n)$ may then be derived by solving equation (3.15). The solution is not defined, of course, if $n < (p-1)$ but the recursive triangularisation process may, nonetheless, be initialised by setting $\mathbf{R}(0) = \mathbf{0}$ and $\mathbf{u}(0) = \mathbf{0}$.

4 Systolic Array Architecture

Gentleman and Kung [10] have shown how the Givens rotation algorithm described above may be implemented in a very efficient pipelined manner using a triangular systolic array. The implementation of a 5-channel adaptive beamforming network using this architecture is shown in figure 4. It may be considered to comprise three distinct sections - the basic triangular array labelled ABC, the right hand column of cells labelled DE and the final processing cell labelled F. The entire array is controlled by a single clock and comprises three types of processing cell. Each cell receives its input data from the directions indicated on one clock cycle, performs the specified function and delivers the appropriate output values to neighbouring cells as indicated on the next clock cycle. Each cell within the basic triangular array stores one element of the recursively evolving triangular matrix $\mathbf{R}(n)$ which is initialised to zero at the outset of the least-squares calculation and then updated every clock cycle. Cells in the right hand column store one element of the evolving vector $\mathbf{u}(n)$ which is also initialised to zero and updated every clock cycle.

Each row of cells performs a basic Givens rotation between one row of the stored triangular matrix and a vector of data received from above so that the leading element of the received vector is eliminated. The reduced data vector is then passed downwards through the array. The boundary cell in each row computes the appropriate rotation parameters and passes them on to the right on the next clock cycle. The internal cells are subsequently used to apply the same rotation to all other elements in the received data vector. Since a delay of one clock cycle per cell is incurred in passing the rotation parameters along the row, it is necessary to impose a corresponding time skew on the input data vectors as indicated in figure 4. This arrangement ensures that as each row $\mathbf{x}^T(t_n)$ of the matrix $\mathbf{X}$ moves down through the array it interacts with the previously stored triangular matrix $\mathbf{R}(n-1)$ and undergoes the sequence of rotations $\hat{\mathbf{Q}}(n)$ described in the earlier analysis. All of its elements are thereby eliminated (one on each row of the array) and an updated triangular matrix $\mathbf{R}(n)$ is generated and stored in the process.

As each element of the vector $\mathbf{y}$ moves down through the right hand column of cells, it undergoes the same sequence of Givens rotations interacting with the previously stored vector $\mathbf{u}(n-1)$ and generating an updated vector $\mathbf{u}(n)$ in the process. The resulting output, which emerges from the bottom cell in the right hand column, is simply the value of the parameter $\alpha(n)$ in equation (3.22).

Each row of cells in figure 4 performs a complex Givens rotation of the form

$$\begin{bmatrix} c & s^* \\ -s & c \end{bmatrix} \begin{bmatrix} 0 & \cdots & 0 & , & \beta r_i & \cdots & \beta r_k & \cdots \\ 0 & \cdots & 0 & , & x_i & \cdots & x_k & \cdots \end{bmatrix} =$$

$$\begin{bmatrix} 0 & \cdots & 0 & , & r_i' & \cdots & r_k' & \cdots \\ 0 & \cdots & 0 & , & 0 & \cdots & x_k' & \cdots \end{bmatrix} \tag{4.1}$$

where the cosine parameter is assumed to be real without loss of generality. As indicated in equation (3.21), incorporating an exponential forget factor simply involves multiplying the elements of $\mathbf{R}(n-1)$ by β before performing each rotation in the sequence of eliminations. The corresponding cell functions are as specified in figure 4. The boundary cell includes an additional parameter γ which is not required within the Givens rotation procedure but will be explained in section 5. Since the matrix $\mathbf{R}$ is initialised to zero, it can be seen that the elements on its leading diagonal (i.e. the values of r within the boundary cells) may be treated as real variables throughout the computation and so the number of real arithmetic operations performed within the boundary cell is, surprisingly, less than that required for an internal cell.

4.1 Square-Root-Free Version

Gentleman [8] and Hammarling [9] have derived extremely efficient QR decomposition algorithms based on modified Givens rotations which require no square-root operation. The essence of these square-root-free algorithms is a factorisation of the form

$$\mathbf{R}(n) = \mathbf{D}^{\frac{1}{2}}(n)\,\overline{\mathbf{R}}(n)\;, \tag{4.2}$$

where

$$\mathbf{D}(n) = diag[r_{11}^2(n),\; r_{22}^2(n)\cdots r_{p-1,p-1}^2(n)]\;. \tag{4.3}$$

and $\overline{\mathbf{R}}(n)$ is a unit upper triangular matrix. The complex Givens rotation in equation (4.1) then takes the form

$$\begin{bmatrix} c & s^* \\ -s & c \end{bmatrix}\begin{bmatrix} 0 & \cdots & 0\;, & \beta\sqrt{d} & , & \cdots & \beta\sqrt{d}\,\overline{r}_k \\ 0 & \cdots & 0\;, & \sqrt{\delta}\,\overline{x}_i & , & \cdots & \sqrt{\delta}\,\overline{x}_k \end{bmatrix} =$$

$$\begin{bmatrix} 0 & \cdots & 0\;, & \beta\sqrt{d'} & , & \cdots & \beta\sqrt{d'}\,\overline{r}'_k \\ 0 & \cdots & 0\;, & 0 & , & \cdots & \sqrt{\delta'}\,\overline{x}'_k \end{bmatrix} \tag{4.4}$$

where $\sqrt{d}$ represents an element in the diagonal matrix $\mathbf{D}^{\frac{1}{2}}(n)$, and $\overline{r}_k$ denotes an associated off-diagonal element from the matrix $\overline{\mathbf{R}}(n)$. Note that each element x_k of the data vector has been expressed in the form

$\sqrt{\delta}\,\overline{x}_k$ where δ is a scaling factor which changes value as a result of the rotation.

The square-root-free algorithm may be implemented using the type of triangular systolic array depicted in figure 5. It operates in a similar manner to the one defined in figure 4 and described above. In this case, however, the boundary cells store and update elements of $\mathbf{D}(n)$ (i.e. the squares of the diagonal elements of $\mathbf{R}(n)$) while the internal cells store and update the off-diagonal elements of $\overline{\mathbf{R}}(n)$. Note that the scale quantities δ are updated only at the boundary cells and passed diagonally from one row to the next. Correct timing of the systolic array requires the δ parameter to be delayed for one clock cycle at the output of each boundary cell and the corresponding storage elements (or latches) are represented by black dots in figure 5. For the purposes of normal least squares processing, the δ parameter is initialised to unity so that on input to the array, $\overline{\mathbf{x}}(n) = \mathbf{x}(n)$. However, as pointed out by Gentleman [9], the δ parameter associated with each input vector may be assigned an arbitrary initial value which serves to weight that row of data accordingly. The systolic array in figure 5 may thus be used to perform a general weighted least squares computation.

As with the conventional Givens rotation algorithm, cells in the right hand column perform the same function as those internal to the triangular array. Factorisation applies for these also and so they update elements of the vector $\overline{\mathbf{u}}(n)$, where

$$\mathbf{u}(n) = D^{\frac{1}{2}}(n)\,\overline{\mathbf{u}}(n)\,. \tag{4.5}$$

It follows from equations (4.2), (4.5), and (3.15) that the weight vector $\mathbf{w}(n)$ is given by

$$\overline{\mathbf{R}}(n)\,\mathbf{w}(n) + \overline{\mathbf{u}}(n) = \mathbf{0}\,, \tag{4.6}$$

and may be obtained, as before, by a simple process of back substitution. Figure 5 specifies the cell functions required for one particular version of the square-root-free algorithm. Note that the parameters d and δ in each boundary cell may be treated as real variables throughout the computation assuming that they are assigned real values initially. This version of the algorithm, which requires 2 multiplications and 2 additions per cycle to be performed in each internal cell, was first suggested by Golub and reported by Gentleman [8]. We have found it to be particularly stable and accurate throughout an extensive programme of finite wordlength adaptive beamforming computations. These observations are supported by the work of Ling, Manolakis and Proakis [11] who derived an equivalent algorithm (the "error-feedback" algorithm) based on a recursive form of the modified Gram-Schmidt orthogonalisation procedure.

5 Direct Residual and Weight Extraction

In many least squares problems, and particularly in adaptive beamforming, the least-squares weight vector $\mathbf{w}(n)$ is not the principal object of interest. Of more direct concern is the corresponding residual, since this constitutes the noise-reduced output signal from a fully adaptive array [12]. In this section, we will show how the "*a-posteriori*" least squares residual

$$e(n,n) = \mathbf{x}^T(t_n)\,\mathbf{w}(n) + y(t_n)\ . \tag{5.1}$$

which depends on the most up-to-date weight vector $\mathbf{w}(n)$, may be obtained directly from the type of systolic array described in section 4 without computing $\mathbf{w}(n)$ explicitly [13].

In order to proceed, it is necessary to consider the structure of the matrix $\hat{\mathbf{Q}}(n)$ which, from equation (3.21), may be expressed in the form

$$\hat{\mathbf{Q}}(n) = \begin{bmatrix} \mathbf{A}(n) & \mathbf{0} & \phi(n) \\ \mathbf{0} & \mathbf{I} & \mathbf{0} \\ \psi^T(n) & \mathbf{0} & \gamma(n) \end{bmatrix} \tag{5.2}$$

where $\mathbf{A}(n)$ is a $p-1 \times p-1$ matrix, $\phi(n)$ and $\psi(n)$ are $p-1 \times 1$ vectors and $\gamma(n)$ is a scalar. Now $\hat{\mathbf{Q}}(n)$ is a product of elementary rotations of the form

$$\hat{\mathbf{Q}}_i(n) = \begin{bmatrix} 1 & & & & & & & \\ & \ddots & & & & & & \\ & & 1 & & & & & \\ & & & c_i(n) & & \cdots & & s_i^*(n) \\ & & & & 1 & & & \\ & & & & & \ddots & & \\ & & & & & & 1 & \\ & & & -s_i(n) & & \cdots & & c_i(n) \end{bmatrix} \qquad (i = 1, 2, \ldots, p-1) \tag{5.3}$$

where all off-diagonal elements are zero except those in the (i, n) and (n, i) locations. It follows directly that

$$\gamma(n) = \prod_{i=1}^{p-1} c_i(n) \tag{5.4}$$

i.e. $\gamma(n)$ is the product of the cosine terms in the $p-1$ rotations represented by $\hat{\mathbf{Q}}(n)$. Multiplying both sides of equation (3.21) by $\hat{\mathbf{Q}}^H(n)$ and noting that the matrix $\hat{\mathbf{Q}}(n)$ is unitary, it is now easy to deduce that

$$\mathbf{x}^T(t_n) = \phi^H \mathbf{R}(n) \tag{5.5}$$

and similarly, from equation (3.22) we have

$$y(t_n) = \phi^H(n)\ \mathbf{u}(n) + \gamma(n)\ \alpha(n) \tag{5.6}$$

Substituting equations (5.5) and (5.6) into equation (5.1) leads to the expression

$$e(n, n) = \phi^H(n)\mathbf{R}(n)\mathbf{w}(n) + \phi^H(n)\mathbf{u}(n) + \gamma(n)\,\alpha(n) \tag{5.7}$$

and, from equation (3.15), it follows immediately that

$$e(n, n) = \gamma(n)\,\alpha(n) \tag{5.8}$$

Now, as noted in section 4, when the conventional Givens rotation algorithm is employed, the parameter $\alpha(n)$ is generated quite naturally within the triangularisation process and simply emerges from the bottom cell of the right hand column in figure 4 eight (in general $2p-2$) clock cycles after the first element of $\mathbf{x}(t_n)$ enters the array. The scalar $\gamma(n)$, as given by equation (5.4), may also be computed very simply. The product of cosine terms is generated recursively by the parameter γ as it passes from cell to cell along the chain of boundary processors. As with the δ parameter in figure 5, correct timing of the systolic array requires this parameter to be delayed by one clock cycle at the output of each boundary cell. The simple product required to form the *a-posteriori* residual as given in equation (5.8) is computed by the final processing cell F in figure 4. Note that the scalar residual is well-defined by the processor computations, even when the data matrix is not of full rank and the weight vector cannot be computed uniquely.

The *a-posteriori* residual may be computed in a similar manner when the square-root-free algorithm is employed. The square-root-free algorithm delivers from the bottom cell in the right-hand column a scalar $\overline{\alpha}(n)$ given by

$$\delta^{\frac{1}{2}}(n)\,\overline{\alpha}(n) = \alpha(n)\,, \tag{5.9}$$

where $\delta^{\frac{1}{2}}$ is the scaling parameter appropriate to the $(p-1)$th row at time t_n and it has been assumed that $\delta(n)$ is initialised to unity on input to the array. From equations (5.8) and (5.9) it follows that

$$e(n,n) = \gamma(n)\,\delta^{\frac{1}{2}}(n)\,\overline{\alpha}(n)\,, \tag{5.10}$$

where $\gamma(n)$ is the product of cosine terms which arise in the conventional Givens rotation algorithm. However, $\delta(n)$, as computed by the boundary processors for the square-root-free algorithm, is simply the product of all the $\overline{c}$ terms and it can easily be shown that this is equivalent to the product of the conventional cosine parameters squared. Thus,

$$\delta(n) = \gamma^2(n)\,, \tag{5.11}$$

and

$$e(n,n) = \delta(n)\,\overline{\alpha}(n)\,. \tag{5.12}$$

Hence the *a-posteriori* residual $e(n,n)$ may also be obtained directly from the square-root-free processor array, using a final multiplier cell F as illustrated in figure 5.

5.1 Weight flushing and freezing

It has been shown that if a data vector $[\mathbf{x}^T(t_n),\ y(t_n)]$ is input to the triangular systolic array in figure 4 or figure 5, the corresponding *a-posteriori* least squares residual e(n,n) emerges from the final cell F after the appropriate number of clock cycles. In order to achieve this result, the array performs two distinct functions: (1) It generates the updated triangular matrix $\mathbf{R}(n)$ and corresponding vector $\mathbf{u}(n)$ (or $\mathbf{D}(n)$, $\overline{\mathbf{R}}(n)$ and $\overline{\mathbf{u}}(n)$ for the square-root-free algorithm) and hence, implicitly, the updated weight vector $\mathbf{w}(n)$: (2) It acts as a simple filter which applies the updated weight vector to the input data according to equation (5.1).

If the array is subsequently "frozen" in order to suppress any further update of the stored values, but allowed to function normally in all other respects, it will continue to perform the filtering operation without affecting the implicit weight vector $\mathbf{w}(n)$. Thus, in response to an input of the form $[\mathbf{x}^T,\ y]$, the frozen network will produce the output value

$$e = \mathbf{x}^T\mathbf{w}(n) + y \tag{5.13}$$

Let ϕ_i denote the $p-1$ element vector whose only non-zero element occurs in the i-th location: It follows that the effect of inputting the sequence of vectors $[\phi_i,\ 0]$ $(i = 1,2....p-1)$ to the frozen array is to produce the sequence of output values $w_i(n)$ $(i = 1,2....p-1)$ and so the weight vector $\mathbf{w}(n)$ may also be obtained from the array without the need for an additional backsubstitution processor. This technique, which amounts to measuring the impulse response of the system, is generally referred to as "weight flushing" [12].

When the square-root-free algorithm is employed, the systolic array in figure 5 may be frozen very simply by setting the forget factor $\beta = 1$ and initialising the parameter δ to zero for any input vector which is to be processed in the frozen mode. As pointed out in section 4, this has the effect of assigning zero weight to that vector within the overall least squares computation and so the processing does not affect any values stored within the array. This property can be verified quite easily by inspecting the square-root-free cells in figure 5. Note that the parameter δ_{in} in for the final processing cell must be set equal to one to avoid suppressing the output residual from the frozen network. Freezing the array is also quite straightforward using the conventional Givens rotation algorithm. However, in this case, an explicit frozen mode of operation must be defined for each type of cell and the method will not be discussed further in this paper.

The operation of the frozen network may also be understood in terms of basic matrix operations [14]. Consider first the basic triangular array ABC in figure 5. In frozen mode, the square-root-free boundary and internal cells effectively perform the reduced processing functions defined in figures 6(a) and 6(b). Now consider the effect of the simplified network upon a row vector $\mathbf{x}^T$, input from above in the usual time-staggered manner. This will give rise to a column vector z which emerges from the right in a similar time-staggered manner after four (in general $p-1$) clock cycles. It is straightforward to verify that the *input* vector $\mathbf{x}$ is related to the *output* vector $\mathbf{z}$ by means of the matrix transformation

$$\mathbf{x} = \overline{\mathbf{R}}^T(n)\,\mathbf{z}\,, \tag{5.14}$$

where $\overline{\mathbf{R}}(n)$ is the unit upper triangular matrix which was defined in equation (4.2) and is stored in the internal cells of the array. Since $\overline{\mathbf{R}}(n)$ is non-singular (no diagonal element of $\overline{\mathbf{R}}(n)$ is zero), it follows that

$$\mathbf{z} = \overline{\mathbf{R}}^{-T}(n)\,\mathbf{x}\,. \tag{5.15}$$

and so the frozen triangular array may be regarded as an $\overline{R}^{-T}$ matrix operator as depicted schematically in figure 6(c).

Now consider the right hand column of cells DE in figure 5. It is easy to show that, in frozen mode, the effect of this array upon a column vector a input from the left in the usual time-staggered manner and a scalar y input from the top (at the same time as the first element of a) is to produce the scalar output $y - \mathbf{a}^T\overline{\mathbf{u}}(n)$ which emerges from the bottom cell after four (in general $p-1$) clock cycles. The vector a also emerges unchanged from the right as depicted in figure 6(d). It follows immediately that, in frozen mode, the effect of the network in figure 5 upon a vector $[\mathbf{x}^T, y]$ input from the top is to produce the scalar output $y - \mathbf{x}^T\overline{\mathbf{R}}^{-1}(n)\mathbf{u}(n)$ which emerges from the internal cell E eight (in general $2p-2$) clock cycles after the first element of x enters the array. But, from equations (4.6) and (5.13), it can be seen that this is precisely the result expected from the frozen network. Assuming that the parameter δ_{in} is set equal to one for the final processing cell F, the same result will emerge as the final output of the array one cycle later.

6 Application of Linear Constraints

So far in this paper, we have only considered the task of minimising the output power from an adaptive combiner of the canonical configuration illustrated in figure 2. In particular, we have shown how a triangular systolic array of the type illustrated in figures 4 and 5 may be used to minimise the norm of a residual vector $\mathbf{e}(n)$ in the form of equation (3.6). This type of processor will henceforth be referred to as a "canonical processor of order p", where $p-1$ is the order of the weight vector $\mathbf{w}$. In this section we show how a canonical processor may also be used to minimise the power of the p-element combined signal in equation (2.2) subject to N simultaneous constraints of the type defined in equation (2.3). Formulated as a least squares problem, this amounts to finding the p-element weight vector $\mathbf{w}(n)$ which minimises the norm of

$$\mathbf{e}(n) = \mathbf{X}(n)\mathbf{w}(n) \tag{6.1}$$

subject to N linear equality constraints. These may be expressed in matrix form as

$$\mathbf{C}\,\mathbf{w}(n) + \mathbf{m} = 0 \tag{6.2}$$

where

$$\mathbf{C} = [\mathbf{c}_1, \mathbf{c}_2, \ldots, \mathbf{c}_N]^T \tag{6.3}$$

is an $N \times p$ matrix of constraint vectors, and

$$\mathbf{m} = -[\mu_1, \mu_2, \ldots, \mu_N]^T \tag{6.4}$$

is a vector of scalar gain factors. It is assumed throughout the following discussion that the principal $N \times N$ submatrix of $\mathbf{C}$ is non-singular. The constraint matrix may then be rendered trapezoidal by the process of orthogonal decomposition of its first N columns. In particular, there exists an $N \times N$ unitary matrix $\mathbf{Q}_0$ such that

$$\mathbf{D}_0^{\frac{1}{2}}[\mathbf{C}', \mathbf{m}'] = \mathbf{Q}_0[\mathbf{C}, \mathbf{m}], \tag{6.5}$$

where $\mathbf{D}_0$ is a diagonal matrix,

$$\mathbf{C}' = [\mathbf{T}, \mathbf{V}] \tag{6.6}$$

and $\mathbf{T}$ is an $N \times N$ unit upper triangular matrix. In effect, multiplication of $\mathbf{C}$ by $\mathbf{Q}_0$ performs a QR decomposition of the principal $N \times N$ submatrix. The constraint equation (6.2) now reads

$$\mathbf{C}'\mathbf{w}(n) + \mathbf{m}' = 0 \tag{6.7}$$

If the vector of weights $\mathbf{w}(n)$ is partitioned in the form

$$\mathbf{w}^T(n) = [\mathbf{w}_a^T(n),\ \mathbf{w}_b^T(n)], \tag{6.8}$$

where $\mathbf{w}_a(n)$ and $\mathbf{w}_b(n)$ are N-element and $(p-N)$-element vectors respectively, then it follows from equation (6.6) that the constraint equation (6.7) may be written as

$$\mathbf{T}\,\mathbf{w}_a(n) + \mathbf{V}\,\mathbf{w}_b(n) + \mathbf{m}' = 0 \tag{6.9}$$

and, as $\mathbf{T}$ is non-singular, it follows that

$$\mathbf{w}_a(n) = -\mathbf{T}^{-1}\,\mathbf{V}\,\mathbf{w}_b(n) - \mathbf{T}^{-1}\mathbf{m}'. \tag{6.10}$$

When the data matrix $\mathbf{X}(n)$ is also conformably partitioned into the $n \times N$ matrix $\mathbf{X}_a(n)$ and the $n \times (p-N)$ matrix $\mathbf{X}_b(n)$ as defined by

$$\mathbf{X}(n) = [\mathbf{X}_a(n),\ \mathbf{X}_b(n)] \tag{6.11}$$

then the constrained residual vector $\mathbf{e}(n)$ in equation (6.1) may be written in the form

$$\mathbf{e}(n) = \mathbf{X}_a(n)\,\mathbf{w}_a(n) + \mathbf{X}_b(n)\,\mathbf{w}_b(n) \tag{6.12}$$

Substituting for $\mathbf{w}_a(n)$ from equation (6.10) into equation (6.12) then gives the form of the residual vector as

$$\mathbf{e}(n) = \mathbf{X}'(n)\,\mathbf{w}_b(n) + \mathbf{y}'(n) \tag{6.13}$$

where

$$\mathbf{X}'(n) = \mathbf{X}_b(n) - \mathbf{X}_a(n)\,\mathbf{T}^{-1}\,\mathbf{V} \tag{6.14}$$

and

$$\mathbf{y}'(n) = -\mathbf{X}_a(n)\,\mathbf{T}^{-1}\mathbf{m}' \tag{6.15}$$

Since the reduced weight vector $\mathbf{w}_b(n)$ is not subject to any further constraints, it follows from equation (6.13) that minimisation of the norm of $\mathbf{e}(n)$ may be achieved using a canonical processor of order $(p - N + 1)$. The reference channel inputs are taken as successive rows $\mathbf{x}'^T(t_i)$ of the transformed matrix $\mathbf{X}'(n)$ and the primary channel inputs are successive elements $\mathbf{y}'(t_i)$ of the vector $\mathbf{y}'(n)$. These input vectors may be expressed in the form

$$\mathbf{x}'^T(t_i) = \mathbf{x}_b^T(t_i) - \mathbf{x}_a^T(t_i)\mathbf{T}^{-1}\mathbf{V}\,, \tag{6.16}$$

and

$$y'(t_i) = \mathbf{x}_a^T(t_i)\mathbf{T}^{-1}\mathbf{m}'\,, \tag{6.17}$$

where $\mathbf{x}_a^T(t_i)$ and $\mathbf{x}_b^T(t_i)$ denote the ith rows of the matrices $\mathbf{X}_a(n)$ and $\mathbf{X}_b(n)$ respectively ie

$$\mathbf{x}^T(t_i) = [\mathbf{x}_a^T(t_i),\ \mathbf{x}_b^T(t_i)] \tag{6.18}$$

The *a posteriori* constrained residual may thus be written in the (unconstrained) canonical form

$$\mathbf{e}(n,n) = \mathbf{x}'^T(t_n)\mathbf{w}_b(n) + y'(t_n)\,. \tag{6.19}$$

The data transformation defined in equations (6.16) and (6.17) may be implemented very efficiently using the type of trapezoidal preprocessor array illustrated schematically in figure 7(a). It constitutes an $N \times N$ triangular array operating in its frozen mode similar to the one depicted in figure 6(c) and an $N \times p - N + 1$ rectangular array of frozen internal cells comprising, in effect, $p - N + 1$ processor columns of the type depicted in figure 6(d). The transformed constraint matrix $\mathbf{C}' = [\mathbf{T}, \mathbf{V}]$ and the corresponding transformed gain vector $\mathbf{m}'$ are stored within the trapezoidal network as indicated. From the discussion in section 5, it can easily be shown that if a data vector $\mathbf{x}^T(t_i)$ is input to this array from the top as shown in figure 7(a) then the appropriate transformed data vector $[\mathbf{x}'^T(t_i),\ \mathbf{y}'^T(t_i)]$ emerges as the corresponding output vector from below[14].

As illustrated in figure 7(b), the combination of the trapezoidal network in figure 7(a) with a canonical adaptive least-squares processor of order $p-N+1$ simply constitutes a canonical least squares processor of order $p+1$, the top N rows of which store the trapezoidal matrix $[\mathbf{C}', \mathbf{m}']$ and operate in their frozen mode. It remains to point out that the trapezoidal matrix $[\mathbf{C}', \mathbf{m}']$ may be obtained from the original rectangular constraint matrix $[\mathbf{C}, \mathbf{m}]$ by using the canonical processor of order $p+1$ in its fully adaptive mode initially. Successive rows of $[\mathbf{C}, \mathbf{m}]$ are input as data to the adaptive network which performs the appropriate square-root-free QR decomposition quite naturally and stores the resulting matrix $[\mathbf{C}', \mathbf{m}']$ within the top N rows. These pre-processor rows are then assigned to their frozen mode of operation while the lower rows remain adaptive and the sequence of data vectors $\mathbf{x}^T(t_i)$ is input to the order p+1 triangular processor in the usual manner. In effect, the entire linearly constrained least squares minimisation procedure has been mapped onto a single triangular processor array designed initially to perform a stable and efficient QR decomposition based on square-root-free Givens rotations [15]. This serves to illustrate the great benefits which can be achieved by developing algorithms and processor architectures together and provides a good example of what the author likes to term "algorithmic engineering".

7 Discussion

In this paper, we have shown how a triangular systolic/wavefront array may be used to perform linearly constrained recursive least squares minimisation with particular application to adaptive beamforming. The array has a number of important advantages such as regular layout and nearest neighbour interconnections which render it very suitable for implementation using VLSI circuit technology. It is also based on a very stable numerical algorithm and therefore provides a good example of "algorithmic engineering". Numerical stability is an important requirement for any algorithm to be used in real time signal processing. The algorithm must produce reliable results irrespective of the data which is received and under the conditions of a finite wordlength number representation. In the design of real time signal processing equipment it is important to reduce the arithmetic wordlength as far as possible since the number of bits has a dramatic effect, not just on the cost of arithmetic processing but, more importantly perhaps, on the cost of interprocessor communication. The advantage, in this respect, of using a good numerical algorithm is clearly illustrated in figure 8. This compares the performance of the square-root-free QR decomposition algorithm represented in figure 5 with the conventional sample matrix inversion method represented in figure 2 when a 24-bit floating point number representation (16-bit mantissa and 8-bit exponent) is used.

The two algorithms were used to process identical data as produced by a computer model which simulated the effect of three equal power jamming signals received individually at levels of 0 dB relative to a thermal noise floor of -50 dB at the antenna array elements. The complex envelope of each jammer was described by an independent, narrow-band Gaussian process. The model also incorporated a desired signal received by the array at a level of 15 dB above the thermal noise floor but approximately 40 dB below the total received jamming. The SNR obtained at the output of an 8 element array is plotted as a function of the number of data vectors $\mathbf{x}(t_i)$ used in the least squares computation. It can be seen that the initial rate of adaptation is extremely rapid for both algorithms. In each case, a good level of jamming cancellation is obtained after about 10 to 20 data samples. However, with sample matrix inversion there is clear evidence of numerical instability as reflected by extreme fluctuations in the adaptive response curve. On the other hand, the QR decomposition algorithm shows no sign of numerical instability and, over the time scale shown on this graph, the SNR continues to improve with time. It was found that the sample matrix inversion technique required a floating point wordlength of 32 bits (24 bit mantissa and 8 bit exponent) to achieve comparable performance with the QR decomposition algorithm in this simulation. It cannot be assumed, of course, that this wordlength would be sufficient for an arbitrary dynamic range environment. One should only conclude that the wordlength required by the sample matrix inversion method will always be significantly greater than that for the QR decomposition algorithm.

Although the QR decomposition algorithm has excellent numerical properties, it should be noted that the constraint preprocessor technique described in section 6 is not unconditionally stable and could break down if the leading sub-matrix of the constraint matrix is not of full rank. However, McWhirter and Shepherd [15] have shown how the preprocessor technique may be generalised in a very simple way to circumvent the problem associated with zeros on the leading diagonal of the reduced constraint matrix. Furthermore, in the important case of a uniformly spaced linear antenna array, the leading sub matrix has a Vandermonde structure and must therefore be of full rank. In more general circumstances, of course, this property cannot be guaranteed but, since the constraint matrix reduction is independent of the data and could therefore be precomputed, it is possible in principle, to avoid any difficulty. A numerically superior technique for incorporating linear constraints has been proposed by Golub [16]. This is based on a column-wise triangular reduction of the constraint matrix but, unfortunately, it does not appear to be compatible with the type of systolic array processor depicted in figures 4 and 5. Bojanczyk and Luk [17] have succeeded in mapping a similar algorithm onto a modified triangular systolic array but the resulting processor appears to be much more complicated and less suitable for real time signal processing.

In this paper, we have concentrated our attention on the square-root-free Givens rotation algorithm and, in particular, the modified version defined in figure 5. The square-root-free algorithm has a number of important advantages quite apart from the simplified arithmetic operations. For example, varying the forget factor β , which determines the duration of the exponential memory, does not affect the internal cells in figure 5 and this greatly simplifies the overall control. Furthermore, the array may be frozen very simply by inputting a the value $\delta_{in} = 0$ to the first boundary cell. Since the parameter δ_{in} constitutes a general weighting factor applied to the input data, it may also be used to perform data downdating in a very simple way. Assuming that a given data vector was originally input to the array with weight $\delta in = 1$, its affect may be removed by inputting it again with weight $\delta_{in} = -\beta^n$ where n is the number of clock cycles which have elapsed. Although this technique corresponds in effect to performing a hyperbolic rotation, it was found to perform quite reliably during our finite wordlength numerical simulations. These observations would appear, in fact, to support the comments made by Van Dooren [18] during his lectures in Leuven.

A six-channel adaptive antenna processor test-bed (the AAPT) incorporating the type of triangular processor array in figures 4 and 5, has recently been constructed by a team of engineers at STL working in collaboration with scientists at Royal Signals & Radar Establishment [19]. The beam-

forming network, which comprises 33 identical processor nodes, was actually implemented as a wavefront array processor (WAP) of the type proposed by S.Y.Kung et al [20]. The overall function of the WAP is identical to that of its systolic counterpart, the only difference being that the individual node processors do not operate synchronously. Instead, the operation of each processor is controlled locally and depends on the necessary input data being available and its previous outputs having been accepted by the appropriate neighbouring processors. As a result, it is not necessary to impose a temporal skew on data input to the triangular WAP; the associated processing wavefront develops naturally within the array. An important advantage of the WAP architecture is that it avoids the problems associated with high speed clock distribution. In order to operate in this mode, every processing element must, of course, incorporate some additional circuitry to implement a bi-directional handshake on each of its input/output links and thereby ensure that the necessary communication protocol is observed. This represents an overhead which is not negligible but was easily absorbed within each node of the AAPT.
The processor node design was based on the Texas Instruments TMS32010 digital signal processing (DSP) chip since the software development system and the chips were both readily available at the time. Like most programmable DSP chips which were then available, the TMS32010 uses a 16-bit fixed point number representation and so the floating point operations required for the test-bed array have to be software programmed. This consumes a surprisingly large number of machine cycles and restricts the throughput rate which can be achieved to $\sim 10^4$ complex samples/sec from each of the six receiver channels. Although this is too slow for most real time applications, it corresponds to an overall processing rate of $\sim 4 \times 10^6$ floating point operations/sec and is more than sufficient to carry out a wide range of laboratory experiments and trials. The AAPT has been used successfully both in the anechoic chamber and in the aeriel field at STL to measure the cancellation performance which may be achieved in various realistic situations.

Since the maximum rate at which the AAPT can process data samples is a limitation not of the architecture, but only the very conservative node processor design, a programmable chip optimised for use as a floating point systolic/wavefront array node processor has recently been designed by

STL. By using one of these chips to implement each node of the AAPT, the maximum processing rate could be increased to at least 2×10^6 samples/sec (equivalent to $\sim 8 \times 10^8$ floating point operations/sec). A very high performance systolic array processor based on the architecture in figure 4 and designed using Weitek floating point chips has recently been reported by the Hazeltine Corporation [21]. This can achieve an overall processing rate in excess of 10^9 floating point operations/sec and serves to highlight the processing power which can be achieved using the architecture described in this paper.

It has not been possible in this paper to cover all of the points which I discussed during my lectures at the NATO meeting. During those lectures, for example, I described a technique for incorporating multiple independent constraints into the least squares minimisation process by using a multiple constraint post-processor in conjunction with the type of triangular systolic/wavefront array described in this paper. The array may then be applied efficiently to the important problem of Minimum Variance Distortionless Response (MVDR) beamforming [22]. I also described how the QR least squares processor could be used in a feedback mode to produce a high performance closed loop processor combining the benefits of rapid adaptation with the self correcting capabilities of a closed loop system [23]. In effect, the systolic QR network is used to implement an accelerated convergence algorithm which is analogous in many respects to the conjugate gradient technique. I trust that the references cited above will enable the interested reader to follow up these aspects

References

[1] Applebaum, S.P., "Adaptive Arrays", IEEE Trans., 1976, AP-24, pp 585-598.

[2] Widrow, B. and McCool, J.M., "A Comparison of Adaptive Algorithms based on the Methods of Steepest Descent and Random Search", IEEE Trans., 1976, AP-24, pp 615-637.

[3] Reed, I.S., Mallett, J.D. and Brennan, L.E., "Rapid Convergence Rate in Adaptive Arrays", IEEE Trans., 1974, AES-10, pp 853-863.

[4] Frost, O.L., "An Algorithm for Linearly Constrained Adaptive Array Processing", Proc. IEEE, 1971, Vol. 60, pp 661-675.

[5] Lawson, G.L. and Hanson, R.J., "Solving Least-Squares Problems", Prentice-Hall, 1974.

[6] Golub, G.H., "Numerical Methods for solving Linear Least-Squares Problems", Num. Math., 1965, 7, pp 206-216.

[7] Givens, W., "Computation of Plane Unitary Rotations Transforming a General Matrix to Triangular Form", J. Soc. Ind. Appl. Math., 1958, 6, pp 26-50.

[8] Gentleman, W.M., "Least-Squares Computations by Givens Transformations without Square-Roots", J. Inst. Maths. Applic., 1973, 12, pp 329-336.

[9] Hammarling, S., "A Note on Modifications to the Givens Plane Rotation", J. Inst. Maths. Applics, 1974, 13, pp 215-218.

[10] Gentleman, W.M. and Kung, H.T. "Matrix Triangularisation by Systolic Arrays", Proc. SPIE, 1981, 298, Real-Time Signal Processing IV.

[11] Ling, F., Manolakis, D. and Proakis, J.G., "A Recursive Modified Gram-Schmidt Algorithm for Least Squares Estimation", IEEE Trans., 1986, ASSP-34, No 4, pp 829-845.

[12] Ward, C.R. Hargrave, P.J. and McWhirter, J.G., "A Novel Algorithm and Architecture for Adaptive Digital Beamforming", IEEE Trans., 1986, AP-34, No 3, pp 338-346.

[13] McWhirter, J.G., "Recursive Least-Squares Minimisation using a Systolic Array", Proc. SPIE, 1983, 431, Real-Time Signal Processing VI, 2983

[14] Shepherd, T.J. and McWhirter, J.G., "A Systolic Array for Linearly Constrained Least Squares Optimisation", in Systolic Arrays, eds W.Moore, A.McCabe, R.Urquhart, Adam Hilger, 1987, pp 151-159.

[15] McWhirter, J.G. and Shepherd, T.J., "A Systolic Array for Linearly Constrained Least Squares Problems", Proc. SPIE, 1986, 696, Advanced Algorithm and Architectures for Signal Processing, pp 80-87.

[16] Golub, G.H. and Van Loan, C. "Matrix Computations", Oxford University Press, 1983.

[17] Bojanczyk, A.W. and Luk, F.T., "A Novel MVDR Beamforming Algorithm", Proc SPIE, 1987, 826, Advanced Algorithms and Architectures for Signal Processing II, pp 12-16.

[18] Van Dooren, P., In this Volume.

[19] McCanny, J.V. and McWhirter, J.G., "Some Systolic Array Developments in the United Kingdom", IEEE Computer, July 1987, pp 51-63

[20] Kung, S.Y., Arun, K.S., Gal-ezer, R.J., Bhaskar Rao, D.V., "Wavefront Array Processor: Language, Architecture and Applications", IEEE Trans., 1982, C-31, No. 11, pp 1054-1066.

[21] Lackey, J.R., Baurle, H.F. and Barlie, J., "Application-Specific Super Computer". Proc SPIE, 1988, 977, Real Time Signal Processing.

[22] McWhirter, J.G. and Shepherd, T.J., "Efficient MVDR Processing using a Systolic Array", Proc SPIE, 1988, 975, Advanced Algorithms and Architectures for Signal Processing III.

[23] Ward, C.R., Hargrave, P.J., McWhirter, J.G. and Shepherd, T.J., "A Novel Accelerated Convergence Technique for Adaptive Antenna Applications" Proc.Int.Conf. on Antennas and Propagation, Warwick, April 1989.

FIGURE CAPTIONS

Figure 1 Schematic of an adaptive beamforming antenna array

Figure 2 Canonical adaptive beamforming configuration

Figure 3 Conventional least squares processing architecture

Figure 4 Systolic array for recursive least squares minimisation using conventional Givens rotations

Figure 5 Systolic array for recursive least squares minimisation using square-root-free Givens rotations

Figure 6 Operation of a frozen least squares processing array

Figure 7 (a) Operation of multiple constraint preprocessor
(b) Multiple constraint preprocessor incorporated into canonical least squares processor array

Figure 8 Output Signal to Noise Ratio obtained using the QR decomposition (QRD) and Sample Matrix Inversion (SMI) algorithms on identical simulated data

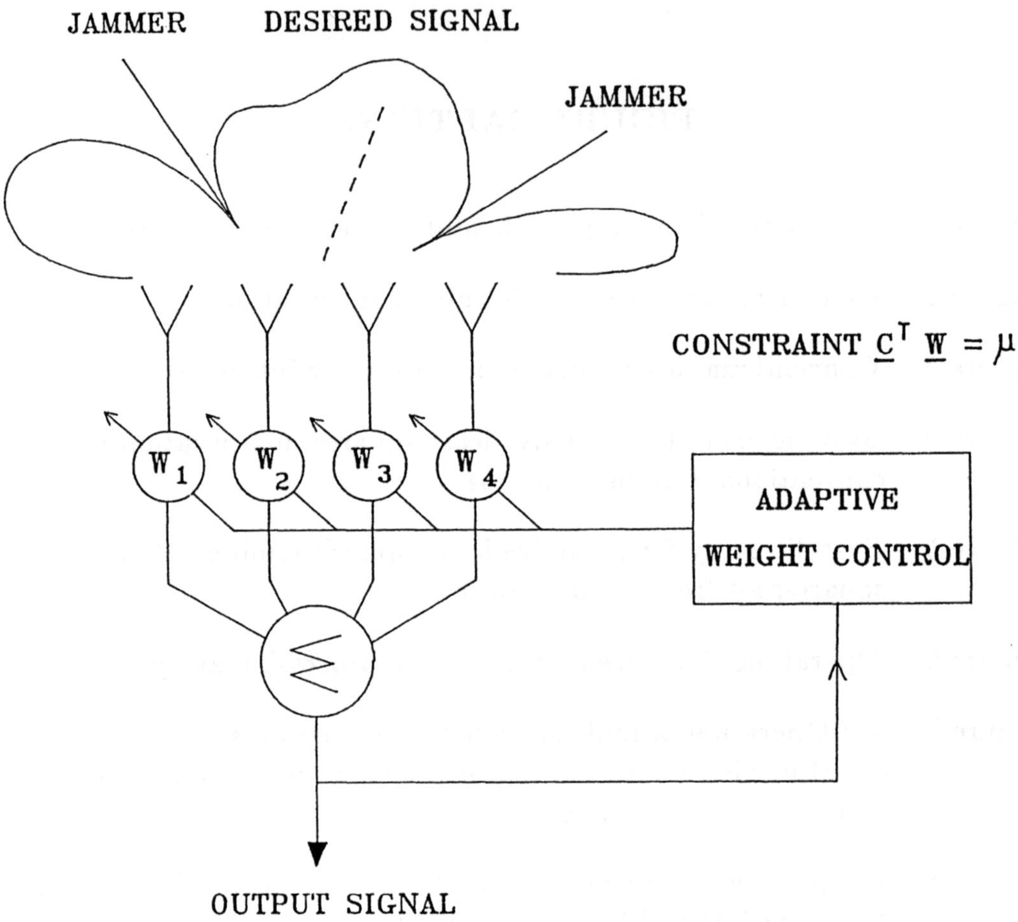
JAMMER
DESIRED SIGNAL
JAMMER
CONSTRAINT $\underline{C}^T \underline{W} = \mu$
W_1
W_2
W_3
W_4
ADAPTIVE
WEIGHT CONTROL
OUTPUT SIGNAL

Figure 1

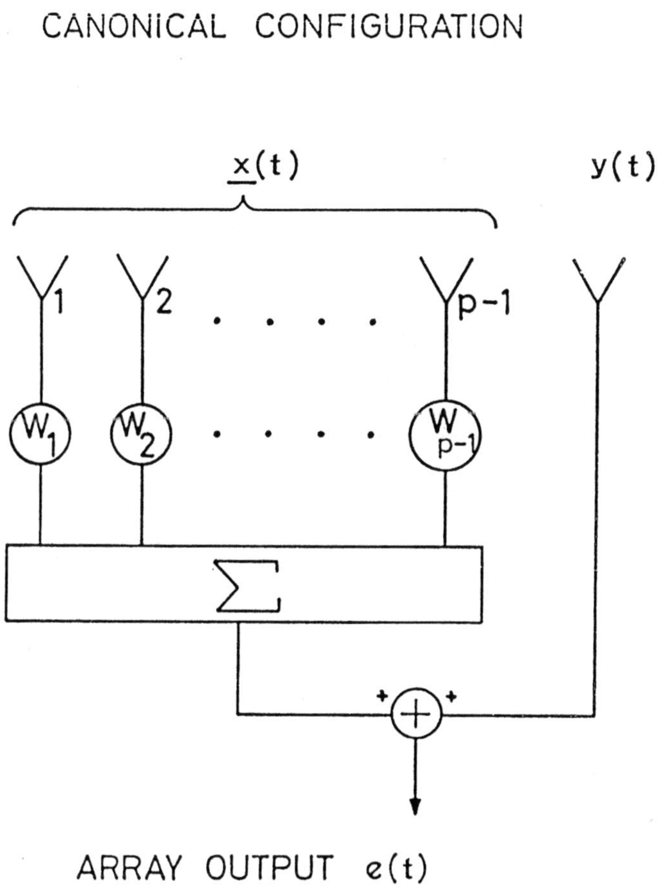
CANONICAL CONFIGURATION
x(t)
y(t)
1
2
p-1
W1
W2
Wp-1
Σ
+
+
ARRAY OUTPUT e(t)

Figurc 2

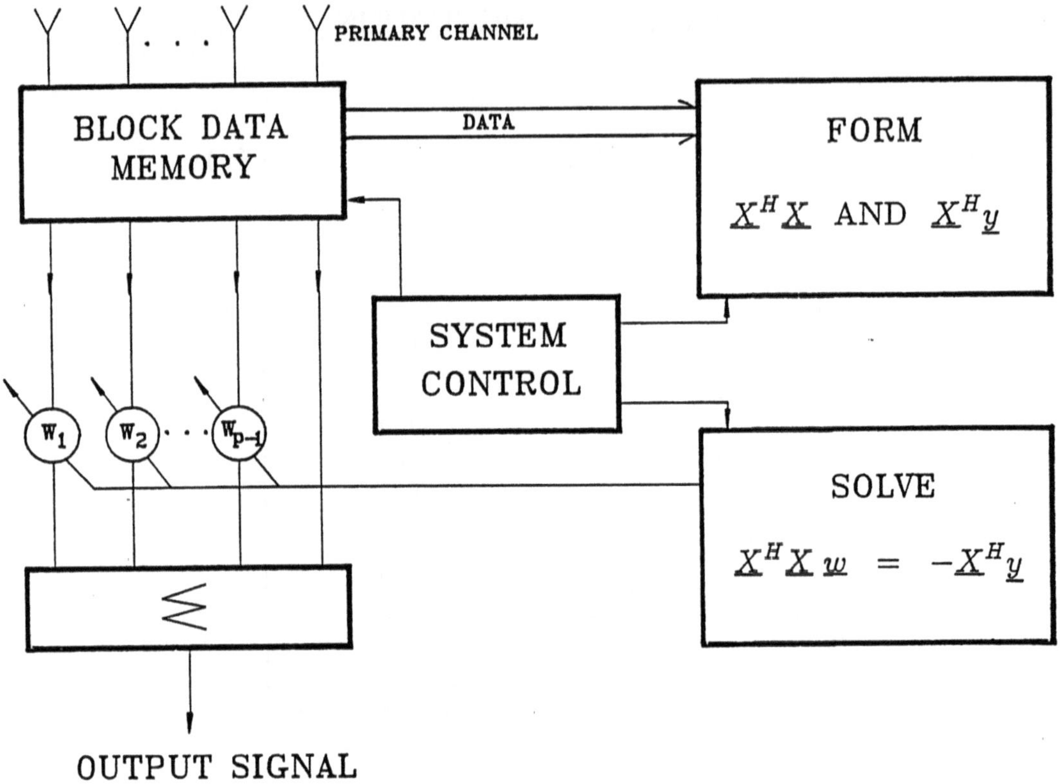
PRIMARY CHANNEL
BLOCK DATA MEMORY
DATA
FORM
X^H X AND X^H y
SYSTEM CONTROL
W1
W2
Wp-1
SOLVE
X^H X w = -X^H y
OUTPUT SIGNAL

Figure 3

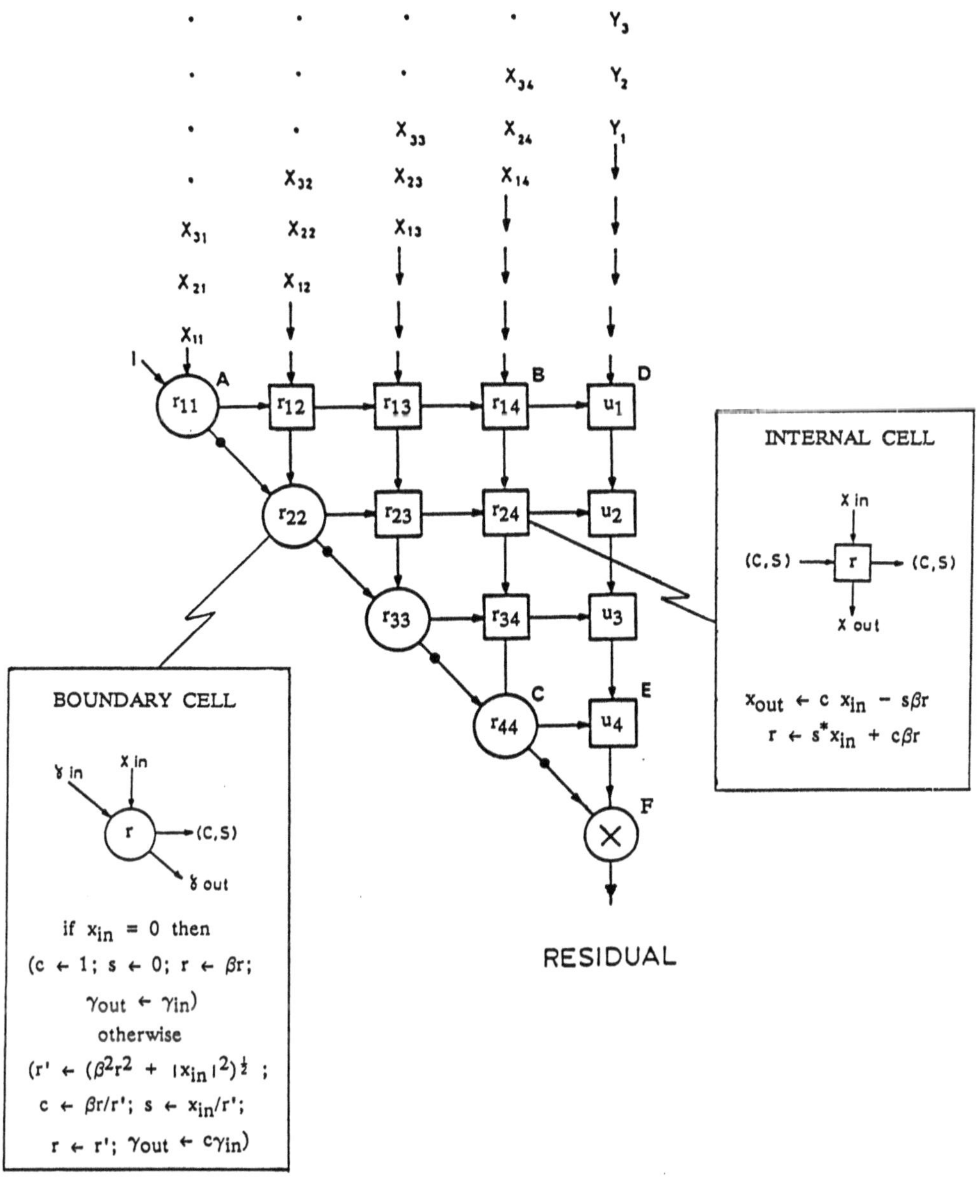

Y3
Y2
Y1
X34
X24
X14
X33
X23
X13
X32
X22
X12
X31
X21
X11
1
A
B
D
C
E
F
r11
r12
r13
r14
u1
r22
r23
r24
u2
r33
r34
u3
r44
u4
RESIDUAL
INTERNAL CELL
X in
(C,S)
r
(C,S)
X out
xout ← c xin − sβr
r ← s*xin + cβr
BOUNDARY CELL
γ in
X in
r
(C,S)
γ out
if xin = 0 then
(c ← 1; s ← 0; r ← βr;
γout ← γin)
otherwise
(r' ← (β²r² + |xin|²)^½ ;
c ← βr/r'; s ← xin/r';
r ← r'; γout ← cγin)

Figure 4

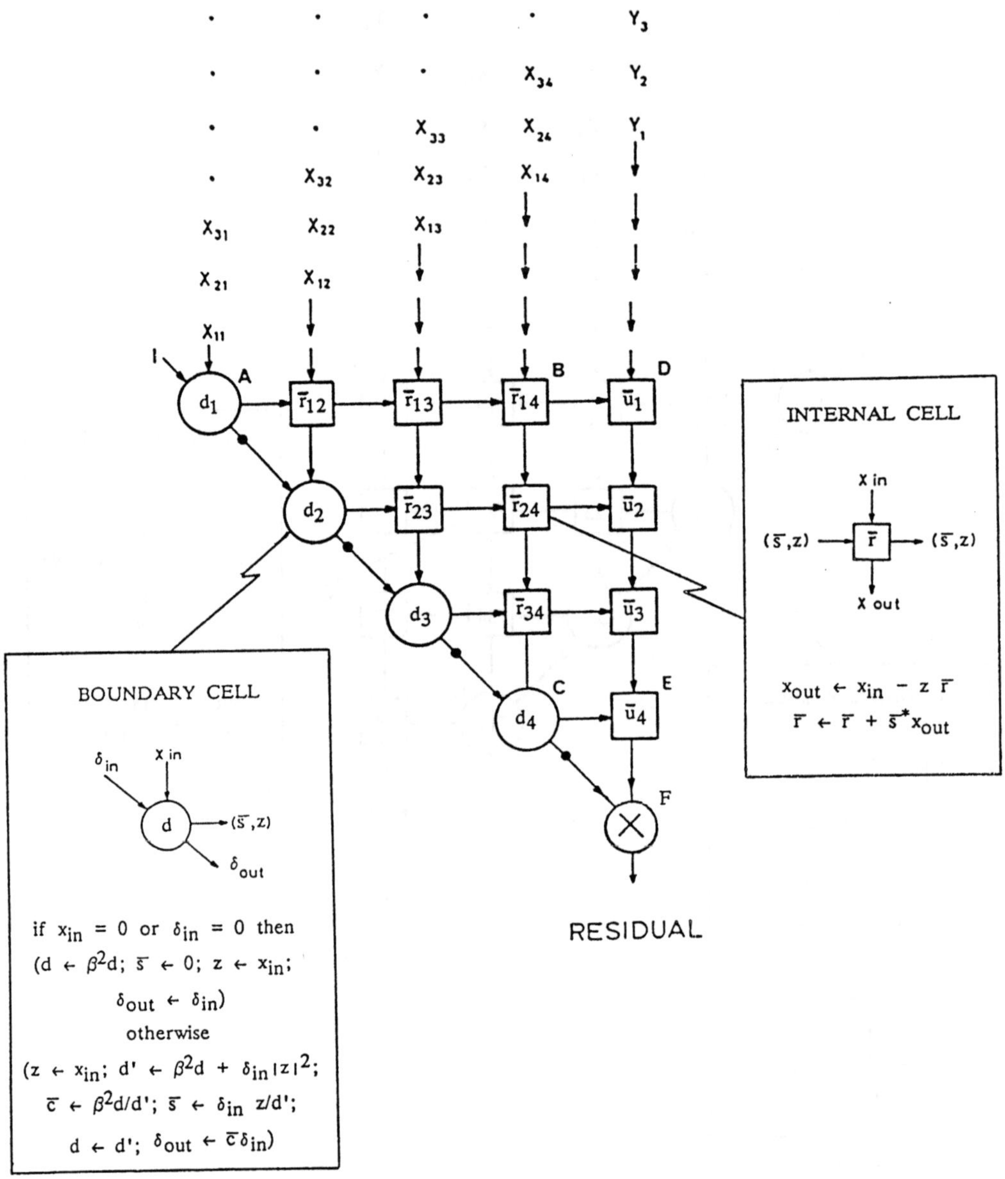

Y3
X34 Y2
X33 X24 Y1
X32 X23 X14
X31 X22 X13
X21 X12
X11
I
A
B
D
d1
r12
r13
r14
u1
d2
r23
r24
u2
d3
r34
u3
C
E
d4
u4
F
RESIDUAL
INTERNAL CELL
X in
(s,z)
r
(s,z)
X out
xout ← xin − z r
r ← r + s*xout
BOUNDARY CELL
δin
X in
d
(s,z)
δout
if xin = 0 or δin = 0 then
(d ← β2d; s ← 0; z ← xin;
δout ← δin)
otherwise
(z ← xin; d' ← β2d + δin|z|2;
c ← β2d/d'; s ← δin z/d';
d ← d'; δout ← c δin)

Figure 5

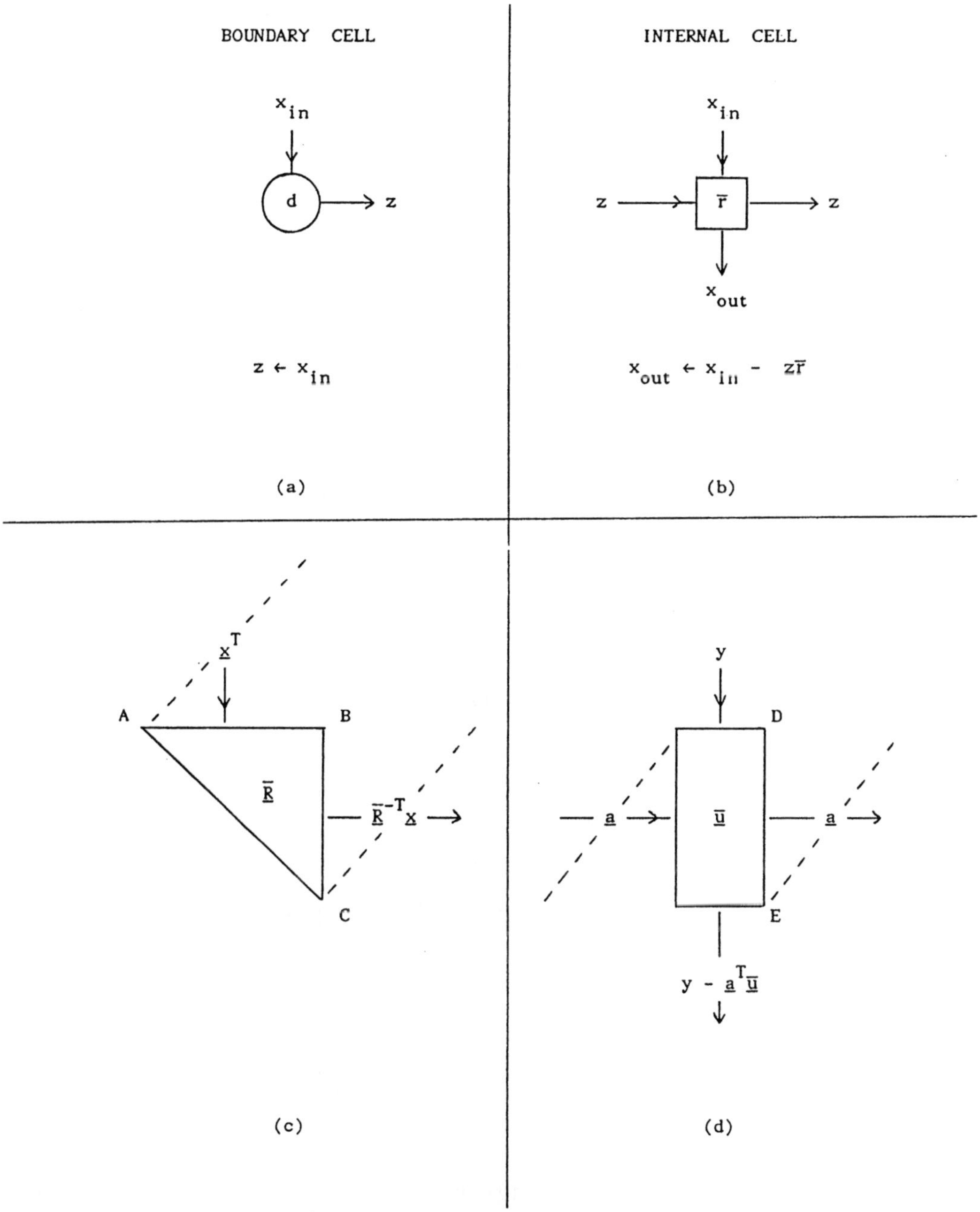
BOUNDARY CELL
x_{in}
d
z
$z \leftarrow x_{in}$
(a)
INTERNAL CELL
x_{in}
z
$\bar{r}$
z
x_{out}
$x_{out} \leftarrow x_{in} - z\bar{r}$
(b)
$\underline{x}^T$
A
B
$\bar{R}$
$\bar{R}^{-T}\underline{x}$
C
(c)
y
D
$\underline{a}$
$\bar{\underline{u}}$
$\underline{a}$
E
$y - \underline{a}^T\bar{\underline{u}}$
(d)

Figure 6

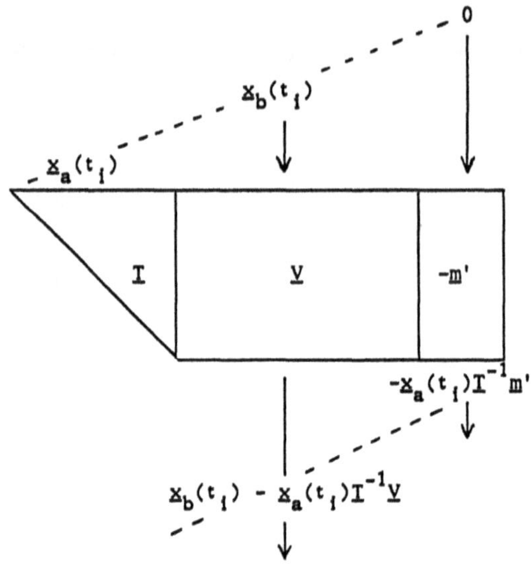
0
$x_b(t_i)$
$x_a(t_i)$
I
V
-m'
$-x_a(t_i)I^{-1}m'$
$x_b(t_i) - x_a(t_i)I^{-1}V$

(a)

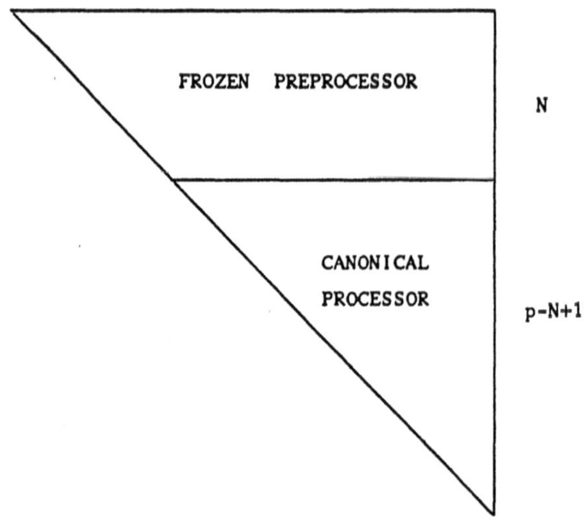
FROZEN PREPROCESSOR
N
CANONICAL
PROCESSOR
p-N+1

(b)

Figure 7

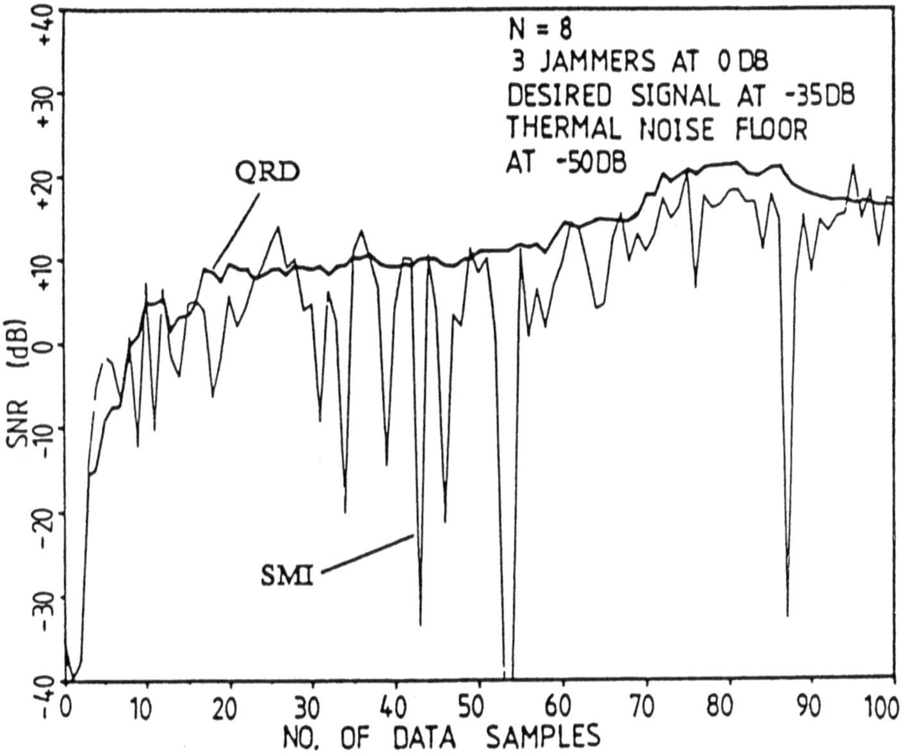
N = 8
3 JAMMERS AT 0 DB
DESIRED SIGNAL AT -35DB
THERMAL NOISE FLOOR
AT -50DB
QRD
SMI
SNR (dB)
+40
+30
+20
+10
0
-10
-20
-30
-40
0
10
20
30
40
50
60
70
80
90
100
NO. OF DATA SAMPLES

Figure 8

Parallel Algorithms for Supercomputers

Gérard Meurant

CEA, Centre d'Etudes de Limeil-Valenton
94190 Villeneuve St. Georges, France

Abstract

In this paper, we describe some of today supercomputers and we give examples of scientific applications that can be efficiently solved on these powerful machines. First, we give a classification of different architectures and we illustrate this description with some specific machines. Then, we discuss the basic issues about vector and parallel computation with examples of performance evaluations. The second part of the paper gives examples of vector and parallel algorithms. We concentrate on solving large linear systems with a parallel fast solver for the Poisson equation and the incomplete Choleski decomposition for the conjugate gradient method.

1 Introduction

During the last fifteen or twenty years there has been a rapid development of fast computers dedicated to large scale scientific computation. To investigate new problems and new algorithms, scientists are asking for computers with more and more computing power and larger and larger memories. Examples of such problems are : weather forecasting, fluid dynamics for the aeronautic industry, detection and simulation of oil reservoirs, simulations for the automobile industry,... This rapid growth of numerical simulation has lead to the design of a new class of machines which are now called Supercomputers. The aim of this paper is to briefly describe this class of machines, their properties and to give some examples of their use.

NATO ASI Series, Vol. F 70
Numerical Linear Algebra, Digital Signal Processing and Parallel Algorithms
Edited by G. H. Golub and P. Van Dooren

The first part of the paper is devoted to vector and parallel supercomputers. Section 2 gives a rough classification of machines that is illustrated in section 3 with some examples (for more details, see Hockney–Jesshope [4]). Section 4 discuss the basic issues in vector and parallel computing, that is to say the relations between the architecture of the machines and the algorithms. Finally, section 5 gives examples of performance evaluation.

The second part describes two applications of supercomputing in numerical linear algebra : a fast solver for the Poisson equation on parallel supercomputers and the use of the incomplete Choleski decomposition with the conjugate gradient method. For the latter, we give performance results on a Cray Y–MP with 8 processors.

2 Classification of supercomputers

Before looking to the different classes of supercomputers, we must define what we mean by "supercomputer". Many definitions have been given of what a supercomputer is, like, for instance,

- the fastest machine at a given time,
- a machine which is an order of magnitude slower than what the users need,
- a machine which cost is between 15 and 20 millions dollars,
- ...

Of course, this is a definition that changes with time, but today (1989) we consider a machine to be a supercomputer if the performances are in the range 500 Mflops–10 Gigaflops, a Mflops being a million of floating point operations per second, with a memory size being in the range 16 Mwords–a few Gigawords. It is likely that these machines will cost around 10–20 M$, sometimes a little more for the most advanced ones.

In this class there are both vector machines with a single processor and multiprocessor vector machines. We will describe some of the machines in more details but examples are : Cray computers (X–MP,Y–MP,2,3), ETA 10 (Control Data), Fujitsu VP, Hitachi S–820, Nec SX, etc... Notice that a machine like the Cray 1, which was one of the first and most famous supercomputers is not anymore in this list.

These machines are used for large-scale scientific computing, solving problems that require thousand and even millions of millions of floating point operations. Supercomputers are used to handle problems with more than 10^{12} –10^{13} floating point operations or to give a short turnaround time for smaller problems. For example, it has been said that a sophisticated model for weather forecasting requires 10^{12}–10^{16} operations that have to be performed within 3 hours of elapsed time; this means a sustained computing speed of 100 Mflops–1 Tflops.

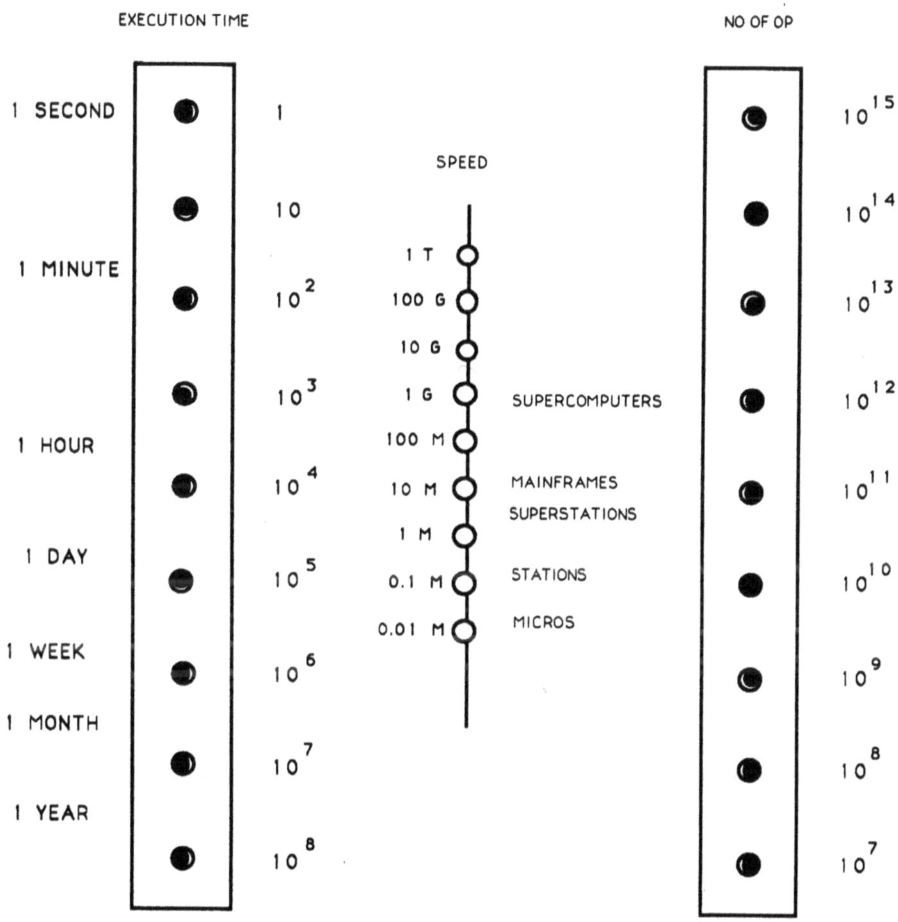

Figure 1

To see the relation between the number of operations and the computing time, it is interesting to look at Figure 1, whose idea was taken from J. Worlton. The left column gives the elapsed time, the middle one the computing speed and the right one the number of operations to be performed. To get the useful information, you simply align three points, one in each column.

It must also be added, that today's scientific computations not only require a fast machine but also a good programming environment with fancy graphic workstations for pre and post–processing, interactivity with the computer, etc... The general trend is towards distributed computing where the supercomputer can be viewed like "a number crunching" server connected to a network. The final goal can be that the supercomputer will be "invisible" for the user, with automatic vectorization and parallelization.

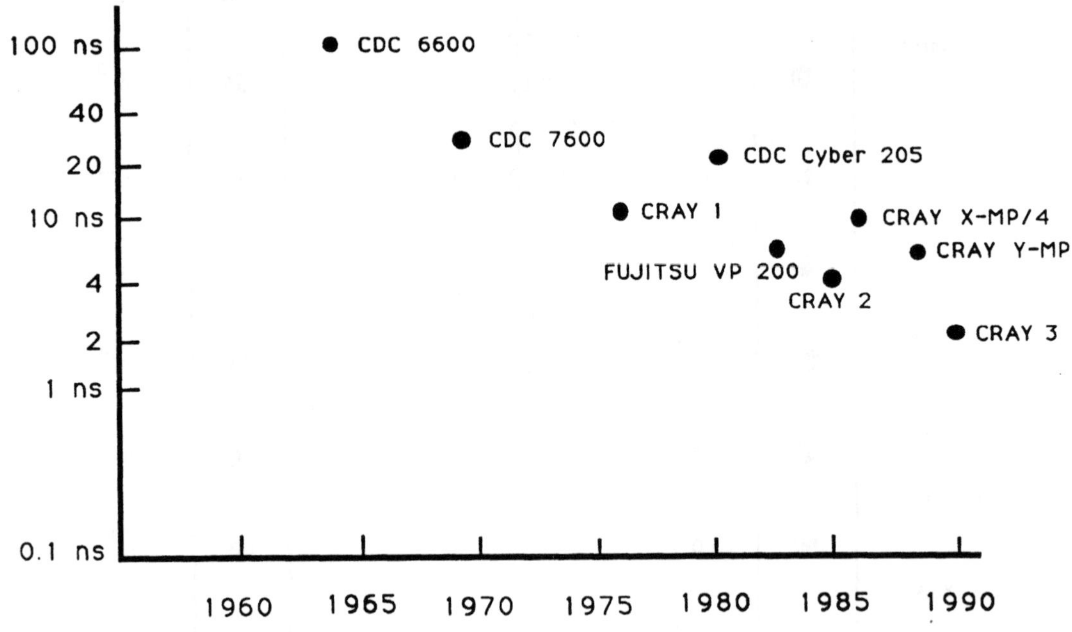

Figure 2

There was always some forms of parallelism in fast scientific computers (multiple functional units in the CDC 6600 (1964), parallelism between additions and multiplications, parallelism between I/O and computations), but this parallelism was more or less hidden from the user. The first supercomputers were single processor machines. For machine designers under the users' pressure, the main question has always been : how to improve the performances, i.e. how to get more Mflops ? For a long time the answer was : use faster electronic circuits (chips) to lower the clock period of the machine and fasten the communication speed between the memory and the computational units. Figure 2 gives the evolution of the clock period (in nanoseconds) as a function of time. It is seen that the progress is slowing down, so it is not possible anymore to rely only on chips progress to improve performance. A good way to produce faster computers is to increase parallelism. During the last five years, we have seen more and more machines with several processors that are, most of the time, tightly coupled. The recent progresses in top of the line supercomputers were given for a small part by technology (factor 1.5 or 2) and for a large part by multiple processors (4 or 8).

Many different ways can be used to classify computers (see Hockney–Jesshope [4]); we will use two different criterias : memory organization and the way processors are controlled.

For multiprocessor machines, we can roughly distinguish three main types of memory organization :

- shared memory,
- local memory,
- hierarchical memory.

In the shared memory organization (see Figure 3), all the processors have equal access to a common address space in a large global memory; this implies that there can be conflicts of accesses to some of the addresses and that these conflicts must be solved, sometimes by the hardware but most of the time by the user (the compiler or the programmer). The advantage of this organization is that there is no questions about where to implement the data.

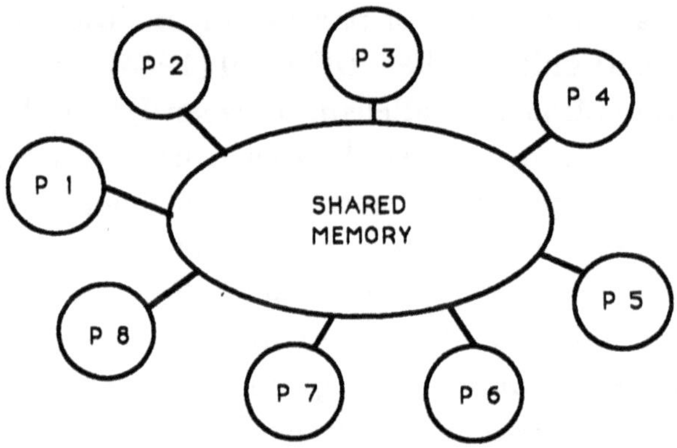

Figure 3

In a local memory organization, each processor has its own private memory that cannot be addressed directly by the other processors. Usually, data is exchanged between processors by message passing through a connection network (see Figure 4).

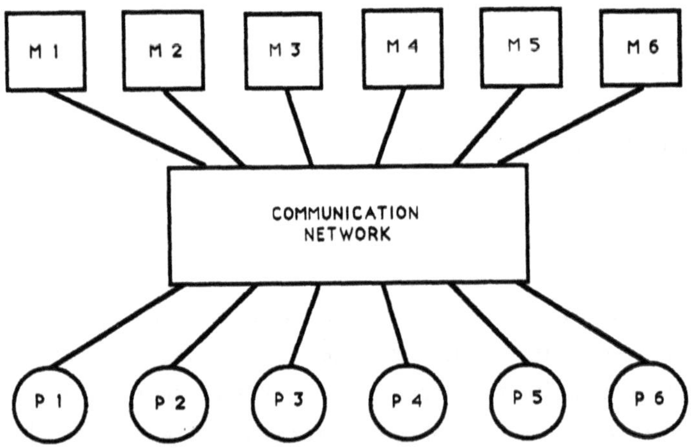

Figure 4

In local memory systems, the bottleneck often is the communication speed between processors, but the synchronizations are data driven.

A hierarchical memory is a combination of the two preceding architectures. Each processor has a local memory and there is also a global memory that can be accessed by each processor. Machines with cache can be, more or less, put into that class although the data movements between caches and memory are driven by the operating system or by hardware.

We will see some examples of machines belonging to these different classes in the next section. Another way of looking at computers is how the flow of information is controlled in the machine. There are two main types of machines : synchronous and asynchronous. For our purposes synchronous machines will be called SIMD (Single Instruction Multiple Data); the asynchronous machines will be denoted MIMD (Multiple Instruction Multiple Data). In a SIMD machine, all processors execute the same instruction (or nothing) on different data under the control of a central authority (either the master processor or the "host" computer); in a MIMD machine, each processor executes its own code, regardless of what the other ones are doing, it is the user's (compiler or programmer) responsability to synchronize the processors.

3 Examples of vector and parallel supercomputers

Vector computers are examples of SIMD machines, at least from the point of view of control. Parallelism is obtained by pipelining basic operations like the addition and multiplication of vectors. Early examples of these machines were the Cray 1 and the CDC Cyber 205. The Cray 1 was a register machine, with 8 vector registers of 64 words each. Data was transferred from memory to the registers and then from the registers to the pipelined functional units (one adder, one multiplier) where the operations were performed. The vector registers can be considered as a vector cache managed by the compiler. In this respect the Cray 1 was a hierarchical memory machine, but this fact was hidden from the user. The CDC Cyber 205 was a memory to memory machine. Data was transferred directly from memory to the functional units. Some other machines with smaller performances (called minisupers) like the SCS 40 or the Convex C2 use the same principles. Many of the parallel machines use vector computers as their building blocks. Below are given some more detailed examples of vector computers (some of them which are not supercomputers anymore). We give the clock cycle, the amount of memory and of vector registers when appropriate; the maximum theoretical speed is the performance you get when all the possible internal parallelism is used; it is the performance you are assured that you will never get. Most of the time, the actual performance is considerably lower than the theoretical peak.

CRAY 1 S :	1 processor 12.5 ns 2 Mwords (8 banks) 8 vector registers (64 words each) 12 functional units max theoretical speed : 160 Mflops
CDC Cyber 205 :	1 processor 20 ns virtual memory 16 Mwords memory to memory architecture 2 or 4 "pipes" scalar unit + vector unit max theoretical speed : 400 Mflops

A "pipe" is a set of functional units in Control data terminology.

Fujitsu VP 200 :	1 processor 7.5 ns (scalar 15 ns) 64 Mwords (256 banks) 8 K of vector registers 2 × 6 functional units max theoretical speed : 534 Mflops
NEC SX 2 :	1 processor 6 ns 128 Mwords (512 banks) 80 K bytes of vector registers

	4 sets of functional units
	max theoretical speed : 1.3 Gflops
Hitachi S–820 :	1 processor
	4 ns (scalar 8 ns)
	64 Mwords + 12 Gbytes of extended storage
	32 vector registers of 512 words
	max theoretical speed : 3 Gflops

Next, we include information about minisupers to give an idea of the differences with supercomputers :

SCS 40 :	1 processor
	45 ns
	Cray compatible
	max theoretical speed : 44 Mflops
Alliant FX/80 :	1 to 8 processors
	164.5 ns
	shared memory with a cache
	max theoretical speed : 188 Mflops
Ardent Titan :	1 to 4 processors
	125 ns
	max theoretical speed : 16 Mflops

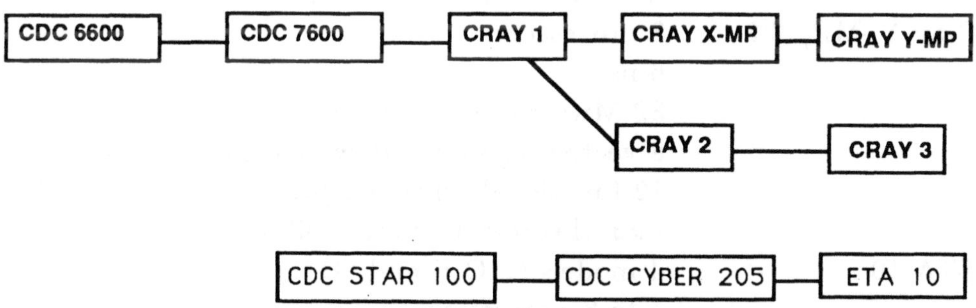

Figure 5

To increase the computing speed the designers had then used multiprocessors machine architectures. The evolution of the different machines of the main manufacturers versus time is illustrated in Figure 5. The machines designed by Seymour Cray are on the top line, although the X–MP and Y–MP were designed by Steve Chen as a continuation of the Cray 1 architecture. Below are the machines marketed by Control Data after the leave of S. Cray. Although, it appeared before the Cray 1, the CDC Star 100 was never successful because the scalar unit was too slow so the machine was unbalanced.

Next, we give descriptions of some multiprocessor vector computers that are of MIMD type.

Cray X–MP/4 :	1 to 4 processors
	8.5 ns
	up to 64 Mwords
	8 vector registers (64 words) per processor
	12 functional units per processor
	max theoretical speed : $235\times4 = 941$ Mflops
Cray 2 :	2 to 4 processors
	4.1 ns
	128, 256 (128 banks) or 512 Mwords
	8 vector registers (64 words) per processor
	local memory 16 Kwords per processor
	9 functional units per processor
	max theoretical speed : 1.9 Gflops
Cray Y–MP/832 :	8 processors
	6 ns
	32 Mwords (256 banks)
	8 vector registers (64 words) per processor
	12 functional units per processor
	max theoretical speed : $333\times8 = 2.6$ Gflops
ETA 10 :	4 models (P,Q,E and G)
	1 to 8 processors
	P: 24 ns, Q: 19 ns, E: 10.5 ns, G: 7 ns (?)
	virtual memory
	shared memory up to 256 Mwords

	local memory 4 Mwords per processor
	4 pipes
	max theoretical speed (G model): 6 Gflops
IBM 3090 VF :	1 to 6 processors
	17.2 ns
	virtual memory + cache
	scalar unit + vector unit (VF)
	max theoretical speed : 696 Mflops

We must also mention some new architectures that are under rapid development. The Connection Machine is an SIMD computer with up to 65,536 simple processors using an hypercube topology. Although this machine was not primarily designed for floating point computations, good performances have been obtained. We see also the emergence of new machines with more modest performances called personal supercomputers of which the Intel hypercube is an example. On the other end, there is a rapid growth of the computing power of super workstations like the Ardent Titan which uses multiprocessors.

4 Basic issues in vector and parallel computing

In this section, we discuss the main issues that have to be taken care of for efficient vector and parallel computing.

The first important problem is how we can describe the parallel algorithm to the compiler. This raises some portability problems, as different manufacturers use different approaches. For instance, Cray has three different ways of expressing the parallelism (synchronization of processors, protected access to shared data) : macrotasking with Fortran calls to a library of routines that call the system primitives; the overhead is large, so this is aimed at large granularity problems (subroutine level) where granularity is roughly defined as the number of operations done between two calls to the library; microtasking aimed at small granularity problems (loop level) with compiler directives to describe the parallelism; autotasking, which is a first attempt to automatic parallelism, the compiler analyses the data dependancies and tries to distribute the workload between the processors. On other machines there are different approaches, for instance IBM is also developing a parallelizing compiler.

An important issue that has to be considered carefully is the load balancing problem. To obtain the maximum efficiency, it is crucial that the processors must be kept as busy as possible. In the design of the algorithm and in coding, we must try to avoid having some idle processors waiting for the other ones to finish some work. All the tasks must be equilibrated.

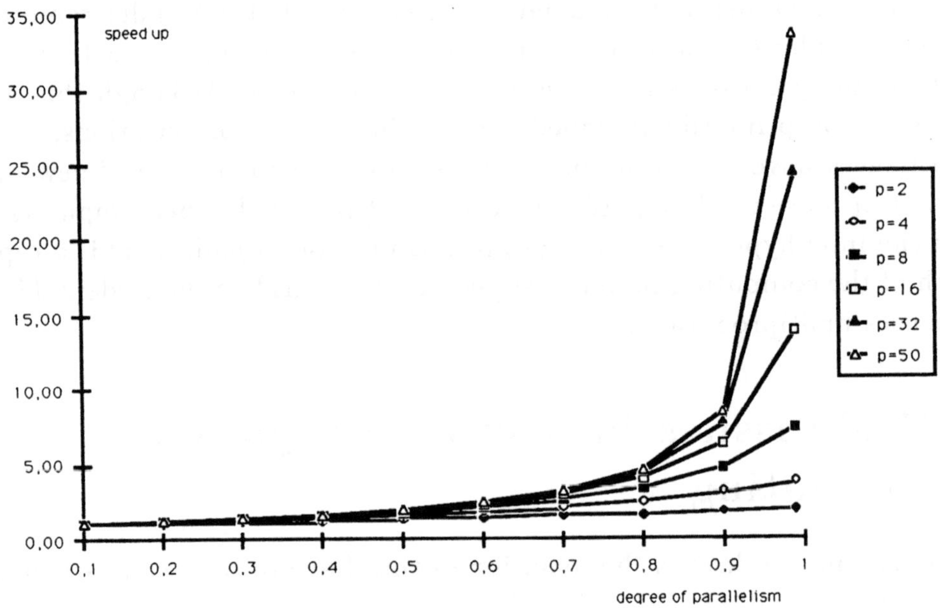

Figure 6

As we have seen, an important problem regarding the algorithm is the granularity of the basic tasks that can be executed in parallel. This is related to the overhead time spent to synchronize the processors. The granularity of the algorithm must be large enough for the computation not being overhead dominated.

The degree of parallelism of the algorithm is also important; for instance, if only 90% of the computation can be parallelized, the speed up will never be greater than 10, whatever the number of processors is. Moreover, in real life the speed up will be decreased by the overhead if the problem size is not large enough. The speed up as a function of the degree of parallelism is given by Amdahl's law whose graph is in Figure 6, p being the number of processors. The rightmost point in the curve is computed for a degree of parallelism of 99% .

5 Performance evaluation

Although, it is important to know what the peak performance of a machine is, in practical applications the computational speed we can reach is usually far from the maximum theoretical performance. This happens for several reasons : time for memory accesses, compiler overheads, the maximum amount of parallelism in the hardware is not optimally used, etc... Hence, it is interesting to assess the machines performances on real applications. Below, we give results for two benchmarks.

The first one is the famous Linpack benchmark, see Dongarra [2], for solving a dense system of linear equations. Table 1 gives the performance results (in Mflops) for a Fortran code and a 100×100 system; they are taken from Dongarra [2], October 1988. This code was not specifically designed for vector computers.

Table 1

Cray Y–MP (1 proc)	79
Cray X–MP (1 proc)	66
ETA 10–E (1 proc)	62
NEX SX–2	43
Cray 2S (1 proc)	41
Hitachi S–820	36
ETA 10–Q (1 proc)	34
ETA 10–P (1 proc)	27
Cray 1S	27
Fujitsu VP–400	20
CDC Cyber 205	17
IBM 3090/180 VF (1 proc)	12
Convex C–210	10
SCS–40	8

To show the differences in performance that can be obtained by paying attention to the characteristics of the architecture, table 2 gives the results of a code solving the same problem but with a vector unrolling technique and a problem size of 300×300.

Table 2

Cray Y–MP (8 proc)	610
Cray 2S (4 proc)	455
Hitachi S–820	440
NEX SX–2	347
Cray 2S (1 proc)	270
Cray Y–MP (1 proc)	250
Fujitsu VP–200	220
Cray X–MP (1 proc)	187
ETA 10–E (1 proc)	182
ETA 10–Q (1 proc)	101
ETA 10–P	80
Cray 1S	66
IBM 3090/180S VF (1 proc)	44
Convex C–210	41
SCS–40	37

Notice the large difference in performance between an "old dusty deck" Fortran code and one which is specially designed for vector machines. It must also be understood that the results are largely compiler dependent; on the same machine compiler progresses can greatly improve the performances : for example, the Cray 1 with the CFT 1.12 compiler gave 12 mflops compared with the 27 Mflops of CFT77 2.1 and it is likely that this will be improved again.

The other example is a conjugate gradient solver for large sparse linear systems, the model of which is given by the Poisson problem in a square domain. The performances are given in Figure 7, for different preconditioners. Going from left to right, the preconditioners are less and less vectorizable. This illustrates the differences in machine performance on codes with different characteristics. In this code, no special optimizations were attempted.

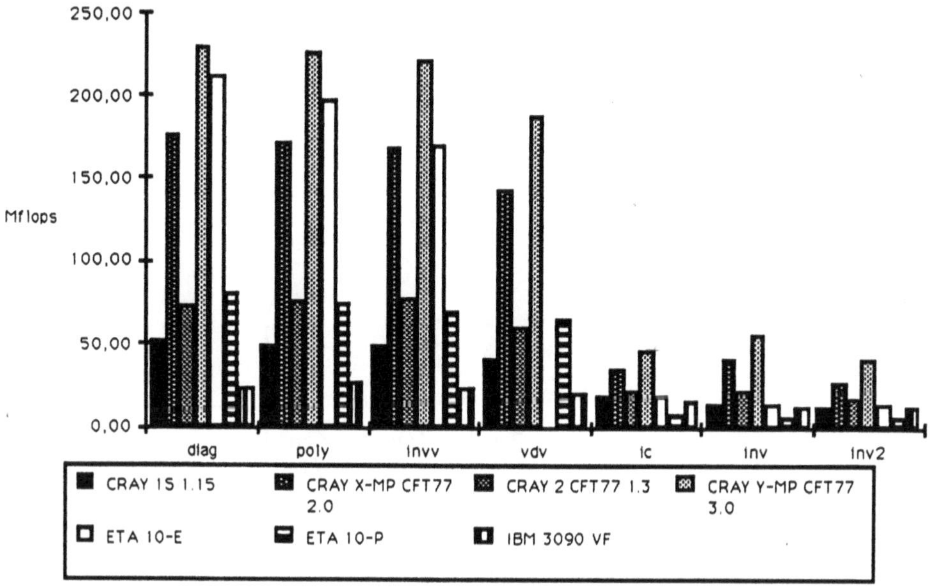

Figure 7

In section 7, we will give examples involving parallel computation on a Cray Y–MP with 8 processors.

6 A parallel fast Poisson solver

In this section we show, as an example, how to derive a parallel algorithm for solving second order separable elliptic equations in a rectangular domain. The general problem we are interested in is

$$-\frac{\partial}{\partial x}(a(x)\frac{\partial u}{\partial x}) - \frac{\partial}{\partial y}(b(y)\frac{\partial u}{\partial y}) + cu = f,$$

$$\text{in } \Omega \subset R^2,$$

$$u|_{\partial\Omega} = 0 \quad \text{or} \quad \frac{\partial u}{\partial n}|_{\partial\Omega} = 0,$$

Ω being a rectangle, although more general boundary conditions can be handled.

With the standard 5 point finite differences scheme and row–wise ordering of the mesh points, we obtain a block tridiagonal linear system :

$$A\,x \;=\; b,$$

with

$$A = \begin{pmatrix} D_1 & A_2^T & & & \\ A_2 & D_2 & A_3^T & & \\ & \ddots & \ddots & \ddots & \\ & & A_{n-1} & D_{n-1} & A_n^T \\ & & & A_n & D_n \end{pmatrix}.$$

D_i is point tridiagonal strictly diagonally dominant, A_i is diagonal. It is well known that A is a positive definite symmetric M-matrix, see Golub–Meurant [3].

One of the main ideas to introduce more parallelism is to reorder the unknowns. We partitioned the domain Ω into subdomains : $\Omega_i, i = 1, \ldots, k$, see Figure 8; we first number the unknowns within each subdomain and then the unknowns on the interfaces.

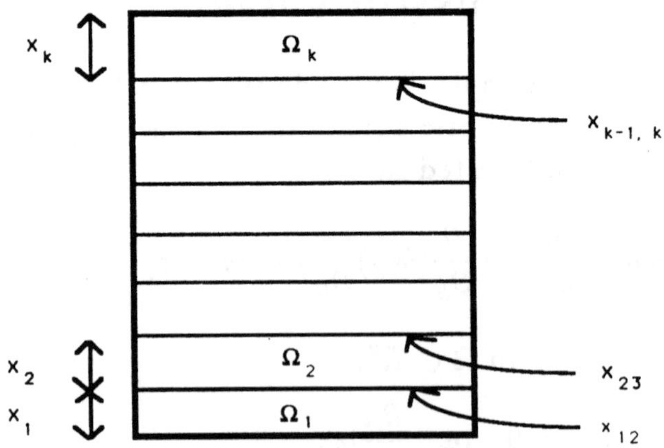

Figure 8

The reordered system is :

$$\begin{pmatrix} B_1 & & & & & C_1 & & & \\ & B_2 & & & & E_2 & C_2 & & \\ & & B_3 & & & & E_3 & \ddots & \\ & & & \ddots & & & & \ddots & C_{k-1} \\ & & & & B_k & & & & E_k \\ C_1^T & E_2^T & & & & B_{1,2} & & & \\ & C_2^T & E_3^T & & & & B_{2,3} & & \\ & & \ddots & \ddots & & & & \ddots & \\ & & & C_{k-1}^T & E_k^T & & & & B_{k-1,k} \end{pmatrix} \begin{pmatrix} x_1 \\ x_2 \\ \cdot \\ \cdot \\ x_k \\ x_{1,2} \\ x_{2,3} \\ \cdot \\ x_{k-1,k} \end{pmatrix} = \begin{pmatrix} b_1 \\ b_2 \\ \cdot \\ \cdot \\ b_k \\ b_{1,2} \\ b_{2,3} \\ \cdot \\ b_{k-1,k} \end{pmatrix}$$

Each B_i is related to a subdomain Ω_i,

$$B_i = \begin{pmatrix} D_i^1 & (A_i^2)^T & & & \\ A_i^2 & D_i^2 & (A_i^3)^T & & \\ & \ddots & \ddots & \ddots & \\ & & A_i^{m_i-1} & D_i^{m_i-1} & (A_i^{m_i})^T \\ & & & A_i^{m_i} & D_i^{m_i} \end{pmatrix}.$$

D_i^j and $B_{i,j}$ are point tridiagonal, m_i is the number of mesh lines in Ω_i. The matrices C_i and E_i have a very special structure, of which we will take advantage of in our methods,

$$C_i = \begin{pmatrix} 0 \\ \vdots \\ 0 \\ C_i^{m_i} \end{pmatrix}, \qquad E_i = \begin{pmatrix} E_i^1 \\ 0 \\ \vdots \\ 0 \end{pmatrix}.$$

$C_i^{m_i}$ and E_i^1 are diagonal matrices.

Then, we eliminate the unknowns $x_1, \ldots, x_k$ corresponding to the subdomains. The reduced matrix for the interfaces is

$$\begin{pmatrix} B'_{1,2} & F_2^T & & & \\ F_2 & B'_{2,3} & F_3^T & & \\ & \ddots & \ddots & \ddots & \\ & & F_{k-2} & B'_{k-2,k-1} & F_{k-1}^T \\ & & & F_{k-1} & B'_{k-1,k} \end{pmatrix}$$

A little algebra shows that we have the following formulas for $i = 1, \ldots, k-1$,

$$B'_{i,i+1} = B_{i,i+1} - C_i^T\, B_i^{-1}\, C_i - E_{i+1}^T\, B_{i+1}^{-1}\, E_{i+1},$$

$$F_i = -C_i^T\, B_i^{-1}\, E_i,$$

The reduced right hand side is

$$b'_{i,i+1} = b_{i,i+1} - C_i^T\, B_i^{-1}\, b_i - E_{i+1}^T\, B_{i+1}^{-1}\, b_{i+1}.$$

We can simplify these expressions by using two factorizations for the matrices B_i, a block LU one (top–down)

$$B_i = (\Delta_i + L_i)\, \Delta_i^{-1}\, (\Delta_i + L_i^T),$$

where,

$$\Delta_i = \begin{pmatrix} \Delta_i^1 & & & \\ & \ddots & & \\ & & \ddots & \\ & & & \Delta_i^{m_i} \end{pmatrix},$$

$$L_i = \begin{pmatrix} 0 & & & \\ A_i^2 & 0 & & \\ & \ddots & \ddots & \\ & & A_i^{m_i} & 0 \end{pmatrix}.$$

By identification, we get

$$\begin{cases} \Delta_i^1 = D_i^1, \\ \Delta_i^j = D_i^j - A_i^j\, (\Delta_i^{j-1})^{-1}\, (A_i^j)^T, \quad j = 2, \ldots, m_i. \end{cases}$$

The other decomposition is a block UL one (bottom–up),

$$B_i = (\Sigma_i + L_i^T)\, \Sigma_i^{-1}\, (\Sigma_i + L_i).$$

Similarly with the previous case, Σ_i is a block diagonal matrix with blocks Σ_i^j given by,

$$\begin{cases} \Sigma_i^{m_i} = D_i^{m_i}, \\ \Sigma_i^j = D_i^j - (A_i^{j+1})^T\, (\Sigma_i^{j+1})^{-1}\, A_i^{j+1}, \quad j = m_i - 1, \ldots, 1. \end{cases}$$

With these notations, we can prove the following results,

Theorem

For $i = 1, \dots, k-1$,

$$B'_{i,i+1} = B_{i,i+1} - (C_i^{m_i})^T \, (\Delta_i^{m_i})^{-1} \, C_i^{m_i} - (E_{i+1}^1)^T \, (\Sigma_{i+1}^1)^{-1} \, E_{i+1}^1.$$

If we define

$$G_i^1 = (\Sigma_i^1)^{-1} \, E_i^1,$$

$$G_i^l = -(\Sigma_i^l)^{-1} \, A_i^l \, G_i^{l-1}, \; l = 2, \dots, m_i,$$

then,

$$F_i = -(C_i^{m_i})^T \, G_i^{m_i}.$$

For the sake of simplicity, we will now concentrate on the model problem,

$$-\Delta u = f \quad \text{in } \Omega =]0,1[\text{ x }]0,1[$$

with Dirichlet boundary conditions $u|_{\partial\Omega} = 0$. Then,

$$A = \begin{pmatrix} T & -I & & & \\ -I & T & -I & & \\ & \ddots & \ddots & \ddots & \\ & & -I & T & -I \\ & & & -I & T \end{pmatrix} \text{ with } T = \begin{pmatrix} 4 & -1 & & & \\ -1 & 4 & -1 & & \\ & \ddots & \ddots & \ddots & \\ & & -1 & 4 & -1 \\ & & & -1 & 4 \end{pmatrix}.$$

T is of order n and A of order n^2. The spectral decomposition of T is $T = Q \, \Lambda \, Q^T$ where Λ is a diagonal matrix and Q is orthogonal. It is easy to see that the Λ_{jj} which are the eigenvalues of A are

$$\Lambda_{jj} = 4 - 2\cos(\frac{j\pi}{n+1}) = 2 + \sigma_j,$$

with $\sigma_j = 4\sin^2(\frac{j\pi}{2(n+1)})$, being the eigenvalues of the one dimensional Laplacian.

Then, we have

$$\Delta_i^j = Q \, \Lambda_i^j \, Q^T$$

$$\Sigma_i^j = Q \, \Omega_i^j \, Q^T$$

Λ_i^j and Ω_i^j are diagonal matrices whose diagonal entries are given by

$$(\Lambda_i^1)_{ll} = \Lambda_{ll}$$

$$(\Lambda_i^j)_{ll} = \Lambda_{ll} - \frac{1}{(\Lambda_i^{j-1})_{ll}} \quad j = 2, \ldots, m_i$$

$$(\Omega_i^{m_i})_{ll} = \Lambda_{ll}$$

$$(\Omega_i^j)_{ll} = \Lambda_{ll} - \frac{1}{(\Omega_i^{j+1})_{ll}} \quad j = m_i - 1, \ldots, 1$$

With these results, we can prove the following theorem for B'_{ij} and F_i.

Theorem

$$B'_{ij} = Q\ (\Lambda - (\Lambda_i^{m_i})^{-1} - (\Omega_j^1)^{-1})\ Q^T$$

Moreover

$$G_i^j = Q\ \Theta_i^j\ Q^T,$$

where Θ_i^j is a diagonal matrix with

$$\Theta_i^1 = -(\Omega_i^1)^{-1}$$

$$\Theta_i^j = (\Omega_i^j)^{-1} \Theta_i^{j-1},$$

then, the spectral decomposition of F_i is

$$F_i = Q\ \Theta_i^{m_i}\ Q^T$$

For the simple case of the model problem, we can analytically compute the elements of Λ's and Ω's i.e. the eigenvalues of B'_{ij} and F_i. Consider the sequence λ_i defined as

$$\lambda_1 = \sigma$$

$$\lambda_i = \sigma - \frac{1}{\lambda_{i-1}},$$

σ being a real number. The solution of this recurrence is for $\sigma \neq 2$

$$\lambda_i = \frac{r_+^{i+1} - r_-^{i+1}}{r_+^i - r_-^i},$$

where r_+ and r_- are the solutions of $r^2 - \sigma r + 1 = 0$. With this result, we have

Theorem

The eigenvalues μ_{ij}^l of B_{ij}^l are

$$\mu_{ij}^l = 2 + \sigma_l - \frac{r_+^{m_i} - r_-^{m_i}}{r_+^{m_i+1} - r_-^{m_i+1}} - \frac{r_+^{m_j} - r_-^{m_j}}{r_+^{m_j+1} - r_-^{m_j+1}}.$$

The eigenvalues δ_i^l of F_i are

$$\delta_i^l = \frac{r_+ - r_-}{r_+^{m_i+1} - r_-^{m_i+1}},$$

with $r_\pm = 1 + \frac{\sigma_l}{2} \pm (\sigma_l + \frac{\sigma_l^2}{4})^{\frac{1}{2}}$.

As we know the eigenvalues and eigenvectors of B_{ij}^l and F_i we are able to efficiently solve the reduced system. Hence, we can define the following parallel algorithm :

- compute in parallel the right hand side of the reduced system (subdomain solves),
- compute the solution of the reduced system,
- compute in parallel the solutions for the subdomains.

To compute the solution of the reduced system, we have several choices that are more or less parallel :

◦ if the order of the system is large enough, we can apply the same method recursively,

◦ otherwise, we can solve the reduced system with the Fourier tridiagonal method that is explained below.

We factor the reduced system as

$$(K + F)\, K^{-1}\, (K + F^T)$$

with

$$K = \begin{pmatrix} K_{12} & & & \\ & K_{23} & & \\ & & \ddots & \\ & & & K_{k-1,k} \end{pmatrix} \quad \text{and} \quad F = \begin{pmatrix} 0 & & & \\ F_2 & 0 & & \\ & \ddots & \ddots & \\ & & F_{k-1} & 0 \end{pmatrix}$$

As before,

$$K_{12} = B'_{12}, \quad K_{ij} = B'_{ij} - F_i \left(B'_{i-1,j-1}\right)^{-1} F_i^T$$

As we already know the eigenvalues of B'_{ij} and F_i, we can compute the eigenvalues ω^l_{ij} of K_{ij}.

$$\omega^l_{12} = \mu^l_{12}, \quad \omega^l_{ij} = \mu^l_{ij} - \frac{\left(\delta^l_i\right)^2}{\omega^l_{i-1,j-1}}$$

All these matrices have the same eigenvectors (given by Q), so the solution of the reduced system is obtained by (see Golub–Meurant [3])

- FFT to transform the right hand side (multiplication by Q^T),
- solving in parallel $k-1$ independent tridiagonal systems to get the Fourier transform of the solution,
- inverse FFT to obtain the solution (mutiplication by Q).

The parallelism in this algorithm is obtained during the FFTs, in the parallel solve of independent tridiagonal systems and solving for the subdomains. Domain Decomposition gives a large grain parallelism (subdomain solves) and the Fourier tridiagonal method a small grain parallelism.

7 Conjugate gradient solvers

In this section we consider another type of vector and parallel algorithm, the conjugate gradient (CG) method for solving large sparse linear systems of which the matrix is symmetric and positive definite. We would like to solve

$$Ax = b,$$

but in fact, we will solve

$$M^{-1}Ax = M^{-1}b,$$

where M is a preconditioning matrix. As a test problem, we will consider the Poisson problem in a square, although as we have seen in the previous section there are more efficient methods to solve it.

Given an initial guess x^0, the algorithm is

$$r^0 = b - Ax^0,$$

for $k = 0, 1, \ldots$ until convergence,

$$Mz^k = r^k,$$

$$\beta_k = \frac{(r^k, z^k)}{(r^{k-1}, z^{k-1})}, \ \beta_0 = 0,$$

$$p^k = z^k + \beta_k p^{k-1},$$

$$\alpha_k = \frac{(r^k, z^k)}{(Ap^k, p^k)},$$

$$x^{k+1} = x^k + \alpha_k p^k,$$

$$r^{k+1} = r^k - \alpha_k Ap^k.$$

If we put aside the problem of finding a good preconditioner M, it is clear that CG is a good method for vector computers as there exist efficient implementations of the basic CG operations :

- 3 saxpys ($y = y + ax$, a being a scalar) or saxpy–like operations
- 2 dot products
- 1 matrix vector product.

One must not forget also, the computation of the stopping criteria and the test for convergence.

The matrix vector product can cause problems depending on how the matrix A is stored. From now on, we will suppose A is stored by diagonals.

As an example, consider in table 3 the asymptotic Mflops rates (vector length of 10000) for these basic operations on one processor of a Cray X–MP, Y–MP (with a 6.4 ns clock) and of the ETA 10–E.

Table 3

operation	CRAY X–MP	CRAY Y–MP	ETA 10–E
saxpy	185	220	378
dot product	209	282	187
matrix vector	169	235	187

Hence, we see that we can obtain very good performances for the basic operations. However, for preconditioned CG, the most important part is solving $Mz^k = r^k$. So, we must develop preconditioners adapted to vector and parallel computation. Of course, this is not too much of a problem (consider, for instance, the Jacobi or diagonal preconditioner); the difficult task is to have both parallelism and efficiency (that is a small number of iterations). The properties that are requested for the preconditioner M are :

- M symmetric positive definite
- M sparse
- M easy to construct
- $Mz = r$ easy to solve on a vector or parallel computer
- "good" distribution of the eigenvalues of $M^{-1}A$.

The simplest preconditioner is the diagonal one (DIAG),

$$M = diag(A).$$

Unfortunately, this preconditioner, even though it is perfectly parallel, is not efficient regarding the rate of convergence as we will see in the numerical experiments.

The most well known and used method is probably the Incomplete Choleski decomposition (IC), see Meijerink and van der Vorst [5]. If we do the standard Choleski decomposition of A, during the elimination we get some fill ins, some elements that are zero in A become non zero in the triangular factors. For the problems under consideration, it can be shown that, within the band of the matrix, the fill ins decrease in magnitude as we go towards the middle of the band. The idea of the Incomplete decomposition is to neglect all or some of the fill ins.

For the row–wise ordering and the most simplest method, things can be formulated very easily in the following way. Let $M = L\ D^{-1}\ L^T$, where L has the same sparsity pattern as the lower triangular part of A. Suppose the five non zero terms on a row of A are

$$A = (c_i, b_i, a_i, b_{i+1}, c_{i+m})$$

and rows of L and D are given by

$$L = (\bar{c}_i, \bar{b}_i, d_i, 0, 0),\ \ D = (0, 0, d_i, 0, 0).$$

By inspection, we can easily see that

$$\bar{c}_i = c_i$$

$$\bar{b}_i = b_i$$

$$d_i = \frac{1}{a_i - b_{i-1}^2 d_{i-1} - c_{i-m}^2 d_{i-m}}.$$

Elements with negative indices are taken to be zero. This method is called IC(1,1). For the Poisson equation, we have a good distribution of the eigenvalues. However, there are some troubles solving the triangular systems. Hence, several modifications to the preconditioner were devised to solve that problem.

One possibility is to compute the unknowns diagonal–wise but, we also get some troubles, short vector lengths, non constant strides, etc... The advantage is that, as we are solving exactly the same system as in the basic method, the convergence rate is the same.

Another possibility is to modify the preconditioner such that we avoid the solve of triangular systems, see van der Vorst [9]. The trick is to replace the inverses of bidiagonal matrices that appear during the triangular solves by truncated Neumann series. It is usual to use three terms in the series, see van der Vorst [9] for a justification.

The interest of this modification is that the solves with bidiagonal matrices are replaced by matrix multiplications that can easily be done in vector or parallel mode. We will call this method ICVDV. However, it must be understood that we modified the preconditioner; so, the problem is to know what the convergence rate of the new method is. With most problems, the number of iterations for ICVDV is within 2 or 3 the same as for IC(1,1); however, examples can be constructed where the matrices involved are not diagonally dominant enough, so the Neumann series converge very slowly and give bad approximations to the inverses. This has the consequence that the number of iterations can be much higher for the modified method.

Next, we consider block preconditioners which, on the problems they are suitable for, give better results that point methods like IC. Let A be block tridiagonal, see section 6. Let L denote the block lower triangular part of A

$$L = \begin{pmatrix} 0 & & & \\ A_2 & 0 & & \\ & \ddots & \ddots & \\ & & A_n & 0 \end{pmatrix}.$$

The (complete) block Choleski decomposition of A can be written as

$$A = (\Sigma + L)\ \Sigma^{-1}\ (\Sigma + L^T)$$

with Σ a block diagonal matrix

$$\Sigma = \begin{pmatrix} \Sigma_1 & & & \\ & \ddots & & \\ & & \ddots & \\ & & & \Sigma_n \end{pmatrix}.$$

By inspection, matrices Σ_i are given by

$$\begin{cases} \Sigma_1 = D_1, \\ \Sigma_i = D_i - A_i\ \Sigma_{i-1}^{-1}\ A_i^T. \end{cases}$$

The matrices Σ_i are dense but, as the D_is are diagonally dominant, the magnitude of the elements of the Σ_is decay away from the diagonal.

The idea introduced in Concus, Golub and Meurant [1] is to approximate these dense matrices with tridiagonal matrices. Let B a matrix and $trid(B)$, a triagonal matrix whose non zero elements are the same as the corresponding ones of B. The preconditioner INV is defined as

$$M = (\Delta + L)\ \Delta^{-1}\ (\Delta + L^T)$$

with

$$\Delta = \begin{pmatrix} \Delta_1 & & & \\ & \ddots & & \\ & & \ddots & \\ & & & \Delta_n \end{pmatrix},$$

and

$$\begin{cases} \Delta_1 = D_1, \\ \Delta_i = D_i - A_i\ trid(\Delta_{i-1}^{-1})\ A_i^T. \end{cases}$$

It has been shown with numerical experiments that this method is very efficient and robust, see Concus, Golub and Meurant [1]. INV is not well suited for vector and parallel machines for two reasons; first of all, in solving $Mz = r$ there is a block recurrence; secondly, for each block we have to solve a tridiagonal system like

$$\Delta_i w = c.$$

A way to vectorize is to replace Δ_i^{-1} by a banded approximation, as we did before. However, instead of using a Neumann series, a better way is to use some diagonals of the inverse Δ_i^{-1}. Taking a 7 diagonal approximation is fine with most problems (with enough diagonal dominance), see Meurant [6]. Then, everything is in vector mode within a block. We call this method INVV.

Let us compare these preconditioners with numerical results for the model Poisson problem on a 50×50 mesh. The figures of table 4 were obtained on one processor of a Cray X–MP/416.

Table 4

precond	no. of it.	Mflops	time (s)
DIAG	110	127	0.041
IC	33	30	0.108
INV	15	19	0.067
VDV	34	79	0.042
INVV	16	97	0.021

In the following, we present another method called the "incomplete twisted factorization" INVkP, that can be seen to be a variation of Domain Decomposition methods, see Meurant [8]. It is easier to explain this factorization with an example. Suppose we have 4 subdomains. Then, as a preconditioner, we take

$$M = \Theta\ \Theta^T,$$

where Θ has the following structure

$$\begin{pmatrix} M_1 & N_2^T & & \\ & M_2 & & \\ & N_3 & M_3 & N_4^T \\ & & & M_4 \end{pmatrix},$$

where M_1, M_3 are block lower bidiagonal, M_2, M_4 are block upper bidiagonal, N_2^T, N_4^T have only a non zero block in the lower left corner, N_3 has only a non zero block in the upper right corner. This means that Θ has alternatively a lower block diagonal, an upper one, a lower one and finally an upper one.

The non–diagonal blocks are equal to the corresponding ones in A. The diagonal blocks can be determined beginning by the top, the bottom and the "middle" elements. The formulas are similar to the ones for the INVV preconditioner; details can be found in Meurant [7], [8].

Figure 9 gives the number of iterations for the model problem and a fixed value of $h = 1/101$ as a function of the number of subdomains. We can see that there is an increase in the number of iterations but, the increase in the degree of parallelism is larger than what we are loosing with the increase in the number of iterations. Figure 10 gives the Mflops rates obtained on a Cray Y–MP/832 using 1 and 8 processors with INVkP as a function of the mesh size; the speed up is given in Figure 11.

We can see that these methods give high Mflops rates and very good speed ups. Moreover, if we look at the computing times, they are faster than more naturally parallel methods like diagonal preconditioners.

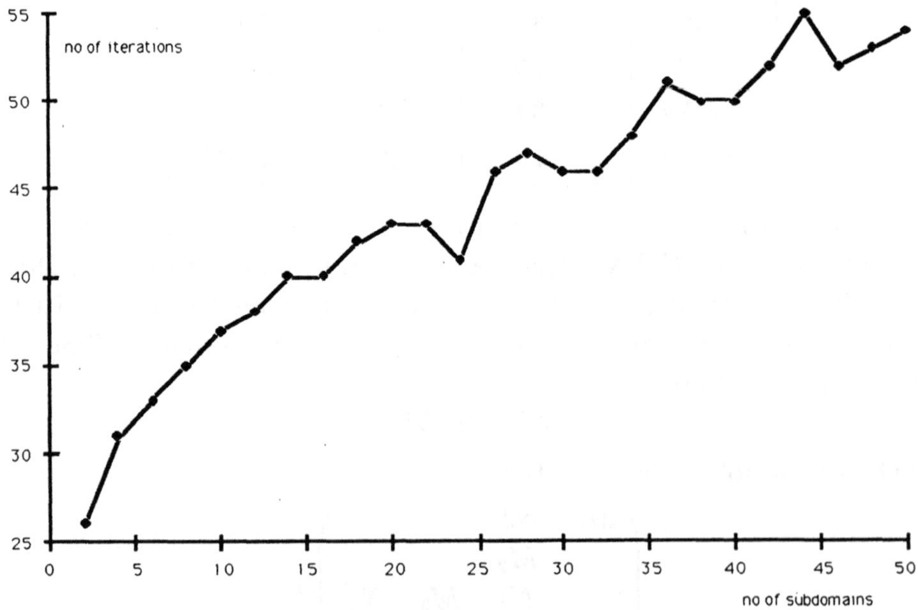

Figure 9

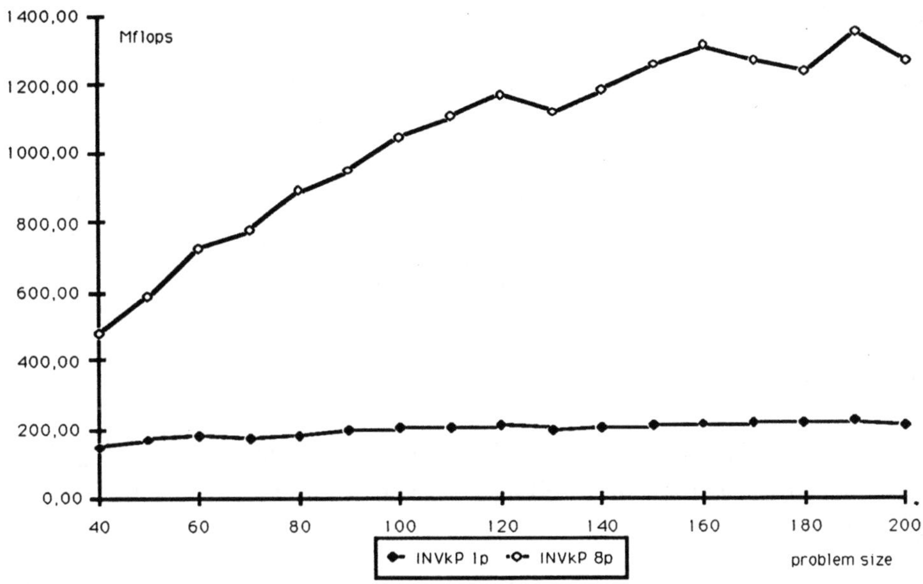

Figure 10

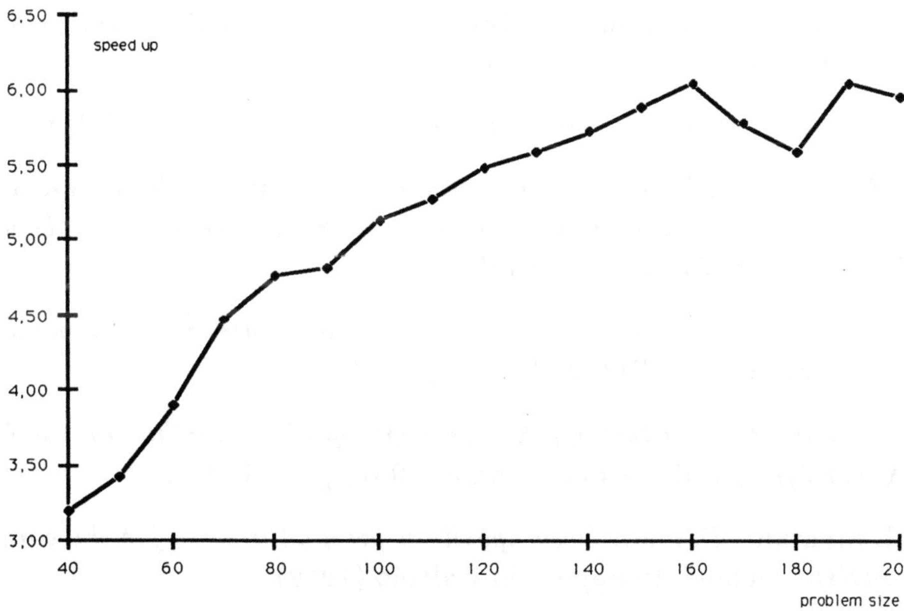

Figure 11

8 Conclusion

In this paper we have described some supercomputers and some applications in numerical linear algebra. We have shown that there are very efficient multiprocessor vector computers with which we can obtain, with suitably parallel algorithms, sustained speed of the order of a Gigaflops. In the next years, the development of these machines will continue allowing to solve new problems but also requiring new algorithms.

References

[1] P. Concus, G.H. Golub & G. Meurant, *Block preconditioning for the conjugate gradient method.* SIAM J. Sci. Stat. Comp., v 6, (1985) pp 220–252.

[2] J.J. Dongarra, *Performance of various computers using standard linear equations software in a Fortran environment.* Argonne National Laboratory, MCS Tech. Mem. n 23 (1988).

[3] G.H. Golub & G. Meurant, *Résolution numérique des grands systèmes linéaires.* Eyrolles, Paris, 1983.

[4] R. Hockney & C. Jesshope, *Parallel Computers 2.* Adam Hilger (1988).

[5] J.A. Meijerink & H.A. van der Vorst, *An iterative solution method for linear systems of which the coefficient matrix is a symmetric M–matrix.* Math. Comp. 31 (1977) pp 148–162.

[6] G. Meurant, *The block preconditioned conjugate gradient method on vector computers.* BIT 24 (1984) pp 623–633.

[7] G. Meurant, *Multitasking the conjugate gradient method on the CRAY X–MP/48.* Parallel Computing 5 (1987) pp 267–280.

[8] G. Meurant, *Domain Decomposition preconditioners for the conjugate gradient method.* to appear in Calcolo (1989).

[9] H.A. van der Vorst, *A vectorizable variant of some ICCG methods.* SIAM J. Sci. Stat. Comput. v 3 (1982) pp 86–92.

Updating Techniques in Parallel Computation

D.C. Sorensen

Mathematics and Computer Science Division
Argonne National Laboratory
9700 South Cass Avenue
Argonne, IL 60439, USA

Abstract

A survey is given of recent techniques for

- updating matrix factorizations. A review is given for updating several standard matrix factorizations and eigen-decompositions
- parallel decomposition of linear systems based on lowrank modifications
- parallel Algorithms for eigen-value problems based on divide and conquer strategies.

NATO ASI Series, Vol. F 70
Numerical Linear Algebra, Digital Signal Processing and Parallel Algorithms
Edited by G. H. Golub and P. Van Dooren

On the Use of the Singular Value Decomposition in Identification and Signal Processing

Joos Vandewalle and Bart De Moor

ESAT (Electronics, Systems, Automation and Technology)
Katholieke Universiteit Leuven
Kardinaal Mercierlaan 94
3030 Heverlee, Belgium

Keywords: singular value decomposition and generalizations, factor analysis, canonical correlations analysis, total linear least squares, realization and identification of state space models, subspace methods

1 Introduction

In recent years, the *'ordinary' singular value decomposition* (**OSVD**) and its generalizations, have become extremely valuable instruments in the analysis and the solution of problems in mathematical engineering. In most applications, the **OSVD** provides a unifying framework, in which the conceptual formulation of the problem, the practical application and an explicit solution that is guaranteed to be numerically robust, are derived at once. In this way, the **OSVD** has become a fundamental tool for the formulation and derivation of new concepts such as angles between subspaces, oriented signal-to-signal ratio's, canonical correlation analysis,... and for the reliable computation of the solutions to problems such as total linear least squares, realization and identification of linear state space models, source separation by subspace methods etc.

NATO ASI Series, Vol. F 70
Numerical Linear Algebra, Digital Signal Processing and Parallel Algorithms
Edited by G. H. Golub and P. Van Dooren

In a first part the perspectives of **SVD** in engineering are situated. In section 3 the **OSVD** and Quotient SVD (**QSVD**) are described. In section 4 several fundamental concepts are defined and their properties discussed in terms of the (generalized) singular value decomposition. A fundamental framework is summarized which allows to formalize 'factor-analysis-like' problems in terms of oriented signal-to-signal ratios. The computational tool is the **OSVD** and one of its generalizations, the **QSVD**, which is also an appropriate instrument to quantify the concept of canonical correlation analysis and the notion of angles between subspaces. We present a survey of several useful original insights in the singular value decomposition structure with respect to backward error analysis, condition numbers, sensitivity and the influence of noise.
In section 5 of the survey, a wide variety of applications from a broad spectrum of scientific disciplines will be cited and illustrated: mechanics (moments of inertia), electrical network analysis (the conditioning of reference node choice), bio-medical engineering (signal source separation of maternal and fetal ECG), realization of systems from impulse response measurements, identification of industrial processes from input-output data, etc....

As will be illustrated, the benefits of using the (generalized) singular value decomposition are most pronounced in those signal processing and identification applications:

- where rank decisions and the computation of the corresponding subspaces determine the complexity and parameters of the model

- where numerical reliability is of primordial importance and the potential loss of numerical accuracy (as caused by the squaring of matrices) is to be avoided.

- where a conceptual framework, such as the notion of oriented signal-to-signal ratio, may provide additional insight, such as in factor-analysis-like problems.

- where the problem can be stated in terms of the (generalized) singular value decomposition, which leads immediately to a reliable and robust solution, such as in a canonical correlation analysis environment.

- where robustness analysis, conditioning and sensitivity optimization are crucial, linked together with geometrical insight and interpretation, for which the **OSVD** and its generalizations may provide meaningful quantification. (condition numbers, principal angles,...)

Moreover, in most engineering applications the number of measurements or the data acquisition poses only minor organisational problems (although the design of a measurement set up causes considerable efforts). The cost of the sensors however increases with higher accuracy and signal-to-noise requirements. In this environment the singular value decomposition is the optimal bridge between limited measurement precision and robust modeling.

As to the computational requirements, the **OSVD** of large matrices poses no considerable difficulties when employing a mainframe computer (matrix size order of magnitude a few hundred). Moderately sized **OSVDs** (order of magnitude 50..70) are nowadays feasible on mini-computers and PC's. In many applications only part of the **OSVD** is needed (e.g. the singular values, the left singular vectors, the smallest singular vector, ...) which results in a reduction of the computational burden. Moreover, it can be expected that the intensive on-going research for parallelized and vectorized algorithms may result in real fast **OSVD** solvers, possibly exploiting the matrix structure which is present in a lot of engineering applications by so-called displacement rank concepts.

2 Opportunities offered by the technology and constraints imposed by the applications

In recent years one can witness an enormous expansion of computational power at all levels : supercomputers, workstations, mini and microcomputers, VLSI, ... In addition the measurements power is increasing by cheap sensors, transducers and data acquisition equipment. However in practice it is much more difficult to increase the accuracy of the measurements than to increase the volume of data. Hence the need for methods and software which can extract more accurate information from the measured data is more acute even at the expense of kflops of computations. In this situation we believe that the singular value decomposition is a very important tool which allows to take profit of the expansion of the computational power in order to improve the accuracy. The **OSVD** is well known in linear algebra for its solid numerical qualities [1]. However it is only recently explored as an important concept in digital signal processing. On the other hand its widespread use is still hampered by the computational burden. Hence it is one of the important targets of intensive research on parallel algorithms. The **OSVD** is thus an important focal point in the three research areas of this Advanced Study Institute.

From an application point of view one can distinguish at least three different environments where these areas meet.

- First of all in the number crunching environment supercomputers are used in order to solve massive problems like weather forecasts, seismic data, optimizations, simulations ... The volume of applications is limited, the value of each application is quite high and the equipment is fixed (e.g. hypercube) and shared with many.

- The design and production environment on the other hand is much more widespread and includes applications like design, laboratory, medicine, plant production, optimization and control. In these situations the equipment consists of a workstation or a personal computer and it can have a mathematical or a signal coprocessor for doing computation intensive operations like **OSVD**.

- The third environment is the consumer environment where dedicated software hardware products are used in sound, video, automotive or telecommunication applications. The high volume of these applications imposes the dedicated nature (e.g. application specific I.C.).

In each of these three environments different trade-offs have to be made between algorithms, parallellization, and architecture (multiprocessor, systolic, bitserial, bitparallel). We are mainly concerned with the design and production and the consumer environment, where the architecture is not fixed. With the advent of VLSI massive amounts of computational power will be available for the consumer and the professional in design and production. The question here is to exploit the computational power in order to perform a number of digital signal processing activities, command, monitoring, data compression, error correction, ... These digital signal processing activities exhibit a number of characteristics which have also an impact on the issues involved in the study of the singular value decomposition. First of all the data stream and the processing of the data should happen in real time. Moreover in recent years the sample rates are increasing from the bps to the kbps and higher rates. On the other hand many variables are continuously being measured or sensored (e.g. automatic control, medicine, ...). The accuracy range is quite different from the accuracy range usually considered in numerical algebra (single or double precision). For measured data 1% of full scale accuracy is usually considered to be very good, which corresponds with 40 dB signal to noise (S/N) ratio. Of course it cannot be tolerated that the algorithms would worsen the S/N ratio substantially. Hence the need for numerically reliable methods is even more acute than in numerical algebra. However the smaller singular triplets often are even less important here. Also it is clear that the **OSVD** is only a part, although often a very important one, of a global system. This global system includes many tasks which can be performed by hardware and software. Typical engineering trade-offs and compromises are made in order to design such systems. The volume of these applications and the dedicated nature of the tasks make these suited for VLSI implementation. It is expected that in this framework **OSVD** will play an important role, in much the same way that FFT has played an important role in digital signal processing and numerical analysis.

3 The (Quotient) Singular Value Decomposition

In this section, the theorems stating the existence and properties of the singular value decomposition are presented. For a proof and computational requirements, the reader is referred to literature [1].

Theorem 1 *The singular value decomposition for real matrices.*
If A is a $m \times n$ real matrix, then there exist real orthogonal matrices

$$U = [u_1 \ u_2 \ \cdots \ u_m], \ V = [v_1 \ v_2 \ \cdots \ v_n]$$

such that

$$U^t \cdot A \cdot V = \begin{bmatrix} diag(\sigma_1, \sigma_2, \cdots, \sigma_r) & 0 \\ 0 & 0 \end{bmatrix}$$

where

$$\sigma_1 \geq \sigma_2 \geq \cdots \geq \sigma_r > \sigma_{r+1} = \cdots = 0$$

The σ_i are the singular values of A and the vectors u_i and v_i are respectively the $i-th$ left and the $i-th$ right singular vector.

The set $\{u_i, \sigma_i, v_i\}$ is called the $i-th$ singular triplet. The singular vectors (triplets) corresponding to large (small) singular values are called large (small) singular vectors (triplets).

The **OSVD** reveals a great deal about the structure of a matrix as evidenced by the following well known corollaries:

Corollary 1 *Let the* **OSVD** *of A be given as in theorem 1 then*

(1) *Rank property*

$$r(A) = r \ \ and \ \ \begin{array}{l} N(A) = span\{v_{r+1}, \cdots, v_n\} \\ R(A) = span\{u_1, \cdots, u_r\} \end{array}$$

(2) *Dyadic decomposition*

$$A = \sum_{i=1}^{r} u_i \cdot \sigma_i \cdot v_i^t$$

(3) *Norms*

$$\| A \|_F^2 = \sigma_1^2 + \cdots + \sigma_r^2 \cdot$$
$$\| A \|_2 = \sigma_1$$

(4) *Rank k approximation.*

$$A_k = \sum_{i=1}^{k} u_i \cdot \sigma_i \cdot v_i^t \quad with \quad k < r$$

then

$$\min_{r(B)=k} \| A - B \|_2 = \| A - A_k \|_2 = \sigma_{k+1}$$

$$\min_{r(B)=k} \| A - B \|_F^2 = \| A - A_k \|_F^2 = \sigma_{k+1}^2 + \cdots + \sigma_p^2$$

This important result is the basis of a lot of concepts and applications such as total linear least squares, data reduction, image enhancement, dynamical system realization theory and in all possible problems where the heart of the solution is the approximation, measured in 2-norm or Frobeniusnorm, of a matrix by one of a lower rank. Many more valuable properties of the **OSVD**, including existence proofs, computational requirements and numerical considerations, sensitivity results, conditioning etc.. can be found in the modern bible of numerical analysis [1] and the references therein.

Theorem 2 *The quotient singular value decomposition for real matrix pairs*
If A is a $m \times n$ matrix with $n \geq m$ and B is a $m \times p$ matrix, then there exist orthogonal matrices U $(n \times n)$ and V $(p \times p)$ and an invertible X $(m \times m)$ such that:

$$X \cdot A \cdot U = D_A = diag(\alpha_i) \;\; \alpha_i \geq 0 \;\; i = 1,..,m$$

and

$$X \cdot B \cdot V = D_B = diag(\beta_i) \;\; \beta_i \geq 0 \;\; i = 1,..,q = min(m,p)$$

where

$$\beta_1 \geq \beta_2 \geq \cdots \geq \beta_r \geq \beta_{r+1} = \cdots = \beta_q = 0 \quad r = rank(B)$$

[1]As there are several other generalizations of the **OSVD**, such as the Product SVD (**PSVD**) and the Restricted SVD (**RSVD**), we propose to call this generalization the *Quotient SVD* **QSVD** because in most applications it are the ratios of the diagonal elements of D_A and D_B that are important. Moreover, under certain conditions, the **QSVD** delivers the **OSVD** of B^+A, which can be considered as a *matrix quotient.* For more details, the reader is referred to [45] and the references therein.

Observe that the **QSVD** reduces to the **OSVD** in the case that $B = I_m$. The elements of the set

$$\sigma(A, B) = \{\alpha_1/\beta_1, ..., \alpha_r/\beta_r\}$$

are referred to as the quotient singular values of A and B. The quotient singular values corresponding to the $\beta_i = 0$ are infinite. They are considered to be equal and come first.

Only in the sixties numerically reliable methods were found to compute the **OSVD** [1]. However in many engineering applications, the general purpose **OSVD** algorithm (complete decomposition, full machine precision) can be replaced by other algorithms with less stringent specifications. This reduces the computational burden and/or storage requirements.

- A lot of applications only require the computation of part of the singular spectrum (realization theory and data reduction: dominant singular triplets, source separation: intermediate triplet, total linear least squares, Pisarenko-type spectral estimation: smallest singular triplets). The storage reduction obtained by storing only the dominant triplets instead of the full matrix can be considerable e.g. in image processing despite the relatively heavy computational requirements.

- Moreover, very frequently the available data are noisy (industrial environment typically 10 % !), indicating that a full precision **OSVD** makes no sense.

- A third specification is the adaptive computation of the **OSVD** of matrices that are time-varying, either elementwise or by adding and deleting columns and/or rows.

- Finally, in a lot of applications the matrices are very structured ((block-) Hankel or - Toeplitz, circulant, ...) so that it is expected that structure exploiting algorithms perform better (faster) than the standard full size **OSVD** algorithm. Moreover, considerable storage gain can be achieved if the elements of the structured matrix could be stored without redundancy, hence requiring matrix-vector multiplication based algorithms.

Results on the *power method and the Chebyshev method* for the **OSVD** can be found in [21,24]. One of the important observations is that the Chebyshev method is nothing else than the eigenvalue power method applied to a certain matrix, constructed from the original one. This allows to translate all results that were derived for the **OSVD** power method in [21] to the Chebyshev algorithm in [24], including convergence criteria, deflation strategies, accelerations algorithms and convergence level control rules that allow to compute the **OSVD** up to the required numerical precision.
An adaptation of Golub's full size **OSVD** for the case that *only the smallest singular triplets are needed*, is described in [32]. By a careful analysis the computational requirements can be reduced in such a way that the partial **OSVD** algorithm can be 3 times faster than the classical one, while obtaining the same accuracy. Moreover, in [36] a comparison can be found between 4 different methods to compute the null space of an approximation of a given matrix in the sense of total linear least squares. Chebyshev iteration [24], inverse iteration, Rayleigh quotient iteration and the Lanczos method have been compared with respect to their computational efficiency in total linear least squares computations. The conclusion is that inverse iteration is the most promising iterative technique for solving generic TLLS problems of known rank. Moreover, the direct methods (Golub's **OSVD** and partial **OSVD**) have also been compared with the iterative ones and several potential applications have been investigated [36] (parameter estimation, subset selection, discrete deconvolution).
Finally, the *one-sided Jacobi method* has been studied in relation to the problem of adaptive computation of the **OSVD** of a matrix containing measurements on the fetal and the maternal ECG [25,26,28,37]. As the measurements (6 to 9 channels) enter at a frequency of 250 Hz, the **OSVD** is updated with orthogonal Givens' rotations as to minimize the Frobenius-norm of the off-diagonal elements. Convergence and speed of a possible implementation are studied. Also the implementation on a TMS-320 signal processor has been tested and has been shown to be feasible for real time applications up to 500 samples per second.

4 Fundamental geometric concepts based on the OSVD and the QSVD

In this section, we discuss several concepts based on the **OSVD** and **QSVD** that are useful in applications such as *oriented signal-to-signal ratios* (section 4.1), *canonical correlations* (section 4.2), *condition numbers* (section 4.3) and *an orthogonality principle for noise and exact data* (section 4.4). In a wide variety of systems and signal processing applications, vector sequences are measured and analysed. Whenever *linear* models are used to describe the measurements, one is interested in their fundamental characteristics, which are their complexity (the rank of certain matrices) and the parameters describing the model. These parameters can often be extracted from certain subspaces, of which the dimension is a measure for the complexity of the model. Hence in identifying linear models from noisy data, one is confronted with two basic non-trivial problems:

- the meaningful estimation of a rank
- the reliable computation of a corresponding subspace.

It is obvious that the technique of the (quotient) singular value decomposition is very appropriate to describe and compute both ranks and subspaces. Examples can be found in a wide variety of applications: Sets of linear equations (total linear least squares [38,1]), the identification of factor-analysis-like models (rotational invariance techniques [12]), separation of MECG/FECG [28,40,50,13], realization of linear state space models from impulse responses [15,22,35] and identification of state space models from noisy input-output measurements via canonical correlation analysis [41-43].

4.1 Oriented energy and signal-to-signal ratios

The recent introduction of the fundamental concepts of oriented energy and oriented signal - to -signal ratio [37] has provided a rational framework in which both the estimation of ranks and subspaces can be formalized in a rigorous way.

Let A and B be two $m \times n$ matrices with $n \gg m$, both containing measurement vector sequences (typically n consecutive sample vectors from m measurements channels). The columns of A and B are denoted by $a_k, b_k \;\; k = 1, .., n$

Definition 1 *The oriented energy of the matrix A, measured in a direction q is defined as:*

$$E_q[A] = \sum_{k=1}^{n} (q^t a_k)^2$$

Definition 2 *The oriented signal-to-signal ratio of the two vector sequences A and B in the direction q is defined as:*

$$E_q[A, B] = E_q[A]/E_q[B]$$

There are straightforward generalizations of these definitions to oriented energy and signal-to-signal ratios in subspaces Q^r. In [37] it is shown that the analysis tool for the oriented energy distribution of a matrix A is the singular value decomposition, while the analysis tool of the oriented signal-to-signal ratio of two vector sequences A and B is the quotient singular value decomposition of the matrix pair $[A, B]$. These well understood matrix factorizations allow to characterize the directions of extremal oriented energy and oriented signal-to-signal ratio:

Theorem 3 *Extremal directions of oriented energy.*

Let A be a $m \times n$ matrix ($n \gg m$) with **OSVD** $A = U \Sigma V^t$ *where* $\Sigma = diag\{\sigma_i\}$. *Then each direction of extremal oriented energy is generated by a left singular vector u_i with extremal energy equal to the corresponding singular value squared σ_i^2.*

Theorem 4 *Extremal directions of oriented signal-to-signal ratio.*

Let A and B be $m \times n$ matrices, with **QSVD***:*

$$A = X^{-1}.D_a.U^t \quad D_a = diag\{\alpha_i\}$$
$$B = X^{-1}.D_b.V^t \quad D_b = diag\{\beta_i\}$$

where the quotient singular values (possible infinite) are ordered such that $(\alpha_1/\beta_1) \geq (\alpha_2/\beta_2) \geq ... \geq 0$. *Then each direction of extremal signal - to*

-signal ratio is generated by a row x_i^t of the matrix X and the corresponding extremal signal-to-signal ratio is the quotient singular value squared $(\alpha_i/\beta_i)^2$.

These two theorems are illustrated for two dimensions in figure 1. Observe that for the oriented energy the maximum and minimum corresponding to the largest resp. smallest singular vectors while a saddle point would correspond to the intermediate singular vector. Observe that the extremal directions of oriented energy are orthogonal while this is not necessarily the case for the signal-to-signal ratio. The underlying tool for the proof of these theorems is nothing else than the Courant-Fisher minimax characterization of the eigenvalues of symmetric operators [37,12]. Now one can proceed by investigating in which directions of the ambient space the vector signal in the matrix A can be best distinguished from the vector signal in the matrix B. This leads to the definition of maximal minimal and minimal maximal signal to signal ratios of two vector sequences [37].

Definition 3 *Maximal minimal and minimal maximal signal-to-signal ratio.*

The maximal minimal signal-to-signal ratio of two m-vector sequences contained in the $m \times n$ *matrices A and B over all possible r-dimensional subspaces* $(r < m < n)$ *is defined as:*

$$MmR[A,B,r] = \underset{Q^r \subset R^m}{Max} \quad \underset{q \in Q^r}{Min} \quad E_q[A,B]$$

Similarly, the minimal maximal signal-to-signal ratio is defined as:

$$mMR[A,B,r] = \underset{Q^r \subset R^m}{Min} \quad \underset{q \in Q^r}{Max} \quad E_q[A,B]$$

The idea behind these definitions is the following: For a given subspace Q^r of the m-dimensional ambient space $(r < m < n)$ there is a certain direction $q \in Q^r$ for which the signal-to-signal ratio of the two vectorsequences A and B is minimal. This direction corresponds to the worst direction q in the sense that in this direction the energy of A is difficult to distinguish from the energy of B. This worst case of course depends upon the precise choice of the subspace Q^r. Among all r-dimensional subspaces, at least one

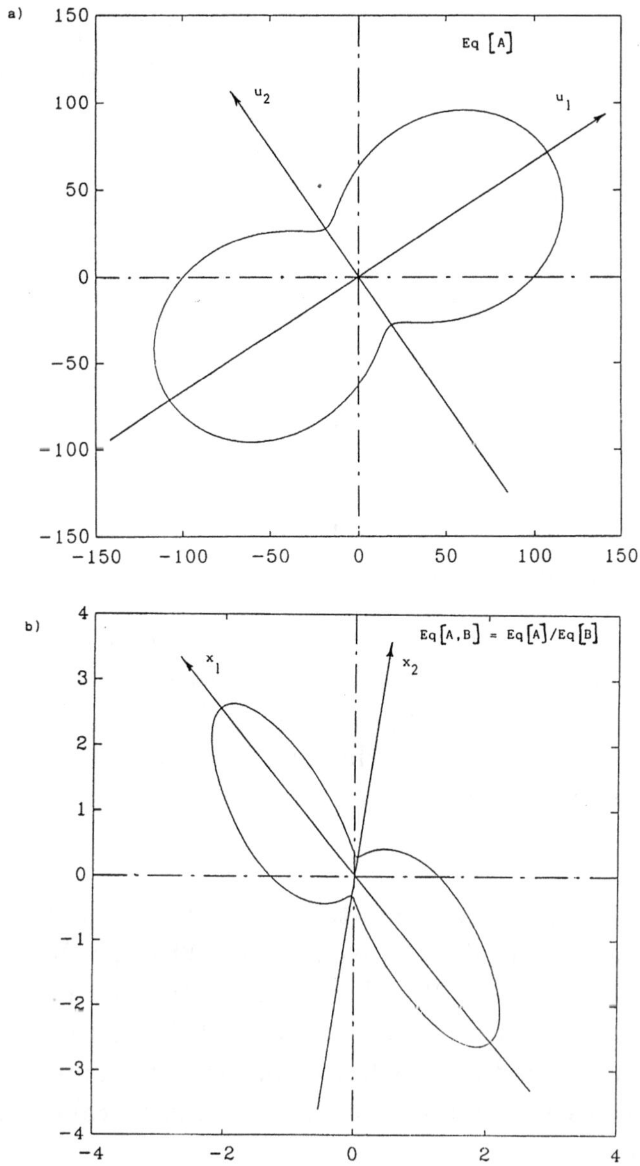

Figure 1: a) Oriented energy of a 2-vector sequence A and b) oriented signal-to-signal distribution of two 2-vector sequences A and B.

r-dimensional subspace has to exist where the worst case is better than all other worst cases. This subspace is the r-dimensional subspace of maximal minimal signal-to-signal ratio. It comes as no surprise that the **QSVD** allows to find this subspace: It is the r-dimensional subspace generated by the first r rows of X, when the quotient singular values are ordered as in theorem 4.

Hence, the concept of oriented signal-to-signal ratio and the **QSVD** allow to formalize all model identification approaches, in which

- the determination of a suitable rank r provides the complexity of the model.
- the model parameters follow from the corresponding subspace of maximal minimal signal-to-signal ratio.

Moreover, it can be shown that when the vector sequence B consists of an unobservable stochastic vector signal with known first and second order statistics (as is the case in most engineering applications), the **QSVD** solution corresponds precisely to the 'classical' Mahalanobis transformation that is commonly used in statistical estimators as a kind of prewhitening filter [37].

Nice applications of the oriented signal-to-signal ratio concept include source separation techniques for fetal ECG extraction [20,13], subspace methods to detect narrowband sources [12] and system identification [41-43]. In general, it is expected that the framework of oriented signal-to-signal ratio will become a powerful analysis tool in what could be called factor-analysis-like modeling and identification methods of which a biomedical example will be presented in section 5.

4.2 Canonical correlation analysis

The **OSVD** provides an important tool in the generalization and characterization of important geometrical concepts. One of these is the notion of *angles between subspaces*, which is a generalization of the angle between two vectors.

Definition 4 *Let F and G be subspaces in R^m whose dimensions satisfy*

$$p = dim(F) \geq dim(G) = q \geq 1$$

The principal angles $\theta_1, \theta_2, .., \theta_q \in [0, \pi/2]$ between F and G are defined recursively by:

$$\cos(\theta_k) = \max_{u \in F} \max_{v \in G} u^t.v = u_k^t.v_k$$

subject to

$$\begin{aligned} &\| u \| = \| v \| = 1; \\ &u^t.u_i = 0 \quad i = 1, ..., k-1 \\ &v^t.v_i = 0 \quad i = 1, ..., k-1 \end{aligned}$$

The vectors $\{u_i\}, \{v_i\}, i = 1, .., q$ are called the principal vectors of the subspace pair [F,G]

If the columns of P $(m \times p)$ and Q $(m \times q)$ define orthonormal bases for the subspaces F and G respectively, then it follows from the minimax characterization of singular values [1] that:

$$\begin{aligned} &[u_1, \cdots, u_p] = P \cdot Y \\ &[v_1, \cdots, v_q] = Q \cdot Z \\ &\cos(\theta_k) = \sigma_k \quad k = 1, \cdots, q \end{aligned}$$

where the **OSVD** of the ('generalized inner') product

$$P^t \cdot Q = Y \cdot diag(\sigma_1, .., \sigma_q) \cdot Z^t$$

From this, it is not difficult to devise an algorithm to compute the intersection of subspaces that are for instance the ranges of two given matrices A $(m \times p)$ and B $(m \times p)$ [1, p.430] This is precisely the idea behind the technique of canonical correlation, which appears to be very fruitful in the identification of linear dynamical state space models from noisy input-output measurements [41-43].

There are several ways to compute the canonical correlation structure of a matrix pair A,B (roughly all possible ways of computing an orthonormal basis for the row spaces): two QR decompositions, two singular value decompositions, one quotient singular value decomposition, all of which are followed by an **OSVD** of the generalized inner product of the orthonormal bases matrices. Another method is the computation of the right null space of the concatenated matrix

$$\begin{bmatrix} A \\ B \end{bmatrix}$$

However, it is expected that depending on the application at hand, one method could be preferable with respect to the others. Finally, let us observe that there exists an intimate relation between total linear least squares and canonical correlation analysis: Let A be an $m \times n$ $(m > n)$ and B an $m \times p$ $(m > p)$ matrix and consider the problem of solving X from $A.X = B$. One can then study the canonical correlation analysis applied to the column spaces of A and B and investigate the solutions to $A.X = B.Y$, which is nothing short of a (generalized) total linear least squares problem.

4.3 Condition numbers

Condition numbers arise naturally in describing the sensitivity of solutions of sets of linear equations to inaccuracies in the data. Consider hereto the following two problems.

Problem 1: For a given $m \times n$ matrix $A(m > n)$, how much is the maximal increase in relative inaccuracy, that may occur when solving x from $A \cdot x = y$, for the worst position of the right hand side y and the unluckiest position of the error dy. This is a question to which the condition number as derived above gives the answer:

$$K(A) = Max_{y,\, dy} \frac{\| dx \| / \| x \|}{\| dy \| / \| y \|} = \sigma_1 / \sigma_n$$

Problem 2 : For a given $m \times n$ matrix A $(m > n)$ and fixed right hand side y, how much is the maximal increase in relative inaccuracy that may occur when solving x from $A.x = y$, for the unluckiest position of the error dy.

The singular value decomposition allows for a direct answer via a kind of 'restricted' condition number:

$$K_y(A) = Max_{dy,A.x=y,A.dx=dy}\frac{\| dx \| / \| x \|}{\| dy \| / \| y \|} = \frac{\| y \|}{\sigma_1 \cdot \| x \|} \cdot K(A)$$

The first condition number only gives information about the matrix A, not taking into account the relative position of the right hand side y. The second condition number takes into account the actual position of y, which may drastically reduce the estimation of the solution sensitivity. First consider the norm amplification $\| x \| / \| y \|$ as a function of the angle θ_x between x and the largest right singular vector and the angle θ_y between y and the largest left singular vector for a simple 2×2 example. One can show [39] that the property of 'being usually close' to either the maximum or the minimum significantly increases with increasing condition number. The minimum corresponds to the inverse of the largest singular value while the maximum equals the inverse of the smallest one.

Some important conclusions, which are general, can be drawn from these observations:

- In many applications, the orientation of the error dy is unknown but one has an idea about the magnitude of the norm $\| dy \|$. Hence, it is meaningful to assume the effect on x most probably to be of a level of $(1/\sigma_r). \| dy \|$, where σ_r is the smallest singular value of A.

- If each angle θ_x is equally probable, then x will have most probably a length equal to $1/\sigma_1. \| y \|$. One can verify that this is the case in discrete deconvolution problems, where the impulse response samples are computed from input-output measurements. The error $\| dx \|$ will most probably be in these applications of the level $K(A) \| dy \|$, which can be rather bad.

- If each angle θ_y is equally probable, then x will have most probably a length close to $1/\sigma_r. \| y \|$. This is the case in applications such as polynomial fitting via Vandermonde matrices. The error $\| dx \|$ will be probably of the level $K_y(A). \| dy \|$.

Now consider the sensitivity measure m_e, which is the error amplification

$$m_e = \frac{\| dx \| / \| x \|}{\| dy \| / \| y \|}$$

for all possible orientations of x, dx, y, dy. For a fixed y, m_e varies from $K_y(A)/K(A)$ to $K_y(A)$ depending on the orientation of dy. The worst case result for dy for fixed y is $m_e = K_y(A)$. Over all possible orientations of y, $K_y(A)$ varies from 1 to K(A). Moreover, if dy has a uniform distribution elementwise, then m_e is usually close to its upperbound $K_y(A)$. If dx has a uniform distribution elementwise, then m_e is usually much smaller than $K_y(A)$. One can conclude that the linear least squares solution x and the error dx tend to be:

- dy-insensitive if y is independent of A (polynomial smoothing)

- dy-sensitive if y is dependent on A (like in deconvolution)

4.4 An orthogonality principle for noisy data

In a lot of mathematical engineering applications, matrices from measured data are constructed. Very often these matrices are rectangular, i.e. they have (say) many more columns than rows. For instance, there are as many rows as measurements channels and each time a measurement on all channels is done, an additional column is added to the matrix. The available data are very frequently rather noisy, i.e. they are perturbed by unobservable errors, of which (in the best case) the statistical properties are known. Moreover, almost always the 'exact data' mathematical model is such (at least in linear applications) that the data matrix would be rank-deficient if the data were noisefree, and that the crucial information about the desired model is contained in the null space of the exact data matrix. Think for instance of the solution of overdetermined linear equations, or dynamical realization of impulse responses from (block)-Hankel matrices or the identification of dynamical state space models from noisy input-output data with canonical correlation analysis. More specifically, we are confronted with the following situation:

$$\underset{m \times n}{C} = \underset{m \times n}{A} + \underset{m \times n}{B} \quad \text{with } m \ll n \text{ where } r(A) = r < m$$

where A represents the exact data matrix of rank r and B represents the matrix containing the perturbations (measurement errors, model mismatch), which are assumed to be additive. A and B as such are unobservable but the entries of their sum C are the measured data. Generically, the matrix B will be of full (row) rank while the matrix A is assumed to be rank-deficient: The crucial model information (in all cited examples) is contained within the orthogonal complement of its columnspace.

The *problem* that will now be investigated is the following: How is the **OSVD** of A modified by the perturbing influence of B and how can properties of A (rank and null spaces) be estimated from computations on C only. The crucial observation to make is contained in the following statement:

Under mild conditions, the canonical angles between the row space of the matrix A and the row space of the matrix B, approach 90^0 *(orthogonality) as the overdetermination* $n \,/\, m$ *increases.*
A general discussion and proof of this statement and the conditions under which it is valid can be found in [44]. An illustration can be found in the Fig.2, where the probability distribution is computed of the angle between a vector and several subspaces in an ambient space of varying dimension. The elements of the vector are independently identically normally distributed with zero mean. It's the authors' belief that this observation is really at the heart of a lot of identification and estimation schemes such as (total) linear least squares, instrumental variables methods, dynamic identification, briefly in all those applications in which

- the estimation of the rank of an exact data matrix through computations on a noisy matrix makes sense,

- the information on the model is contained in the null space of the measurement matrix.

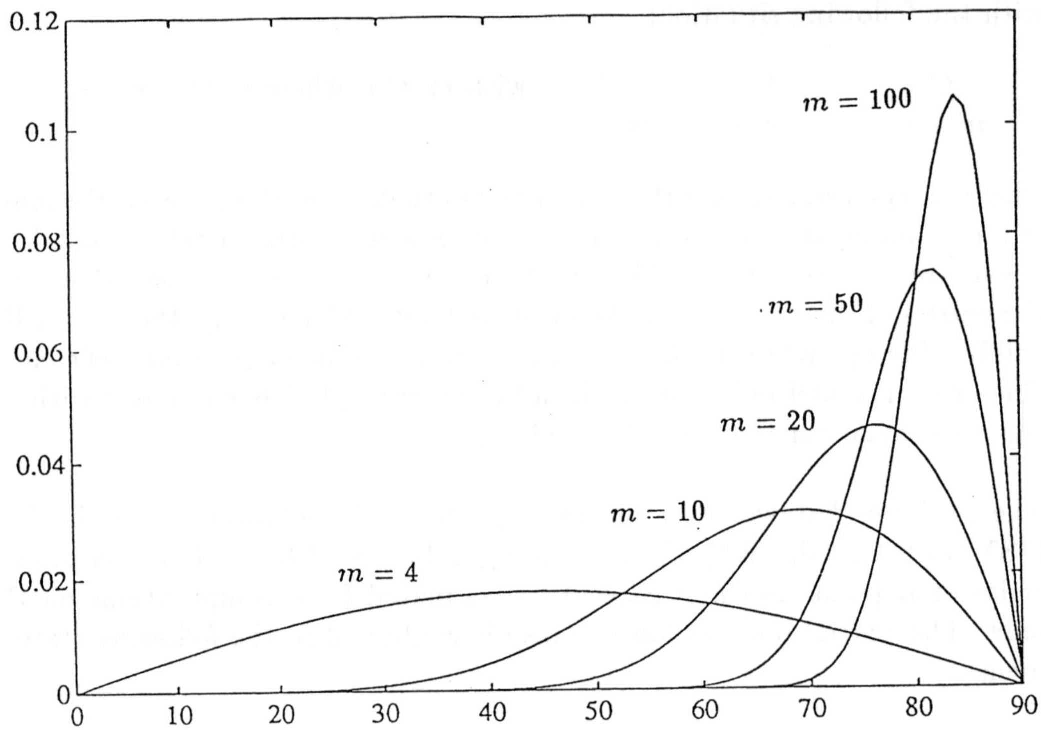

Figure 2: Probability distribution of the angle between an arbitrary direction in R^m (uniformly distributed) and a fixed 2-dimensional plane for m=4 (1), m=10 (2), m=20 (3), m=50 (4), m=100 (5). Observe the increasing probability that the direction is orthogonal to the plane.

In [41-44] this statement is even taken to be an axiomatic basis for a new conceptual identification framework.

Let us demonstrate how the preceding statement can be applied to the modification analysis of the **OSVD** of A and C. We will derive the conditions under which the column space of the matrix A, including its dimension (rank) can exactly be recovered from a subspace of the columnspace of the matrix C.

Let A and B have the **OSVDs**:

$$\underset{m\times n}{A} = \underset{m\times r}{U_a} \; \underset{r\times r}{.\Sigma_a.} \; \underset{r\times n}{V_a^t}$$

$$\underset{m\times n}{B} = \underset{m\times m}{U_b} \; \underset{m\times m}{.\Sigma_b.} \; \underset{m\times n}{V_b^t}$$

where $r < m < n$. Denote by $U_a^\perp$ any $m \times (m-r)$ orthogonal matrix satisfying

$$U_a^t.U_a^\perp = 0$$

and assume that the row spaces of A and B are orthogonal:

$$V_a^t.V_b = 0$$

Then, it is easy to show that the matrix C can be written as:

$$C = [U_a \;\; U_a^\perp]\begin{bmatrix} P_1^t \\ P_2^t \end{bmatrix} \quad with \quad \begin{array}{l} P_1^t = \Sigma_a V_a^t + U_a^t.U_b.\Sigma_b.V_b^t \\ P_2^t = (U_a^\perp)^t.U_b.\Sigma_b.V_b^t \end{array}$$

Let P_1^t and P_2^t have the **OSVDs**:

$$\begin{array}{l} P_1^t = X_1 \cdot S_1 \cdot Y_1^t \\ P_2^t = X_2 \cdot S_2 \cdot Y_2^t \end{array}$$

then we have the result that C can be written as:

$$C = [U_a.X_1 \;\; U_a^\perp.X_2]\begin{bmatrix} S_1 & 0 \\ 0 & S_2 \end{bmatrix}\begin{bmatrix} Y_1^t \\ Y_2^t \end{bmatrix}$$

which is (up to a reordering of the singular values in S_1 and S_2) a singular value decomposition if:

$$Y_1^t.Y_2 = 0$$

or equivalently:

$$P_1^t.P_2 = 0$$

Because $V_a^t.V_b = 0$ this is true iff:

$$U_a^t.[U_b.(\Sigma_b.\Sigma_b^t).U_b^t].U_a^\perp = 0$$

The factor between square brackets can be recognized to be the sample covariance matrix of the vector signal contained in the matrix B. Now let Q_1 and Q_2 be such that:

$$U_a = U_b.Q_1 \quad U_a^\perp = U_b.Q_2$$

then the following set of matrix equations is to be satisfied:

$$\begin{aligned} Q_1^t.Q_1 &= I_r \\ Q_2^t.Q_2 &= I_{m-r} \\ Q_1^t.Q_2 &= 0 \\ Q_1^t.(\Sigma_b.\Sigma_b^t).Q_2 &= 0 \end{aligned}$$

This poses no problem if $(\Sigma_b.\Sigma_b^t)$ is a multiple of the identity matrix. This is then equivalent with the fact that the oriented energy of the matrix B is isotropic. For general $(\Sigma_b.\Sigma_b^t)$ it is not so difficult to see that a solution looks as follows: Q_1 may contain any set of r different columns of the identity matrix I_m (or a linear combination of these) while Q_2 may contain the remaining m-r columns (or a linear combination of them). Hence, Q_1 and Q_2 must have a complementary zero pattern. The conclusion is that U_a must be generated by r singular vectors of U_b.

Hence, we have demonstrated that under the orthogonality condition of the row spaces of A and B, the subspace spanned by the vectors of U_a is not mixed with vectors of the subspace spanned by $U_a^\perp$ if either the oriented energy distribution of B is isotropic or if the subspace U_a is generated by r left singular vectors of the matrix B. However, it can only be recognized as such from the singular values, if one has a priori information about the relative magnitude of the singular values contained in S_1 and S_2. If one knows for instance a priori that all singular values in S_1 are larger than those in S_2, then the first r left singular vectors of C generate the subspace spanned by the vectors of U_a. As an illustration consider the following:

Example: If the entries of B are identically normally distributed with zero mean and variance σ^2, then with increasing probability for increasing overdetermination n/m the **OSVD** of C will approach the following 'limit' **OSVD**:

$$[U_a \;\; U_a^\perp]\begin{bmatrix}(\Sigma_a^2+n.\sigma^2I_r)^{\frac{1}{2}} & 0\\ 0 & \sqrt{n}.\sigma.I_{m-r}\end{bmatrix}\begin{bmatrix}(\Sigma_a^2+n.\sigma^2I_r)^{-\frac{1}{2}}(\Sigma_aV_a^t+\sqrt{n}.\sigma.V_1^t)\\ V_2^t\end{bmatrix}$$

where $V_1 = B^t.U_a/(\sqrt{n}\sigma)$ and $V_2 = B^t.U_a^\perp/(\sqrt{n}\sigma)$.

Note that the noise covariance matrix is a multiple of the identity matrix $E(B.B^t) = n\sigma^2.I_m$. Moreover, remark the Pythagoras-like squaring of the scaled orthogonal matrices in the upper part of the right singular matrix. This implies that one may not hope to recover approximately the row space of the matrix A from the analysis of the matrix C. The columnspace of A however can be recovered, by inspecting the singular values of C: As is obvious, a 'noise threshold' $\sqrt{n}\sigma$ will be recognized in the singular values (for sufficiently large overdetermination n/m).

For a detailed account of the application of this orthogonality principle in system identification, the interested reader is referred to [44].

5 Applications of the Quotient Singular Value Decomposition

In this section, we discuss some applications of the (quotient) singular value decomposition that are closely related to some of its well known algebraic, geometric and numerical properties. We have included a mechanical example (section 5.1), a biomedical signal processing application (section 5.2) and a summary of results on the application of the **OSVD** in realization theory and system identification (sections 5.3 and 5.4). Other applications can be found in the references.

5.1 Moments of inertia

First the close connection between the **OSVD** and the Principal Axes and Moments of Inertia (PAMI) of a rigid body is established. Only the discretized case will be considered. Consider a rigid body consisting of K point masses $m(k)$, described with coordinates $x_i(k)$ in a N-dimensional Euclidean space. With respect to a certain reference O (often the center of gravity), one defines the moment of rotational inertia I_t around an axis t as:

$$I_t = \sum_{k=1}^{K} r_{t,m(k)}^2 m(k)$$

where $r_{t,m(k)}$ is the distance from the k-th point mass to the axis. I_t is a positive function of the orientation t, with one minimum I_{t1}, one maximum I_{tN} and $N-2$ saddle points $I_{t2}, \cdots, I_{t(N-1)}$. The orientations $t_1, .., t_N$ of extremal rotational inertia form a set of orthogonal directions in the N-dimensional Euclidean space and are called the principal axis (PA) of the rigid body and the associated moments of inertia are called the principal moments of inertia (MI) with respect to O. Classically, the PAMI are obtained as the right eigenstructure of a so-called inertia tensor T, i.e. the PA are the right eigenvectors and the MI are the eigenvalues of T, where:

$$T = \begin{bmatrix} t_{11} & -t_{12} & -t13 & \cdots & -t_{1N} \\ -t_{21} & t_{22} & -t23 & \cdots & -t_{2N} \\ \cdot & \cdot & \cdot & \cdots & \cdot \\ -t_{N1} & -t_{N2} & -t_{N3} & \cdots & t_{NN} \end{bmatrix}$$

with

$$t_{ii} = \sum_{k=1}^{K} m(k) \cdot [\sum_{\substack{n=1 \\ n \neq i}}^{N} x_n^2(k)] \quad i = 1, \cdots, N$$

$$t_{ij} = \sum_{k=1}^{K} m(k) \cdot x_i(k) \cdot x_j(k) \quad \substack{i,j=1,\cdots,N \\ i \neq j}$$

T is easy to obtain from the mass distribution of the considered body, described with respect to a chosen basis. It is a symmetric, positive semi-definite matrix. However, the crucial point we want to emphasize is that in forming T explicitly, important numerical accuracy may be lost! Computing the PAMI via the eigenstructure of the inertia tensor T requires ϵ^2 precision computations in order to handle properly ϵ-precision data [39].

However the degeneration of precision is not essential to the PAMI problem itself, but to its formulation as an eigenstructure problem. The loss of accuracy is caused by the mere fact of using the tensor T.

Theorem 5 *On the relation between* **OSVD** *and PAMI.*

Consider a rigid body B of K discrete point masses $m(k)$, located at $x(k) = [x_1(k), \cdots, x_N(k)]^t$ in an N-dimensional coordinate system with reference O. For such a rigid body, a $N \times K$ matrix M is constructed with elements $M(i,j) = \sqrt{m(j)} \cdot x_i(j)$.

- *Then a rigid body B' can be constructed with identical PAMI with N unit pointmasses, located at*

$$\sigma_i \cdot u_i \quad i = 1, \cdots, N$$

where σ_i and u_i are the i-th singular value and i-th left singular vector in the **OSVD** *of M.*

- *The principal axis of inertia of B are the left singular vectors of M*
- *The principal moments of inertia are obtained from the singular values of M as:*

$$I_{t_i} = \sum_{\substack{n=1 \\ n \neq i}}^{N} \sigma_n^2$$

The interested reader is encouraged to investigate the benefits of this relation between **OSVD** and the computation of the PAMI of a rigid body, by applying this result to the matrix M

$$M = 1/\sqrt{2} \begin{bmatrix} 1 & \mu \\ 1 & -\mu \end{bmatrix}$$

Computing the **OSVD** of M in ϵ machine precision, delivers now the second singular value μ with a precision of ϵ/μ, whereas the computation via the tensor T only has precision ϵ/μ^2.

Corollary 2 *If a body is composed of two subbodies B_1 and B_2 with equivalent PAMI decompositions $U_1 \cdot \Sigma_1$ and $U_2 \cdot \Sigma_2$, then an M-matrix for B is obtained as:*

$$M_B = [U_1 \cdot \Sigma_1 \quad U_2 \cdot \Sigma_2]$$

and its associated equivalent PAMI structure is $U \cdot \Sigma$ where the **OSVD** *of $M = U \cdot \Sigma \cdot V^t$.*

5.2 Fetal ECG extraction

The measurements of this biomedical application are obtained from cutaneous electrodes placed at the heart and the abdomen of the mother. If there are p measurement channels (typically 6 to 8), the sampled data are stored in a pxq matrix M_{pq} where q denotes the number of consecutive samples that are processed. The p observed signals $m_i(t)$ (the rows of M_{pq}) are modeled as unknown linear combinations (modeled by a static $p \times r$ matrix T) of r source signals $s_j(t)$, corrupted by additive noise signals $n_i(t)$ with known (or experimentally verified) second order statistics. Hence the model has the well known factor-analysis-like structure:

$$M_{pq} = T_{pr}.S_{rq} + N_{pq}$$

where the rows of S_{rq} are the source signals. The problem now consists of a rank decision to estimate r and of a subspace determination problem to determine the subspace generated by the columns of the matrix T, which are the so-called lead vectors. Since the second order statistics are assumed known, the conceptual framework of oriented signal-to-signal ratio (Mahalanobis transformation [37]) could be applied. However, it has been verified [20,40] that for this specific application with an appropriate position of the electrodes, the subspace spanned by the lead vectors of the mother heart is three dimensional and orthogonal to the three-dimensional subspace generated by the lead vectors of the fetal heart transfer. Moreover, the source signals of mother and fetal are orthogonal vectors if considered over a sufficiently long time wherein the contribution of the mother heart is much stronger than that of the fetal heart. For all this reasons, one single **OSVD** suffices to identify the subspace corresponding to the fetal ECG and by projecting the measurements on this subspace, the MECG can be eliminated almost completely. For more details on this separation based on the strength of the signals we refer to [20,25,26,28,40].

Besides the single **OSVD** approach for FECG/MECG separation, which is based on some restrictive source orthogonality requirements (though fulfilled under mild conditions) it is interesting to note that another **OSVD** based method, described in [13], for the same problem, lends itself very naturally to an interpretation and computation in terms of oriented signal-to-signal ratio and **QSVD**. Basically, the method consists of constructing (by visual inspection) 2 matrices. The first one contains only FECG complexes while the second one is built from maternal ECG complexes. The method then reduces to the determination of that subspace in which the FECG signal can best be distinguished (from the point of view of oriented energy), from the mother ECG. This is equivalent with the determination of the maximal minimal signal-to-signal ratio subspace of a fixed dimension (for instance dimension 3 for the FECG, a choice which can be based upon a physical electro-magnetic model). The measurements are then projected into this subspace, which results in a maternal ECG filtering effect. This is a source separation based on the relative strength.

It is interesting to note that recently high resolution subspace methods have been introduced and analysed [12] to detect the number and the location of narrowband sources. Essentially, these methods reduce to the oriented signal-to-signal ratio framework.

5.3 Realization and exponential fitting

The problem of realization of state space models is the following (stated here for discrete time systems):

Given (possibly noise corrupted) Markov parameters H_k of a linear system. Find a minimal state space representation of the form

$$\begin{aligned} x_{k+1} &= A.x_k + B.u_k \\ y_k &= C.x_k \end{aligned}$$

As is well known, the Markov parameters satisfy $H_k = C.A^{k-1}.B$.

The problem was solved in its full generality by Ho-Kalman and the **OSVD** was introduced in its solution in [9,14]. The algorithm is by now almost classical:

- construct the (block) Hankel matrix H of Markov parameters H_k and choose its dimensions sufficiently large.
- Factorize it using **OSVD**: $H = U.\Sigma.V^t$ This allows to estimate the order of the system from inspection of the rank. Moreover, C and B can be read off immediately from certain (block) rows and (block) columns respectively.
- Exploit the so-called shift structure of the (block) Hankel matrix in order to estimate the state transition matrix A. This reduces to the solution of an overdetermined set of linear equations in [9], in which it differs from [14] where an additional (block) Hankel matrix is to be constructed.

The use of the **OSVD** in these has the almost 'classical' advantages: robust rank estimation, noise insensitivity, high resolution and accuracy. This solution is applied to the estimation of amplitudes, dampings, frequencies and phases in a series of papers [15,22,35]. Furthermore, it has been applied for high resolution spectral analysis in the separation and localisation of narrow-band sources, the analysis and classification of electromyograms and as a second step in a two step identification procedure, in which first the impulse response is estimated using TLLS deconvolution. Several practical problems have been studied:

- the relation between the number of measurements to be used and the accuracy of the estimated parameters (poles, ..). It is obvious that the rank decision becomes easier and the accuracy improves considerably as the number of measurements is increased.

- In [22], one can find some results on the application of Hankel - **OSVD** based methods for high resolution spectral analysis. It has been verified with some simple experiments that the method is extremely robust and performs as well as algorithms that are claimed to be optimal.

- In [35], the sensitivity of the method is investigated for the estimation of coefficients c_i and exponents b_i from noisy observations f(t):

$$f(t) = \sum_{i=1}^{n} c_i.exp(b_i.t) + n(t)$$

As a result, it is shown that, under mild conditions, the error in the computed exponents is of the order:

$$O(\sigma_{n+1}/(\sigma_n - \sigma_{n+1}))$$

the quotient of the largest 'noise' singular value and the gap between the smallest 'signal' and the largest 'noise' singular value. This error estimate is much better than Kung's [9], since the constant in the order term does not contain the (rather large) norm of the pseudo-inverse of the Hankel matrix as in [9]. The result is a rigorous demonstration of some facts that are already intuitively obvious: The more the measurements are corrupted by noise, the smaller will be the gap between σ_n and σ_{n+1} and hence the less accurate will be the estimates. Moreover, there is a one-to-one relation between the exponents and the subspace that is spanned by 'corresponding' singular vectors [34]. When singular values get close, the corresponding singular vectors are no longer well conditioned and 'noise' and 'signal' subspaces get mixed. In [35] one can find also some experimental verification of the fact that 'square' Hankel matrix are best suited for the estimation of the rank (the number of exponentials).

5.4 Identification of state space models from noisy input-output measurements

Starting from the mid-seventies, the **OSVD** and to a lesser extent, the **QSVD** have made their appearance in the systems and control literature, where they have become the cornerstone of numerically reliable implementation of algorithms for Kalman decomposition [7], controllability and observability questions [2], the concept of balanced realization, realization theory [9,14] and numerically reliable computation of the generalized eigenstructure of matrices [5].

The use of **OSVD** and **QSVD** in identification and modeling problems starting from noisy input-output measurements of linear systems, has been and still is the subject of intensive research at the ESAT laboratory. Considerable experience has been gained by applying the derived methods to several data sets from industrial processes. The first results concerned a 'brute force' approach [15,18,19] that consisted of a deconvolution algorithm (TLLS) for the computation of the multivariable impulse response. This was then realized into a state space model using an **OSVD** based realization algorithm. Finally, a certain tail correction iteration procedure was applied in order to ameliorate the estimates. However, in [29,33,41,44], a fundamental structured matrix input - output equation is derived, which provides a much more elegant framework for the formulation and solution of the multivariable identification problem. Moreover, the new approach fits perfectly well into the conceptual framework of oriented signal-to-signal ratio and canonical correlation analysis.

If a linear system, with m inputs and l outputs is described by the above mentioned state space equations: then by straightforward substitutions, the following input-output matrix equation can be derived:

$$Y_h = \Gamma_i.X + H.U_h$$

Here, Y_h (U_h) is a block Hankel matrix with block dimensions $i \times j$, containing i+j-1 consecutive output (input) vectors. There are several good reasons to choose these dimensions in such a way that $max(li, mi) \ll j$. Γ_i is the extended observability matrix. X contains j consecutive state space vectors and H is a lower triangular block Toeplitz matrix containing Markov parameters [29]. In a realistic identification environment, only input-output observations are available. Hence, only the matrices Y_h and U_h are known, up to additive noise. In a series of papers [29,41-44] the following results have been obtained from the geometrical representation of this input output equation:

$$rank \begin{bmatrix} Y_h \\ U_h \end{bmatrix} = rank(U_h) + n$$

where n is the dimension of the (excited) observable part of state space. Hence, under mild conditions [29], one can estimate n from the singular value decomposition of the concatenation of the input and output block Hankel matrix.

- The singular values of U_h serve as quantitative measures for the degree of persistency of excitation of the input sequence. Loosely speaking, the input sequence has to be persistently exciting in order to 'excite' all modes of the systems. When the matrix U_h is (nearly) rank deficient (some singular values are small) the input sequence is 'poor' in that it (almost) consists of a finite number of complex exponentials. When the singular values are all (almost) equal, the input sequence tends to be 'white' noise. Also for an impulsive input, the singular values are all equal (SISO).

Three different identification approaches can now be derived from this input-output equation: a linear least squares approach, a total linear least squares approach and a canonical correlation approach.

Linear Least Squares:

Let $U^\perp$ be any $j \times (mi - rank(U_h))$ matrix satisfying $U_h.U^\perp = 0$. Consider the **OSVD** of $Y_h.U^\perp = P.S.Q^t$. Under mild conditions [29], rank(S) = n and there exists a non-singular $n \times n$ matrix R such that:

$$P = \Gamma_i.R$$

This implies that a realization of the state transition matrix and the output matrix of the form $R^{-1}.A.R, C.R$ can be performed in a similar way as in Kung's realization algorithm. The matrices $R^{-1}.B$ and D follow from a set of linear equations [29,33]. It is shown that this identification approach corresponds to a *linear least squares version* for identification problems where the input is noisefree while the output is noisy. The row space of the output block Hankel matrix is orthogonalized with respect to the input block Hankel row space.

Total Linear Least Squares:

Let the **OSVD** of

$$\begin{bmatrix} Y_h \\ U_h \end{bmatrix} = \begin{bmatrix} P_{11} & P_{12} \\ P_{21} & P_{22} \end{bmatrix} \begin{bmatrix} S_1 & 0 \\ 0 & 0 \end{bmatrix} Q^t$$

where $rank(S_1) = rank(U_h) + n$ and the partitioning of the left singular matrix is such that P_{11} is a $(li) \times (mi + n)$ matrix. Then, there exists a non-singular nxn matrix T such that:

$$P_{11}.P_{21}^{\perp} = \Gamma_i.T$$

where $P_{21}^{\perp}$ is any $(mi + n) \times n$ matrix satisfying $P_{21}.P_{21}^{\perp} = 0$. This implies that a realization of the state transition matrix and the output matrix of the form $T^{-1}.A.T, C.T$ can be performed in a similar way as in Kung's realization algorithm. The matrices $T^{-1}.B$ and D follow from a set of linear equations [29,33]. Contrary to the previous versions, this corresponds to a total linear least squares approximation of the multivariable identification problem, which applies when both input and output are corrupted by the same amount of noise. Considerable insight has been gained into the behavior of the algorithm in noisy industrial applications. More details are found in [41-44].

Canonical Correlations:

The canonical correlation approach to the identification of a state space model, is based upon the following fundamental observation [41-44]:

Let Y_1, U_1 be a output - input block Hankel pair (block dimensions $i \times j$) containing output- input measurements on a linear dynamical system up to time k and let Y_2, U_2 be another output input block Hankel pair of block dimensions $i \times j$, containing measurements from time k+1 on. If the rows of the matrix Z (with j columns) form a basis for the intersection of the row spaces of

$$\begin{bmatrix} Y_1 \\ U_1 \end{bmatrix} \quad and \quad \begin{bmatrix} Y_2 \\ U_2 \end{bmatrix}$$

then:

- $dim(row\ space\ Z) = rank(Z) = n$
- there exists a non-singular $n \times n$ matrix R such that

$$Z = R.[x_{k+1} \quad x_{k+2} \ldots x_{k+j}]$$

Hence, the matrix Z is nothing but a state vector sequence realization. Once such a sequence is available, the model matrices A, B, C, D follow from the set of linear equations:

$$\begin{bmatrix} x_{k+1} \\ y_k \end{bmatrix} = \begin{bmatrix} A & B \\ C & D \end{bmatrix} \begin{bmatrix} x_k \\ u_k \end{bmatrix}$$

that can be solved with TLLS or one of its variations.

Hence, the 'difficult' problem of identification of a linear state space model has now been reduced to 2 **OSVD** steps, that may be implemented in a very streamlined identification algorithm. Adaptive versions for updating and downdating the QR- and **OSVD** factorizations via a gliding window approach are actually being implemented, taking into account the specific structure of the matrices. For more detail, the reader may wish to consult [41-44].

The above summarized identification algorithms have been tested with success on a lot of industrial processes including glass furnaces, power plants, chemical reactors, biological systems, heating and ventilation of confined spaces, and the identification of a flexible arm, the results of which can be found in [29] [41] [42] [43] [44].

6 Conclusions

In this paper, it is claimed that the singular value decomposition and the quotient singular value decomposition have a great potential for signal processing in much the same way as the FFT had a great impact on digital signal processing in the seventies and eighties. Several applications were presented or referred to. The benefits of using the (quotient) singular value decomposition are most pronounced in those applications:

- where essentially rank decisions and the computation of the corresponding subspaces determine the complexity and parameters of the model
- where numerical reliability is of crucial importance and the potential loss of numerical accuracy is to be avoided.
- where a conceptual framework, such as the notion of oriented signal-to-signal ratio, may provide unrevealed additional insight, such as in factor-analysis-like problems.
- where the problem can be stated directly in terms of the (quotient) singular value decomposition, which leads immediately to a reliable and robust solution, such as in a canonical correlation analysis environment.
- where robustness analysis, conditioning and sensitivity optimization are crucial, linked together with geometrical insight and interpretation, for which the **OSVD** and the **QSVD** may provide meaningful quantifications (condition numbers, principal angles,...).

Moreover, in most engineering applications the number of measurements or the data acquisition poses only minor organisational problems (although the design of a measurement set up causes considerable efforts). The cost of the sensors however increases with higher accuracy and signal-to-noise requirements. In this environment the (quotient) singular value decomposition is the optimal bridge between limited measurement precision and robust modeling.
As to the computational requirements, the **OSVD** of large matrices poses no considerable difficulties when employing a mainframe computer (matrix size order of magnitude a few hundred). Moderately sized **OSVDs** (order of magnitude 50..70) are nowadays feasible on mini-computers and PC's. However, it can be expected that the intensive on-going research for parallellized and vectorized algorithms may result in real fast **OSVD** solvers, possibly exploiting the matrix structure which is present in many engineering applications.

Acknowledgement: The authors greatly appreciate the stimulating interaction between all the members of the research team at ESAT.

References

[1] Golub G.H., Van Loan C.F., *"Matrix Computations"*. North Oxford Academic / Johns Hopkins University Press. 1983.

[2] Klema V.C., Laub A.J., "The singular value decomposition:its computation and some applications", *IEEE Trans. Automatic Control*, Vol.AC-25, no.2,pp.164-167, 1980.

[3] Lawson C.L., Hanson R.J., *"Solving least squares problems"*, Prentice Hall Series in Automatic Computation, Englewood Cliffs 1974.

[4] Parlett B.N., *"The symmetric eigenvalue problem"* Prentice Hall Series in Computational Mathematics, Prentice Hall Inc., Englewood Cliffs N.J. 1980.

[5] Van Dooren P., "The generalized eigenstructure in linear systems theory", *IEEE Trans. Automatic Control*, Vol.AC-26, no.1, pp.111-130, 1981.

[6] Hammarling S., "The singular value decomposition in multivariate statistics", *AGM Signum Newsletter*, Vol.20, no.3, pp.2-25, July 1985.

[7] Fernando K.V., Hammarling S.J., "A product induced singular value decomposition for two matrices and balanced realisation", *NAG Technical report TR8/87*, The Numerical Algorithms Group Limited, 1987

[8] Kung S.Y., "A new identification and model reduction algorithm via singular value decomposition", *Proc. 12-th Asilomar Conf. Circ. Syst. comp.*, Nov. 1978.

[9] Moore B.C., "Principal component analysis in linear systems: Controllability, Observability and Model reduction", *IEEE Trans. Automatic Control*, Vol.AC-26, pp.17-32, Feb. 1981.

[10] Paige C.C., Saunders M.A. *Towards a generalized singular value decomposition.* SIAM J. Numer. Anal., 18, pp.398-405, 1981.

[11] Paige C.C., "Computing the generalized singular value decomposition", *SIAM J. Sci.Statist.Comput.*, 7, pp.1126-1146,1986.

[12] Roy R.H., "Estimation of signal parameters via rotational invariance techniques", Ph.D. thesis, Stanford University, August 1987.

[13] van Oosterom A., Alsters J., "Removing the maternal component in the fetal ECG using singular value decomposition", In *Electrocardiography '83*, I.Ruttkay-Nedecky and P.MacFarlane, Eds. Amsterdam, The Netherlands: Excerpta Medica, 1984,pp.171-176.

[14] Zeiger H.P., Mc Ewen A.J., "Approximate linear realizations of given dimensions via Ho's algorithm", *IEEE Trans. Automatic Control*, Vol.AC-19, pp.153, 1974.

[15] Staar J., Vandewalle J., "Comparison of multivariable MBH realization algorithms in the presence of multiple poles and noise disturbing the Markov sequence",*4th Int. Conf. on Anal. and Optimization of Systems.* A.Bensoussan and J.L.Lions (Ed.), Springer, pp.141-160, Berlin, 1980.

[16] Staar J., Vandewalle J., "Numerical implications of the choice of a reference node in the nodal analysis of large circuits", *Int. J. of Circuit Theory and Applications*, Vol.9,pp.488-492,1981.

[17] Staar J., Vandewalle J., "Singular value decomposition: A reliable tool in the algorithmic analysis of systems", *Journal A*, Vol.23, pp.69-74, 1982.

[18] Vandewalle J., Staar J., "Modeling of linear systems: critical examples, problems of numerically reliable approaches", *Proc. IEEE Int. Symp. on Circuits and Systems*, ISCAS-82, Rome, pp.915-918, 1982.

[19] Staar J., Vandewalle J., Claeys F., De Bock H., "On the recursive identification of multivariable state space models", *Symposium on Application of multivariable systems theory*, The Institute of Measurements and Control, Plymouth, Great Britain, 26-28 October 1982, pp.49-57.

[20] Vanderschoot J., Vandewalle J., Janssens J., Sansen W., Vantrappen G., "Extraction of weak bioelectrical signals by means of singular value decomposition", *Proc. of Sixth Int. Conf. on An. and Opt. of Systems*, Nice, June 1984, Springer Verlag, Berlin, pp.434-448.

[21] Vandewalle J., Staar J., De Moor B., Lauwers J., "An adaptive singular value decomposition algorithm and its application to adaptive realization", *Proc. of Sixth Int. Conf. on An. and Optimization of Systems*, Nice, June 1984, Springer Verlag, Berlin, pp.33-47.

[22] De Moor B., Vandewalle J., "A numerically reliable algorithm for fitting a sum of exponentials or sinusoids to noisy data", *Proc. IFAC Symp. on Comp. Aided Control Systems Design.* Copenhague, CADCE '85, pp.434-439., June 1985.

[23] Van Huffel S., Vandewalle J., "The use of total least squares techniques for identification and parameter estimation", *7th IFAC Conference on Identification and Parameter Estimation.* York 3-7 July 1985. Vol.2, pp.1167-1172.

[24] De Moor B., Vandewalle J., "An adaptive singular value decomposition algorithm based on generalized Chebyshev recursions", in: Mathematics in signal processing, T.S. Durrani, J.B. Abbiss, J.E. Hudson, R.N. Madan, J.G. McWhirter and T.A. Moore (ed.), Clarendon Press, Oxfordm 1987, pp. 607-635.

[25] Callaerts D., Vanderschoot J., Vandewalle J., Sansen W., Vantrappen G., Janssens J., "An adaptive on-line method for the extraction of the complete fetal electrocardiogram from abdominal multilead recording", *Journal of Perinatal Medicine*, Vol.14, pp.421-433, 1986.

[26] Callaerts D., Vanderschoot J., Vandewalle J., Sansen W., "An online adaptive algorithm for signal processing applications using SVD", *EUSIPCO-86*, The Hague, 2-5 September 1986, Signal Processing III: Theories and Applications Eurasip, pp.953-956.

[27] Van Huffel S., Vandewalle J., De Roo M., Willems J., "Reliable and efficient deconvolution technique based on total least squares for calculating the renal retention function", *Medical and Biological Engineering and Computing.* Vol. 25, pp.26-33 January 1987.

[28] Vanderschoot Jan, Callaerts Dirk, Sansen W., Vandewalle J., Vantrappen G., Janssens J., "Two methods for optimal MECG elimination and

FECG detection from skin electrode signals", *IEEE Transactions on Biomedical Engineering*, Vol. BME-34, no.3, March 1987.

[29] De Moor B., Vandewalle J., "A geometrical strategy for the identification of state space models of linear multivariable systems with singular value decomposition", *Proc. of the 3rd Int.Symp. on Appl. of Multivariable System Techniques*, Inst. of Measurements and Control, pp.59-68, 1987.

[30] Van Huffel S., Vandewalle J., "Subset selection using the total least squares approach in Collinearity problems with errors in the variables", *Lin. Alg. and its Applications*, Vol.88/89, pp.695-714, April 1987.

[31] Van Huffel S., Vandewalle J., "Algebraic relationships between classical regression and total least squares estimation", *Lin. Alg and its Applications*, Vol. 93, pp.149-160, August 1987.

[32] Van Huffel S., Vandewalle J., Haegemans A., "An efficient and reliable algorithm for computing the singular subspace of a matrix, associated with its smallest singular values", *J.of Comput. and Applied Mathematics*, Vol.19,pp.313-330, 1987.

[33] Swevers J., Adams M., De Moor B., "A new direct time domain identification method for linear systems using singular value decomposition", *Proc. of the 12th International Seminar on Modal Analysis*, Katholieke Universiteit Leuven, Leuven 21-24 September 1987.

[34] De Moor B., Vandewalle J., "Non-conventional matrix calculus in the analysis of rank deficient Hankel matrices of finite dimensions", *Systems and Control Letters*,1987, Vol. 9, pp. 401-410.

[35] De Groen P., De Moor B., "The fit of a sum of exponential to noisy data", *Journal of Computational and Applied Mathematics*, 1987.

[36] Van Huffel, S., Vandewalle, J., "The partial Total Least Squares Algorithm", *Journal Comput. and Applied Math.*, Vol. 21, pp. 333-341, 1988.

[37] De Moor, B., Staar, J., Vandewalle, J., "Oriented energy and oriented signal-to-signal ratio concepts in the analysis of vector sequences and time series", in *SVD and Signal Processing* E. Deprettere (ed.), North Holland, 1988, pp. 209-232.

[38] Vanhuffel, S., Vandewalle, J., "The total least squares technique: computation, properties and applications", in *SVD and Signal Processing* E. Deprettere (ed.), North Holland, 1988, pp. 189-207.

[39] Vandewalle, J., De Moor, B., "A variety of applications of singular value decomposition in identification and signal processing"", in *SVD and Signal Processing* E. Deprettere (ed.), North Holland, 1988, pp. 43-91.

[40] Callaerts, D., Vandewalle, J., Sansen, W., Moonen, M., "On-line algorithm for signal separation based on SVD", in *SVD and Signal Processing* E. Deprettere (ed.), North Holland, 1988, pp. 269-276.

[41] De Moor, B., Moonen, M., Vandenberghe, L., Vandewalle, J., "Identification of linear state space models with singular value decomposition using canonical correlation concepts", in *SVD and Signal Processing* E. Deprettere (ed.), North Holland, 1988, pp. 161-169.

[42] De Moor, B., Moonen, M., Vandenberghe, L., Vandewalle, J., "The application of the canonical correlation concept to the identification of linear state space models", in A. Bensousan, J.L. Lions, (Eds.) *Analysis and Optimization of Systems* Springer Verlag, 1988, pp. 1103-1114.

[43] De Moor, B., Moonen, M., Vandenberghe, L., Vandewalle, J., "A geometrical approach for the identification of state space models with singular value decomposition", *Proceedings of the International Conference on Acoustics, Speech and Signal Processing 1988*, New york, April 1988, pp. 2244-2247.

[44] De Moor B., *Mathematical concepts and techniques for modelling of static and dynamic systems*, Ph. D. Thesis, Katholieke Universiteit Leuven, Electrical Eng. Dept., 1988.

[45] De Moor B., Golub G.H. *Generalized Singular Value Decompositions: A proposal for a standardized nomenclature.* Report, Dept. Computer Science, January 1989, Stanford University.

Structured Linear Algebra Problems in Digital Signal Processing

Paul M. Van Dooren

Philips Research Laboratory Brussels
B-1170 Brussels, Belgium

Abstract

In this paper we give a survey of a number of linear algebra problems occurring in digital signal processing, where the structure of the matrices involved is crucial. Although the problems one wants to solve for these matrices are rather classical, one can not make use anymore here of standard linear algebra tools, since the structure of the matrices has to be taken into account. We discuss in this paper some of these problems and show how structure affects the sensitivity of the problem at hand and how algorithms should be adapted in order to cope with the structure constraint.

1 Introduction

Structured matrices have been around for a long time and are encountered in various application fields. In linear algebra several algorithms have been derived for dealing with such matrices but one is then mainly concerned with exploiting the structure of the matrices in order to improve the complexity of the problem, or in other words to speed up the algorithm. In digital signal processing problems often have to be solved in real time and it is then imperative to compute solutions at a very high speed. For this reason one prefers to use fast algorithms, either by simplifying the assumptions made on the problem, or by exploiting the structure of the problem. Yet several of these so-called fast algorithms suffer from loss of accuracy

NATO ASI Series, Vol. F 70
Numerical Linear Algebra, Digital Signal Processing and Parallel Algorithms
Edited by G. H. Golub and P. Van Dooren

during their (real time) execution which then results in complete divergence of the algorithm from the correct answer.

The importance of a correct understanding of the sensitivity of structured linear algebra problems and of the error propagation in algorithms dealing with them, is being recognized more and more these days. In the next Section we mention a number of structured matrices and indicate in which problem of digital signal processing they occur. When there are fast algorithms available for these problems we briefly discuss them. In Section 3 we then analyze if the constraint of structure on a matrix may affect the sensitivity of a problem defined for such a matrix. We also look at the effect that structure may have on the stability of an algorithm and how structure could be exploited in general. In the last Section we mention a number of open problems for which there are no *structured* algorithms available yet, and show how crucial they can be for such examples.

The paper certainly does not claim to be a complete survey of structured linear algebra problems in digital signal processing. It merely tries to pin down a number of problems in which structure is present and for which it is also crucial to deal with structure in an appropriate manner. The problems may not be very similar at a first glance. It is precisely the aim of this paper to show that in fact they are, and to convince people in digital signal processing to design algorithms that do take structure into account if numerical reliability is to be guaranteed.

2 Structured matrix problems in DSP

In this section we go over a number of structured matrices that are often encountered in digital signal processing applications. For each of them we mention the typical applications and the efficient algorithms that deal with them.

Toeplitz matrices are certainly the most commonly occurring structured matrices in digital signal processing. They can be non-square in general, but when they are square, they are usually symmetric (or Hermitian in the complex case) and positive definite. A symmetric $n \times n$ Toeplitz matrix

T_n :

$$T_n \doteq \begin{bmatrix} t_0 & t_1 & \dots & t_{n-1} \\ t_1 & \ddots & \ddots & \vdots \\ \vdots & \ddots & \ddots & t_1 \\ t_{n-1} & \dots & t_1 & t_0 \end{bmatrix} \tag{1}$$

may e.g. represent the covariance matrix of a stochastic signal $\{s_i\}$:

$$t_i = E\{s_j \cdot s_{j-i}\} \tag{2}$$

or the autocorrelation matrix of a finite data sequence $\{s_i,\ i = 1, \dots, N\}$:

$$t_i = \frac{1}{N} \sum_j s_j \cdot s_{j-i} \tag{3}$$

In both cases T_n is positive definite and one typically wants to compute its Cholesky factorization :

$$T_n = LDL^T. \tag{4}$$

From this decomposition one then computes the first row/column of T_n^{-1}, which yields e.g. the whitening filter of the signal $\{s_i,\ i = 1, \dots, N\}$ [25]. Fast methods to compute such factorizations are the Levinson-Durbin algorithm [25] (which in fact computes L^{-1} instead of L) and the Schur-Cholesky algorithm [12],[7]. Both require $O(n^2)$ floating point operations (also called *flops*) while the standard Cholesky decomposition for a general positive definite matrix would require $O(n^3)$ flops. Moreover, these two algorithms have been shown to have reasonable properties of numerical reliability [3] [8] [4].

In spectral estimation one is rather interested in computing one or a few of the extremal eigenvalues of such matrices. The Pisarenko method requires the smallest eigenvalue of T_n :

$$T_n x_{min} = \lambda_{min} x_{min} \tag{5}$$

in order to construct the closest singular Toeplitz matrix $T_n^* \doteq T_n - \lambda_{min} I_n$ to T_n. The singular Toeplitz matrix T_n^* now represents the covariance matrix of a signal $\{s_i^*,\ i = 1, \dots, N\}$ which consists of pure harmonics [14] :

$$t_j^* = \sum_k \rho_k e^{\mathrm{i} j \theta_k}, \tag{6}$$

where the ρ_k and θ_k can be retrieved from the vector x_{min}. Algorithms for constructing this vector/value pair are e.g. explained in [10], [31].

In circular convolution of two periodic signals with periods $\{s_i,\ i = 0,\ldots,n-1\}$ and $\{c_i,\ i = 0,\ldots,n-1\}$, one wants to multiply the vector $[\ s_0\ s_1\ \ldots\ s_{n-1}\ ]'$ with the circulant matrix C_n :

$$C_n \doteq \begin{bmatrix} c_0 & c_1 & \ldots & c_{n-1} \\ c_{n-1} & \ddots & \ddots & \vdots \\ \vdots & \ddots & \ddots & c_1 \\ c_1 & \ldots & c_{n-1} & c_0 \end{bmatrix} \tag{7}$$

If n is a power of 2, this can be done via FFT techniques. One then goes to the frequency domain via the FFT of the two sequences :

$$\hat{s}_i \doteq \sum_{j=0}^{n-1} w^{j+i} s_j, \quad \hat{c}_i \doteq \sum_{j=0}^{n-1} w^{j+i} c_j, \quad w \doteq e^{-\mathrm{i}\frac{2\pi}{n}} \tag{8}$$

and one merely multiplies the spectra $\hat{s}_i$ and $\hat{c}_i$ to get the spectrum $\hat{s}_i\hat{c}_i$ of the convolved signal. The transformations require $O(n \log n)$ flops instead of the normal $O(n^2)$ flops needed for the product of C_n with the vector of s_i's.

For recursive least squares filtering one typically has to deal with a non-square $m \times n$ Toeplitz matrix T_{mn} , representing the *data matrix* of the least squares problem :

$$T_{mn} \doteq \begin{bmatrix} t_0 & t_{-1} & \ldots & t_{-(n-1)} \\ t_1 & \ddots & \ddots & \vdots \\ \vdots & \ddots & \ddots & t_{-1} \\ \vdots & & \ddots & t_1 \\ t_{m-n} & & & t_0 \\ \vdots & \ddots & & \vdots \\ \vdots & & \ddots & \vdots \\ t_{m-1} & \ldots & \ldots & t_{m-n} \end{bmatrix} \tag{9}$$

Problems in connection to this matrix typically require its QR-factorization :

$$T_{mn} = Q \begin{bmatrix} R \\ 0 \end{bmatrix} \tag{10}$$

which is closely linked with the Cholesky factorization of the matrix $T_{mn}^T T_{mn}$. Indeed, it is easily seen that the latter matrix is Toeplitz and positive definite and that its Cholesky factor L^T is equal to R. This is again used in linear prediction and in fast Kalman filtering [9], [24]. Algorithms providing such factorizations require $O(mn)$ flops instead of the classical $O(mn^2)$ flops for $m \times n$ matrices without specific structure. In signal processing, one is usually interested in having recursive algorithms for "updating" the QR decomposition, i.e. computing

$$T_{m+1,n} = Q_+ \begin{bmatrix} R_+ \\ 0 \end{bmatrix} \tag{11}$$

where $T_{m+1,n}$ is the matrix T_{mn} with one additional row appended to it. The "updating" consists in constructing Q_+ and R_+ directly from Q and R, without passing via the factorization (11). Algorithms for doing this have been found [23] [30] and typically require $O(n)$ for updating R_+ from R (updating Q_+ is often not requested). Unfortunately, these algorithms may suffer from error build-up and may eventually diverge completely.

In the multi-channel analogues of the above problems, one has to deal with block Toeplitz matrices and several "block"-versions have been developed of the above scalar algorithms. Some other matrices can e.g. also be rewritten as a block Toeplitz matrix. This is the case with the Sylvester matrix :

$$S_{2n} \doteq \begin{bmatrix} a_0 & a_1 & \dots & a_n & 0 & \dots & 0 \\ 0 & a_0 & a_1 & \dots & a_n & \ddots & \vdots \\ \vdots & \ddots & \ddots & \ddots & & \ddots & 0 \\ 0 & \dots & 0 & a_1 & a_2 & \dots & a_n \\ 0 & \dots & 0 & b_1 & b_2 & \dots & b_n \\ \vdots & \cdot^{\cdot^{\cdot}} & \cdot^{\cdot^{\cdot}} & \cdot^{\cdot^{\cdot}} & & \cdot^{\cdot^{\cdot}} & 0 \\ 0 & b_0 & b_1 & \dots & b_n & & \vdots \\ b_0 & b_1 & \dots & b_n & 0 & \dots & 0 \end{bmatrix} \tag{12}$$

Such matrices appear in adaptive control, where it is used for GCD extraction or also, to estimate how close one is to having a non-trivial GCD. The typical algorithm is the Euclidean algorithm, which is known to be unstable in general.

A $m \times n$ *Hankel (or block Hankel) matrix* H_{mn} :

$$H_n \doteq \begin{bmatrix} h_0 & h_1 & h_2 & \cdots & h_{n-1} \\ h_1 & h_2 & & \cdot^{\cdot^{\cdot}} & \vdots \\ h_2 & & & & \vdots \\ \vdots & \cdot^{\cdot^{\cdot}} & & & h_{m+n-3} \\ h_{m-1} & \cdots & \cdots & h_{m+n-3} & h_{m+n-2} \end{bmatrix} \tag{13}$$

where the entries h_i are possibly matrices themselves. The scalar version is typically encountered as the impulse response matrix of a transfer function, while the block case occurs e.g. as the input-output pair matrix where each h_i is now a 2×1 block $\begin{bmatrix} y_i \\ u_i \end{bmatrix}$. Both matrices are encountered in problems of identification and/or model reduction of transfer functions. The multichannel cases give rise to block hankel matrices as well. Typical algorithms used for such matrices are variants of the Padé algorithm, or more reliable QR and SVD methods.

All the above examples are what could be called *patterned matrices*. The matrices are *linear* in some coefficients, which occur according to some *pattern* in the matrices. The result of this is that these matrices can be written as a linear combination of certain *basis matrices*. For T_4 this would e.g. be :

$$T_4 \doteq c_0 \begin{bmatrix} 1 & 0 & 0 & 0 \\ 0 & 1 & 0 & 0 \\ 0 & 0 & 1 & 0 \\ 0 & 0 & 0 & 1 \end{bmatrix} + c_1 \begin{bmatrix} 0 & 1 & 0 & 0 \\ 1 & 0 & 1 & 0 \\ 0 & 1 & 0 & 1 \\ 0 & 0 & 1 & 0 \end{bmatrix} + c_2 \begin{bmatrix} 0 & 0 & 1 & 0 \\ 0 & 0 & 0 & 1 \\ 1 & 0 & 0 & 0 \\ 0 & 1 & 0 & 0 \end{bmatrix} + c_3 \begin{bmatrix} 0 & 0 & 0 & 1 \\ 0 & 0 & 0 & 0 \\ 0 & 0 & 0 & 0 \\ 1 & 0 & 0 & 0 \end{bmatrix} \tag{14}$$

If one consider an $m \times n$ matrix as an mn-tuple and one introduces the standard inner product for these mn-tuples, then the set of patterned matrices form a linear subspace for which the basis matrices form an orthogonal basis. Other matrices for which this is true are the *sparse matrices* (e.g. band matrices) :

$$A_n \doteq \begin{bmatrix} a_{11} & a_{12} & 0 & \ldots & 0 \\ a_{21} & a_{22} & \ddots & \ddots & \vdots \\ 0 & \ddots & \ddots & \ddots & 0 \\ \vdots & \ddots & \ddots & \ddots & a_{n-1,n} \\ 0 & \ldots & 0 & a_{n,n-1} & a_{nn} \end{bmatrix} \tag{15}$$

This property may sometimes be exploited nicely as is shown in the next section.

More involved matrices are those in which the parameters enter in a non-linear fashion. Examples of this are now given below.

A $m \times n$ *Vandermonde matrix* V_{mn} :

$$V_{mn} \doteq \begin{bmatrix} 1 & x_1 & \ldots & x_1^{n-1} \\ 1 & x_2 & \ldots & x_2^{n-1} \\ \vdots & \vdots & & \vdots \\ 1 & x_m & \ldots & x_m^{n-1} \end{bmatrix} \tag{16}$$

These matrices occur in the construction of a least squares polynomial fit of degree $n-1$ through m data points. If the polynomial is $P(x) = a_0 + a_1 x + \ldots + a_{n-1}x^{n-1}$ and the data points are the pairs $(x_i, y_i), i = 1, \ldots, m$ then one has to solve in least squares sense (and often recursively) the equation $V_{mn}a = y$, where y and a are vectors containing the coefficients a_i and y_i, respectively.

A $n \times n$ *Nevanlinna-Pick* matrix N_n :

$$N_n \doteq \begin{bmatrix} \frac{1-\bar{s}(x_1)s(x_1)}{1-\bar{x}_1 x_1} & \frac{1-\bar{s}(x_1)s(x_2)}{1-\bar{x}_1 x_2} & \cdots & \frac{1-\bar{s}(x_1)s(x_n)}{1-\bar{x}_1 x_n} \\ \frac{1-\bar{s}(x_2)s(x_1)}{1-\bar{x}_2 x_1} & \frac{1-\bar{s}(x_2)s(x_2)}{1-\bar{x}_2 x_2} & \cdots & \frac{1-\bar{s}(x_2)s(x_n)}{1-\bar{x}_2 x_n} \\ \vdots & \vdots & & \vdots \\ \frac{1-\bar{s}(x_n)s(x_1)}{1-\bar{x}_n x_1} & \frac{1-\bar{s}(x_n)s(x_2)}{1-\bar{x}_n x_2} & \cdots & \frac{1-\bar{s}(x_n)s(x_n)}{1-\bar{x}_n x_n} \end{bmatrix} \tag{17}$$

is typically a square matrix entering in various approximation problems. For the choice $s(x) = x^n$ one obtains $\frac{1-\bar{s}(x_i)s(x_j)}{1-\bar{x}_i x_j} = \sum_{k=0}^{n-1} \bar{x}_i^k x_j^k$ and the Nevanlinna-Pick matrix is then the gramian of the Vandermonde matrix : $P_n = V_{mn}^* V_{mn}$. Other choices of $s(x)$ are found in approximation problems with Blashke products [13].

A $m \times n$ *Loewner matrix* L_{mn} :

$$L_{mn} \doteq \begin{bmatrix} \frac{\bar{y}_1 - y_1}{\bar{x}_1 - x_1} & \frac{\bar{y}_1 - y_2}{\bar{x}_1 - x_2} & \cdots & \frac{\bar{y}_1 - y_n}{\bar{x}_1 - x_n} \\ \frac{\bar{y}_2 - y_1}{\bar{x}_2 - x_1} & \frac{\bar{y}_2 - y_2}{\bar{x}_2 - x_2} & \cdots & \frac{\bar{y}_2 - y_n}{\bar{x}_2 - x_n} \\ \vdots & \vdots & & \vdots \\ \frac{\bar{y}_m - y_1}{\bar{x}_m - x_1} & \frac{\bar{y}_m - y_2}{\bar{x}_m - x_2} & \cdots & \frac{\bar{y}_m - y_n}{\bar{x}_m - x_n} \end{bmatrix} \tag{18}$$

is a matrix entering in interpolation problems with rational functions [1]. A closely related matrix is the *Cauchy matrix* C_{mn} with elements $\frac{1}{x_i - y_j}$ (see [17]).

A *unitary Hessenberg* matrix U_n :

$$U_n \doteq \begin{bmatrix} -c_1 & -s_1 c_2 & -s_1 s_2 c_3 & \cdots & -s_1 s_2 \cdots s_{n-1} e_n \\ s_1 & -c_1 c_2 & -c_1 s_2 c_3 & \cdots & -c_1 s_2 \cdots s_{n-1} e_n \\ 0 & s_2 & -c_2 c_3 & & \vdots \\ \vdots & \ddots & \ddots & \ddots & \vdots \\ 0 & \cdots & 0 & s_{n-1} & -c_{n-1} e_n \end{bmatrix} \tag{19}$$

is in fact completely determined by the $n-1$ pairs $\{c_i, s_i\}$ (satisfying $c_i^2 + s_i^2 = 1$) and the number e_n (satisfying $| e_n |= 1$). This matrix has all its eigenvalues on the unit circle and the phases ω_i of these eigenvalues can be used as approximations of the resonance frequencies of the ladder filter with reflection coefficients c_i [14].

A *controllability matrix* $C(A, b)$ of an $n \times n$ matrix A and an n vector b is a Krylov matrix :

$$C(A, B) = \begin{bmatrix} b & Ab & A^2b & \ldots & A^m b \end{bmatrix} \tag{20}$$

It is a structured matrix whose rank is full if and only if the underlying state space model $x_{k+1} = Ax_k + bu_k$ is controllable. Therefore, the distance of $C(A, b)$ to a singular matrix is sometimes used as "measure of controllability" of this system [32].

3 Sensitivity and stability

Numerical linear algebra is probably the area where the issues of *sensitivity of a problem* and *stability of an algorithm* are best understood and most precisely described. An important result in this area is that for a considerable class of algorithms, rigorous bounds have been derived for the error propagation during the execution of the algorithm on a computer with relative precision ϵ. When these bounds can be interpreted as "equivalent" (but fictitious) errors on the data – so-called *backward errors* –, this yields a powerful tool for analyzing and comparing algorithms and for predicting their behaviour for various problems. We briefly introduce here these two general concepts first, before applying them to the "structured" problems described in the previous section.

3.1 Conditioning

We follow here the analysis of Rice [26]. Let us consider a mapping $y = f(x)$ from a vector space $\mathcal{F}_d^k$ to a vector space $\mathcal{F}_s^\ell$, where $\mathcal{F}_d$ and $\mathcal{F}_s$ are the fields of either real numbers $\mathcal{R}$ or complex numbers $\mathcal{C}$. We call $\mathcal{F}_d^k$ the *data* space and $\mathcal{F}_s^\ell$ the *solution* space. Examples from numerical linear algebra would e.g. be the function $\Lambda = f(H)$ that maps an $n \times n$ Hermitean matrix $H \in \mathcal{C}^{n\times n}$ to its n-tuple of eigenvalues $\Lambda \in \mathcal{R}^n$, or the function $x = f(A,b)$ that maps a real system of equations $(A,b) \in \mathcal{R}^{n\times(n+1)}$ to its solution $x = A^{-1}b \in \mathcal{R}^n$.

The sensitivity of such a function is now characterized as follows. First choose norms $\| \cdot \|_d$ and $\| \cdot \|_s$ (independently) in the *data* space, respectively, *solution* space. Then the *absolute condition number* $\kappa_{abs}(f,x_0)$ of the function f at the point x_0 is defined as :

$$\kappa_{abs}(f,x_0) = \lim_{\epsilon\to 0} \sup_{x,\|x-x_0\|_d\le\epsilon} \|f(x)-f(x_0)\|_s/\|x-x_0\|_d \tag{21}$$

If we exclude the origin in the data and solution space (i.e. $x_0 \neq 0$ and $f(x_0) \neq 0$), then we can define a *relative condition number* $\kappa_{rel}(f,x_0)$ of f at the point x_0 in a similar manner :

$$\kappa_{rel}(f,x_0) = \lim_{\epsilon\to 0} \sup_{x,\|x-x_0\|_d\le\epsilon} \frac{\|f(x)-f(x_0)\|_s}{\|f(x_0)\|_s} \Big/ \frac{\|x-x_0\|_d}{\|x_0\|_d} \tag{22}$$

The condition number $\kappa_{abs}(f,x_0)$ (resp. $\kappa_{rel}(f,x_0)$) thus gives a bound for absolute (resp. relative) perturbations in the solution $f(x_0)$ when applying infinitesimally small perturbations to x_0 :

$$\|f(x)-f(x_0)\|_s \le \kappa_{abs}(f,x_0)\|x-x_0\|_d + o(\epsilon) \tag{23}$$

$$\frac{\|f(x)-f(x_0)\|_s}{\|f(x_0)\|_s} \le \kappa_{rel}(f,x_0)\frac{\|x-x_0\|_d}{\|x_0\|_d} + o(\epsilon) \tag{24}$$

Since $\kappa_{abs}(f,x_0)\cdot\|x_0\|_d = \kappa_{rel}(f,x_0)\cdot\|f(x_0)\|_s$, we restrict ourselves in the sequel to $\kappa_{abs}(f,x_0)$ and drop the suffix *abs*. The condition number $\kappa(f,x_0)$ is illustrated in Fig. 1, where an ϵ-ball around x_0 is mapped into a region around $f(x_0)$ of smallest radius $k\epsilon$. For infinitesimally small ϵ, the ratio k of both radii tends of course to the condition number $\kappa(f,x_0)$.

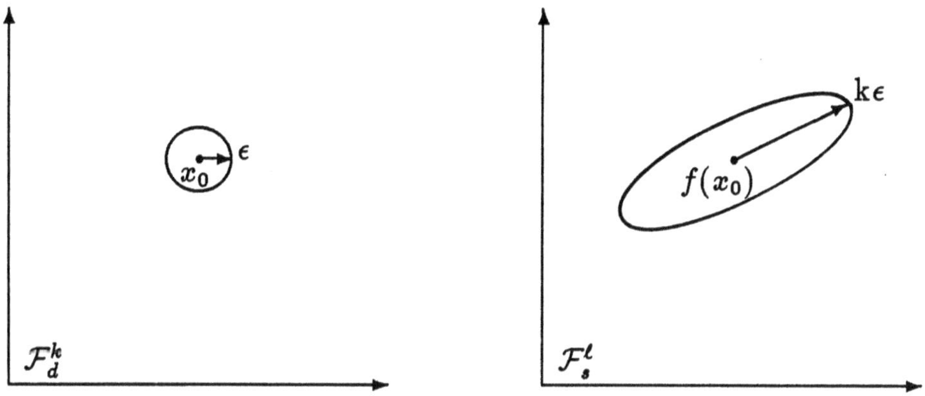

Fig. 1 : Condition number $\kappa(f, x_0)$

3.2 Structured conditioning

Let us now consider a structured A matrix in the date space $\mathcal{F}_d^k$, i.e. whose entries $a_{i,j}$ are related via certain functions as for the examples of Section 2. Since the entries $a_{i,j}$ are solutions of a set of polynomial equations in several variables, the set of matrices A is called an *algebraic set* $\mathcal{V}$. A crucial property needed in the sequel is that $\mathcal{V}$ is *connected* – i.e. that any two matrices with that particular structure can be connected via a continuous path of matrices belonging to $\mathcal{V}$ – and *unbounded* – i.e. there is always an element of $\mathcal{V}$ of arbitrarily large norm. These properties are trivially satisfied for patterned matrices since they form a linear subspace in $\mathcal{F}_d^k$, but they can also easily be proved for the other examples given in Section 2. Indeed this follows from the fact that the parameters defining the matrices, can have arbitrary values (possibly, except for a finite number of points). For each parameter the allowed values are thus connected and unbounded and this then obviously holds as well for the set $\mathcal{V}$.

A consequence of this is that if A_0 is a *point* in $\mathcal{F}_d^k$ which also belongs to $\mathcal{V}$, then the *intersection* of an ϵ-ball $\mathcal{B}(A_0, \epsilon)$ around A_0 with $\mathcal{V}$ always contains a point which is ϵ distant from A_0. This intersection $\mathcal{V} \cap \mathcal{B}(A_0, \epsilon)$ is illustrated in Fig. 2 by the dashed area in the data space.

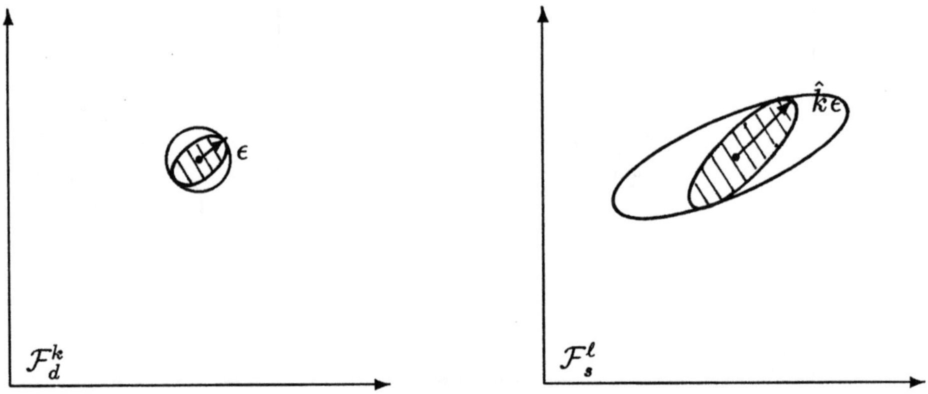

Fig. 2 : Restricted condition number $\hat{\kappa}(f, x_0)$

In the solution space $\mathcal{F}_s^\ell$ we indicated the mapping of $\mathcal{V} \cap \mathcal{B}(A_0, \epsilon)$ as a dashed area as well. Obviously, the largest distance $\hat{k}\epsilon$ from $f(A_0)$ is smaller now than the largest distance $k\epsilon$ in Fig. 1. Since $\hat{k}$ tends to $\hat{\kappa}(f, x_0)$ for $\epsilon \to 0$ we thus have by a continuity argument that the following inequality always holds :

$$\hat{\kappa}(f, x_0) \leq \kappa(f, x_0) \tag{25}$$

if the same norms are used in both definitions of condition numbers. This restriction is important since in the structured case, one is tempted to use a norm defined on the *parameter vector* instead of on the *matrix* defined by these parameters (e.g. for a Toeplitz matrix T_n this would be the parameter vector $[\ t_0\ , \ldots, t_n\ ]$). When different norms are used for the definitions of $\kappa(f, x_0)$ and $\hat{\kappa}(f, x_0)$, the above inequality has to be corrected with appropriate scale factors which, unfortunately, are not always easy to evaluate or even to bound (think e.g. of the example of the Nevanlinna-Pick matrix N_n).

3.3 Stability

Let us now see if structure may also affect definitions of numerical stability. For this we first recall the usual forms of numerical stability. These are :

forward stability :

$$\overline{f}(x_0) = f(x_0) + \Delta_1, \qquad \|\Delta_1\|_s \leq c_1\epsilon \tag{26}$$

backward stability :

$$\overline{f}(x_0) = f(x_0 + \Delta_2), \qquad \|\Delta_2\|_d \leq c_2\epsilon \tag{27}$$

mixed stability :

$$\overline{f}(x_0) = f(x_0 + \Delta_3) + \Delta_4, \qquad \begin{array}{l} \|\Delta_3\|_d \leq c_3\epsilon \\ \|\Delta_4\|_s \leq c_4\epsilon \end{array} \tag{28}$$

and *weak stability* :

$$\overline{f}(x_0) = f(x_0) + \kappa(f, x_0)\Delta_5, \qquad \|\Delta_5\|_s \leq c_5\epsilon \tag{29}$$

As ordered above the "type" of stability becomes weaker and weaker. Indeed, we can rewrite (23) into $f(x_0 + \Delta_d) = f(x_0) + \kappa(f, x_0)\Delta_{s'}$ with $\|\Delta_{s'}\|_s \leq \|\Delta_d\|_d + o(\epsilon)$. Using this, one then obtains for backward stability :

$$\overline{f}(x_0) = f(x_0) + \kappa(f, x_0)\Delta_{2'}, \qquad \|\Delta_{2'}\|_s \leq c_2\epsilon + o(\epsilon) \tag{30}$$

and mixed stability :

$$\overline{f}(x_0) = f(x_0) + \kappa(f, x_0)\Delta_{3'} + \Delta_4, \qquad \begin{array}{l} \|\Delta_{3'}\|_s \leq c_3\epsilon + o(\epsilon) \\ \|\Delta_4\|_s \leq c_4\epsilon \end{array} \tag{31}$$

When looking at the bound for the "equivalent" forward errors, it is clear that the bounds are less restrictive (if we assume that $\kappa(f, x_0) \geq 1$). Indeed, (26) implies (27) because of (30), (27) obviously implies (28), and (28) implies (29) because of (31).

3.4 Structured stability

Let us now suppose that we can guarantee that an algorithm is such that the part of the error $\hat{\Delta}_d$ interpreted in the data space respects the structure of given matrices, then the following bounds are obtained for what we would

like to call :
structured backward stability :

$$\overline{f}(x_0) = f(x_0 + \hat{\Delta}_2), \qquad \|\hat{\Delta}_2\|_d \leq c_2\epsilon \tag{32}$$

structured mixed stability :

$$\overline{f}(x_0) = f(x_0 + \hat{\Delta}_3) + \Delta_4, \qquad \begin{array}{l} \|\hat{\Delta}_3\|_d \leq c_3\epsilon \\ \|\Delta_4\|_s \leq c_4\epsilon \end{array} \tag{33}$$

Using now $f(x_0 + \hat{\Delta}_d) = f(x_0) + \hat{\kappa}(f, x_0)\Delta_{s'}$ with $\|\Delta_{s'}\|_s \leq \|\hat{\Delta}_d\|_d + o(\epsilon)$ one then obtains for *structured backward stability* :

$$\overline{f}(x_0) = f(x_0) + \hat{\kappa}(f, x_0)\Delta_{2'}, \qquad \|\Delta_{2'}\|_s \leq c_2\epsilon + o(\epsilon) \tag{34}$$

and for *structured mixed stability* :

$$\overline{f}(x_0) = f(x_0) + \hat{\kappa}(f, x_0)\Delta_{3'} + \Delta_4, \qquad \begin{array}{l} \|\Delta_{3'}\|_s \leq c_3\epsilon + o(\epsilon) \\ \|\Delta_4\|_s \leq c_4\epsilon \end{array} \tag{35}$$

which are clearly better bounds than (30) and (31), since the structured condition number $\hat{\kappa}(f, x_0)$ can only be smaller than the unstructured condition number $\kappa(f, x_0)$. Similarly we can define *structured weak stability* for an algorithm when the following holds :

$$\overline{f}(x_0) = f(x_0) + \hat{\kappa}(f, x_0)\Delta_5, \qquad \|\Delta_5\|_s \leq c_5\epsilon \tag{36}$$

where again, the structured condition number appears and a better bound is obtained.

4 How much can structure affect this ?

In the previous section we showed that structure can only mean an improvement for the sensitivity of the problem. But how big an improvement can it be ? Is this not negligible in most cases ? We give a number of examples here illustrating what range of improvements can be obtained.

4.1 Structure improving distance measures

The problem of improving distance measures by imposing structure is closely related to that of improving the condition number. Indeed let $\mathcal{S}$ be some set to which one wants to determine the distance from a given point x_0. The set $\mathcal{S}$ could be e.g. the singular matrices ($\mathcal{S}$ is then a *variety*) or the symmetric positive definite matrices ($\mathcal{S}$ is then a *cone*). In the picture below, we represent $\mathcal{S}$ more or less as a variety. The *unstructured* distance Δ of x_0 to $\mathcal{S}$ is then just the radius of the smallest ball around x_0 which has at least one point in common with $\mathcal{S}$. The intersection of this Δ ball with the structured matrices $\mathcal{V}$ may very well not have a point in common with $\mathcal{S}$. The *structured* distance $\hat{\Delta}$ of x_0 to $\mathcal{S}$ is then the distance to the closest point on $\mathcal{S} \cap \mathcal{V}$ and this may be much larger than Δ.

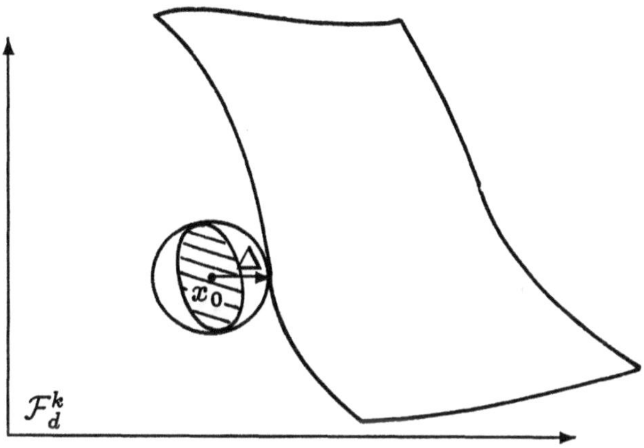

Fig. 3 : Distances Δ and $\hat{\Delta}$ to $\mathcal{S}$

We now give two extreme examples. Consider first the subspace of Toeplitz matrices T_n and the closeness problem to the singular matrices. If we use the spectral norm as distance measure then it is easy to show that both the unstructured distance Δ and the structured distance $\hat{\Delta}$ are equal to λ_{min}, and the closest matrix in both cases is equal to $T_n - \lambda_{min} I$. The closest unstructured matrix happens indeed to be Toeplitz !

A second example illustrates the other extreme. Consider the patterned matrices

$$\mathcal{V} = \left\{ \begin{bmatrix} 0 & a & b \\ a & 0 & c \\ b & c & 0 \end{bmatrix} \text{ for } a, b, c \text{ arbitrary} \right\} \tag{37}$$

and the matrix M_0 with parameter values $a = c = 0$ and $b = 1$. This matrix has eigenvalues equal to $\lambda_1 = -1$, $\lambda_2 = 0$, $\lambda_3 = 1$. Suppose we want to find the smallest perturbation that moves eigenvalue λ_2 to the value ϵ while the other two eigenvalues remain unchanged, and let us use the spectral norm again as distance measure. For an unstructured perturbation the minimum distance Δ is in fact ϵ and the matrix that achieves this is

$$M_\epsilon = \begin{bmatrix} 0 & 0 & 1 \\ 0 & \epsilon & 0 \\ 1 & 0 & 0 \end{bmatrix} \tag{38}$$

The closest *structured* matrix, on the other hand, is at distance $\hat{\Delta} = \infty$ since there is no structured matrix that will achieve this. Indeed, the structured matrices (37) have trace 0 and hence the prescribed set of eigenvalues is unattainable.

4.2 The importance of choosing a measure

In the two above example we chose the spectral norm as distance measure both for structured and unstructured perturbations. How the choice of a norm can affect certain answers is shown below by some examples again.

Consider the following system :

$$A = \begin{bmatrix} 1/2 & \sqrt{\epsilon} \\ 0 & 1/2 \end{bmatrix}, \; b = \begin{bmatrix} 0 \\ \sqrt{\epsilon} \end{bmatrix} \tag{39}$$

then its controllability matrix $C_2(A, b)$ equals

$$C_2(A, b) = \begin{bmatrix} b, & Ab \end{bmatrix} = \begin{bmatrix} 0 & \epsilon \\ \sqrt{\epsilon} & \sqrt{\epsilon}/2 \end{bmatrix} \tag{40}$$

This is a structured matrix of which the column rank is full if and only if the underlying state space model $x_{k+1} = Ax_k + bu_k$ is controllable. Let us use $\delta_1 = \|C_2(A,b) - C_2(A_\epsilon, b_\epsilon)\|_2$ as distance measure and let $\hat{\Delta}_1 = \min\delta_1$ be the distance to the "uncontrollable set", where $(A_\epsilon,\ b_\epsilon)$ is uncontrollable. Then $\hat{\Delta}_1$ is smaller than ϵ since the system (A_ϵ, b) with $A_\epsilon = \frac{1}{2}I$ is already uncontrollable and has distance $\delta_1 = \epsilon$. Take on the other hand the measure $\delta_2 = \|[A,\ b] - [A_\epsilon,\ b_\epsilon]\|_2$ then $\hat{\Delta}_2 = \min\delta_2$ can be shown to be of the order of $\sqrt{\epsilon}$ [32]. Clearly the previous measure can be seriously distorted from this one.

A second example illustrates the computability of the solution. Consider the set of arbitrary matrices and the cone $\mathcal{P}$ of positive definite Toeplitz matrices. Suppose one wants to find the closest matrix T in $\mathcal{P}$ to a given (arbitrary) matrix A. If one measures distances in Frobenius norm : $\delta_F = \|A - A_\epsilon\|_F$, then the problem can be solved as follows. Let $\mathcal{T}$ be the subspace of Toeplitz matrices. The closest Toeplitz matrix $T \in \mathcal{P}$ to A is easy to find. Indeed, since the Frobenius norm is just the 2-norm in the vector space of matrices, and since $\mathcal{T}$ is a linear subspace, one just projects A on $\mathcal{T}$ to find, say T_A. This projection is trivial since (14) is an orthogonal basis for $\mathcal{T}$; the projection then merely amounts to averaging the elements of A along the 1's in each basis vector. Since $\mathcal{P}$ is a subset of $\mathcal{T}$, we can now split our original problem in two orthogonal ones. Let T be the closest matrix in $\mathcal{P}$ to T_A. Then the *vectors* $A - T_A$ and $T_A - T$ are orthogonal to each other (in the above norm). Hence, T is also the closest matrix in $\mathcal{P}$ to A. Moreover, we can also say the following. If T_A happens to be in $\mathcal{P}$ then $T = T_A$ is our solution. Otherwise, there is no solution since $\mathcal{P}$ is an open set in $\mathcal{T}$. The boundary of $\mathcal{P}$ in $\mathcal{T}$ are indeed all singular matrices, and the closest $T \in \mathcal{P}$ to a matrix T_A outside $\mathcal{P}$ therefore is also singular. The above problem is thus solved by merely projecting A onto $\mathcal{T}$ to get T_A and checking whether this lies also in $\mathcal{P}$.

Notice that the above reasoning does not work anymore as soon as one uses another norm. The Frobenius norm induces a inner product in the vector space of matrices, which is crucial here. A similar technique can be used in the approximation of Hankel matrices by singular ones. In [27] this is exploited to give a simple (but not optimal) approximation, while the same problem using spectral norms is shown to lead to a complicated generalization of a fixed point problem in [6].

4.3 Structure improving sensitivity

In the definition of conditioning, the choice of measures is again crucial. It is quite difficult to find closed formulas for the condition number of a structured matrix problem unless one chooses an appropriate norm which makes the task somewhat easier. Even then, one often can only find bounds rather than exact formulas. Below we give two examples of such closed formulas, which were only recently obtained.

Consider first the sensitivity of the polar decomposition of a nonsingular $n \times n$ matrix X (here $.^*$ denotes Hermitian conjugate) :

$$X = U \cdot H, \quad U^*U = I, \; H = H^* > 0. \tag{41}$$

Suppose X is a real matrix but we want to analyze the effect of both real and complex perturbations on the unitary factor U. (The analysis for the sensitivity of H follows from the one of U.) If we use the spectral norm in the data space of matrices X and the Frobenius norm in the solution space U, then it is shown in [22] that the condition number for arbitrary perturbations of a complex or real matrix is given by :

$$\kappa(U) = 1/\sigma_n(X). \tag{42}$$

Since X is a real matrix, we might consider real perturbation only. In the space of complex matrices this is in fact a structured perturbation. It is shown in [22] again that this restricted condition number is now

$$\hat{\kappa}(U) = 1/[\sigma_n(X) + \sigma_{n-1}(X)] \tag{43}$$

It is obvious that one can find a real matrix X with any prescribed singular values $\sigma_n(X)$ and $\sigma_{n-1}(X)$, and that the difference between the condition numbers κ and $\hat{\kappa}$ can thus be arbitrarily large.

Our second example concerns a classical example of badly conditioned matrices with respect to inversion : the $n \times n$ Hilbert matrix H, with elements $H_{i,j} = \frac{1}{i+j-1}$. When using the spectral or Frobenius norm for the data space H as well as the solution space H^{-1}, the unstructured condition number is (essentially) exponential in n, the order of H [18]:

$$\kappa(H^{-1}) \approx 10^n. \tag{44}$$

Recently, it was observed in [17] that much better numerical results were obtained than what could be expected from the above condition number, when using a specialized algorithm for the inversion of Cauchy matrices. The algorithm in [17] considers in fact the entries of H to be $H_{i,j} = \frac{1}{x_i - y_j}$ (with $x_i = i-1$ and $y_j = -j$) and the vectors of elements x_i and y_j are processed by the algorithm rather than the elements of H. Under the supposition that such an algorithm would perturb the elements x_i and y_j only, one ought to consider the *structured* condition number $\hat{\kappa}(H^{-1})$ when analyzing error propagation. It is proved in [17] that this condition number is now (essentially) quadratic in n :

$$\hat{\kappa}(H^{-1}) \approx n^2 \tag{45}$$

when choosing appropriate norms in both data space and solution space. We remark here that in fact [17] use a relative condition number, which affects to some extent the expressions (44) and (45). Nevertheless, the difference between κ and $\hat{\kappa}$ is indeed striking !

4.4 Structure improving stability

We have seen in previous sections that structured matrices should lead also to better stability results provided the accumulated errors correspond exactly to perturbed *structured* data (or are at least very close to such data).

In practice, one observes a completely different phenomenon. Algorithms exploiting structure of the processed matrix often do that in order to reduce the number of operations. These so-called *fast* algorithms violate basic principles of numerically stable algorithms (such as pivoting in LU decompositions) and turn out to be *unstable*, even in an unstructured sense. Typical examples of this are the fast Hankel factorization algorithms based on the Padé algorithm and its variants, as was shown in [11]. Is it a general principle that one has to sacrifice stability for speed in these fast algorithms, or can we hope for algorithms that exploit structure in such a manner that both speed and numerical stability are obtained ? Ideally one would even wish stability to hold in a structured sense (see section

3.4) since this yields improved bounds for the error propagation. We now mention a number of results that point in that direction.

Two of the first analyses made for popular signal processing algorithms are related to the Toeplitz matrices. In [3] [8] an error analysis is given showing that the well-known Levinson-Durbin method for positive definite Toeplitz matrices is stable in a weak and unstructured sense. It is indeed proven that the *computed* triangular factors contain *forward* errors that are of the order of the it unstructured condition number of the problem. It is not known if the *structured condition number* of this problem can be much better than the unstructured one, and hence no structured stability result is obtained there. The same holds for the analysis of the Lattice Algorithm analyzed in [9] : weak stability is proven in an unstructured sense and no results are derived in the structured sense. Another fast method for rectangular Toeplitz matrices is described in [23] and is known to be unstable (even in the unstructured sense, which is worse). Recently a stabilized version was proposed in [30], but there the analysis is only statistical. A more rigorous proof for this algorithm would be welcome. An analysis of *structured errors* appeared in for Vandermonde matrices in [19] and for both Vandermonde and Cauchy matrices in [17]. For Vandermonde matrices structured stability is proved for special cases only, while for Cauchy matrices *weak structured stability* is proved for a special relative condition number.

5 Concluding remarks

From the above discussions it appears that concrete results are only very rare. But it appears from the literature that the awareness of the problem of structured stability is growing and that results in this direction are slowly being obtained [5] [19] [17]. Early work on stability [20] [11] lead to definitions of different types of stability as e.g. given in section 3.3. Also for structured problems one finds results about various types of stability and conditioning [21] [4] [15]. Yet we believe that the list of possible types of numerical stability as presented here in section 3, is more complete than what appeared before. Moreover we tried to illustrate with several examples that each one of these definitions has a different meaning for the obtained errors.

We should also point out the existence of new results for norms which are proportional to the size of the individual elements of the matrix [28] [2]. It is interesting to notice that there a simple step of iterative refinement lead to much stronger results of stability. Is it also possible to apply such a technique here ? That this might be possible is indicated by the error correction mechanism presented in [30]. A general idea would be to try to guarantee that at each step of a structured algorithm one would try to keep the numerical errors on the variety of structured matrices (or at least very close to it) by some projection mechanism or iterative refinement. By doing so, one would obtain an algorithm with numerical errors that satisfy the constraints of structure in the matrix, and would therefore yield better error bounds based on the structured condition number. For closeness problems the structure constraint leads to minimization problems with constraints which are often much harder to solve than the unconstrained problem. Similar results in a related area indicate that performant algorithms may be very hard to find indeed [16] [29]. Let us terminate this paper with a word of good hope. When problem is well described, one is already half way to the solution; since clearly there is an increasing activity in this area, solution can not be far away anymore.

Acknowledgement

This material was presented in parts at several conferences (*NATO ASI* in Leuven, *Mathematics in Signal Processing* in Warwick and *Future Directions in Matrix Theory* in Auburn). People attending these talks have made valuable comments, incorporated in this text. They include D. Slock, A.J.E.M. Janssen, G. Golub, J. Demmel and H. Lev-Ari.

References

[1] A. Antoulas and B. Anderson, State space and polynomial approaches to rational interpolation, Proceedings MTNS, 1989.

[2] M. Arioli, J. Demmel and I. Duff, Solving sparse linear systems with sparse error bound, *SIAM J. Matrix Anal. Appl.* **10** : 165-190 (1989).

[3] A. Bultheel, Error analysis of incoming and outgoing schemes for the trigonometric moment problem, *in : Lecture Notes on Mathematics* **888**, Springer Verlag, 1981.

[4] J. Bunch, The weak and strong stability of algorithms in numerical linear algebra, *Lin. Alg. & Appl.* **88** : 49-66 (1987).

[5] R. Byers, A. Bunse-Gerstner and V. Mehrmann, A chart of numerical methods for structured eigenvalue problems, *SIAM J. Matrix Anal. Appl.*, to appear.

[6] J. Cadzow, Signal enhancement : a composite property mapping algorithm, *IEEE Trans. ASSP* **10** : 49-62 (1988).

[7] J. Chun and T. Kailath, Generalized displacement structure for block-Toeplitz, Toeplitz-block and Toeplitz-derived matrices, *these Proceedings*, pp. 215-236.

[8] G. Cybenko, The numerical stability of the Levinson-Durbin algorithm for Toeplitz systems of equations, *SIAM J. Sci. & Stat. Comp.* **1** : 303-310 (1980).

[9] G. Cybenko, The numerical stability of the lattice algorithm for least-squares linear prediction problems, *BIT* **24** : 441-455 (1984).

[10] G. Cybenko and C. Van Loan, Computing the minimum eigenvalue of a symmetric positive definite Toeplitz matrix, *SIAM J. Sci. & Stat. Comp.* **7** : 123-131 (1986).

[11] L. S. de Jong, Towards a formal definition of numerical stability, *Numer. Math.* **28** : 211-219 (1977).

[12] J.-M. Delosme and I. Ipsen, Parallel solution of symmetric positive definite systems with hyperbolic rotations, *Lin. Alg. & Appl.* **77** : 75-111 (1986).

[13] Ph. Delsarte, Y. Genin and Y. Kamp, On the role of the Nevanlinna-Pick problem in circuit and system theory, *Circuit Theory and Applications* **9** : 177-187 (1981).

[14] Ph. Delsarte, Y. Genin, Y. Kamp and P. Van Dooren, Speech modelling and the trigonometric moment problem, *Philips J. of Research* **37** : 277-292 (1982).

[15] J. Demmel, On condition numbers and the distance to the nearest ill-posed problem, *Numer. Math.* **51** : 251-289 (1987).

[16] M. Fan and A. Tits, Characterization and efficient computation of the structured singular value, *IEEE Trans. Aut. Contr.* **AC-31** : 734-743 (1986).

[17] I. Gohberg and I. Koltracht, Error analysis for Cauchy and Vandermonde matrices, *Proceedings MTNS*, Amsterdam, 1989.

[18] G. H. Golub and C. Van Loan, *Matrix Computations*, Johns Hopkins University Press, 1989.

[19] N. J. Higham, Error analysis of the Björck-Pereyra algorithms for solving Vandermonde matrices, *Numer. Math.* **50** : 613-632 (1987).

[20] W. M. Kahan, A survey of error analysis, *Information Processing 71* : 1214-1239, North Holland, 1972.

[21] W. M. Kahan, *Conserving confluence curbs ill-condition*, Computer Sci. Techn. Rept. 6, Univ. of California at Berkeley, 1972.

[22] C. Kenney and A. Laub, *Condition estimates for the matrix sign function and polar decomposition*, Sci. Comput. Lab. Techn. Rept., Univ. of California at Santa Barbara, 1989.

[23] L. Ljung, M. Morf and D. Falconer, Fast calculations of gain matrices for recursive estimation schemes, *Int. J. Contr.* **27** : 1-19 (1978).

[24] S. Ljung and L. Ljung, Error propagation of recursive least-squares adaptation algorithms, *Automatica* **20** : 157-167 (1985).

[25] J. Markel and A. Gray, *Linear Prediction of Speech*, Springer Verlag, 1976.

[26] J. Rice, A theory of condition, *SIAM J. Num. Anal.* **3** : 287-310 (1966).

[27] A. Shaw and R. Kumaresan, Some structured matrix approximation problems, *Proceedings IEEE Conf. on ASSP* : 2324-2327 (1988).

[28] R. D. Skeel, Scaling for numerical stability in Gaussian elimination, *J. Assoc. Comput. Mach.* **26** : 494-526 (1979).

[29] S. Skogestad and M. Morari, Some new properties of the structured singular value, *IEEE Trans. Aut. Contr.* **AC-33** : 1151-1154 (1988).

[30] D. Slock and T. Kailath, Numerically stable fast transversal filters for recursive least squares adaptive filters, *IEEE Trans. ASSP*, to appear.

[31] W. Trench, Numerical solution of the eigenvalue problem for Hermitian Toeplitz matrices, *SIAM J. Matrix Anal. Appl.* **10** : 135-146 (1989).

[32] P. Van Dooren, The generalized eigenstructure problem in linear system theory, *IEEE Trans. Aut. Contr.* **AC-26** : 111-129 (1981).

Constructing a Unitary Hessenberg Matrix from Spectral Data*

Gregory Ammar†

Northern Illinois University
Dept. of Mathematical Sciences
DeKalb, IL 60115, USA

William Gragg‡

Naval Postgraduate School
Department of Mathematics
Monterey, CA 93943, USA

Lothar Reichel§

Bergen Scientific Centre IBM
Allégaten 36 N-5007 Bergen, Norway

Abstract

We consider the numerical construction of a unitary Hessenberg matrix from spectral data using an inverse QR algorithm. Any upper unitary Hessenberg matrix H with nonnegative subdiagonal elements

*In memory of Peter Henrici

†This research was supported in part by the National Science Foundation under grant DMS-8704196, and by the Foundation Research Program of the Naval Postgraduate School

‡on leave from University of Kentucky, Department of Mathematics, Lexington, KY 40506, USA

§Norway on leave from University of Kentucky, Department of Mathematics, Lexington, KY 40506, USA

NATO ASI Series, Vol. F 70
Numerical Linear Algebra, Digital Signal Processing and Parallel Algorithms
Edited by G. H. Golub and P. Van Dooren

can be represented by $2n - 1$ real parameters. This representation, which we refer to as the *Schur parametrization* of H, facilitates the development of efficient algorithms for this class of matrices. We show that a unitary upper Hessenberg matrix H with positive subdiagonal elements is determined by its eigenvalues and the eigenvalues of a rank-one unitary perturbation of H. The eigenvalues of the perturbation strictly interlace the eigenvalues of H on the unit circle.

AMS(MOS) Subject Classification: 15A18, 65F15.

Keywords: inverse eigenvalue problem, unitary matrix, orthogonal polynomial.

1 Introduction

In this paper we focus on an inverse eigenvalue problem for unitary Hessenberg matrices with positive subdiagonal elements. Throughout this paper all Hessenberg matrices are *upper* Hessenberg matrices. This class of matrices bears many similarities with the class of Jacobi matrices, i.e. real symmetric tridiagonal matrices with positive subdiagonal elements. Any matrix in either class is normal and has distinct eigenvalues. Both $n \times n$ Jacobi matrices and $n \times n$ unitary Hessenberg matrices with positive subdiagonal elements can be parametrized by $2n - 1$ real parameters. This is obvious for Jacobi matrices; for unitary Hessenberg matrices this parametrization is described below. Since unitary Hessenberg matrices with positive subdiagonal elements are determined by $O(n)$ parameters, one can develop efficient algorithms for this class of matrices. These algorithms are analogous with algorithms for Jacobi matrices. For example, the unitary QR algorithm, described in [6], has many similarities with the QR algorithm for Jacobi matrices. Another example is provided by the divide-and-conquer methods that have been developed for both the tridiagonal and unitary eigenproblems [3], [4], [8], [9].

In this paper we show another analogy of unitary Hessenberg matrices with Jacobi matrices, namely, that a unitary Hessenberg matrix H with positive subdiagonal elements is uniquely determined by its eigenvalues and the eigenvalues of a unitary rank-one perturbation of H. The matrix H can be constructed using $O(n^2)$ arithmetic operations using an inverse unitary

QR algorithm. Similar results for Jacobi matrices are well-established [1], [2] , [7]. The inverse unitary QR algorithm is analogous with the algorithm described in [7] for Jacobi matrices.

2 Unitary Hessenberg Matrices and Szegö Polynomials

We refer to a *finite Schur parameter sequence* of length n as a sequence of complex numbers $\{\gamma_j\}_{j=1}^n$ with $|\gamma_j| < 1$ for $1 \le j < n$ and $|\gamma_n| = 1$. Also define the *complementary Schur parameters* $\{\sigma_j\}_{j=1}^{n-1}$ by $\sigma_j := \sqrt{1-|\gamma_j|^2}$. Associated with the finite Schur parameter sequence $\{\lambda_j\}_{j=1}^n$ is a unitary Hessenberg matrix H with *positive* subdiagonal elements

$$H = H(\gamma_1, \ldots, \gamma_{n-1}, \gamma_n) := G_1(\gamma_1)G_2(\gamma_2)\ldots G_{n-1}(\gamma_{n-1})\tilde{G}_n(\gamma_n), \quad (2.1)$$

where the Givens reflector $G_j(\gamma_j)$ is the identity matrix of appropriate size except for the 2×2 principal submatrix

$$G_j\begin{bmatrix} j & j+1 \\ j & j+1 \end{bmatrix} = \begin{bmatrix} -\gamma_j & \sigma_j \\ \sigma_j & \overline{\gamma}_j \end{bmatrix},$$

with the bar denoting complex conjugation. The matrix on the right in the product (2.1) is defined by $\tilde{G}_n(\gamma_n) :=$ diag$[1, 1, \ldots, 1, -\gamma_n]$. The nonzero entries of $H = [\eta_{j,k}]_{j,k=1}^n$ are then given by $\eta_{j+1,j} := \sigma_j$ and $\eta_{j,k} := -\bar{\gamma}_{j-1}\sigma_j\sigma_{j+1}\ldots\sigma_{k-1}\gamma_k$ for $1 \le j \le k$, where $\gamma_0 := 1$.

It is easy to see that every $n \times n$ unitary Hessenberg matrix $H = [\eta_{j,k}]_{j,k=1}^n$ with $\eta_{j+1,j} > 0$ is uniquely determined by a finite Schur parameter sequence of length n. In fact, the Schur parameters $\{\gamma_j\}_{j=1}^n$ and the complementary Schur parameters $\{\sigma_j\}_{j=1}^{n-1}$ can be determined from H by

$$\sigma_j = \eta_{j+1,j}, \qquad 1 \le j < n; \qquad \gamma_j = -\eta_{1,j}/\sigma_1\sigma_2\cdots\sigma_{j-1}, \qquad 1 \le j \le n.$$

Hence, we have a one-to-one correspondence between Schur parameter sequences of length n and $n \times n$ unitary Hessenberg matrices with positive subdiagonal elements. This *Schur parametrization of unitary Hessenberg matrices* with positive subdiagonal elements shows that these matrices are determined by $2n-1$ real parameters: the real and imaginary parts of γ_j for $1 \leq j < n$, and the argument of γ_n. To avoid numerical instability, however, we also retain the complementary Schur parameters.

Let $H = H_n := H(\gamma_1, \ldots, \gamma_{n-1}, \gamma_n)$, and let $H_k := G_1(\gamma_1)G_2(\gamma_2)\ldots G_{k-1}(\gamma_{k-1})\tilde{G}_k(\gamma_k)$ be the leading principal submatrix of $H = H_n$ of order k. Introduce the functions

$$\begin{aligned} \phi_k(\lambda) &:= e_1^T(\lambda I - H_k)^{-1}e_1, \\ \psi_k(\lambda) &:= \det(\lambda I - H_k), \\ \pi_k(\lambda) &:= \det(\lambda I - H_{k-1}''), \end{aligned}$$

where $e_1 := [1, 0, \ldots, 0]^T$, and H_{k-1}'' denotes the trailing principal submatrix of H_k of order $k-1$. Thus, $\phi_k(\lambda) = \pi_k(\lambda)/\psi_k(\lambda)$. The following proposition can be verified by induction.

Proposition 2.1. The polynomials $\psi_k(\lambda)$ and $\pi_k(\lambda)$, $k > 0$, satisfy

$$\begin{bmatrix} \psi_k(\lambda) & \pi_k(\lambda) \\ \tilde{\psi}_k(\lambda) & \tilde{\pi}_k(\lambda) \end{bmatrix} = \begin{bmatrix} \lambda & \gamma_k \\ \bar{\gamma}_k\lambda & 1 \end{bmatrix} \begin{bmatrix} \psi_{k-1}(\lambda) & \pi_{k-1}(\lambda) \\ \tilde{\psi}_{k-1}(\lambda) & \tilde{\pi}_{k-1}(\lambda) \end{bmatrix};$$

$$\begin{bmatrix} \psi_0(\lambda) & \pi_0(\lambda) \\ \tilde{\psi}_0(\lambda) & \tilde{\pi}_0(\lambda) \end{bmatrix} = \begin{bmatrix} 1 & 1/\lambda \\ 1 & 0 \end{bmatrix}$$

It follows that for each k, the polynomials $\tilde{\psi}_k(\lambda) := \lambda^k\bar{\psi}_k(1/\lambda)$ are obtained by reversing and conjugating the coefficients of $\psi_k(\lambda)$. From the initial conditions we recognize $\psi_k(\lambda)$ to be the kth *Szegö polynomial* determined by the Schur parameter sequence $\{\gamma_j\}_{j=1}^n$. □

The Szegö polynomials $\{\psi_k\}_{k=0}^n$ are orthogonal with respect to a discrete measure on the unit circle. This measure assigns a positive weight ω_k to each zero λ_k of $\psi_n(\lambda)$. These weights are the numerators in the partial fraction decomposition

$$\phi_n(\lambda) =: \sum_{k=1}^{n} \frac{\omega_k}{\lambda - \lambda_k},$$

see [5] for details.

Let $H = U\Lambda U^*$, with $\Lambda = \operatorname{diag}[\lambda_1, \lambda_2, \ldots, \lambda_n]$ and U unitary, be the spectral resolution of $H = H(\gamma_1, \ldots, \gamma_{n-1}, \gamma_n)$, where $*$ denotes transposition and complex conjugation. Let $u := U^T e_1$ be the vector containing the first components of the eigenvectors of H. Since H has nonzero subdiagonal elements, every entry of $u = [v_1, v_2, \ldots, v_n]^T$ is nonzero, and we normalize U so that each $v_k > 0$. Then

$$\phi_n(\lambda) = u^*(\lambda I - \Lambda)^{-1}u = \sum_{k=1}^{n} \frac{v_k^2}{\lambda - \lambda_k}.$$

Hence, the weights $\omega_k = v_k^2$ are guaranteed to be positive, and $\sum_{k=1}^{n} \omega_k = 1$.

3 The Inverse Unitary QR Algorithm

Given n distinct unimodular complex numbers $\{\lambda_k\}_{k=1}^{n}$ and associated positive weights $\{v_k^2\}_{k=1}^{n}$, we can construct a unitary Hessenberg matrix H with the λ_k and v_k equal to the eigenvalues and first components of the corresponding eigenvectors of H, respectively. This construction is achieved using an inverse QR algorithm, which is analogous with the procedure of [7] for real symmetric tridiagonal matrices.

The required Hessenberg matrix is obtained by performing a sequence of elementary unitary similarity transformations to transform the matrix $\begin{bmatrix} \delta & u* \\ u & \Lambda \end{bmatrix}$ to a Hessenberg matrix $\begin{bmatrix} \delta & e_1* \\ e_1 & H \end{bmatrix}$ without using or changing the arbitrary entry δ. Then $H = H(\gamma_1, \ldots, \gamma_{n-1}, \gamma_n)$ has the desired eigenvalues and associated eigenvectors.

The idea is to build the Hessenberg matrix by adding weight-abscissa pairs one at a time. Suppose that we have constructed the unitary Hessenberg matrix $H_m := H(\gamma_1, \ldots, \gamma_{m-1}, \gamma_m)$, for some $m < n$, corresponding with the weight-abscissa pairs $\{(\omega_k, \lambda_k)\}_{k=1}^{m}$. Let $\sigma_0 := (\sum_{k=1}^{m} \omega_k)^{1/2}$ and assume that the first components of the eigenvectors of H_m are $\{(v_k/\sigma_0)\}_{k=1}^{m}$. In order to add the weight-abscissa pair (v^2, λ) and construct the corresponding $(m+1) \times (m+1)$ unitary Hessenberg matrix $H'_{m+1} := H(\lambda'_1, \lambda'_2, \ldots, \lambda'_{m+1})$, we perform a sequence of unitary similarity transformations to put the $(m+2) \times (m+2)$ matrix

$$\widetilde{H}^{(1)} := \begin{bmatrix} \delta & v & \sigma_0 e_1{*} \\ v & \lambda & 0^* \\ \sigma_0 e_1 & 0 & H_M \end{bmatrix} = \begin{bmatrix} \delta & v & \sigma_0 & & & & \\ v & \lambda & & & & & \\ \sigma_0 & & \times & \times & \times & \times & \times \\ & & \times & \times & \times & \times & \times \\ & & & \times & \times & \times & \times \\ & & & & \times & \times & \times \\ & & & & & \times & \times \end{bmatrix}$$

into Hessenberg form without changing δ. Let $\sigma_0' := \sqrt{\sigma_0^2 + v^2}$ and $\alpha_0 := -v/\sigma_0'$. Then

$$G_2(\alpha_0)\widetilde{H}^{(1)} =: \widetilde{H}^{(2)} = \begin{bmatrix} \delta & v & \sigma_0 & & & & \\ \sigma_0' & \times & \times & \times & \times & \times & \times \\ 0 & \times & \times & \times & \times & \times & \times \\ & & \times & \times & \times & \times & \times \\ & & & \times & \times & \times & \times \\ & & & & \times & \times & \times \\ & & & & & \times & \times \end{bmatrix}$$

is a Hessenberg matrix with a trailing principal $(m+1)\times(m+1)$ submatrix, which is both unitary and of Hessenberg form. On the completion of the similarity transformation of $\tilde{H}^{(1)}$, we obtain

$$\widetilde{H}^{(2)}G_2^*(\alpha_0) = \begin{bmatrix} \delta & \sigma_0' & & & & & \\ \sigma_0' & \times & \times & \times & \times & \times & \times \\ & \times & \times & \times & \times & \times & \times \\ & \oplus & \times & \times & \times & \times & \times \\ & & & \times & \times & \times & \times \\ & & & & \times & \times & \times \\ & & & & & \times & \times \end{bmatrix}. \tag{3.1}$$

The circled element in (3.1) forms a "bulge", which is to be chased down along the subdiagonal in order to obtain a matrix of Hessenberg form. Define $G_3(\alpha_1)$ so that $G_3(\alpha_1)\tilde{H}^{(2)}G_2^*(\alpha_0) =: \tilde{H}^{(3)}$ is a Hessenberg matrix. Then $\tilde{H}^{(3)}G_3^*(\alpha_1)$ has a bulge, which we annihilate by multiplying from the left with $G_4(\alpha_2)$. Proceeding in this manner, we ultimately chase the bulge off the bottom of the matrix, and obtain the Hessenberg matrix $\tilde{H}^{(m-1)}G^*_{m-1}(\gamma_{m-3})$, which is unitarily similar to $\tilde{H}^{(1)}$. The trailing principal $(m+1)\times(m+1)$ submatrix of $\tilde{H}^{(m-1)}G^*_{m-1}(\gamma_{m-3})$ is unitarily similar to a unitary Hessenberg matrix with positive subdiagonal elements. The latter matrix is the desired Hessenberg matrix $H'_{m+1} = H(\gamma'_1, \gamma'_2, \ldots, \gamma'_{m+1})$.

This procedure for adding a weight-abscissa pair to H_m, if implemented by directly manipulating the elements of the matrices $\tilde{H}^{(k)}$, for $1 \leq k < m$, would require $O(m^2)$ arithmetic operations. However, we note that for each k the trailing $(m+1)\times(m+1)$ principal submatrix of $\tilde{H}^{(k)}$ is unitary and of Hessenberg form, and, therefore, is unitarily similar to a unitary Hessenberg matrix, denoted by $\hat{H}^{(k)}_{m+1}$, with positive subdiagonal elements. Hence, we can carry out the similarity transformations by manipulating the Schur parameters of the matrices $\hat{H}^{(k+2)}_{m+1} = H(\gamma'_1, \gamma'_2, \ldots, \gamma'_k, \bar{\lambda}^k\alpha_k, \gamma_{k+1}, \ldots, \gamma_m)$. This gives rise to a method that requires only $O(m)$ arithmetic operations in order to add a weight-abscissa pair to H_m. We refer to this method as the *inverse unitary QR algorithm* because of its relationship with the unitary QR algorithm presented in [6].

Inverse Unitary QR Algorithm: adding a weight-abscissa pair (v^2, λ), where $|\lambda| = 1,\ v > 0$.

$$\sigma'_0 := \sqrt{\sigma_0^2 + v^2};$$
$$\alpha_0 := -v/\sigma'_0;\ \beta_0 := \sigma_0/\sigma'_0;$$

for $k := 1, 2, \ldots, m$

$$\begin{aligned}
\sigma'_k &:= \beta_{k-1}\sqrt{\sigma_k^2 + |\alpha_{k-1} + \gamma_k\lambda^{k-2}\bar{\alpha}_{k-1}|^2};\\
\alpha_k &:= \beta_{k-1}\lambda(\alpha_{k-1} + \gamma_k\lambda^{k-2}\bar{\alpha}_{k-1})/\sigma'_k;\\
\beta_k &:= \beta_{k-1}\sigma_k/\sigma'_k;\\
\gamma'_k &:= \beta_{k-1}^2\gamma_k - \bar{\lambda}^{k-2}\alpha_{k-1}^2;
\end{aligned}$$

$$\gamma'_{m+1} := \lambda\gamma_m;$$

Thus, the inverse unitary QR algorithm can be used to construct the unitary $n \times n$ Hessenberg matrix $H = H(\gamma_1, \ldots, \gamma_{n-1}, \gamma_n)$ from its eigenvalues and the first components of its normalized eigenvectors in $O(n^2)$ arithmetic operations.

4 An Inverse Spectral Problem

Let $H = H(\gamma_1, \ldots, \gamma_{n-1}, \gamma_n)$ and $\phi_n(\lambda) = \dfrac{\pi_n(\lambda)}{\psi_n(\lambda)} = \sum_{k=1}^{n} \dfrac{\omega_k}{\lambda - \lambda_k}$. Let $\alpha := e^{i\tau}$ for some $0 < \tau < 2\pi$, and consider the polynomials

$$\chi_k(\lambda) := (1-\alpha)\lambda\pi_k(\lambda) + \alpha\psi_k(\lambda), \quad 0 \le k \le n. \tag{4.1}$$

Proposition 4.1. The polynomials $\{\chi_k(\lambda)\}_{k=0}^{n}$ are the monic Szegö polynomials corresponding with the Schur sequence $\{\alpha\gamma_k\}_{k=1}^{n}$.
Proof. The definition (4.1) and Proposition 2.1 yield

$$\begin{bmatrix} \chi_k(\lambda) \\ \xi_k(\lambda) \end{bmatrix} = \begin{bmatrix} \lambda & \gamma_k \\ \overline{\gamma}_k\lambda & 1 \end{bmatrix} \begin{bmatrix} \chi_{k-1}(\lambda) \\ \xi_{k-1}(\lambda) \end{bmatrix}, \begin{bmatrix} \chi_0(\lambda) \\ \xi_0(\lambda) \end{bmatrix} = \begin{bmatrix} 1 \\ \alpha \end{bmatrix},$$

where $\xi_k(\lambda) := (1-\alpha)\lambda\tilde{\pi}_k(\lambda) + \alpha\tilde{\psi}_k(\lambda)$. On setting $\tilde{\chi}_k(\lambda) = \bar{\alpha}\xi_k(\lambda)$, we obtain

$$\begin{bmatrix} \chi_k(\lambda) \\ \tilde{\chi}_k(\lambda) \end{bmatrix} = \begin{bmatrix} \lambda & \alpha\gamma_k \\ \overline{\alpha}\,\overline{\gamma}_k\lambda & 1 \end{bmatrix} \begin{bmatrix} \chi_{k-1}(\lambda) \\ \tilde{\chi}_{k-1}(\lambda) \end{bmatrix}, \begin{bmatrix} \chi_0(\lambda) \\ \tilde{\chi}_0(\lambda) \end{bmatrix} = \begin{bmatrix} 1 \\ 1 \end{bmatrix}.$$

□

An immediate consequence of Proposition 4.1 is that each zero of $\chi_n(\lambda)$ has unit modulus. When $\alpha = -1$ the polynomials $\{\chi_k(\lambda)\}_{k=0}^{n}$ are known as the Szegö polynomials of the second kind corresponding with $\{\gamma_k\}_{k=1}^{n}$.

Proposition 4.2. The zeros $\{\mu_k\}_{k=0}^{n}$ of $\chi_n(\lambda)$ strictly interlace the zeros $\{\lambda_k\}_{k=0}^{n}$ of $\psi_n(\lambda)$ on the unit circle. Moreover, $\prod_{k=1}^{n}(\mu_k/\lambda_k) = \alpha$.
Proof. Let $\lambda := e^{i\theta}$ for some $0 \le \theta < 2\pi$, and let $\alpha =: e^{i\tau}$ for some $0 < \tau < 2\pi$. Then

$$\begin{aligned} \Phi_n(\lambda) &:= \tfrac{\chi_n(\lambda)}{\psi_n(\lambda)} = \textstyle\sum_{k=1}^{n} \omega_k \tfrac{\lambda - \alpha\lambda_k}{\lambda - \lambda_k} \\ &= \textstyle\sum_{k=1}^{n} \omega_k \tfrac{1 - \lambda\bar{\lambda}_k + \alpha(1 - \bar{\lambda}\lambda_k)}{|\lambda - \lambda_k|^2} \\ &= \textstyle\sum_{k=1}^{n} 2\, e^{i\tau/2} \omega_k \tfrac{\mathrm{Re}(e^{i\tau/2}(1 - \bar{\lambda}\lambda_k))}{|\lambda - \lambda_k|^2}. \end{aligned}$$

Thus, $\operatorname{Im}(e^{-i\tau/2}\Phi_n(e^{i\theta})) \equiv 0$ for $0 \le \theta < 2\pi$. Let $\lambda_k =: e^{i\theta_k}$. We may assume that $0 \le \theta_1 < \theta_2 < \cdots < \theta_n < 2\pi$. Then

$$\begin{aligned} e^{-i\tau/2}\Phi_n(e^{i\theta}) &= \textstyle\sum_{k=1}^n 2\,\omega_k \frac{\cos(\tau/2)-\cos(\tau/2+\theta_k-\theta)}{(1-\cos(\theta_k-\theta))^2+\sin^2(\theta_k-\theta)} \\ &= \textstyle\sum_{k=1}^n \omega_k \frac{\cos(\tau/2)-\cos(\tau/2)\cos(\theta_k-\theta)+\sin(\tau/2)\sin(\theta_k-\theta)}{1-\cos(\theta_k-\theta)} \\ &= \textstyle\sum_{k=1}^n \omega_k \left(\cos(\tau/2)+\sin(\tau/2)\cot((\theta_k-\theta)/2)\right). \end{aligned}$$

From $0 < \tau < 2\pi$ and $\omega_k > 0$ for all k, it follows that

$$e^{-i\tau/2}\frac{d}{d\theta}\Phi_n(e^{i\theta}) = \frac{1}{2}\sin(\tau/2)\sum_{k=1}^{n}\omega_k/\sin^2((\theta_k-\theta)/2) > 0 \qquad \text{for } \theta \neq \theta_k,\ 1 \le k \le n.$$

Thus, $\Phi_n(e^{i\theta}) \to -\infty$ as $\theta \searrow \theta_k$, and $\Phi_n(e^{i\theta}) \to \infty$ as $\theta \nearrow \theta_{k+1}$. This shows that $\theta \to \Phi_n(e^{i\theta})$ has precisely one zero in $]\theta_k, \theta_{k+1}[$. Consequently, the zeros μ_j of $\Phi_n(\lambda)$ strictly interlace the λ_j on the unit circle. The second statement of the Proposition follows from the fact that $\chi_n(0)/\psi_n(0) = \alpha$. □

Let $\lambda_k =: e^{i\theta_k}$ and $\mu_k =: e^{i\nu_k}$ for $0 \le \theta_k, \nu_k < 2\pi$. Then we have $\sum_{k=1}^n(\nu_k - \theta_k) = \tau$. We may assume that the arguments have been ordered so that

$$0 \le \theta_1 < \nu_1 < \theta_2 < \cdots < \nu_{n-1} < \theta_n < \nu_n < \theta_1 + 2\pi. \tag{4.2}$$

Proposition 4.3. With the above notation, and the ordering (4.2), the weights ω_k are given by

$$\omega_k = \frac{1}{\sin(\tau/2)}\frac{\prod_{j=1}^{n}\sin((\nu_j-\theta_k)/2)}{\prod_{\substack{j=1\\ j\neq k}}^{n}\sin((\theta_j-\theta_k)/2)}.$$

Proof. We have

$$(1-\alpha)\lambda_k\omega_k = \lim_{\lambda\to\lambda_k}(\lambda-\lambda_k)\frac{\chi_n(\lambda)}{\psi_n(\lambda)} = \frac{\chi_n(\lambda_k)}{\psi_n'(\lambda_k)} = \frac{\prod_{j=1}^{n}(\lambda_k-\mu_j)}{\prod_{\substack{j=1\\ j\neq k}}^{n}(\lambda_k-\lambda_j)}$$

since $\chi_n(\lambda)$ and $\psi_n(\lambda)$ are monic. Thus,

$$(1-\alpha)\omega_k = e^{-i\theta_k} \frac{\prod_{j=1}^{n} (e^{i\theta_k} - e^{i\nu_j})}{\prod_{\substack{j=1 \\ j\neq k}}^{n} (e^{i\theta_k} - e^{i\theta_j})} = \frac{\prod_{j=1}^{n} (1 - e^{i(\nu_j-\theta_k)})}{\prod_{\substack{j=1 \\ j\neq k}}^{n} (1 - e^{i(\theta_j-\theta_k)})}$$

$$= -2ie^{i\tau/2} \frac{\prod_{j=1}^{n} \sin((\nu_j - \theta_k)/2)}{\prod_{\substack{j=1 \\ j\neq k}}^{n} \sin((\theta_j - \theta_k)/2)}, \tag{4.3}$$

because

$$1 - e^{i\beta} = -2i\sin(\beta/2), \tag{4.4}$$

and $\sum_{j=1}^{n}(\nu_j - \theta_j) = \tau$. The formula for the weights now follows by substituting (4.4) with $\beta := \tau$ into (4.3). □

We can now state the inverse spectral problem and its solution.

Problem: Given two sets of n mutually interlacing points on the unit circle $\{\lambda_k\}_{k=1}^n$ and $\{\mu_k\}_{k=1}^n$, determine the unique unitary Hessenberg matrix $H = H(\gamma_1, \ldots, \gamma_{n-1}, \gamma_n)$ and $|\alpha| = 1$ such that the spectrum of H is $\{\lambda_k\}_{k=1}^n$ and the spectrum of $H(\alpha\gamma_1, \ldots, \alpha\gamma_{n-1}, \alpha\gamma_n)$ is $\{\mu_k\}_{k=1}^n$.

Solution: Let $\alpha := \prod_{k=1}^n(\mu_k/\lambda_k)$, calculate the weights by Proposition 4.3, and use the inverse unitary QR algorithm to construct $H = H(\gamma_1, \ldots, \gamma_{n-1}, \gamma_n)$.

References

[1] D. Boley and G.H. Golub, *Inverse Eigenvalue Problems for Band Matrices,* in Lecture Notes in Mathematics 630, Numerical Analysis, Proceedings of the Biennial Conference Held at Dundee, 1977, G.A. Watson, ed., Springer-Verlag, New York, 1978, pp. 23-31.

[2] D. Boley and G.H. Golub, *A Survey of Matrix Inverse Eigenvalue Problems,* Inverse Problems 3, (1987) 595-622.

[3] J.J.M. Cuppen, *A Divide and Conquer Method for the Symmetric Tridiagonal Eigenproblem,* Numer. Math. 36, (1981) 177-195.

[4] J.J. Dongarra and D.C. Sorensen, *A Fully Parallel Algorithm for the Symmetric Eigenproblem,* SIAM J. Sci. Stat. Comput. 8, (1987) s139-s154.

[5] W.B. Gragg, *Positive Definite Toeplitz Matrices, the Arnoldi Process for Isometric Operators, and Gaussian Quadrature on the Unit Circle* (in Russian), in Numerical Methods in Linear Algebra, E.S. Nikolaev, ed., Moscow University Press, Moscow, 1982, pp. 16-32.

[6] W.B. Gragg, *The QR Algorithm for Unitary Hessenberg Matrices,* J. Comput. Appl. Math. 16, (1986) 1-8.

[7] W.B. Gragg and W.J. Harrod, *The Numerically Stable Reconstruction of Jacobi Matrices from Spectral Data,* Numer. Math. 44, (1984) 317-355.

[8] W.B. Gragg and L. Reichel, *A Divide and Conquer Algorithm for the Unitary Eigenproblem,* in Hypercube Multiprocessors 1987, M.T. Heath, ed., SIAM, Philadelphia, 1987, pp. 639-647.

[9] W.B. Gragg and L. Reichel, *A Divide and Conquer Method for the Unitary and Orthogonal Eigenproblems,* BSC Report 88/31, Bergen Scientific Centre, Bergen, 1988.

Parallel Computation of the Generalized Singular Value Decomposition*

Zhaojun Bai

Courant Institute of Mathematical Sciences
New York University
251 Mercer Street
New York, NY 10012, USA

Abstract

The generalized singular value decomposition (GSVD) is the simultaneous reduction of any two matrices having the same number of columns to diagonal forms by premultiplying by two different orthogonal matrices and postmultiplying by the same nonsingular matrix. The recent advent of real time signal processing and other problems have given impetus to the development of the parallel computation of the GSVD [3,7]. Brent *et al.* [2] have developed a parallel algorithm on systolic array. However, their approach requires the use of different configurations of systolic arrays for singular value decomposition (SVD), QR decomposition and matrix-matrix multiplication. In addition to some difficulties in numerical treatment, how to avoid costly interfacing between these arrays and devise a single array that is capable of performing all these basic matrix computations is still an open problem.

Recently, Luk [5] proposed a parallel implementation of Paige's Jacobi-like algorithm[6]. But in his work, only the simplest case of

*Prepared for an Interdisciplinary Meeting on Numerical Linear Algebra, Digital Signal Processing, and Parallel Algorithms, August 1988. This work was supported in part by the U.S. Air Force Office of Scientific Research under Grant AFOSR-82-0078.

NATO ASI Series, Vol. F 70
Numerical Linear Algebra, Digital Signal Processing and Parallel Algorithms
Edited by G. H. Golub and P. Van Dooren

computing the GSVD of two $n \times n$ triangular matrices, one of which is nonsingular, is considered. The scheme cannot treat the general case.

Our parallel direct GSVD algorithm will treat the general case. The algorithm falls into two parts: first, the input matrices are preprocessed in parallel by computing their upper trapezoidal forms, for which we develop the parallel QR decomposition with column pivoting. And second, the GSVD of two upper trapezoidal matrices is computed in parallel using generalized Jacobi-Kogbetliantz-iteration scheme. The 2-by-2 GSVD is developed to be used in the the forward and backward sweeps to keep the trapezoidal structure. We also consider in detail how to organize the computation, how to partition the matrix and how to map these pieces of data and computations onto the network of processors. The whole algorithm is designed to be implemented efficiently on distributed-memory parallel computer architectures. The time is $O(n^2)$ for parallel-preprocessing, and $O(n^2/p)$ for the GSVD of two upper trapezoidal matrices, where p is the dimension of the triangular array of processors. In fact, the computation of the GSVD of two upper trapezoidal matrices can also be implemented directly on a triangular systolic array. How to efficiently compute the QR decomposition with column pivoting on systolic array is still unknown, although there are many papers concerning the QR decomposition without column pivoting on systolic array, e.g. see [4].

References

[1] Z. Bai, *The direct GSVD algorithm and its parallel implementation.* University of Maryland, Dept. of Computer Science, TR-1901, 1987.

[2] R. P. Brent, F. T. Luk and C. Van Loan, *Computation of the generalized singular value decomposition using mesh-connected processing*, TR-CS-83-563, Dept. Computer Science, Cornell Univ. Ithaca, NY 1983

[3] K. Bromley and J. M. Speiser, *Singular processing algorithm, architectures and applications*, Tutorial 31, SPIE 27th Annual Internat. Thch. Symp., San Diego, 1983.

[4] W. M. Gentleman and H. T. Kung, *Matrix triangularization by systolic array*, Proc. SPIE Vol. 298, Real Time Signal Processing IV, pp.19-26.

[5] F. T. Luk, *A Parallel Method for computing the Generalized Singular Value Decomposition*, J. of Para. and Dist. Comput. 2(1985), pp.250-260.

[6] C. C. Paige, *Computing the generalized singular value decomposition*, SIAM J. Sci. Stat. 7(1986), pp.1126-1146.

[7] S. Van Huffel and J. Vandewalle, *Analysis and properties of the generalized total least squares problem* $AX \approx B$ *when some or all columns in* A *are subject to error*, submitted to SIAM J. Matrix Anal Appl.

Correcting Interface Errors Arising from Segmentation in a Parallel Iterative Algorithm

J.M. Barry, J.P. Pollard

Australian Nuclear Science and Technology Organisation
Private Mail Bag 1
Menai NSW 2234
Australia

E.L. Wachspress

Department of Mathematics
University of Tennessee
Knoxville, Tennessee
37796, USA

Abstract

Several parallel iterative strategies based on domain segmentation were previously developed by the authors. While each is effective, the error distribution after every global iteration is not uniform within each segment, the error being greater at the extremities. A correction to the solution is developed at the interfaces between the disjoint segments which accelerates convergence.

1 Introduction

Numerous strategies exist for the iterative solution of large systems of linear equations on parallel computer architectures. The most appropriate choice

NATO ASI Series, Vol. F 70
Numerical Linear Algebra, Digital Signal Processing and Parallel Algorithms
Edited by G. H. Golub and P. Van Dooren

is far from clear cut. This work was commenced without reference to a particular type of hardware, however, it was assumed that a small number of relatively sophisticated processors would be available. Each processor should at least have vector facilities, but it is even more preferable that they have lower levels of parallelism. A completely suitable computer for the approach studied here does not exist. The algorithm testing is restricted to a sequential machine, and in discussion of its performance, approximate measures of extent of work are employed.

This study was prompted by the solution of neutron diffusion equations over three-dimensional grids as large as 300×300×100. Due to limited facilities, the largest system tested to date corresponds to a 64×64×64 grid. The parallel strategies described are not restricted to neutron diffusion problems.

2 Parallel iteration based on polynomial correction

Large system of linear equations

$$A\phi = s \tag{2.1}$$

arising in the discretisation of a p.d.e. over a regular three-dimensional mesh are divided as evenly as possible in the (z) direction into a number of segments (n). The segments are assigned to individual processors so that each iteratively solve at the tth global iteration a reduced linear system

$$A_{i,j}\phi_i^{(t)} = s_i - A_{i,i-1}\phi_{i-1}^{(t-1)} - A_{i,i+1}\phi_{i+1}^{(t-1)}, \quad (i = 1, 2, \ldots, n). \tag{2.2}$$

If the outer block scheme was used on its own, very slow convergence would result. Instead at each stage of the global iteration, a polynomial correction is computed and applied to each place [3] before a further iterative solution is obtained for each block (2.2). In the correction we seek to "rebalance" the solution of the underlying equation (2.1), thus accelerating the slower Gauss-Jacobi convergence.

A polynomial in two variables $p(x, y)$ is chosen to multiplicatively rebalance the solution for each plane $\phi_k^{(t)}$ of the current t^{th} iterate. The correction

$$\phi_k^{(t)} \leftarrow p_k(x,y)\phi_k^{(t)}, \quad (k = 1, 2, \ldots, N_z),$$

is applied, where $\phi_k^{(t)}$ represents the current iterate for the kth plane. N_z polynomials of the form

$$p(x,y) = a_0 + a_1 x + a_2 y + a_3 xy + a_4 x^2 y + a_5 xy^2 + a_6 x^2 + a_7 y^2$$

are used as rebalancing functions. The $8N_z$ polynomial coefficients required are globally determined from a weighted residual or variational formulation. With Galerkin weighting, a greatly reduced linear system is obtained which is symmetric and positive definite. It is a block tridiagonal system and solved by Cholesky decomposition. The bandwidth is 24 and although the direct sequential process is adequate for studies to date, parallel approaches are possible. More important that its solution is the cost of its generation. Here vector facilities at least, are essential for adequate performance. Lower levels of parallelism within each processor would be even more advantageous.

3 Improving solutions on segment boundaries

When equations (2.2) are solved, the error in the solution for each plane is significantly higher for planes that form the interfaces between each segment. This situation persists even after the polynomial corrections are applied. Ideally, if the solution were known at the interface planes, the global process would converge in one iteration. This prompted us to seek an improvement to the estimate of the solution on the adjacent interface planes for each processor. The improvement is computed for less effort than necessary for a global iteration of the above parallel iterative strategy.

The method of improvement evolved from an alternative parallel iterative approach [1 and 2] based on coarse mesh rebalancing.. In this, a correction is computed at any stage of the iterative solution of the original system of equations (2.1) from a new system that is of significantly reduced order. It may be viewed as "collapsing" many points of the fine physical grid to a single new coarse grid point. Rather than solve equations (2.1) in terms of neutron flux, an alternative system is generated in which multiplicative correction factors are sought for the estimated flux at each new grid point. These are used to correct the flux at the original fine mesh grid points that "collapse" to the new grid point. With multiplicative correction the tth trial solution $\phi^{(t)}$ for ϕ of the original system (2.1), is improved by

$$\phi^{(t)} \leftarrow \sum_{k=1}^{K} c_k \, P_k \, \phi^{(t)} \,, \tag{3.1}$$

where there are K coarse grid points and P_k is a diagonal matrix with ones on the diagonal corresponding to grid points to be corrected by the constant multiplicative factor c_k. K may vary between 1 and $N_x N_y N_z$. The factors c_k of (3.1) are determined from solution of the Galerkin weighted residual

$$\sum_{k=1}^{K} < P_\ell \phi^{(t)}, \, AP_k \phi^{(t)} > c_k = < P_\ell \phi^{(t)}, s >, \; \ell = 1, 2, \ldots, K \tag{3.2}$$

where $<,>$ denotes inner product. In this work $(n-1)$ systems of coarse mesh equations similar to (3.2) are generated, corresponding to the number of segment interfaces as defined in Section 2. On each processor the grid is generated by a halving rule defined as

i. for the two interfaces planes and their immediate neighbours, no collapsing takes place, i.e. a total of $4N_xN_y$ correction factors are sought corresponding to the original fine grid points,

ii. for the planes adjacent to the above $N_xN_y/4$ coarse mesh points on each plane,

iii. for the planes adjacent to the above $N_xN_y/16$ coarse mesh points on each plane,

iv. and so on until a lower limit, usually of 16 is met for each plane.

The matrix system (3.2) generated for each of the processors is symmetric and positive definite, and as for Section 2, the ICCG method of solution is used. The seven-point finite difference stencil of the original system becomes a ten-point stencil when the number of coarse grid points varies between the planes. Like the original there is no need to compute the off-diagonal components for the incomplete Cholesky decomposition. The new structures are solved in parallel on $(n-1)$ processors. Trial solutions of unity for the components of c_k are used. Upon solution of (3.2) not only are the flux estimates at the segment interfaces corrected, but all the other points between them are adjusted where appropriate, albeit by a coarser grid correction. The solution of (3.2) involves the determination of a number of unknowns whose order is approximately the equivalent of five (x, y) planes of the original points.

4 Results

Two trial problems are used to test the algorithm. These are defined respectively as Problems 1 and 3 in [1]. Results are reported for Problem 1 with $N_x = N_y = N_z = 32$. The number of iterations with ICCG on this problem is 39 which is equivalent to 32×39 (1248) iterations through a typical (x, y) plane. Evaluation of the parallel algorithm is difficult as a completely adequate architecture is not available, consequently the method is not examined in terms of hardware performance. On a serial computer generation of the Galerkin relationship in Section 2 is costly. The Galerkin relationship (3.2) is less expensive, and there is a significant degree of commonality between the processors which may be exploited. In other research [2] a hybrid correction procedure involving multiplicative and additive correction is reported. This is as effective as the pure multiplicative form used here but involves less overhead. Evaluation of both Galerkin forms can be done in parallel. With appropriate hardware at a sub-processor level, this cost is reduced to insignificant proportions, and this is assumed in this study. The assumption leads to optimistic results for present day hardware-even so the cost of generation is not considered in the evaluation of the algorithm.

It is difficult to evaluate the algorithm adequately in terms of iteration or operation counts owing to its complexity. Instead the method is compared with the "best" general algorithm for such problems, namely ICCG. In the form used, no additional infill is allowed in the incomplete Cholesky decomposition. Because ICCG is used as the inner-pass iteration driver in both Sections 2 and 3 and as the benchmark for the "best" sequential method, the criteria for evaluation selected is the number of plane equivalent iterations. This is the number of conjugate gradient passes through a full (x, y) plane (or its equivalent in the case of the interface correction, because the iteration count is then adjusted for the varying sizes of the planes). This gives a rough measure for studying the parallel algorithm. The number of conjugate gradient passes for the first stage of the algorithm (Section 2) are shown in addition to the number of global iterations. The solution of the LDL^T Cholesky phase is ignored in these estimates. At present, the LDL^T solution is performed sequentially; multicoloured ordering undoubtedly increase the level of parallelism but at the expense of the iteration count.

In Table (1a) results are given for the basic algorithm of Section 2. Compared with ICCG (1248 plane iterations) the method is not attractive for any number of processors. The enhancements [1] alter this opinion but that approach is not as effective as this method. In Table (1b) results for the parallel algorithm, as described in Sections 2 and 3 are reported. For a small number of processors there is a dramatic improvement in performance over ICCG. The cost of the interface enhancement is a relatively low proportion of the calculation, adding the equivalent of only 140 plane equivalent iterations in the case of two processors. The improvement over ICCG shows in the global iteration count and the number of plane equivalent iterations. As the number of processors increases, the cost of the interface correction becomes higher due to the relatively excessive number of planes represented in fine mesh. For 16 processors for example, approximately $75(15 \times 5)$ planes of information are needed, substantially more than the original problem. A natural constraint on the number of processors is suggested.

In Table (1c) a modification is made arising from experience with ADI methods. On subsequent global iterations the segments are defined to follow a cycle of (x, y), (y, z) and (x, z) orientations. This allows for greater global communication throughout the grid and a reduction in global and planar iterations.

In Table (1d), the correction of Section 3 is computed on a coarse grid of half size, that is $(N_x/2, N_y/2)$ grid points on each of the interface group of planes. The cost of the interface correction is reduced and its effect is noticeable when the number of processors is large. Few harmful effects appear from the less accurate interface correction, other than an increase of one global iteration.

Results for Problem 2 are reported in Table 2, ICCG requires 59 global (2360 plane) iterations for the problem. Again the value of the parallel approach is apparent, however, its performance with the coarse mesh interface without alternating pancakes is not as impressive as it was for the previous problem. In Table 3, results for Problem 1 on finer grids are presented. On a 48×48×48 grid ICCG requires 62 global (2976 plane) iterations, while on the larger 64×64×64 grid, 84 global (5376 plane) iterations are necessary.

5 Conclusion

By correcting the solution estimate for the interface planes, the performance of the parallel algorithm is improved vastly. The restricted measure of performance defined in Section 4 (the optimistic improvement factor of the method over sequential ICCG) is shown in Table 4 for three cases. (The results obtained so far for the finer grid version of Problem 1 are not for the form with coarse mesh interface correction. A slightly greater improvement is anticipated when these calculations are completed.) Ignoring the overheads, (which are significant on the sequential machine used), the method shows the potential of promising performance and a linear improvement is suggested for a limited number of processors. There are limitations with this approach for present day "supercomputers". The problem of generating the Galerkin relationships efficiently is the most important, but this could vanish with suitable parallel sub-architectures on the processors. The cost of solution of the reduced system of Section 2 by direct methods is not significant. The alternating segment enhancement which is desirable for good performance is not well suited to machines with cache memory, or vector processors that are unable to handle large strides.

The absence of a completely suitable machine is of little consequence insofar as algorithm research is concerned. Success with the approach warrants further investigation. Instead of the divide and conquer concept of massive parallelism the idea of limited division with the subsequent application of secondary parallelism holds a promise that deserves close attention.

TABLE 1

Results for Problem 1, 32×32×32 grid

	# Processors	Global Iterations	Plane Iterations	Plane Equivalent Iterations
(a)	Polynomial correction, no interface correction			
	2	20	3504	
	4	23	3286	
(b)	Polynomial correction, interface correction			
	2	8	1136	1276
	4	10	1000	1460
	8	12	664	1894
	16	11	402	3143
(c)	Polynomial correction, interface correction, alternating pancakes			
	2	6	800	915
	4	8	640	985
	8	7	344	1114
	16	7	260	2043
(d)	Polynomial correction, interface correction, coarse mesh interface, alt. pancakes			
	2	7	784	824
	4	8	632	756
	8	8	424	714
	16	8	372	1025

TABLE 2

Results for Problem 2, 40×40×40 grid

	# Processors	Global Iterations	Plane Iterations	Plane Equivalent Iterations
(a)	Polynomial correction, no interface correction			
	2	21	7680	
	4	20	6460	
(b)	Polynomial correction, interface correction			
	2	7	1840	2086
	4	7	1450	2003
(c)	Polynomial correction, interface correction, alternating pancakes			
	2	8	1900	2079
	4	10	1740	2288
(d)	Polynomial correction, interface correction, coarse mesh interface			
	2	11	2880	3009
	4	11	2520	2817
(e)	Polynomial correction, interface correction, coarse mesh interface, alt. pancakes			
	2	6	1700	1761
	4	10	1910	2094

TABLE 3

Results for Problem 1, 48×48×48 grid

	# Processors	Global Iterations	Plane Iterations	Plane Equivalent Iterations
(a)	Polynomial correction, interface correction			
	2	9	2280	2496
	4	13	2772	3521
(b)	Polynomial correction, interface correction, alternating pancakes			
	2	8	1872	2020
	4	8	1476	1913
	Results for Problem 1, 64×64×64 grid			
(a)	Polynomial correction, interface correction, alternating pancakes			
	2	8	3872	4071
	4	9	2784	3397
	8	9	2392	4041

TABLE 4

Optimistic improvement factor over sequentiel ICCG

# Processors	Optimistic Improvement Factor
Problem 1, polynomial correction, interface correction, coarse mesh interface, alternating pancakes	
2	2.9
4	6.3
8	13.2
16	18.6
Problem 2, polynomial correction, interface correction, coarse mesh interface, alternating pancakes	
2	2.6
4	4.4
Problem 1, polynomial correction, interface correction, alternating pancakes, 64×64×64 grid	
2	2.4
4	6.0
8	10.0

References

[1] Barry, J.M., Pollard, J.P. and Wachspress E.L. (1988a) - Some parallel iterative schemes for large systems of linear equations arising in three dimensional modelling - to appear.

[2] Barry, J.M., Pollard, J.P. and Wachspress, E.L. (1988b) - A method of parallel iteration - to appear.

[3] Barry, J.M. and Wachspress, E.L. (1988) - A method of parallel iteration based on polynomial correction in Noye J. and Fletcher C., Computational techniques and applications CTAC-87, North Holland, pp. 93-98.

[4] Mahoney, M.A. and Wachspress, E.L. (1985) - Numerical solution of 3D neutron group diffusion equations on the CYBER-205 computer, in Proceedings of international meeting on advances in nuclear engineering computational methods, American Nuclear Society (1), pp. 66-71.

Multidimensional Internally Lossless Filters of Fully Recursive Half-Plane Type

Sankar Basu

Electrical Engineering Department
Stevens Institute of Technology
Hoboken, NJ 07030, USA

Recently, motivated by needs for parallel processing of $2-D$ signals a scheme known as the fully recursive half-plane scheme has been proposed, and a method of designing transfer functions of filters having this recursive structure has been outlined in [1]. The recursion equation describing the relation between the input x and output y of a filter of this type is given by:

$$\beta_0[\{y_n(m)\}] = \sum_{i=1}^{L_D} \beta_i[\{y_{n-i}(m)\}] + \sum_{i=1}^{L_N} \alpha_i[\{x_{n-i}(m)\}] \tag{1}$$

where $x_n(m)$, $y_n(m)$ denote the n-th row of the input and the output signal, and the (row) operations $\alpha_i[\cdot]$ and $\beta_i[\cdot]$ respectively, denote $1-D$ linear shift invariant convolution operations with the $1-D$ sequences $a_i(m)$ and $b_i(m)$. Considering the $2-D$ Z-transform of (1), and assuming that the operations $\alpha_i[\cdot]$ and $\beta_i[\cdot]$ are all rational, we then have $H(z_1, z_2)$ in (2) for the transfer function of the filter, where $A_i(z_1)$, $B_i(z_1)$ are rational transfer functions representing the convolutional row operations just mentioned.

$$H(z_1, z_2) = \sum_{i=0}^{L_N} A_i(z_1) z_2^i \Big/ \sum_{i=0}^{L_D} B_i(z_1) z_2^i \tag{2}$$

On the other hand, it is now well known that for the successful operation of a digital filter structural considerations (in addition to an input-output

NATO ASI Series, Vol. F 70
Numerical Linear Algebra, Digital Signal Processing and Parallel Algorithms
Edited by G. H. Golub and P. Van Dooren

description) must be taken into account. The class of structurally passive filters variously known as the wave digital filters [2], orthogonal filters [3] or the lossless bounded real filters [4], are known to satisfy the essential properties of insensitivity to coefficient perturbation and non-linear arithmetic conditions resulting from overflow, finite precision arithmetic etc. Although much progress has been documented in the synthesis and design of $1-D$ structurally passive filters, methods for two and higher dimensions are still evolving.

In the present paper, $2-D$ fully recursive half-plane scheme is first presented. Next, (pseudo) passive or (pseudo) lossless fully recursive half-plane $2-D$ filters are introduced and then a method of their structurally passive synthesis and subsequently that of their design is discussed for the first time. The problem of synthesis of quarter plane causal structurally passive multidimensional filters can be equivalently viewed as the classical network theoretic problem of synthesizing a lossless but otherwise arbitrarily prescribed multidimensional transfer function as an interconnection of elementary building blocks such as capacitors and inductors. This latter problem does not have a solution in dimensions higher than 2, whereas in $2-D$ synthesis is feasible only in an unconstrained topological structure. In contrast, the present work provides us with a synthesis of arbitrary lossless fully recursive half-plane multidimensional filters. Additionally, unlike the quarter plane case referred to, earlier the synthesis is obtained in a fixed predetermined structure potentially useful for fast implementation. The synthesis method takes advantage of a recent algorithm for the design of structurally passive $1-D$ filters advanced by Rao and Kailath [5] as an extension of the celebrated Schur algorithm. Although the details of our method differ nontrivially from $1-D$ due to considerations characteristic of multidimensional problems (e.g., those utilizing techniques from elementary algebraic curve theory), the synthesis to be outlined can be considered, at least at a conceptual level, to be a generalization of the result in [5] to two-port transfer functions, the coefficients of numerator and denominator polynomials of which are rational functions. The class of multidimensional rational functions which completely describe the transfer functions of the recursive filters of the type under consideration is characterized.

A note regarding the stability of the filter is in order. The region of analyticity of the transfer function of our filter will be found to marginally differ from those previously considered in the $2-D$ half-plane literature. This is primarily due to the fact that the results such as those in [1] are motivated by bounded-input-bounded-output considerations, whereas, in contrast, our results are driven by passivity considerations. The fact that this difference in consideration does indeed lead to diverging formulations of stability in multidimensions ($m > 1$), but not in $1-D$, is now known. Thus, there is no contradiction between our stability results and those existing in the half-plane literature so far.

References

[1] J. Lee and J.W. Woods, Design and implementation of two-dimensional fully recursive digital filters, IEEE Trans. on Acoustics, Speech and Signal Processing, vol. 34, No. 1, pp. 178-191, Feb. 1986.

[2] A. Fettweis, Multidimensional digital filters with closed loss behavior designed by complex network theory approach, IEEE Trans. on Circuits and Systems, vol. 34, No 4, pp. 338-345, April 1987.

[3] E.F. Deprettere and P. Dewilde, Orthogonal cascade realization of multiport digital filters, Int. J. Circuits, Th. and Appl., vol. 8, pp. 254-272, July 1980.

[4] P. Vaidyanathan, S.K. Mitra, Passivity properties of low sensitivity digital filter structures, IEEE Trans. on CAS, vol. 32, pp. 217-224, March 1985.

[5] S.K. Rao and T. Kailath, Orthogonal digital filters for VLSI implementation, IEEE Trans. on Circuits and Systems, vol. 31, pp. 933-945, Nov. 1984.

Rank-One Extensions of the Generalized Hermitian Eigenvalue Problem for Adaptive High Resolution Array Processing

C. Beattie

Department of Mathematics
Virginia Polytechnic Institute and State University
Blacksburg, VA 24061, USA

A. Beex, and M. Fargues

Department of Electrical Engineering
Virginia Polytechnic Institute and State University
Blacksburg, VA 24061, USA

The determination of number, location, and movement of radiating sources using passive arrays of sensors is an important problem in radar imaging, medical imaging, and seismology. High resolution array processing, methods utilize spectral information contained in the observed signal correlation matrix in order to decompose incoming signals into signal and noise components. This allows extraction of various characteristic parameters such as source intensity and direction of arrival. Treatment of nonstationary signals in a colored noise environment necessitate the analysis of generalized Hermitian pencils. In order to conserve computational resources and permit real time processing of information, one may wish to process these matrix pencils adaptively only up to the smallest order necessary for full estimation of the desired signal parameters.

NATO ASI Series, Vol. F 70
Numerical Linear Algebra, Digital Signal Processing and Parallel Algorithms
Edited by G. H. Golub and P. Van Dooren

The method we describe successively updates the spectral decomposition of signal correlation matrices subject to rank-one extensions that correspond to inclusion of additional sensor data. We draw substantial inspiration from the analysis of Bunch, Nielsen, and Sorensen for rank-one matrix modifications and develop analogous root-finding and deflation strategies that allow treatment of rank-one extensions to generalized Hermitian pencils. In turn, new variations on the root-finding process lend some insight into possible improvements of the original Bunch-Nielsen-Sorensen algorithm SESUPD.

A Hybrid Scheme for the Singular Value Decomposition on a Multiprocessor

Michael Berry* and Ahmed Sameh*

Center for Supercomputing Research and Development
University of Illinois at Urbana-Champaign
Urbana, IL, 61801, USA

Abstract

We present a hybrid multiprocessor scheme for determining clusters of singular values and corresponding singular vectors of rectangular matrices in which the number of rows is substantially larger or smaller than the number of columns. In this scheme, we perform an initial orthogonal factorization on the tall matrix (either $\mathbf{A}$ or $\mathbf{A}^{\mathrm{T}}$) using a multiprocessor block Householder algorithm. A parallel one-sided Jacobi method is then applied to the resulting upper triangular $\mathbf{R}$ to reduce the matrix $\mathbf{S} = \mathbf{R}^{\mathrm{T}}\mathbf{R}$ to block diagonal form. This is followed by applying either a one-or two-sided Jacobi method to obtain the eigenvalues of these diagonal submatrices (which are the clusters of singular values of $\mathbf{A}$) in parallel. These hybrid methods are well suited for the hierarchical memory architectures of the Alliant FX/8, CEDAR, and CRAY 2. Some initial results on the Alliant FX/8, CRAY X-MP/416, and CRAY 2 are presented along with a comparison in the performance of the classical bi-diagonalization technique used in EISPACK and LINPACK.

*Research supported by the U.S. Department of Energy under Grant No. US DOE DE-FG02-85ER25001.

NATO ASI Series, Vol. F 70
Numerical Linear Algebra, Digital Signal Processing and Parallel Algorithms
Edited by G. H. Golub and P. Van Dooren

1 Introduction

The singular value decomposition (SVD) is commonly used in determining the solution of unconstrained linear least squares problems. In applications such as real-time signal processing, the solution to these problems is needed in the shortest possible time. Given the growing availability of multiprocessor computer systems, there has been great interest in the development of parallel implementations of the singular value decomposition. We present a multiprocessor scheme for determining the SVD of an m by n matrix $\mathbf{A}$, in which either $m \gg n$ or $n \gg m$, that has a clustered spectrum of singular values.

Without loss of generality, suppose $\mathbf{A}$ is a real m by n matrix with $m \gg n$, and rank$(\mathbf{A}) = r$. The singular value decomposition of $\mathbf{A}$ can be defined as

$$\mathbf{A} = \mathbf{U\Sigma V^T}, \tag{1}$$

where $\mathbf{U^TU} = \mathbf{V^TV} = \mathbf{I_n}$ and $\mathbf{\Sigma} = diag(\sigma_1, ..., \sigma_n)$, $\sigma_i > 0$ for $1 \leq i \leq r$, $\sigma_j = 0$ for $j \geq r+1$. The first r columns of the orthogonal matrices $\mathbf{U}$ and $\mathbf{V}$ define the orthonormalized eigenvectors associated with the r nonzero eigenvalues of $\mathbf{AA^T}$ and $\mathbf{A^TA}$, respectively. The singular values of $\mathbf{A}$ are defined as the diagonal elements of $\mathbf{\Sigma}$ which are the nonnegative square roots of the n eigenvalues of $\mathbf{AA^T}$. The multiprocessor scheme we are proposing consists of three stages which are outlined below.

2 Block Householder Reduction

Given the m by n matrix $\mathbf{A}$, where $m \gg n$, we perform a block generalization of Householder's reduction

$$\mathbf{A} = \mathbf{QR}, \tag{2}$$

where $\mathbf{Q}$ is m by n and has orthogonal columns. We note that $\mathbf{R}$, an n by n upper-triangular matrix, will be singular for rank deficient matrices $\mathbf{A}$(i.e. $r < n$). Block formulations of Householder transformations were proposed by Brenlund and Johnsen [2], Dietrich [3], and more recently by Bischof and Van Loan [1]. We adopt the Bischof and Van Loan Method which is based on the observation that the product of k Householder transformations $\mathbf{H_k} = \mathbf{P_k}...\mathbf{P_2}\mathbf{P_1}$ can be written in the form

$$\mathbf{H_k} = \mathbf{I} - \mathbf{V_k}\mathbf{U_k^T}, \tag{3}$$

where $\mathbf{U_k}$, $\mathbf{V_k}$ are m by k matrices so that $\mathbf{U_k} = (u_1, u_2, ..., u_k)$ and $\mathbf{V_k} = (\mathbf{P_k}\mathbf{V_{k-1}}, u_k)$. In [1] it was shown that such a block scheme has the same numerical properties as the classical Householder factorization method.

One major motivation for using this block generalization method on a machine such as an Alliant FX/8 lies in its extensive use of matrix-matrix multiplication modules. As demonstrated in [4] and [8], optimal performance of classical linear algebra algorithms on hierarchical memory systems, such as that of the Alliant FX/8 and *CEDAR* [11], can be achieved by reformulating the algorithm in terms of such modules (BLAS3) which achieve high data locality. The efficiency of this particular block Householder method (3) is best realized in the speed improvements obtained over vectorized implementations of the classical Householder factorization [7]. For a 1000 by 100 matrix $\mathbf{A}$, speed improvements as high as 9 over the classical scheme have been achieved on the eight processors of the Alliant FX/8.

3 One-Sided Jacobi Method

Applying the one-sided Jacobi method ([9], [10], [12]) to the n by n upper triangular matrix $\mathbf{R}$ (which may be singular) from the block Householder factorization of $\mathbf{A}$, we determine an orthogonal matrix $\mathbf{V} = [\tilde{\mathbf{V}}, \tilde{\mathbf{W}}]$, where $\tilde{\mathbf{V}}$ is n by r, as a product of plane rotations such that

$$\mathbf{R}\tilde{\mathbf{V}} = \tilde{\mathbf{Q}} = (\tilde{\mathbf{q}}_1, \tilde{\mathbf{q}}_2, ..., \tilde{\mathbf{q}}_\mathbf{r}), \tag{4}$$

and

$$\tilde{\mathbf{q}}_\mathbf{i}^\mathbf{T}\tilde{\mathbf{q}}_\mathbf{j} = {\sigma_i}^2\delta_{ij},$$

where the columns of $\tilde{\mathbf{Q}}$ are orthogonal, and δ_{ij} is the Kronecker delta. We then may write $\tilde{\mathbf{Q}}$ as

$$\tilde{\mathbf{Q}} = \tilde{\mathbf{U}}\tilde{\mathbf{\Sigma}} \text{ with } \tilde{\mathbf{U}}^\mathbf{T}\tilde{\mathbf{U}} = \mathbf{I_r}, \text{ and } \tilde{\mathbf{\Sigma}} = diag(\tilde{\sigma}_1, ..., \tilde{\sigma}_r), \tag{5}$$

and hence combining (4) and (5) we obtain

$$\mathbf{R} = \tilde{\mathbf{U}}\tilde{\mathbf{\Sigma}}\tilde{\mathbf{V}}^{\mathbf{T}}. \tag{6}$$

From (2), (6) we obtain

$$\mathbf{A} = \mathbf{QR} = \mathbf{Q}\tilde{\mathbf{U}}\tilde{\mathbf{\Sigma}}\tilde{\mathbf{V}}^{\mathbf{T}} = \mathbf{U}\tilde{\mathbf{\Sigma}}\tilde{\mathbf{V}}^{\mathbf{T}}, \tag{7}$$

where $\mathbf{U} = \mathbf{Q}\tilde{\mathbf{U}}$. Thus we can realize the decomposition in (1) after determining the matrix $\tilde{\mathbf{V}}$ in (4).

We construct the matrix $\tilde{\mathbf{V}}$ via the (i,j) plane rotations discussed in [5] so that

$$\mathbf{r_i^T r_j} = 0, \text{ and } ||\mathbf{r_i}||_2 > ||\mathbf{r_j}||_2, \tag{8}$$

where $\mathbf{r_i}$ designates the i-th column of $\mathbf{R} \equiv \mathbf{R}\tilde{\mathbf{V}}$. Note that (8) requires the columns of $\tilde{\mathbf{Q}}$ to decrease in norm from left to right, and hence the resulting σ_i to be in monotonic non-increasing order. Several parallel schemes for ordering the (i,j) plane rotations may be used in orthogonalizing the columns of $\mathbf{R}$. One possible orthogonalization scheme is given in [5] and [6]. A sweep consists of $n(n-1)/2$ plane rotations and at the end of any particular sweep s_i we have

$$\mathbf{V_{s_i}} = \mathbf{V_1 V_2 ... V_n},$$

and hence

$$\mathbf{V} = \mathbf{V_{s_1} V_{s_2} ... V_{s_t}}, \tag{9}$$

where t is the number of sweeps required for convergence. This process is iterative since orthogonality of several columns established by one transformation may be destroyed in subsequent transformations.

4 Numerical Decoupling

We can define the one-sided Jacobi iteration on the upper-triangular matrix $\mathbf{R} \equiv \mathbf{R_1}$, by

$$\mathbf{R_{k+1}} = \mathbf{R_k V_k}, k = 1, 2, ..., \tag{10}$$

where

$$\mathbf{R_k} = \mathbf{D_k} + \mathbf{E_k} + \mathbf{E_k^T},$$

and

$$\mathbf{S_k} = \mathbf{R_k^T R_k} = \tilde{\mathbf{D}}_\mathbf{k} + \tilde{\mathbf{E}}_\mathbf{k} + \tilde{\mathbf{E}}_\mathbf{k}^\mathbf{T},$$

where $\mathbf{D_k}$, $\tilde{\mathbf{D}}_\mathbf{k}$ are diagonal matrices and $\mathbf{E_k}$, $\tilde{\mathbf{E}}_\mathbf{k}$ are strictly upper-triangular. From (6), we note that the square roots of the eigenvalues of $\mathbf{S_k}$ are the singular values of the original matrix $\mathbf{A}$. As discussed in [6] for matrices $\mathbf{A}$ having clustered spectra of singular values, $\mathbf{S_k}$ will converge to a diagonal form through an intermediate *block diagonal* form, where each of the principal submatrices (positioned along $\tilde{\mathbf{D}}_\mathbf{k}$) have diagonal elements which comprise one cluster of singular values of $\mathbf{A}$. To obtain such a form for $\mathbf{S_k}$, we must have the diagonal elements of $\mathbf{S_k}$ corresponding to clustered singular values occur in adjacent positions on the diagonal, [6]. This adjacency is guaranteed by the enforcement of the monotonically non-increasing column norm ordering in $\mathbf{R_k}$, (8). Thus, after a particular number of critical sweeps, c, we obtain

$$\mathbf{S_c} = \begin{pmatrix} \mathbf{T_1} & & & \\ & \mathbf{T_2} & & \\ & & \ddots & \\ & & & \mathbf{T_p} \end{pmatrix} \tag{11}$$

so that the eigenpairs of each $\mathbf{T_i}, (1, 2, ..., p)$, where p is the number of clusters, can be computed in parallel by either the one- or two-sided Jacobi method. Each symmetric positive definite matrix $\mathbf{T_i}$ will, in general, be dense of order q_i, representing the number of singular values of $\mathbf{A}$ contained in the i-th cluster. Given the quadratic convergence property of the two-sided Jacobi method for symmetric matrices having clustered spectra ([15]), we may obtain a faster global convergence for $k > c$ if the two-sided rather than the one-sided Jacobi method is applied to each of the $\mathbf{T_i}$. Thus, a *hybrid* method consisting of an initial phase of several one-side Jacobi iterations followed by the two-sided Jacobi method on the resulting subproblems would combine the optimal parallelism of the one-sided Jacobi method and the fast convergence of the two-sided Jacobi method for clustered spectra. Of course the difficulty in implementing such a method lies in the determination of the critical number of sweeps, c. We note that such a hybrid SVD method would be quite suitable for implementation on multiprocessors such as the CRAY X-MP/416, CRAY 2 and *CEDAR*, [11].

For *CEDAR*, each cluster of processors can determine the SVD of one of the $\mathbf{T_i}$'s and hence compute one cluster of singular values and corresponding left and right singular vectors of $\mathbf{A}$, as specified in (7) and (9). Optimal load balancing on any multiprocessor, of course, will depend upon the number of clusters of singular values, p, and the order, q_i, of each $\mathbf{T_i}$.

5 Gershgorin Disk Separation

In order to determine the number of critical sweeps, c, required to decouple $\mathbf{S_k}$ into the $\mathbf{T_i}$'s, we resort to the separation in radii of the n Gershgorin disks associated with the spectrum of $\mathbf{S_k}$. We note that the $\mathbf{S_k}$'s are completely determined via the inner products in (8), which are by-products of the rotation angle computation in the one-sided Jacobi method described in Section 3 (see [5] or [6]). Consider the case where $c = 2$. Let us define the ith Gershgorin radius associated with $\mathbf{S_k}$, g_i^k, as

$$g_i^k = \sum_{j=1(\neq i)}^{n} |s_{ij}^k|, \text{ for } i = 1, 2, ..., n,$$

and the partial Gershgorin radius, $\hat{g}_i^k$, as

$$\hat{g}_i^k = \sum_{j=i+1}^{n} |s_{ij}^k|, \text{ for } i = 1, 2, ..., n-1.$$

We choose a prescribed minimum separation ϵ of the $\hat{g}_i^k$'s, and determine the set $\Im_k$ (for each k) given by

$$\Im_k = \{i | \hat{g}_i^k - \hat{g}_{i-1}^k| > \epsilon\}. \tag{12}$$

Then, for each $i \in \Im_k$ we determine g_i^k and g_{i-1}^k. If

$$g_i^k + g_{i-1}^k < |s_{ii}^k - s_{i-1,i-1}^k|, \tag{13}$$

then the two clusters in the spectrum of $\mathbf{A}$ are approximated by s_{jj}^k, $j = [1, i-1]$, and s_{ll}^k, $l = [i, n]$. We note that while (12) constitutes a necessary condition for the separation of the two unions of Gershgorin disks, we must always have (13) to guarantee that the two clusters of ordered singular values can be separated.

6 Performance

In Table 1, we present timing results (wall-clock execution times) for the computation of clustered singular values (no singular vectors) by a hybrid SVD scheme (HYJAC), and an optimized version of the classical bi-diagonalization SVD algorithm (SVD2) implemented in [7] and [13]. SVD2 utilizes efficient matrix-vector (BLAS2) kernels for the application of subsequent Householder transformations in the reduction to bi-diagonal form. HYJAC consists of two phases: orthogonal factorization (Section 2), and one-sided Jacobi iterations (Section 3). Upon detection of the block diagonal form in (11) via (12) and (13), HYJAC then applies a two-sided Jacobi method (see [5], [6]) to each of the $\mathbf{T_i}$'s (concurrently) to resolve the corresponding independent clusters of singular values. The $n \times n$ ($n = 64$ and 128) matrices used in the experiments in Table 1 are chosen to have four clusters of singular values, σ_i, given by

$$\sigma_i = \begin{cases} 1 & + & (i-1) \times \eta, & i = 1, ..., n/4, \\ 10 & + & (i - n/4 - 1) \times \eta, & i = n/4 + 1, ..., n/2, \\ 10^2 & + & (i - n/2 - 1) \times \eta, & i = n/2 + 1, ..., 3n/4, \\ 10^3 & + & (i - 3n/4 - 1) \times \eta, & i = 3n/4 + 1, ..., n, \end{cases}$$

where $\eta = 10^{-10}$. HYJAC and SVD2 are used to accurately compute all the singular values of the test matrices to 14 decimal digits in 64-bit arithmetic on an Alliant FX/8[1] (128 KB Cache Memory), an Alliant FX/80[2] (512 KB Cache Memory), a CRAY X-MP/416[3], and a CRAY 2[3].

After the decoupling phase in HYJAC, all 8 computational elements (CE's) on the Alliant computer systems are used to resolve each of the clusters of singular values, whereas each CPU on the CRAY systems is dedicated to only one of the four clusters of singular values. The microtasking (Micro) implementation (see [14]) of HYJAC on both the Alliant

[1]Located at the University of Illinois Center for Supercomputing Research and Development (CSRD), Urbana, IL.

[2]Special thanks to University of Illinois National Center for Supercomputing Applications (NCSA), Urbana, IL. for providing us access to an Alliant FX/80.

[3]Special thanks to the Cray Research Inc. for providing us access to both a CRAY X-MP/416 and CRAY 2 at the Mendota Heights Computation Center, Mendota Heights, MN.

HYJAC						
n	Alliant		Cray X-MP/416		Cray 2	
	FX/8	FX/80	Micro	Macro	Micro	Macro
64	.50	.33	.09	.16	.26	.18
128	3.0	2.0	.60	.65	.85	.75

SVD2				
n	Alliant		Cray X-MP/416	Cray 2
	FX/8	FX/80	1CPU	1CPU
64	.36	.22	.11	.13
128	1.4	1.0	2.2	1.3

Table 1: Performance of Conventional and Hybrid Methods for Model Problem in Wall-Clock Times.

and CRAY computer systems uses each CPU or CE to compute and apply a single independent Jacobi rotation in the four necessary *global* one-sided Jacobi sweeps prior to decoupling (11). Using macrotasking (Macro) (see [14]) on the CRAY computer systems, each CPU is used to compute $\lfloor n/2 \rfloor /4$ or $\lfloor (n-1)/2 \rfloor /4$ independent Jacobi rotations prior to numerical decoupling. We note that SVD2 is implemented on only one CPU of the CRAY X-MP/416 and CRAY 2.

From Table 1, we note that for $n = 128$ HYJAC can be ~ 2 to ~ 4 times faster than SVD2 on the CRAY 2 (macrotasking) and the CRAY X-MP/416 (microtasking), respectively. Although HYJAC is not competitive with SVD2 on the Alliant computer systems, we do gain some reduction in execution time from the larger cache memory (which can enhance data locality for the decoupling) and faster processor clock speed of the Alliant FX/80 compared to the Alliant FX/8.

References

[1] C. Bischoff and C. Van Loan, "The WY representation for products of Householder Matrices", TR 85-681, Cornell University, Department of Computer Science, 1985.

[2] O. Bronlund and T. Johnsen, "QR-factorization of partitioned matrices", Computer Methods in Applied Mechanics and Engineering, 1974, 3, 153–172.

[3] G. Dietrich, "A new formulation of the hypermatrix Householder-QR decomposition", Computer Methods in Applied Mechanics and Engineering, 1976, 9, 273–280.

[4] M. Berry and others, "Parallel Numerical Algorithms on the CEDAR System", Lecture Notes in Computer Science, 1986, 237, 25–39

[5] M. Berry and A. Sameh, "Multiprocessor Jacobi schemes for dense symmetric eigenvalue and singular value decompositions", 546, University of Illinois at Urbana-Champaign, Center for Supercomputing Research and Development, 1986.

[6] M. Berry and A. Sameh, "Parallel algorithms for the singular value and dense symmetric eigenvalue problems", 761, University of Illinois at Urbana-Champaign, Center for Supercomputing Research and Development, 1988.

[7] J. Dongarra and J. Bunch and C. Moler and G. W. Stewart, "LINPACK Users' Guide", SIAM, Philadelphia, 1979.

[8] K. Gallivan and W. Jalby and U. Meier, "The Use of BLAS3 in Linear Algebra on a Parallel Processor with a Hierarchical Memory", SIAM J. Sci. Stat. Comput, 1987, 8, 6, 1079–1084.

[9] M. R. Hestenes, "Inversion of matrices by biorthogonalization and related results", J. Soc. Indust. Appl. Math., 1958, 6, 51–90

[10] H. F. Kaiser, "The JK method: a procedure for finding the eigenvectors and eigenvalues of a real symmetric matrix", The Computer Journal, 1972, 15, 33, 271–273.

[11] D. Kuck and E. Davidson and D. Lawrie and A. Sameh, "Parallel Supercomputing Today and the CEDAR Approach", Science, 1986, 231, 967–974,

[12] J. C. Nash, "A one-sided transformation method for the singular value decomposition and algebraic eigenproblem", The Computer Journal, 1975, 18, 1, 74–76.

[13] B. Smith and others, "Matrix Eigensystem Routines - EISPACK Guide", second Edition, Springer-Verlag, Berlin, 1976.

[14] CRAY X-MP Multitasking Programmer's Reference Manual, Cray Research Inc., Mendota Heights, MN, 1987, SR-0222.

[15] J. Wilkinson, "Almost diagonal matrices with multiple or close eigenvalues", Linear Algebra Appl., 1968, 1, 1–12.

The Weak Stability of Algorithms for Matrix Computations

James R. Bunch

Department of Mathematics
University of California, San Diego
La Jolla, CA 92093, USA

Abstract

The weak stability of an algorithm (a small relative error in the computed solution for every well-conditioned system) is shown to be equivalent to a small relative residual for every well-conditioned system or to the solution of a nearby problem (in the relative sense) for every well-conditioned system.

We say that an algorithm for solving systems of linear equations in finite precision arithmetic is *stable* for a class of matrices $\mathcal{A}$ if for all A in $\mathcal{A}$ and for all b the computed solution $\hat{x}$ to $Ax = b$ satisfies $\hat{A}\hat{x} = \hat{b}$ for some $\hat{A}$ close to A and $\hat{b}$ close to b. This does not say that $\hat{x}$ is close to x! However, if A is well-conditioned ($\kappa(A) \equiv \|A\| \, \|A^{-1}\|$ is not too large), then $\hat{x}$ is close to x [5, p. 27; 7, pp. 194-5; 9, p. 190].

In [2] a modified definition of stability was introduced, called *weak stability*. We say that an algorithm for solving systems of linear equations is *weakly stable* for a class of matrices $\mathcal{A}$ if for each well- conditioned A in $\mathcal{A}$ and for each b the computed solution $\hat{x}$ to $Ax = b$ has small relative error, i.e., $\|x - \hat{x}\|/\|x\|$ is small. In other words, an algorithm is weakly stable for a class of matrices $\mathcal{A}$ if it computes an accurate solution for every well-conditioned A in $\mathcal{A}$.

In [2] the concept of *strong stability* was also introduced and it was extended in [4]. However, in this paper we shall consider some equivalent definitions of weak stability.

NATO ASI Series, Vol. F 70
Numerical Linear Algebra, Digital Signal Processing and Parallel Algorithms
Edited by G. H. Golub and P. Van Dooren

The following theorem is well-known, e.g., see [1], p. 458.

Theorem 1: If $Ax = b$, $r = Ay - b$, A nonsingular, and $b \neq 0$, then

$$\frac{1}{\kappa(A)}\frac{\|r\|}{\|b\|} \leq \frac{\|x-y\|}{\|x\|} \leq \kappa(A)\frac{\|r\|}{\|b\|} .$$

Theorem 1 says that if A is well-conditioned, then the relative error $\|x - y\|/\|x\|$ in an approximate solution y is small iff the relative residual $\|r\|/\|b\|$ is small.

However, if A is ill-conditioned, then a small relative residual does not necessarily imply that the relative error is the approximate solution is small.

Example 1: Consider $Ax = b$, where

$$A = \begin{bmatrix} 1+\varepsilon & 1 \\ 1 & 1 \end{bmatrix} \quad \text{and} \quad b = \begin{bmatrix} 1 \\ 1 \end{bmatrix} . \qquad \text{Then} \quad x = \begin{bmatrix} 0 \\ 1 \end{bmatrix} .$$

$$\text{Let} \quad y = \begin{bmatrix} 2 \\ -1 \end{bmatrix} . \ \text{Then} \quad r = \begin{bmatrix} 2\varepsilon \\ 0 \end{bmatrix} \quad \text{and} \quad \|r\|_\infty/\|b\|_\infty = 2\varepsilon ,$$

$$\text{but} \quad \|x - y\|_\infty/\|x\|_\infty = 2 .$$

The next theorem shows that a small relative residual always implies that the approximate solution is the exact solution to a nearby system (whether the matrix well-conditioned or ill-conditioned).

Theorem 2: Suppose $y \neq 0$, is an approximate solution to $Ax = b, b \neq 0$, $A \neq 0$. Let $r = Ay - b$. Suppose $\alpha \equiv \|r\|_2/\|b\|_2$ is small (i.e., the relative residual is small). Then y is the exact solution to a problem close to the original, i.e., there exists an E such $(A+E)y = b$, where $\|E\|_2/\|A\|_2$ is small, namely $E = -ry^T/y^Ty$ and $\|E\|_2/\|A\|_2 \leq \alpha/(1-\alpha)$.

Proof: Let $E = -\frac{ry^T}{y^Ty}$. $(A+E)y = Ay - \frac{ry^T}{y^Ty} = Ay - r = Ay - (Ay - b) = b$.

$\|E\|_2 \leq \frac{\|r\|_2 \ \|y\|_2}{\|y\|_2^2} = \frac{\|r\|_2}{\|y\|_2}$. $r + b = Ay$. $\|b\|_2 - \|r\|_2 \leq \|r+b\|_2 \leq \|A\|_2\|y\|_2$.

$\frac{1}{\|b\|_2 - \|r\|_2} \geq \frac{1}{\|A\|_2\|y\|_2}$ since $\|b\|_2 > \|r\|_2$. So

$\frac{1}{\|y\|_2} \leq \frac{\|A\|_2}{\|b\|_2 - \|r\|_2}$ and $\frac{\|E\|_2}{\|A\|_2} \leq \frac{\|r\|_2}{\|b\|_2 - \|r\|_2} = \frac{\|r\|_2/\|b\|_2}{1 - \|r\|_2/\|b\|_2}$. qed

Note that if $b = 0$ and $y = 0$, we may take $E = 0$. In fact, the matrix $E = -ry^T/y^Ty$ has the smallest 2-norm of all such systems [2, p. 16].

Is the converse true? Suppose $(A + E)y = b$, $Ax = b$, and $\|E\|/\|A\|$ is small. Let $r = Ay - b$. Is $\|r\|/\|b\|$ small? This is true if A is well-conditioned, but false if A is ill-conditioned, e.g., see [3].

Theorem 3: If $r = Ay - b$, $(A + E)y = b$, A nonsingular, and $\|A^{-1}\| \, \|E\| < 1$, then

$$\frac{\|r\|}{\|b\|} \leq \frac{\|A^{-1}\| \, \|E\|}{1 - \|A^{-1}\| \, \|E\|} = \frac{\kappa(A) \, \|E\|/\|A\|}{1 - \kappa(A)\|E\|/\|A\|}.$$

Example 2: Let $A = \begin{bmatrix} 1+\epsilon & 1 \\ 1 & 1-\epsilon \end{bmatrix}$, $E = \begin{bmatrix} 0 & 0 \\ 0 & \epsilon^2 \end{bmatrix}$, and $b = \begin{bmatrix} 1 \\ -1 \end{bmatrix}$, where

$0 < \epsilon << 1$. Then $\kappa_\infty(A) = (2+\epsilon)^2/\epsilon^2 \cong 4/\epsilon^2$ and $\|E\|_\infty/\|A\|_\infty = \epsilon^2/(2+\epsilon) \cong \epsilon^2/2$.

Let $y \equiv (A + E)^{-1}b = \frac{1}{\epsilon^3} \begin{bmatrix} 2 - \epsilon + \epsilon^2 \\ -2 - \epsilon \end{bmatrix} \cong \begin{bmatrix} 2/\epsilon^3 \\ -2/\epsilon^3 \end{bmatrix}$. Then $r = Ay - b = \begin{bmatrix} 0 \\ \frac{2}{\epsilon} + 1 \end{bmatrix}$. So $\|r\|_\infty/\|b\|_\infty = \frac{2}{\epsilon} + 1$, which is not at all small.

Theorems 2 and 3 and Example 2 show that a small *relative* residual $\|r\|/\|b\|$, $r = Ay - b$, always implies that the approximate solution y is the exact solution of a nearby system, $(A + E)y = b$, where "nearby" means $\|E\|/\|A\|$ is small. However, the converse is *false* unless A is well-conditioned. I emphasize this because many people confuse the situation concerning *relative* residuals and solving a *nearby* system in the *relative* sense with a theorem by Oettli and Prager [6] (see also [8], pp. 181-183) which concerns the *absolute* residual and solving a *nearby* system in the *absolute* sense. We can state their theorem as follows, where $\leq$ between vectors and matrices is to be understood as holding componentwise.

Theorem 4 (Oettli and Prager): Let $\Delta A \geq 0$, $\Delta b \geq 0$; let $D=\{E:|E|\leq \Delta A\}$ and $G=\{f:|f|\leq \Delta b\}$. Then there is an E in D and f in G such that $(A+E)y = b+f$ iff $|r| \leq \Delta A|y| + \Delta b$, where $r = Ay - b$.

In words, this says that if $|y|$ is not too large, then the approximate solution y is the exact solution to a *nearby* system in the *absolute* sense iff the *absolute* residual is small. See also [9], pp. 251-254. However, as Theorems 3 and 4 and Example 2 show, the situation is quite different when discussing *relative* residuals and *relative* perturbations of systems. Furthermore, when discussing matrix computations, it is always more relevant to discuss *relative* error, *relative* residuals, and *relative* perturbations than to discuss *absolute* error, *absolute* residuals, and *absolute* perturbations.

Of course, it is well known [5, p. 27; 7, pp. 194-195; 9, p. 190] that if we have solved a nearby system (in the relative sense) which is well conditioned, then we have a good approximate solution (i.e., the relative error is small):

Theorem 5: If $Ax = b$ and $(A+E)y = b+f$, where A is nonsingular and $b \neq 0$, and if $\|E\|/\|A\| < 1/\kappa(A)$, then $A+E$ is nonsingular and

$$\frac{\|x-y\|}{\|x\|} \leq \frac{\kappa(A)}{1-\kappa(A)\|E\|/\|A\|}\left[\frac{\|E\|}{\|A\|} + \frac{\|f\|}{\|b\|}\right].$$

If we put together Theorems 2, 3, and 5 we see that we have the following equivalences for weak stability.

Theorem 6: An algorithm for solving systems of linear equations in finite precision arithmetic is weakly stable for a class of matrices $\mathcal{A}$ if for all well-conditioned A in $\mathcal{A}$ and for all b the computed solution $\hat{x}$ to $Ax = b$ satisfies any of the following:

(1) $\frac{\|x-\hat{x}\|}{\|x\|}$ is small;

or

(2) $\frac{\|r\|}{\|b\|}$ is small where $r = A\hat{x} - b$;

or

(3) there is an E such that $(A+E)\hat{x} = b$, where $\|E\|/\|A\|$ is small.

In other words, an algorithm is *weakly stable* if for each well-conditioned matrix, the computed solution has small *relative* error or the *relative* residual is small or we solved a problem close (in the *relative* sense) to the original problem.

References

[1] K.E. Atkinson, *An Introduction to Numerical Analysis*, Wiley, 1978.

[2] J.R. Bunch, "The Weak and Strong Stability of Algorithms in Numerical Linear Algebra", *Linear Algebra and Appl.* 88/89 (1987): 49-66.

[3] J.R. Bunch, "Stability Considerations for Toeplitz Matrix Computations", UCSD Technical Report.

[4] J.R. Bunch, J.W. Demmel, and C.F. Van Loan, "The Strong Stability of Algorithms for Solving Symmetric Linear Systems", SIMAX, submitted.

[5] G.H. Golub and C.F. Van Loan, *Matrix Computations*, Johns Hopkins University Press, 1983.

[6] W. Oettli and W. Prager, "Compatibility of Approximate Solution of Linear Equations with Given Error Bounds for Coefficients and Right-Hand Sides", *Numerische Mathematik*, 6(1964):405-409.

[7] G.W. Stewart *Introduction to Matrix Computations*, Academic Press, 1973.

[8] J. Stoer and R. Bulirsch, *Introduction to Numerical Analysis*, Springer-Verlag, 1980.

[9] J.H. Wilkinson, *The Algebraic Eigenvalue Problem*, Clarendon Press, 1965.

Parallelism in Dynamic Programming and Control

J. L. Calvet, J. Dantas de Melo and J. M. Garcia

LAAS du CNRS
7, ave. du Colonel-Roche
31077 Toulouse Cedex, France

Abstract

Dynamic programming using parallel processing techniques is discussed for optimal control. An implementation on a CRAY-2 for optimizing the station acquisition of geostationary satellites is mentioned.

Keywords: parallel processing, dynamic programming, optimal control

1 Introduction

It is well known that Dynamic Programming [DP] constitutes a very general method for optimization and that it is very useful in Control where it gives feedback solutions. However, it suffers from the "curse of dimensionality" when the state space is an infinite set. The complexity of the basic computational procedure is twofold. Exponential algorithmic complexity in states and controls. Complexity of the tasks to be performed: quantization and data ordering, calculus of transition costs and state-transition functions, optimization of cost-to-go functions [CTG], interpolation of the optimal CTG and controls.

Parallel computing is undoubtedly an issue to push back [to some extent] such limits.

NATO ASI Series, Vol. F 70
Numerical Linear Algebra, Digital Signal Processing and Parallel Algorithms
Edited by G. H. Golub and P. Van Dooren

Due to space limitation, the paper focuses on the CTG optimization. Parallelism in states and controls is first considered. Based on the pioneer work by Casti et al. [1], our approach then considers works on parallel DP applied to search optimal path in a multistage graph [2-4]. This leads through the use of linear algebra paradigms to define basic parallel algorithms for optimal control.

Likewise, parallelism in stages makes use of the work by Li and Wah [5] and gives an appropriate formulation.

The paper ends by mentioning the implementation on a CRAY-2 of a parallel state algorithm for optimizing the station acquisition of a geostationary satellite.

2 Outline of the basic DP computational procedure

Let us consider the problem of optimal control:

Solve for $u \in \{u(0),\ldots,u(K\text{-}1)\}$

$$J = Opt_u \quad \left\{ \sum_{k=0}^{K-1} r_k(x_k, u_k) + r_K(x_K) \right\} \tag{1}$$

$$s.t. \quad x_{k+1} = f_k(x_k, u_k); \quad k = 0, 1, \ldots, K-1; \quad x_0 = x_0^*$$

$$where \qquad x_k \in \mathcal{X}_k \subset \mathcal{R}^n; \quad u_k \in \mathcal{U}_k(x_k) \subset \mathcal{R}^m$$

The DP functional equations issued from the application of the optimality theorem are:

for k = K-1 down 0, solve

$$I_k(x_k) = Opt_{u_k} \{ r_k(x_k, u_k) + I_{k+1}[f_k(x_k, u_k)] \} \tag{2}$$

$$with \qquad I_K(x_K) = r_K(x_K)$$

This gives $J = I_0(x_0^*)$ and $u_k opt = u_k(x_k)$.

The formulation (2) expresses explicitely the optimality principle by Bellman (1957), i.e. "whatever the initial state and control are, the remaining decision must constitute an optimal policy, with regard to the state resulting from the first decision".

For the general case the complexity of DP lies in the fact that solving the functional equations amounts to an implicit enumeration, numerically processed after discretizing the state space and very often the control space, which moreover implies to use some interpolation techniques.

1) Quantization and data ordering

The CTG $I_k(x_k)$ are defined on a grid designed by the quantization of states [and controls]. Hence, the range of the quantized state variables at step k can be stored in the form of a $n_s(k) \times n$ matrix-array, where:

$$n_s(k) = \prod_{l=1}^{n} \# x_{l,k}$$

and $\# x_{l,k}$ means the number of quantization levels of the lth state-vector component. Thus, the ith row of this matrix, denoted by α_i^T, contains the coordinates of the particular grid point to which is assigned the label i.

Likewise, a label $j[j \in \{1, 2, ..., n_c(k)\}]$ will be used to code the values of the control variables.

Basic algorithm

For sake of conciseness, one can summarize the general backward optimization part of the DP algorithm by three nested loops, as follows:

i)Iteration over all stages $k = K\text{-}1, \ldots, 1$

ii) Iteration over all states $j = 1, \ldots, n_s(k)$

iii) Iteration over all controls $i = 1, \ldots, n_c(k)$

Parallel computing is associated with the suppression of at least one of these loops.

3 Parallelism in states and controls

By using somewhat abusive notations such as:

$$I_k(i) \equiv I_k(x_k)\,|_{x_k=\alpha_i} \qquad \textit{and} \quad r_k(i,j) \equiv r_k(x_k, u_k)|_{x_k=\alpha_i, u_k=\beta_j}$$

one can rewrite, in matrix-form, the functional equations (2), as follows:

$$\begin{bmatrix} I_k(1) \\ \cdot \\ I_k(i) \\ \cdot \\ I_k(n_s(k)) \end{bmatrix} = \underline{Opt}_j \left\{ \begin{bmatrix} r_k(1,1) & \cdot & r_k(1,j) & \cdot & r_k(1,n_c(k)) \\ \cdot & \cdot & \cdot & \cdot & \cdot \\ r_k(i,1) & \cdot & r_k(i,j) & \cdot & r_k(i,n_c(k)) \\ \cdot & \cdot & \cdot & \cdot & \cdot \\ r_k(n_s(k),1) & \cdot & r_k(n_s(k),j) & \cdot & r_k(n_s(k),n_c(k)) \end{bmatrix} + \right.$$

$$+\left[\begin{array}{ccccc} J_{k+1}(1,1) & . & J_{k+1}(1,j) & . & J_{k+1}(1,n_c(k)) \\ . & . & . & . & . \\ J_{k+1}(i,1) & . & J_{k+1}(i,j) & . & J_{k+1}(i,n_c(k)) \\ . & . & . & . & . \\ J_{k+1}(n_s(k),1) & . & J_{k+1}(n_s(k),j) & . & J_{k+1}(n_s(k),n_c(k)) \end{array}\right]\Bigg\} \quad (3)$$

where:

- $\underline{Opt}_j$ means the optimization operator, row by row;
- $J_{k+1}(i,j) \equiv I_{k+1}[f_k(i,j)]$

The computational task, viewed as the sum of two matrices, exhibits several parallelisms:

1) Parallelism in states

It corresponds to the concurrent sum [in a vector-like parallel processing] of all elements of two respective columns of the matrices "r_k" and "J_{k+1}".

2) Parallelism in controls

Change columns by rows in the previous description.

3) Parallelism in states and controls

It corresponds to the concurrent sum [in a matrix-like parallel processing] of all elements (i,j).

Remark 1. If considering the computational task in terms of both the sum of two matrices and the subsequent optimization of each row, it results that only the first algorithm can be completely parallelized.

By denoting respectively $\underline{I}_k$ as the $n_s(k)$-dimensioned vector with elements $I_k(i)$, $\underline{r}_k^j$ and $\underline{J}_{k+1}^j$ as the jth columns of the matrices "r_k" and "J_{k+1}", we can rewrite the equation (3) in vector-form, as follows:

$$\underline{I}_k = \underline{Opt}_{j=1,\ldots,n_c(k)}[\underline{r}_k^j + \underline{J}_{k+1}^j]$$

Note that this algorithm merely generalizes the parallel DP algorithm considered in [2-5] for solving the classical search of an optimal path in a multistage graph. The only difference lies in the arguments of the transition function: $f_k(i,j)$ in Control versus $f_k(j)$ in the mentioned problem. The matrix "J_k" reduces in the later case to a vector column and the algorithm can be viewed as a product matrix-vector.

Remark 2. For the parallel algorithms 2) and 3), the optimization task can be partially parallelized by grouping recursively the results of the sums in a binary optimization tree then performing a divide and conquer approach.

Remark 3. The efficiency of these algorithms depends obviously on the relation between n_s and n_c.

4 Parallelism in stages

It is explicitely exhibited when considering another and probably the most ancient formulation of the optimality principle (Bernoulli ,1706) which states that all subsequences of an optimal trajectory are also optimal. The alternative form of the functional equations (2), can be expressed as:

$$J \quad = \quad \mathcal{J}(x_0, x_K) \quad = \quad Opt_x \sum_{\gamma=0}^{N-1} \mathcal{J}[x_{k(\gamma)}, x_{k(\gamma+1)}] \tag{4}$$

where:

- $x \in \{x_k(0), ..., x_k(N-1)\}$;

- $k(\gamma)$ means an increasing function in γ , with integer values, which determines what stages the decomposition of the horizon-time will be considered $[k(0) = 0, k(N) = K]$;

- $\mathcal{J}[x_{k(\gamma)}, x_{k(\gamma+1)}]$ denotes the optimal CTG from one state at time [stage] $k(\gamma)$ to one state at time $k(\gamma+1)$.

The possibility of a parallel computing of (4) issues from the property that all operands $\mathcal{J}$ in the sum are independent ones to the others. Further, the global solution can be performed by a divide and conquer technique.

According to the data ordering previously considered, we summarize as follows the parallel stages algorithm:

1) Compute in parallel all optimal subsequences and store the results in matrices $M_{\gamma,\gamma+1}$ defined by:

$$M_{\gamma,\gamma+1} \quad = \quad \begin{bmatrix} \mathcal{J}(1,1) & . & \mathcal{J}[1, n_s(k(\gamma+1))] \\ . & . & . \\ \mathcal{J}[n_s(k(\gamma)), 1] & . & \mathcal{J}[n_s(k(\gamma)), n_s(k(\gamma+1))] \end{bmatrix}$$

where $\mathcal{J}(i,j)$ roughly denotes the optimal CTG from the state with label i at step $k(\gamma)$ to the state j at step $k(\gamma+1)$.

2) Group the matrices in a binary tree and "multiply" them in a divide and conquer manner; the multiplication of the matrices

$$M_{\gamma,\gamma+2} = M_{\gamma,\gamma+1} \otimes M_{\gamma+1,\gamma+2}$$

being defined such that

$$M_{\gamma,\gamma+2}(i,j) = Opt_{l=1,\dots,n_s(k(\gamma+1))} \left[M_{\gamma,\gamma+1}(i,l) + M_{\gamma+1,\gamma+2}(l,j)\right]$$

$$i = 1,\dots,n_s(k(\gamma)) \quad and \quad j = 1,\dots,n_s(k(\gamma+2))$$

Remark 1. This algorithm is clearly well suited, within a compromize for N, if the number K of stages is large;

Remark 2. However substantial storage and computing requirements may be needed.

5 A state parallel algorithm implementation on a CRAY-2 and conclusions

One important feature of parallel algorithms is the machine dependency. Whence the interest of developing concrete applications. Indeed, little is known about the performance of standard schemes on existing parallel machines. Let us quote here the experiments by Ribeiro [3], Kindervater and Trienekens [4] with parallel DP algorithms for solving shortest paths problems.

With the support of the "Centre National d'Etudes Spatiales", a Fortran based program has been developed to solve on a parallel machine [presently a CRAY-2] the problem of optimizing the orbit transfer of geostationary satellites. The present state of this work is reported in [6]. Roughly speaking, one can say that the complexity of the problem [non differentiable objective function, inequality constraints on state and control variables, unconnected state space domain ...] prevents from using standard Nonlinear Programming methods. DP is a priori well-suited for this case and one way considered to circumvent the algorithm complexity is supercomputing. A modular software has been designed for adressing to a large class of optimal control problems. The exploitation of parallelism concerns as well the CTG optimization [a state parallel vector processing] as other functions like the data base generation, the computation of cost and transition functions, the multivariable interpolation procedures, the recovery of the optimal trajectory. Moreover, a partitioning of the vectors in the whole program enables multitasking on the four processors of the CRAY-2.

This program has been validated on a simplified problem formulation. The first performance we have evaluated is a speed-up of 40.058 [11.284 if vector processing only] which has been obtained for a quantization of variables implying the solution of 58394463 functional equations. A more detailed description of the program, the problem and the performance evaluation will be presented in a forthcoming Report.

References

[1] J. Casti, M. Richardson and R. Larson, *Dynamic Programming and Parallel Computers*, J.O.T.A., Vol. 12, no. 4, 1973.

[2] A. Bossavit, *Comment j'ai vectorisé certains de mes programmes*, Publication E.D.F. HI 4271-00, Paris, 1982.

[3] C. Ribeiro, *Performance evaluation of vector implementations of combinatorial algorithms*, Parallel Computing 1, 1984.

[4] G.A.P. Kindervater and H.W.J.M. Trienekens, *Experiments with parallel algorithms for combinatorial problems*, European J. of Oper. Res. (33),1988.

[5] G.J. Li and B.W. Wah, *Systolic processing for dynamic programming problems*, Proc. Int. Conf. on Parallel Processing, IEEE, 1985.

[6] J.M. Garcia, J. Dantas de Melo and J.L. Calvet, *Vectorisation et Parallelisation de l'Algorithme de Programmation Dynamique*, LAAS Report no. 88231, 1988.

Design Issues for Parallel Matrix Algorithms

Vladimir Cherkassky and Ross Smith

Department of Electrical Engineering
University of Minnesota
Minneapolis, MN 55455, USA

Abstract

It is well known that parallelism by itself does not lead to higher speeds. For example, communication or I/O overheads can lead to situations where adding processors actually increases total execution time. It is important, therefore, to understand and to classify properly the factors that affect the optimal design and implementation of parallel algorithms. In order to make our discussion meaningful and specific, we illustrate these factors and trade-offs for:

- a given class of problems (matrix algorithms)
- a particular architecture type (distributed-memory, message-passing MIMD hypercube)
- a particular machine (NCUBE)

We conclude that analytic performance modeling and optimal design of algorithms are usually possible only for simple matrix algorithms and under simplifying assumptions concerning implementation and machine dependent factors. It does not seem feasible or practical to develop exact (analytic) performance models for algorithms that take into account all problem-dependent, architecture-dependent and implementation (machine) - dependent factors. Therefore, for practical purposes, it is important to develop approximate performance models for specific parallel machines(s), and to verify them experimentally.

NATO ASI Series, Vol. F 70
Numerical Linear Algebra, Digital Signal Processing and Parallel Algorithms
Edited by G. H. Golub and P. Van Dooren

INDEX TERMS: communication overhead, hypercube, load balancing, mapping, NCUBE, parallel matrix algorithms, partitioning, performance evaluation.

1 Introduction

In this paper we are concerned with various issues that affect design and implementation of parallel algorithms. These include both theoretical issues such as the choice of an algorithm, problem partitioning, mapping, and more practical considerations of the specific parallel machine, e.g., communication overheads, I/O costs etc. More specifically, we seek to construct analytic models to evaluate the execution time for various problems (algorithms) taking into account all significant design factors as parameters (variables). We can then use problem execution time as an optimality criterion, so that an optimal algorithm design, or the optimal choice of design parameters, guarantees the minimum-time solution for a given problem on a given (real) parallel machine. We are not concerned with the asymptotic complexity or speedup that can be achieved simply by increasing the problem size and fixing the number of processors. Instead, we would like to quantify (or at least to understand) the relationships between problem size, number of processors, partitioning and mapping strategies, and details of specific machine implementation.

In the scientific community there is a wide range of opinions on the viability of massive parallelism. For example, a popular point of view states that good performance can be achieved simply by increasing the problem size, while a more negative view, Amdahl's law [1967], asserts that the serial part of an algorithm will dominate execution time for a large number of processors. Yet recently, researchers at Sandia National Laboratories have achieved an almost linear speedup for several important scientific applications [Gustafson et al. 1988] by using inherently parallel applications and, most importantly, by scaling the problem size in proportion to the number of processors. We believe this confusion of opinions is due to the fact that each point of view takes into account only a few relevant factors and ignores the rest.

2 Taxonomy of Factors for Algorithm Design

Developing and implementing of a parallel algorithm is a complex design problem which involves the choice of an algorithm based on the analysis of the inherent parallelism in the problem, problem partitioning into smaller subproblems, mapping subproblems onto processors, and algorithm realization on a specific machine. Therefore, efficient algorithm design is affected by a variety of *interrelated* factors that can be classified as:

1. *Problem* and *algorithm dependent* factors. These include:
problem type, problem size, the choice of a parallel algorithm and programming style/software overhead.

In this paper, we study parallel matrix algorithms. Efficient algorithm designs must consider several factors at once (multivariable optimization). For example, consider the matrix size and the number of processors for a given matrix algorithm. One extreme is mapping the whole matrix onto a single processor (fully sequential), the other is to map each element onto a different processor. The latter case achieves maximum parallelism, but also incurs maximum interprocessor communication overhead. Clearly, there is an optimal correspondence between the problem size and the number of processors. The choice of an algorithm is often based on purely scientific grounds (e.g. maximum speedup, minimum memory requirements). However, pragmatic considerations are equally important. For example, Moler [1985] focused on a LINPACK-style implementation of matrix algorithms on a hypercube. He stated the necessity of developing a single storage scheme for all algorithms, and proposed a common scheme to map matrix columns onto the processors. Finally, there are certain (usually small) software overheads dependent on how one actually writes the parallel code. For example, a parallel algorithm usually requires more indexing than the sequential one, leading to the so called 'software decomposition overhead' [Fox et al. 1987].

2. *Architecture dependent* factors. These include:

- multiprocessor system type, i.e., shared-memory (tightly coupled) or message-passing (loosely coupled) systems.

- Interconnection network topology, i.e. multistage interconnection network, pyramid, hypercube, regular mesh, etc. In this paper we use message-passing parallel machines with a hypercube interconnection (NCUBE).

- Number of processors, which can be fixed or variable. For hypercube computers, this number is variable since a large hypercube can be partitioned into several smaller ones.

- Problem partitioning and mapping. These factors are both problem and architecture-dependent, i.e., problem partitioning is directly related to the number of processors. Typical matrix partitioning strategies are: square partitioning, rectangular block partitioning, row (or column) partitioning.

- Importance of mapping for balanced algorithm design. A good mapping should distribute computational workload evenly among the processors. Since matrix algorithms usually have regular structure, static mapping is used to achieve balanced execution.

3. *Implementation machine-dependent* factors. These include:

- I/O and host bottleneck. In a typical hypercube multiprocessor the host, having a communication path to each of the nodes (processors), performs program development, data downloading, I/O control, etc. For some applications, the host and the host-to-node interconnection may become a bottleneck. Ni et al. [1987] discussed this problem for evaluating the sum of a large array of numbers so that each node in the hypercube calculates a partial sum and the host does the final summation. They concluded that the problem (array) must be partitioned properly, otherwise the host will become a bottleneck.

- Communication system (for message passing multiprocessors):

  - is communication concurrent with computation
  - full or half duplex mode
  - overhead of send/receive primitives

  - on-board vs. off-board communication costs. For example, for the NCUBE, up to 64 nodes fit on one board. Therefore, the performance estimates for more than 64 processors should reflect higher off-board communication costs.

- Processor system design:
  - existence of the specialized hardware (e.g., floating-point unit)
  - data-dependent timing for floating-point operations. For example, a floating point addition in the NCUBE processor is data-dependent, since the operands are shifted to line up their binary points prior to the actual addition. Also, the processor checks whether either operand is zero and the floating point operation takes less time if this is true.

- effects of massive parallelism. [Gustafson et al. 1988]. There are several new effects that become increasingly important as the number of processors grows large (i.e. more than a thousand). First, large ensembles of processors are likely to contain defective nodes which slow down their operation without causing incorrect results, (i.e., soft communication errors with successful retries, error corrections on memory, etc.). These defective nodes introduce a load imbalance which can reduce the efficiency of a (theoretically) perfectly balanced algorithm from 99% to 90% [Gustafson et al. 1988]. Second, variation caused by data-dependent timing for basic arithmetic results in a slight load imbalance and becomes visible on large applications.

3 Design of Matrix Algorithms: Matrix Multiplication

Much work on parallel matrix algorithms was done before parallel machines were readily available [Sameh and Kuck. 1978; Dekel et al. 1981]. These studies focused mainly on asymptotic estimates of the order of complexity of parallel algorithms and often made assumptions not currently true. Many of these theoretical estimates are not accurate for smaller matrices and/or fine grain parallel processors when communication overheads and details of machine implementation have a significant effect on the performance. We discuss next analytic and experimental performance modeling of a simple algorithm, matrix multiplication, in order to illustrate the interplay between the problem partitioning/mapping and the NCUBE communication model.

3.1 NCUBE experimental model

Computational model – The NCUBE is a commercial MIMD message passing parallel processor [Colley et al. 1986; NCUBE 1986]. A host computer is used to control the hypercube. An n-dimensional hypercube has 2^n processors, and n communication links per processor. While the largest NCUBE machine has 1024 processors, the machine we used has 64 processors. Each processor, or node, consists of a microprocessor and 128 Kbytes of local memory. The microprocessor is a custom built 32-bit chip with a floating point unit, a memory interface unit, and 11 bidirectional direct memory access (DMA) channels. Dunigan [1986] showed that the chip can execute up to 125 thousand floating point operations per second or 233 thousand integer instructions per second.

Communication model – to send a message, a processor sets up the DMA, which then runs independently sending messages serially at about 8 Mbps. The transfer of messages is accomplished with two subroutines: send and receive. For sending, the message is moved from the program's memory to the operating system's memory. Once the DMA is set up to send the message, the program can continue. For receiving, the program is interrupted to set up the DMA so that it can transfer the message to the operating system's memory. When the program is ready to receive the message it is moved from the operating system's memory to the program's memory.

These subroutines have three time components:

1. The time spent by the DMA transmitting a message to a neighboring node. We can ignore the DMA transmission time because the CPU is performing arithmetic while the DMA is transmitting its messages, and the CPU arithmetic operations will always take longer than the DMA transmission for each iteration of our matrix algorithm.

2. The time spent transferring a message to and from the operating system's buffer. We denote the time required to transfer one 32-bit word to or from the operating system's buffer as T_{mov}. Note that T_{mov} is typically small.

3. The overhead involved in invoking the send and receive subroutines and setting up the DMA. We denote the time required to set up a message for sending as T_{send} and for receiving as T_{rcv}. On the NCUBE, the overhead of invoking the send and receive subroutines is the largest part of the communication cost when sending a single 32-bit word.

Thus, by combining these components the time to send a message of w words is

$$T_{send} + w \times T_{mov} \qquad (1-a)$$

and the time to receive w words is

$$T_{rcv} + w \times T_{mov} \qquad (1-b)$$

Letting $T_{comm} = T_{send} + T_{rcv}$ and $T'_{mov} = 2T_{mov}$, an entire message takes $T_{comm} + w \times T'_{mov}$. From [Dunigan 1986], $T_{comm} = 440\mu s$ and $T'_{mov} = 10\mu s$. Let us briefly compare the NCUBE with the Cosmic Cube [Fox et al. 1987]. Since $T_{send} >> T_{mov}$ on the NCUBE, the order of complexity to send (or receive) a message of a few words is approximately $O(1)$. In other words, short messages (i.e. less than 20 words) take approximately the same amount of time to send. In contrast, for the Cosmic Cube, $T_{comm} = 80\mu s$ and $T'_{mov} = 40\mu s$. Thus, an $O(w)$ model is used to model communication overheads, i.e. the overhead of invoking the send and receive subroutines is ignored and communication time is proportional to the number of words, w. In addition to these communication differences, our experimental results show that the NCUBE can multiply four times as fast as the Cosmic cube, primarily because the node processor in the NCUBE is four times faster.

3.2 Analytic and Experimental Modeling

The problem graph for matrix algorithms discussed in this paper is a toroidal mesh, which corresponds to square matrix partitioning. An m by m torus is embedded onto the hypercube, with processor P_{ij} located in row i and column j. Each processor runs an identical algorithm. The matrices are N by N dense, square matrices which are partitioned into n by n submatrices, where $n = N/m$. Given matrix **A**, Figure 1 shows a partitioning for $m = 2$, where $\mathbf{A}_{ij}$ refers to the submatrix in the *ith* row and *jth* column. Our analysis is limited to square matrices and submatrices to simplify the presentation. Thirty-two bit floating point numbers are used. The time spent loading and unloading the matrices is ignored. This is a valid assumption in many applications where matrices come from previous computation in the hypercube.

$\mathbf{A}_{11}$	$\mathbf{A}_{12}$
$\mathbf{A}_{21}$	$\mathbf{A}_{22}$

Figure 1. Partitioning of Matrix **A** for $m = 2$.

We now examine in detail matrix multiplication: $\mathbf{C} = \mathbf{A}\times\mathbf{B}$. We use a modification of Cannon's algorithm described by Dekel et al. [1984] where each node holds a submatrix of each matrix **A**, **B**, and **C**, instead of holding an element of each matrix. For example, the initial mapping of the matrices onto a 4 by 4 torus is shown in Figure 2. This skewed mapping of **A** and **B** requires no more time to load than a straightforward mapping, yet during multiplication the communication between nodes is minimized, since **A** and **B** only need to be shifted circularly one position for each iteration of the algorithm. Once matrices A and B are loaded, the algorithm in each node proceeds as shown in Figure 3. A node receives a message and passes it along before performing multiplication. Hence, calculations and message passing are overlapped. Noted on the right in Figure 3 is the time to execute each step. The times T_{send}, T_{rcv} and T_{mov} were explained in Section 3.1. T_{calc_1} is the time required to multiply and add two 32-bit floating point numbers.

$\mathbf{A}_{11}$ $\mathbf{B}_{11}$ $\mathbf{C}_{11}$	$\mathbf{A}_{12}$ $\mathbf{B}_{22}$ $\mathbf{C}_{12}$	$\mathbf{A}_{13}$ $\mathbf{B}_{33}$ $\mathbf{C}_{13}$	$\mathbf{A}_{14}$ $\mathbf{B}_{44}$ $\mathbf{C}_{14}$
$\mathbf{A}_{22}$ $\mathbf{B}_{21}$ $\mathbf{C}_{21}$	$\mathbf{A}_{23}$ $\mathbf{B}_{32}$ $\mathbf{C}_{22}$	$\mathbf{A}_{24}$ $\mathbf{B}_{43}$ $\mathbf{C}_{23}$	$\mathbf{A}_{21}$ $\mathbf{B}_{14}$ $\mathbf{C}_{24}$
$\mathbf{A}_{33}$ $\mathbf{B}_{31}$ $\mathbf{C}_{31}$	$\mathbf{A}_{34}$ $\mathbf{B}_{42}$ $\mathbf{C}_{32}$	$\mathbf{A}_{31}$ $\mathbf{B}_{13}$ $\mathbf{C}_{33}$	$\mathbf{A}_{32}$ $\mathbf{B}_{24}$ $\mathbf{C}_{34}$
$\mathbf{A}_{44}$ $\mathbf{B}_{41}$ $\mathbf{C}_{41}$	$\mathbf{A}_{41}$ $\mathbf{B}_{12}$ $\mathbf{C}_{42}$	$\mathbf{A}_{42}$ $\mathbf{B}_{23}$ $\mathbf{C}_{43}$	$\mathbf{A}_{43}$ $\mathbf{B}_{34}$ $\mathbf{C}_{44}$

Figure 2. The initial mapping of the submatrices onto a 4 by 4 mesh.

Iteration Steps	Time fo step
for $k = 1$ to $m - 1$	
1. *Send* **A** *to the east*	$T_1 = T_{send} + T_{mov} \times n^2$
2. *Send* **B** *to the south*	$T_2 = T_{send} + T_{mov} \times n^2$
3. $\mathbf{C} = \mathbf{C} + \mathbf{A} \times B$	$T_3 = T_{calc_1} \times n^3$
4. *Receive* **A** *from the west*	$T_4 = T_{rcv} + T_{mov} \times n^2$
5. *Receive* **B** *from the north*	$T_5 = T_{rcv} + T_{mov} \times n^2$
next	
6. $\mathbf{C} = \mathbf{C} + \mathbf{A} \times \mathbf{B}$	$T_6 = T_{calc_1} \times n^3$

Figure 3. Matrix Multiplication Algorithm.

The time to execute this algorithm is found by summing these six time components.

$$\begin{aligned} T = (m-1) \quad \times & \quad \{T_{send} + T_{mov} \times n^2 + T_{send} + T_{mov} \times n^2 \\ + & \quad T_{rcv} + T_{mov} \times n^2 + T_{rcv} + T_{mov} \times n^2 \\ + & \quad T_{calc_1} \times n^3\} + T_{calc_1} \times n^3 \end{aligned}$$

We can simplify this equation by letting $T_{comm_1} = T_{send} + T_{rcv}$

$$T = 2(m-1)(T_{comm_1} + (2T_{mov})n^2) + (m \times T_{calc_1})n^3 \tag{2}$$

Because matrix multiplication is a simple algorithm, the analytic model (2) is quite accurate and it is possible to obtain parameter values by timing the program execution. The parameter values were calculated using a few experimental times by applying the least square curve fitting technique. Using $T_{comm_1} = 4613$, $T_{calc_1} = 210$, and $T_{mov} = 17$ (values are in clock cycles) our model accurately predicts (within 1% for all points) execution time, T, for arbitrary values of m and n. These values are similar to those found by Dunigan [1986] and Gustafson and Montry [1988].

Figure 4 shows the execution time on different sized grids versus the size of the matrix. The processor clock cycle is used as a time unit so that all experimental results are independent of the actual processor clock rate. The execution time in seconds is found by dividing the number of cycles by the clock rate, where a K refers to 1024 cycles. The points marked on the graphs indicate experimental times while the continuous lines are derived from the analytical expression (2).

For a given N, the optimal submatrix size, n, can be found from Figure 4 by finding the optimal grid size, m, and then calculating $n = N/m$. The optimal submatrix size can also be found analytically by taking the derivative of T with respect to n and setting it to zero.

$$\frac{dT}{dn} = (2T_{calc_1} - \frac{8}{N} \times T_{mov})n^3 + (4T_{mov})n^2 - 2T_{comm_1} = 0$$

Since $T_{calc_1} >> T_{mov}$ for the NCUBE, n can be found approximately from

$$n \approx \left[\frac{T_{comm_1}}{T_{calc_1}}\right]^{\frac{2}{3}} \tag{3}$$

Note that the optimal size of the submatrix does not depend on the original matrix size. For the NCUBE $n = 2.80$. Thus, the fastest execution time is guaranteed by choosing the n closest to 2.80. Obviously, the best choice is $n = 3$. However, when N is not divisible by 3 the choices are $n = 2$ or $n = 4$. From (2) it can be shown that the fastest execution time is obtained when $n = 2$.

Figure 4. Execution Time vs. Size of Large Matrix for Matrix Multiplication

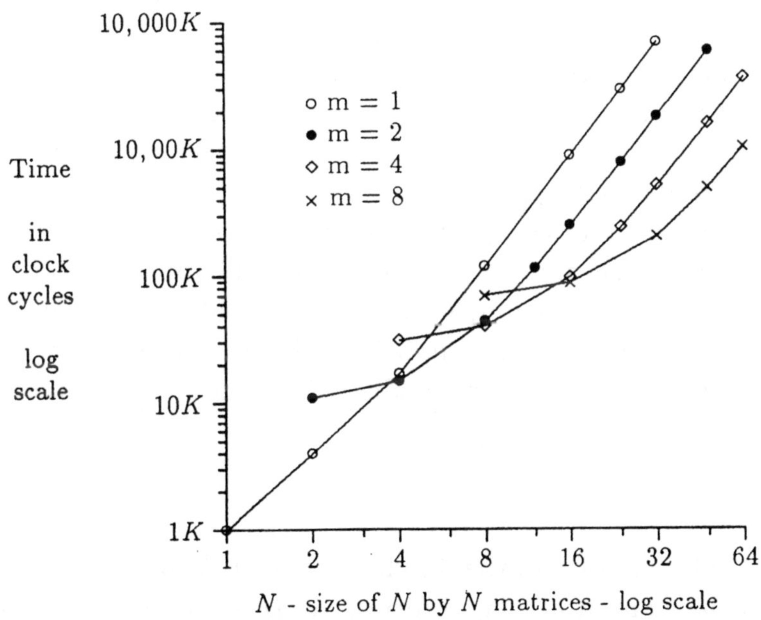

4 Conclusions

We discussed efficient design and implementation of matrix algorithms on the NCUBE parallel machine. Unlike many previous theoretical studies which focus on asymptotic performance of parallel processors, our analytic modeling approach also takes into account communication overheads and the details of the machine implementation. For matrix multiplication, it is shown that the optimal submatrix size assigned to each node of the NCUBE should be typically small to guarantee minimum execution time. However, if the time to set up the programs and load and unload the matrices is accounted for, then the optimal size of a submatrix may change. Comparison of our results on matrix multiplication with Fox et al. [1987] leads to the conclusion that details of machine implementation can greatly affect communication overheads and overall performance. The matrix multiplication algorithm was very straightforward, lending itself well to a detailed analysis. However, the same approach can be used to implement more complex algorithms. For example, for the balanced Gaussian elimination algorithm, Cherkassky and Smith [1988] have recently developed an (approximate) analytic model that takes into account the problem size, hypercube size and the NCUBE communication system parameters. Approximations were necessary since Gaussian elimination exhibits different synchronization dependencies for different-sized hypercubes.

Our experience with more complex algorithms (e.g. Gaussian elimination, Givens rotations) suggests that the best practical approach is to combine analytic performance modeling with experimental verification. Such an approach allows one to capture important design factors that significantly affect the performance of parallel algorithms on real machines.

References

[1] Amdahl, G. 1967. Validity of the Single Processor Approach to Achieving Large-Scale Computer Capabilities. *AFIPS Conference Proceedings*,30, 483-485.

[2] Cherkassky, V., and Smith, R. 1988. Efficient Mapping and Implementation of Matrix Algorithms on a Hypercube. To appear in *The Journal of Supercomputing.*

[3] Colley, S., Hayes, J., Mudge, T., Palmer, J., and Stout, Q. 1986. Architecture of a Hypercube Supercomputer. In *Proc. ICPP*, 653-660.

[4] Dekel, E., Nassimi, E., and Sahni, S. 1981. Parallel Matrix and Graph Algorithms. *SIAM J. of Computing*, Vol. 10, no. 4, (Nov.) 657-675.

[5] Dunigan, T. H. 1986. Hypercube Performance. In *Proceedings of the Second Conference on Hypercube Multiprocessors*, Knoxville, (August) 178-192.

[6] Fox, G. C., Otto, S. W., and Hey, A. J. 1987. Matrix Algorithms on a Hypercube I: Matrix Multiplication. *Parallel Computing*, 4, 17-31.

[7] Gustafson, J. Montry, G., and Benner, R. 1988. Development of Parallel Methods for a 1024-Processor Hypercube. *SIAM J. on Scientific and Statistical Computing*, Vol. 9, No. 4, (July).

[8] Moler, C. 1985. Matrix Computation on a Distributed Memory Multiprocessor. In *Proceedings of the First Conference on Hypercube Multiprocessors*, Knoxville, (August) 181-195.

[9] NCUBE Corp. 1986. *NCUBE Handbook*, version 1.0, Beaverton, Ore., (April).

[10] Ni, L. M., King, C. T., and Prins, P. 1987. Parallel Algorithm Design Considerations for Hypercube Multiprocessors. In *Proc. ICPP (Aug.)*, 717-720.

[11] Sameh, A. H., and Kuck, D. J. 1978. On Stable Parallel Linear System Solvers. *J. of the ACM*, Vol. 25, No. 1, (January) 81-91.

Fast Computation of a Restricted Subset of Eigenpairs of a Varying Hermitian Matrix

Pierre Comon

Thomson Sintra, ASM, BP 53
F-06801 Cagnes sur mer, France

Abstract

Standard algorithms require an order of dm^2 flops to compute d eigenpairs of a $m \times m$ semi-definite positive hermitian matrix, R. A larger amount of computations (of same order) is required to find d left singular pairs of a square matrix, D, such that $R = DD^H$; this is known to be numerically more stable. Note that D is not unique, the best choice in the sequel turns out not to be the Cholesky factor of R. The algorithm proposed in this contribution will compute d left singular pairs of D in a faster way, such that it requires an order of $\alpha m d^2$ flops. This algorithm takes advantage of the fact that the matrix considered is nothing else but a matrix whose d eigenpairs are known, perturbed by a rank-one modification. It is pointed out in the sequel why the technique of Bunch et al [1] is not relevant for $O(md^2)$ algorithms, neither for square-root implementations.

1 Introduction

The approximation of a matrix by another of lower rank is desired in many situations such as data analysis, seismics, image processing and coding, pattern recognition, system theory, mechanics, spectral analysis, sonar, radar, geophysics... In some of these applications, for instance active sonar, the

NATO ASI Series, Vol. F 70
Numerical Linear Algebra, Digital Signal Processing and Parallel Algorithms
Edited by G. H. Golub and P. Van Dooren

matrix considered, $R(t)$, is varying with time, and it is necessary to track the variations of its reduced rank approximate (say rank d) in real time [3] (note that in other similar applications, the smallest eigenpair is desired [11] instead of the largest ones.) Therefore, it is suitable to use iterative algorithms that are able to:

1. Compute only a restricted subset of d eigenpairs, in order to decrease the complexity.

2. Use the estimate of the previous time step $(t-1)$ as initial guess at step t.

3. If possible, compute the d left singular pairs of a matrix D instead, such that $DD^H = R$, in order to improve the rounding errors.

4. If $d = O(1)$, have a global complexity of $O(m)$.

Consider the following sequence of rank-one random matrices, whose expectation is almost surely positive definite, and slowly varying with time t:

$$A(t) = r(t)r(t)^H,\ \overline{R}(t) = E\{r(t)r(t)^H\}$$

where $r(t)$ is an observed m-variate random process. In practice, the number of samples is very limited, especially in non stationary environments, so that the estimation of the local covariance matrix $\overline{R}(t)$ cannot be very much averaged. This yields the fact that the performance of the processing in physical applications, in which these eigenpairs are required, will be eventually limited by the memory length and not by the roughness of the calculations. Thus, the aim of the paper is to provide an adaptive method for computing approximately d eigenpairs of a $m \times m$ hermitian random matrix varying with time. In this problem, we can have $d << m$ or $d = O(m)$ as well, but the smaller d, the more attractive the algorithm compared to standard ones.

Since the so-called "Square-Root" algorithms have received attention because of their ability to always find positive eigenvalues to semi-definite positive matrices (in other words they are numerically more stable), we shall focus on the computation of the left singular pairs of a square matrix, D, such that $R = DD^H$. Matrix D can be the Cholesky factor of R, but this is not necessarily the best solution to assume, regarding the computational burden.

As shown in the literature, it is rather easy to satisfy condition 2) in iterative algorithms, but standard algorithms from EISPACK, such as the symmetric QR or the Golub-Reinsch [7] algorithm do not satisfy condition 1). The method of Bunch et al [1] does not satisfy condition 1). either. On the other hand, block-Lanczos methods, including a block-size reduction as soon as eigenvalues have converged, satisfy conditions 1) to 3), but require $O(m^2 d)$ operations, which is too much according to goal 4) [2] [5] [6]. One of the heavy tasks is the explicitation of matrices R or D, and it should be also avoided. Finally, in "square root" algorithms, it is less obvious to take into account the fact that the matrix $R(t)$ is modified by a rank-one matrix, as show the contents of this paper.

2 The context of adaptive computation

2.1 Observation model

For practical value, the sequence of matrices $\overline{R}(t)$ is replaced by a sequence of estimates defined recursively by:

$$R(t) = b(t)R(t-1) + a(t)r(t)r(t)^H \tag{1}$$

where $b(t)$ and $a(t)$ are positive reals, both smaller than one. Generally, $b(t) = 1 - a(t)$ in order to have consistent estimates, and also to avoid divergence. Though the eigenpairs of $R(t)$ are the quantities of interest, the matrix $R(t)$ itself is rarely available, so that its possible explicitation (requiring usually more than m^3 flops) must be taken into account in the global complexity.

2.2 The Technique of Bunch et al

The algorithm proposed by Bunch et al in [1] allows to compute d eigenpairs of a matrix $R(t-1)$ added by a rank-one perturbation, $a(t)r(t)r(t)^H$. In this algorithm, all the eigenpairs of $R(t-1)$ need to be known, since the first step of the algorithm is to diagonalize the matrix $R(t-1)$. If we used this algorithm with the purpose of computing d eigenpairs of $R(t)$ for each value of t, we would be forced to compute all of them at each time step t. In such a case, the technique does not deserve much attention any more. Moreover, it applies to the computation of eigenpairs of a given real symmetric (or hermitian complex) matrix, and has not the advantages displayed by the square-root algorithms. As far as we know, it can be roughly adapted to compute singular triplets of a matrix modified by the addition of a column (its square is perturbed by a rank-one matrix) [3], but the rounding errors are not better than in the standard version of [1].

3 A square-root Subspace Iteration with Ritz acceleration (SIR)

Let $L(t)$ denote the Cholesky factor of $R(t)$, as $R(t) = L(t)L(t)^H$. It is suitable to utilize the subspace iteration in the adaptive context for two reasons. First, the choice of the estimated dominant eigenvectors $U(t-1)$ as starting guess at step t is evident; second, the eigenpairs vary sufficiently slowly to allow convergence in one step, especially if we use the Ritz acceleration. We shall describe here a square-root version of the Subspace Iteration with Ritz acceleration (SIR), introduced in [2]. The adaptive algorithm may be written as:

main

Assume $L(t)$ is $m \times n$ (possibly $m = n$, in particular if L is the Cholesky factor of R) start with an arbitrary $m \times d$ matrix $U(0)$,
and orthonormalize it.
As time increases, $t = 1, \ldots$
Update $L(t)$; $[U(t), \Sigma(t)] = \text{sir}\ (U(t-1), L(t))$;

function [Unew, Σ] = **sir**(Uold, Lnew)
Wleft = Lnew^H*Uold;
Compute the $n \times d$ matrix X in the SVD of Wleft : Wleft $= X\Sigma Y^H$;
Set Vnew $= X$.
Wright = Lnew*Vnew;
Compute the $m \times d$ matrix X in the SVD of Wright : Wright $= X\Sigma Y^H$;

Set Unew $= X$.

Note that in the iteration, the singular values are updated twice. Though it may be not obvious at first glance, the obtained outputs have taken advantage of the Ritz acceleration. In fact, it can be seen clearly that if $LV = X\Sigma Y^H$, then $\Sigma = X^H LVY$ are the d dominant Ritz values, and X the corresponding left Ritz vectors. When $n = m$, this algorithm requires about $2dm^2 + 6d^2m + O(md)$ - $O(d^3)$ flops, plus $O(2m^2)$ flops to update the Cholesky factor $L(t)$ [3].

4 A fast subspace iteration, OK

The previous algorithm satisfies the condition 1) through 3). A faster algorithm has been proposed by [9], for computing d eigenpairs of a matrix slowly varying with time, $R(t)$. This algorithm satisfies conditions 1), 2) and 4); its convergence has been proved in the almost sure sense in [10]. The purpose of the next section, and eventually of the whole paper, is to find a square-root version to it. Suppose that, at step t, we have:

$$R(t)\, v_i(t) \approx v_t(t)\, \lambda_i(t), \quad 1 \leq i \leq d .$$

Then, from (1), the following matrix-vector product can be approximately computed as:

$$R(t+1)\, v_i(t) \approx b(t+1)\, \lambda_i(t)[v_i(t) + \beta_i(t+1)\, r(t+1)] ,$$

where

$$\beta_i(t+1) = a(t+1)\, r(t+1)^H\, v_i(t)/b(t+1)\, \lambda_i(t) .$$

This product requires $O(3m)$ flops whereas, in general, it requires $O(m^2)$ flops. In [9], $\lambda_i(t+1)$ is approximated by $v_i(t)^H R(t+1) v_i(t)$. The reason to that is that further simplifications occur, yielding the subroutine below, to be run at each time update. The following function gives the new d dominant eigenpairs in function of the old ones, and requires an order of $5m$ flops ($4m$ if the Gram-Schmidt procedure is utilized instead of Householder QR):

function $[U,\Lambda] = \mathbf{ok}(U(o),\Lambda(o),a,b,r)$ **Complexity** $\pm O(d)$

For $i = 1$ to d

$\beta_i = a\, r^H u_i(o)/b\lambda_i(o);\ w_i = u_i(o) + \beta_i r;$ $3md$

$\lambda_i = b\,\lambda_i(o) + a|r^H u_i(o)|^2.$

$U = \text{orth}\{w_j;\ 1 \le j \le d\}$, via economic QR factorization. $2md^2 - 2d^3/3$

5 Fast square-root algorithms, FASIR

In this section, we give the chronological stages in the design of a fast square-root algorithm, FASIR, namely an algorithm having an $O(m)$ complexity. Now, conditions 1) to 4) are thus all satisfied. FASIR stands for FAst Subspace Iteration with Ritz acceleration; the algorithms of the FASIR family are approximations to the SIR algorithm previously described. In SIR, the dominant tasks were (i) the updating of $L(t)$ (ii) the computation of the matrix-vector products $L(t)^H U(t-1)$ and $L(t)V(t)$, both of order m^2. As explained in the sequel, it is possible to approximate these two tasks at once and get a solution within $[3md^2 + 3md + 5d^3/3] + O(d^2)$ flops. It may be viewed as a square root version of the fast algorithm proposed by Karhunen [9] [10], and includes also the Ritz acceleration.

Suppose $R(t) = D(t)D(t)^H$, where $D(t)$ is a square $m \times m$ matrix (not necessarily triangular); then the approximation is based on the following:

$$D(t-1)^H U(t-1) \approx V(t-1)\,\Sigma(t-1) \text{ and } D(t-1)\,V(t) \approx U(t-1)\,\Sigma(t) \quad (2)$$

where $U(t-1)$ and $V(t-1)$ denote the matrices formed by the d left and right estimated singular vectors of $D(t-1)$ respectively, and $\Sigma(t-1)$ the $d \times d$ matrix of corresponding estimated singular values. The consequences of this approximation are (i) it is very fast to compute approximatively the above mentioned matrix-vector products without updating $D(t)$ explicitely (ii) if the approximation is well satisfied at step $t-1$, then it will be also well satisfied at step t if the perturbation is small enough (iii) if the approximation is not well satisfied for some t, it may take longer to catch up the true eigenpairs of $R(t)$. Convergence has not been proved yet, but it has been proved for the algorithm of Karhunen [10] which utilizes similar approximations.

For the sake of compactness in the description of the matrices involved in the algorithm, semicolumns denote carriage returns (i.e. next row) and periods denote spaces (i.e. next column). Moreover, $A(i:j,:)$ denotes the submatrix of A formed by its rows i through j; similarly, $A(:,i:j)$ is the matrix made of A's columns i to j. These are standard notations in the *Matlab* package. The matrix $D(t)$ is defined (implicitely) by:

$$D(t) = \text{the } m \text{ first columns of } \{[\sqrt{b}(t)\, D(t-1), \quad \sqrt{a}(t)\, r(t)]Q^H\} \qquad (3)$$

where Q is a unitary $m+1$ by $m+1$ matrix, choosen such that it cancels the last column of $F^H = u(t-1)^H\ [\sqrt{b}(t)\, D(t-1),\ \sqrt{a}(t)\, r(t)]$. Clearly, it suffices that Q modifies only the $d+1$ last rows of F such that they form a $d+1$ by d upper triangular matrix; this requires $O(2d^3/3)$ multiplications [4]. As detailed in the algorithm FASIR below, the product $D(t)^H u(t-1)$ can now be run within $O(md^2)$ multiplications with the help of (2). The new singular values $\Sigma(t)$ are then estimated in the same manner as in SIR. Next, it is more complicated to compute the product $D(t)v(t)$. First, note that Q is made of two diagonal blocks: One is the $m-d$ by $m-d$ identity matrix, and the other is a unitary $d+1$ by $d+1$ matrix, Qf. Define Qt by the first m rows of Q, $Qt \triangleq Q(1:m,:)$. Then, $D(t) = [\sqrt{b}(t)D(t-1),$ $\sqrt{a}(t)r(t)]Qt^H$. Furthermore, Qt is made of a first m by m block which can be written as: $Qt(:,1:m) = I(m,m) + [0\ 0;\ 0\ g]$, where g is the d by d block $g \triangleq Qf(1:d,\ 1:d)$ - $I(d,d)$, and a second block, m by 1, $Qt(:,m+1) \triangleq [0;\ Qf(:,m+1)]$. These remarks allow to compute the product $D(t)v(t)$ in function of $D(t-1)v(t)$, so that (2) may be used; the other remaining matrix-vector products need only $O(md^2)$ multiplications. The complete algorithm can be summarized as follows:

main

$U(0)$ arbitrary $m \times d$; $U(0)^H U(0) = I$; $D(0) = [r(1-m),\ldots,r(0)]/\sqrt{m}$;
$V(0) = \text{orth}\{D(0)^H U(0)\}$;
Ddold = the d last columns of $D(0)$;
For $t = 0, 1, 2, \ldots$

  $[U(t),\ \Sigma(t),\ V(t),\ Ddnew] = \text{fasir1}(U(t-1),\ \Sigma(t-1),\ V(t-1),$
  $\sqrt{a}(t),\ \sqrt{b}(t),\ Ddold,\ r(t))$
  Ddold = Ddnew;

	Complexity $\pm$ O(d²)
function $[U, \Sigma, V, Ddnew]$ = **fasir1**(Uold, Σold, Vold, ra, rb, Ddold, r)	
Compute $F = [rb\, Vold\, \Sigma old;\ ra\ r^H\ Uold]$, F is $(m+1) \times d$	$2md$
Find the $(d+1) \times (d+1)$ matrix Qf which triangularizes the south $(d+1) \times d$ block of F (the last row of F is cancelled), and explicit Qf	$2d^3/3$
Wleft = The $m \times d$ upper triangular block of Diag $\{1, Qf\}F$;	
Compute the $n \times d$ matrix X in the SVD of Wleft: Wleft $= X\Sigma Y^H$;	$3md^2 - d^3/3$
Set $V = X$.	
$g = Qf(1:d, 1:d)$ -$I(d,d)$; $bD = rb\ Ddold$	
$T = bD\ g + ra\ r\ Qf(d+1, 1:d)$;	$md + md^2$
Wright = rb Uold $\Sigma + T\ V(m-d+1:m, :)$;	$2md$
Compute the $m \times d$ matrix X in the SVD of Wright:	
Wright $= X\Sigma Y^H$; Set $U = X$;	$3md^2 - d^3/3$
Ddnew $= bD + T$;	
return	

According to the complexities given in *LINPACK*'s guide [4], the total number of multiplications required in this algorithm at each time step is $[7md^2 + 5md] + O(d^2)$, viz, of order $11m$. It turns out, after a more careful study, that the $m - d$ first columns of D tend exponentially to zero when time tends to infinity. Thus, there exists a simpler and more economic algorithm, which coincides with the above when the steady-state is reached; here is its matlab code:

	Complexity $\pm$ O(d)
function $[U, \Sigma]$ = **fasir2**(Uo, Σo, ra, rb, r)	
$[m, d] = \text{size}(Uo)$; $rauo = ra^* uo$; $rbso = rb^* \Sigma o$;	md
$wu = [rbso';\ x'^* rauo]$;	md
$[V, \Sigma, toto] = svd(wu, 0)$;	$2d^3 + O(d^2)$
$rbaso = rb/ra^* \Sigma o$; $baus = Uo^* rbaso$;	md
$Wv = [baus,\ x]$;	
$[U,\ \Sigma,\ titi] = svd(Wv, 0)$;	$3md^2 - d^3/3$
$\Sigma = ra^* \Sigma(1:d,\ 1:d)$; $U = U(:, 1:d)$;	

Now, this simplified version of *FASIR* requires only an order of $6m$ multiplications.

6 Equivalence to a simpler algorithm

In this section, we show why the assumptions (2) are equivalent to looking for the eigenpairs of another matrix, $\tilde{R}(t)$. Consider the approximations:

(a) $R(t) \approx \tilde{R}(t) \triangleq b(t)\, U(t-1)\Sigma(t-1)^2 U(t-1)^H + a(t) r(t) r(t)^H$.

(b) $R(t-1)\, U(t-1) \approx U(t-1)\, \Sigma(t-1)^2$.

If the subspace iteration is utilized, the matrix $R(t)$ is used only through matrix-vector products, so that assumption (a) is equivalent to:

$$R(t)\, U(t-1) \approx b(t)\, U(t-1)\Sigma(t-1)^2 + a(t) r(t) r(t)^H U(t-1).$$

From definition (1), it is clear that this relation is nothing else but (b). This shows the equivalence of assumptions (a) and (b) if we restrict ourselves to the use of the subspace iteration. This remark exhibits the existence of an obvious square-root algorithm, able to compute exactly the left singular pairs of a matrix $\widetilde{D}(t)$, where $\widetilde{D}(t)\widetilde{D}(t)^H = \tilde{R}(t)$. In fact, defining $\widetilde{D}(t) = [\sqrt{b}(t) U(t-1)\Sigma(t-1),\ \sqrt{a}(t) r(t)]$ amounts to the algorithm:

Complexity ± O(d)

function $[U, \Sigma] =$ **fasir3**$(Uo,\ \Sigma o,\ ra,\ rb,\ r)$	
$rab = ra/rb;\ w = [Uo^* \Sigma o,\ rab^* r];$	md
$[U, \Sigma, titi] = svd(w, 0);$	$3md^2 - d^3/3$
$\Sigma = rb^* \Sigma;$	

This last algorithm performs twice as slow as *FASIR2* and better than *OK*, but requires only an order of $4m$ multiplies to compute approximately $O(1)$ dominant eigenpairs of $R(t)$ in a recursive and square-root fashion.

References

[1] J.R. Bunch, C.P. Nielsen and D.C. Sorenson, "Rank One Modification of the Symmetric Eigenproblem", *Numerische Math.*, vol. 31, 1978, 31-48.

[2] P. Comon, "Adaptive Computation of a Few Extreme Eigenpairs of a Positive Definite Hermitian Matrix", *Proceedings of Eusipco 88*, Sept. 5-8, Grenoble, France.

[3] P. Comon and G.H. Golub, "Tracking of a Few Extreme Singular Values and Vectors in Signal Processing", to be submitted.

[4] J.J. Dongara, C.B. Moler, J.R. Bunch and G.W. Stewart, *LINPACK User's Guide*, SIAM, Philadelphia, 1979.

[5] G.H. Golub and W. Kahan, "Calculating the Singular Values and Pseudo Inverse of a Matrix", *SIAM Journal Num. Anal.*, ser. B, vol. 2, no 2, 1965, 205-224.

[6] G.H. Golub, F.T. Luk and M.L. Overton, "A Block Lanczos Method for Computing the Singular Values and Corresponding Singular Vectors of a Matrix", *ACM Trans. on Mathematical Software*, vol 7, no 2, june 1981, 149-169.

[7] G.H. Golub and C. Reinsch, "Singular Value Decomposition and Least Squares Solutions", *Numerische Math.*, 1970, vol. 14, 403-420, or *Handbook of Automatic Computation*, 1971, Wilkinson and Reinsch editors, Springer Verlag.

[8] G.H. Golub and C.F. Van Loan, *Matrix Computations*, 1983, Hopkins.

[9] J. Karhunen, "Adaptive Algorithm for Estimating Eigenvectors of Correlation Type Matrices", *IEEE International Conference on ASSP 84*, San Diego, 1984.

[10] E. Oja and J. Karhunen, "On Stochastic Approximation of Eigenvectors and Eigenvalues of the Expectation of a Random Matrix", *Journal of Mathematical Analysis and Applications*, vol. 106, no 1, Feb 1985, 69-84.

[11] V.U. Reddy, B. Egardt and T. Kailath, "Least Squares Type Algorithm for Adaptive Implementation of Pisarenko's Harmonic Retrieval Method", *IEEE Trans. on ASSP*, vol 30, no 3, june 1982, 399-405.

[12] D.C. Sorenson, "Updating techniques in Parallel Computation", this volume.

Parallelization of Toeplitz Solvers

E. de Doncker and J. Kapenga

Computer Science Department
Western Michigan University
Kalamazoo MI 49008, USA

Abstract

We shall describe parallel algorithms for solving Toeplitz and block Toeplitz systems, based on the Levinson type methods implemented in serial in the Toeplitz Package (by Arushanian et al. [1]). A matrix A, with blocks $A_{i,j}$, has block Toeplitz structure if the blocks satisfy $A_{i,j} = A_{j-i}$ for all i, j such that $0 \leq i, j \leq n - 1$ and each block is a general matrix. The implementation and performance of these algorithms on MIMD shared memory machines will be discussed. It will be shown that the contribution of the inner product calculations to the time complexity is negligible for typical applications.

1 Introduction

Toeplitz and block Toeplitz matrices play an important role in such areas as time series analysis and the numerical solution of integral equations with difference kernels. It is our goal to deal efficiently with the numerical solution of linear systems involving very large matrices with Toeplitz structures on MIMD shared memory machines. In this situation the order of the systems considered is typically much larger than the number of processors available.

Consider the system

$$Ax = b \tag{1}$$

NATO ASI Series, Vol. F 70
Numerical Linear Algebra, Digital Signal Processing and Parallel Algorithms
Edited by G. H. Golub and P. Van Dooren

where

$$A = \begin{bmatrix} A_0 & A_1 & A_2 & \cdots & A_{n-1} \\ A_{-1} & A_0 & A_1 & \cdots & A_{n-2} \\ \vdots & \vdots & \vdots & & \vdots \\ A_{-n+1} & A_{-n+2} & A_{-n+3} & \cdots & A_0 \end{bmatrix} \tag{2}$$

is a block Toeplitz matrix with general blocks A_i, $0 \leq i \leq n-1$ each of order $m \geq 1$. In the case where $m = 1$, A will be called Toeplitz, and the blocks A_i will represent scalars a_i.

The parallel Toeplitz solver will be described in section 2. We shall cover the block Toeplitz case in section 3. Timing results will be given in section 4, based on the solution of systems required in applications involving one and two-dimensional singular integral equations with difference kernels.

2 The Toeplitz solver

The method is based on a Levinson type algorithm [10],[11],[12],[13],[14], with a computational complexity of $O(n^2)$ (for $m = 1$). A serial version is implemented as algorithm $TSLD$ in the Toeplitz Package [1],[19]. The algorithm comprises a recursion for $k = 0, 1, \ldots, n-1$. Using the notation of [1], at step k a system

$$C_k y_k = d_k \tag{3}$$

is solved, where C_k is the leading principal submatrix of order $k+1$ of A and $d_k = [b_0, b_1, ..., b_k]^T$.

At step 0, $y_0 = b_0/a_0$. At step k, the vector y_k (of length $k+1$) is derived from y_{k-1} by

$$y_k = \begin{bmatrix} y_{k-1} \\ 0 \end{bmatrix} + z_k, \tag{4}$$

where z_k is the solution of the system

$$C_k z_k = \begin{bmatrix} 0 \\ f_{k,k} \end{bmatrix}, \tag{5}$$

and $f_{k,k}$ is a scalar defined in Figure 1. That is, z_k corresponds to the last column of the inverse matrix ${C_k}^{-1}$, to within a factor $f_{k,k}$. The solution to the last system is the principal computation performed in each recurrent

Figure 1: The Recursive Parallel Toeplitz Algorithm

Step 0: Initialize;
for $k = 1$ to $n-1$ do
    Step k: Phase 1: $f_1 = 0; f_2 = 0; f_{k,k} = b_k;$

    Phase 2: for $l = 1$ to k do all
        $f_1 \mathrel{+}= a_{-l} g_{k-l,k-1};$
        $f_2 \mathrel{+}= a_l h_{l-1,k-1};$
        $f_{k,k} \mathrel{+}= -a_{-l} y_{k-l,k-1};$
    enddo;

    Phase 3: $v = -f_1/q_{k-1}; r = -f_2/q_{k-1};$
    $q_k = q_{k-1} - f_1 f_2/q_{k-1}; f_{k,k} = f_{k,k}/q_k;$

    Phase 4: for $l = 0$ to k do all
        $g_{l,k} = g_{l,k-1} + v h_{l-1,k-1};$
        $h_{l,k} = r g_{l,k-1} + g_{l-1,k-1};$
        $y_{l,k} = y_{l,k-1} + f_{k,k} h_{l,k};$
    enddo;
enddo;

step of the algorithm and strongly dominates all other operations. The algorithm assumes that the inverse ${C_k}^{-1}$ exists at each step.

Figure 1 represents the concurrent computation of y_k for $k \geq 1$. The quantities f_1, f_2 and $f_{k,k}$, and the vectors h_k (the last column of the inverse matrix ${C_k}^{-1}$ scaled by a factor q_k), g_k (the correspondingly scaled first column of ${C_k}^{-1}$) and y_k are each calculated in parallel.

Note that the representation assumes $h_{0,0} = g_{0,0} = 1$; $q_0 = a_0$ and furthermore $h_{-1,k-1} = g_{-1,k-1} = h_{k,k-1} = g_{k,k-1} = y_{k,k-1} = 0$ for $k \geq 1$. In the 'do all' construct of *Phase 2*, the $\mathrel{+}=$ operator represents a 'fetch and add' operation. This implies the correct accumulation of the sum $s = s + r$ in a serial algorithm is given by $s \mathrel{+}= r$ in the parallel do all. This may be implemented by a lock on the shared operand in specific implementations on shared memory MIMD machines (such as the Sequent Balance). Most parallel machines have some intrinsic support for the fetch and add within a do all loop. On a machine like the Alliant FX-8, where processors have vector capabilities, the fetch and add operations should be implemented with an intrinsic sum, which helps vectorize the summation.

Assume that a total of P processes are used for the execution of *Step* k on P or more processors. Synchronization is needed for concurrent execution of *Phase 2* and of *Phase 4*. Pre-scheduling can usually be done as a result of code inserted by the compiler via the use of compiler directives. Alternatively, the scheduling can be achieved through the use of monitors [17],[18],[20] here in the form of the simple self-scheduling do-loop primitive or, more appropriately, the askfor monitor. The latter will allow each process to request a work unit consisting of a set $\mathcal{L}$ of indices l from the shared task pool. Each task acquired by the process then requires the calculations within the corresponding set of loop iterations. This reduces the synchronization overhead for systems of interest on any shared memory machine ($P \ll n$) to a negligible quantity and allows portable code. For example, the Argonne askfor monitor macro is supported on the Alliant, Sequent, Cray-XMP and Encore Multimax, and can be ported to similar architectures.

In the parallel algorithm for the recursive step only *Phase 2* and *Phase 4* are parallelized. When $P \ll n$ the serial code executed in *Step* k during *Phase 1* and *Phase 3* is of $\mathcal{O}(1)$, the accumulation of the independent sums from the P processes at the end of each do all requires $\mathcal{O}(\log P)$ time (although a $\mathcal{O}(P)$ time sum would be fine and faster on some machines) and *Phase 2* and *Phase 4* each require order $\mathcal{O}(k/P)$ time .

Thus, the total time spent over all the n *Steps* calculating independent sums is $\mathcal{O}(n^2/P)$, the total time spent combining these sums is $\mathcal{O}(n \log P)$ (or $\mathcal{O}(nP)$) and all other overheads are $\mathcal{O}(n)$. The speedup of the parallel algorithm over the serial one is close to P.

This is very different than the case where Toeplitz solutions are sought on systolic arrays or MIMD processor arrays, with perhaps $P = \mathcal{O}(n^2)$ processors, where the goal is to find a solution in order n time. It should not be expected that such an algorithm would be effective when simulated on a shared memory MIMD machine where $P \ll n$.

3 The block Toeplitz solver

The algorithm, a serial version of which is implemented in routine *TGSLD* in the Toeplitz Package [1],[19], is the block analogue of the *TSLD* algorithm. The overall algorithm is similar to the Toeplitz case with the *Step* k in Figure 1 generalized to the algorithm given in Figure 2.

In each step of the recursion, for $k = 0, 1, \ldots, n-1$, the last block column H_k scaled by an order m matrix Q_k of ${C_k}^{-1}$ is calculated, where C_k is the leading principal submatrix of order $(k+1)m$ of A. The existence of the inverse is assumed by the algorithm. A vector z_k is determined for $k \geq 1$ as a linear combination of the last m columns of ${C_k}^{-1}$, and is used to obtain y_k as in (4), the solution of a system of the form (3) where now $d_k = [b_0, b_1, \ldots, b_{(k+1)m-1}]^T$.

At *Step 0*, y_0 is determined by solving $A_0 y_0 = d_0$. Figure 2 represents the parallelization of *Step k* for the block algorithm. F_1, F_2, V, R, P_k and Q_k denote matrices of order m; $f_{k,k}$ is a vector of length m.

The representation assumes initializations which are the block analogues of those in Figure 1 and furthermore $P_0 = A_0$. Each of the phases, as well as *Step 0*, is performed concurrently. In *Step 0*, the order m system is solved by Gaussian elimination with a parallel decomposition step, in a modification by Brewer [6] of the Linpack routine *DGEFA* [15]. Parallelism is present in *Phase 1* and in *Phase 3* through the required multiplications and additions of matrices of order m. Furthermore, where a set of m systems with the same system matrix of order m have to be solved, this is done concurrently by parallel decomposition of the matrix as in *Step 0*, followed by the m backsubstitutions in parallel. Note that the granularity of the do all constructs in *Phase 2* and *Phase 4* is far larger than that of their correspondents in Figure 1.

We implemented the parallel block Toeplitz solver on the Sequent by using compiler directives for pre-scheduling of the loop structures. This also required modifications of the original serial code by, for example, eliminating the need of reduction variables where possible, or recoding do loops in order to change the order of the nesting.

Figure 2 : The Recursive Step of the Parallel Block Toeplitz Algorithm

Step k: Phase 1: $F_1 = 0; F_2 = 0; f_{k,k} = [b_{km}, b_{km+1}, \ldots, b_{km+m-1}]^T;$

Phase 2: for $l = 1$ to k do all

$F_1 \mathrel{+}= A_{-l}G_{k-l,k-1};$

$F_2 \mathrel{+}= A_l H_{l-1,k-1};$

$f_{k,k} \mathrel{+}= -A_{-l}y_{k-l,k-1};$

enddo;

Phase 3: $V = -{Q_{k-1}}^{-1}F_1; R = -{P_{k-1}}^{-1}F_2;$

$P_k = P_{k-1} + F_2 V; Q_k = Q_{k-1} + F_1 R; f_{k,k} = {Q_k}^{-1} f_{k,k};$

Phase 4: for $l = 0$ to k do all

$G_{l,k} = G_{l,k-1} + H_{l-1,k-1}V;$

$H_{l,k} = G_{l,k-1}R + G_{l-1,k-1};$

$y_{l,k} = y_{l,k-1} + H_{l,k}f_{k,k};$

enddo;

4 Applications and timing

In this section we will give timing results for the solution of Toeplitz and block Toeplitz systems arising in the numerical solution of one and two-dimensional Fredholm integral equations of the 2nd kind, with singular difference kernels. It is well-known that the convergence of standard numerical techniques for one-dimensional problems is rather slow in cases of derivative singularities in the solution. We used collocation with piecewise polynomial basis functions, together with a nonlinear extrapolation on sequences of results obtained at specific points for a geometrically decreasing mesh size. This method requires no analytical information about the behaviour of the solution function at a given point nor knowledge of the forms of early terms in the corresponding asymptotic error functional expansion. For 1-D problems where this type of information is available, Spence [5] suggests a linear extrapolation; Schneider [7] uses a non-uniform mesh approach.

Table 1: Times (in sec.) for Solving Toeplitz Systems (Alliant FX-8)

n	Number of Processors					
	1*	1	2	4	6	8
64	.056	.019	.016	.015	.016	.016
128	.207	.060	.042	.033	.034	.034
256	.670	.200	.127	.087	.081	.077
512	2.492	.760	.430	.260	.220	.200
1024	9.867	2.290	1.550	.890	.690	.610
2048	39.386	11.400	5.900	3.200	2.390	2.060
4096	159.454	45.200	23.100	12.200	9.600	7.600

* without vectorization

Table 2: Times (in sec.) for Solving Block Toeplitz Systems (Sequent)

n=m	order of system	Number of Processors					
		1	2	4	8	12	16
4	16	.22	.15	.12	.12	.12	.12
8	64	4.30	2.30	1.37	.87	.90	.92
16	256	107.30	54.60	28.20	15.21	14.90	8.90
32	1024	3053.00	1531.80	776.00	397.00	301.60	205.30

We consider a one-dimensional integral equation, from Spence [5], of the form

$$\lambda x(s) = \int_{-1}^{1} g(s,t)h(s,t)x(t)dt + y(s), \tag{6}$$

where $h(s,t)$ and $y(s)$ are continuous functions and $g(s,t)$ is termed 'weakly singular'. In our example, $\lambda = 1.5$, $\gamma = -\frac{1}{2}$ and $y(s) = 1.5(1-s^2)^{\frac{3}{4}} - 0.375\pi\sqrt{2}(2-s^2)$. Times for solving the Toeplitz systems required for a collocation with piecewise constant elements, on an Alliant FX-8 (with 8 processors and vector capabilities) are given in Table 1 for the indicated numbers of processors used (P) and orders of the systems (n).

Two-dimensional integral equations of the type

$$x(s,s') = c_0 + c_1 \int_0^d \int_0^1 g(s,s',t,t')x(t,t')dtdt', \tag{7}$$

where c_0, c_1 and d are constants and $g(s,s',t,t') = \log\sqrt{(s-t)^2 + (s'-t')^2}$, which arise from electrical engineering problems, can be solved numerically by collocation at the centroids of the rectangles in a uniform grid, using piecewise constant basis functions [8],[9]. The resulting system matrix, of order mn (= the number of rectangles in the grid), is block Toeplitz and easily large. In Table 2 we give timings obtained on a Sequent Balance 21000, for such a system from an integral equation, using $m = n$.

Note that, in the case of a complex integral equation (e.g. with complex c_1 in Equation 7 [8],[9]), one can rely on the complex arithmetic version of the block Toeplitz solver.

The timings given both in Table 1 and Table 2 were obtained with light loads on the machines.

5 Acknowledgement

This project was supported in part by a Summer grant from the Faculty Research and Sponsored Programs Division at Western Michigan University.

We also thank ACRF at Argonne National Laboratories for providing excellent support and use of the machines.

References

[1] Arushanian O.B., Samarin M.K., Voevoedin V.V., Tyrtyshnikov E.E., Garbow B.S., Boyle J.M., Cowell W.R. and Dritz K.W., The Toeplitz Package Users' Guide, Argonne National Laboratory, Report MCS ANL-83-16, 1983.

[2] Kapenga J. and de Doncker E., "Concurrent management of priority queues", Proceedings of the Third SIAM Conference on Parallel Processing for Scientific Computing, Dec. 1987, Los Angeles, to appear, 1988.

[3] de Doncker E. and Kapenga J.A., "A portable parallel algorithm for multivariate numerical integration and its performance analysis", Proceedings of the Third SIAM Conference on Parallel Processing for Scientific Computing, Dec. 1987, Los Angeles, to appear, 1988.

[4] de Doncker E., Kapenga J.A. and Spence A., "Parallel solution of singular integral equations", Presentation at the SIAM 35th Anniversary Meeting, 1987, Paper in preparation, 1987.

[5] Spence A., "Product integration for singular integrals and singular integral equations", Numerische Integration, Hämmerlin, G., Birkhäuser Verlag, Basel, 288-300, 1979.

[6] Brewer O., "A Parallel DGEFA on the Sequent", personal communication, 1987.

[7] "Product integration for weakly singular integral equations", Schneider C., Mathematics of Computation, Vol. 36, pp. 207-213, 1981.

[8] Graham I.G., "Some application areas for Fredholm integral equations of the second kind", The application and numerical solution of integral equations, Sijthoff and Noordhoff, Alphen aan den Rijn, Anderssen R.S., de Hoog F.R. and Lukas M.A., 1980.

[9] Graham I.G., "Collocation methods for two dimensional weakly singular integral equations", Journal of the Australian Mathematical Society, B, Vol. 22, pp. 456-473, 1981.

[10] Levinson N., "The Wiener RMS error criterion in filter design and prediction", Journal of Mathematical Physics, Vol. 25, pp. 261-278, 1947.

[11] Trench W.F., "An algorithm for the inversion of finite Toeplitz matrices", J. SIAM, Vol. 12, pp. 515-522, 1964.

[12] Zohar S., "Toeplitz matrix inversion, the algorithm of W.F. Trench", J. ACM, Vol. 16, pp. 592-601, 1969.

[13] Zohar S., "The solution of a Toeplitz set of linear equations", J. ACM, Vol 21, pp. 272-276, 1974.

[14] Cybenko G., "The numerical stability of the Levinson-Durbin algorithm for Toeplitz systems of equations, SIAM Journal of Scientific and Statistical Computations, Vol.1, 303-319, 1980,

[15] Dongarra J.J., Bunch J.R., Moler C.B. and Stewart G.W., "LINPACK users' guide", SIAM, Philadelphia, 1979.

[16] de Doncker E., Kapenga J.A. and Spence A., "Parallel solution of singular integral equations", Presentation at the SIAM 35th Anniversary Meeting, 1987, Paper in preparation, 1987

[17] Lusk E.L. and Overbeek R.A., "Implementation of monitors with macros: a programming aid for the HEP and other parallel processors", Argonne National Laboratories, Note MCS ANL-83-97, 1983

[18] Lusk E.L. and Overbeek R.A., "Use of monitors in FORTRAN: A tutorial on the barrier, self-scheduling do-loop and askfor monitors", Argonne National Laboratory, Report MCS ANL-84-51, 1984.

[19] Garbow B.S., "The Toeplitz Package Implementation Guide", Argonne National Laboratory, Report MCS ANL-83-17, 1983.

[20] Boyle J., Butler R., Disz T., Glickfeild B., Lusk E., Overbeek R., Patterson J. and Stevens R., "Portable programs for parallel processors", Holt, Rinehart and Winston, 1987.

Parallel Gaussian Elimination, iPSC/2 Hypercube versus a Transputer Network

L. Beernaert, D. Roose

Department of Computer Science, K.U. Leuven
Celestijnenlaan 200A, B-3030 Leuven, Belgium

S. Van Praet and P. de Groen

Department of Mathematics and Computer Science
V.U. Brussel, Pleinlaan 2
B-1050 Brussel, Belgium

Abstract

Gaussian elimination with partial pivoting can be implemented in several ways on a multitude of parallel hardware. In this note we give a comparison of implementations of the algorithm for banded matrices on an iPSC/2 hypercube and on a square grid of Transputers. For both implementations a scattered square allocation scheme is used. The comparison shows that machine characteristics strongly influence the attainable efficiency.

AMS(MOS) classification : 65F05

Keywords : band matrices, Gaussian elimination, parallel algorithms, distributed-memory multiprocessor.

NATO ASI Series, Vol. F 70
Numerical Linear Algebra, Digital Signal Processing and Parallel Algorithms
Edited by G. H. Golub and P. Van Dooren

1 Introduction

The performance of an algorithm in an environment of independent parallel processors with only local memory heavily depends on the distribution of the data over the processing elements and on the communication patterns. Several authors, e.g. Fox e.a. [3], Saad & Schultz [8], Beernaert & Roose [1] and Van Praet & de Groen [9], describe banded linear system solvers on distributed-memory multiprocessors. Most important network topologies are the hypercube, the two-dimensional grid and the ring-structure. The first topology, used in the iPSC/2, is richest in communication possibilities and both other structures can be simulated on it. In a network of Transputers only four links per processor are available and those are exploited most efficiently in a two-dimensional grid.

The aim of this paper is to describe implementations of Gaussian elimination for band matrices on the iPSC/2 hypercube and on a two-dimensional square grid of T414 Transputers, and to compare the relative efficiencies that can be obtained on these types of hardware. The allocation scheme wraps the matrix around the square grid in both directions, as in Fox e.a. [3], which seems most efficient for this type of problems.

2 A glance at the hardware

a) A **Transputer** consists of a processor with local (cache) memory and four data-links on chip. The Transputers we used in the network are of the type T414 without floating point hardware, operating at 30 Kflops for emulated 32-bit reals. Per processor a local memory of 256 Kbytes is available. The communication-links operate at the standard rate of 10 Mbits/sec. The start-up time for a message is 6.3 μsec (see Fig. 1). Communication is fast in comparison to computation.

The basic language is OCCAM [6], which provides Pascal-like structures for describing a sequential process, constructors for putting processes in parallel and low-level features for describing process-communication. The only communication facility is sending data to and receiving data from one of the four neighbours. The processor suspends a process until the neighbour is ready for the exchange of data.

In a network-configuration one Transputer functions as the 'host', by which the network is directed. This 'host' is installed in a PC and does not take part in the computational process itself. The network used in our experiments is a two-dimensional grid of processors that are connected circularly column- and rowwise. The 'host' is interconnected in the first column of the network.

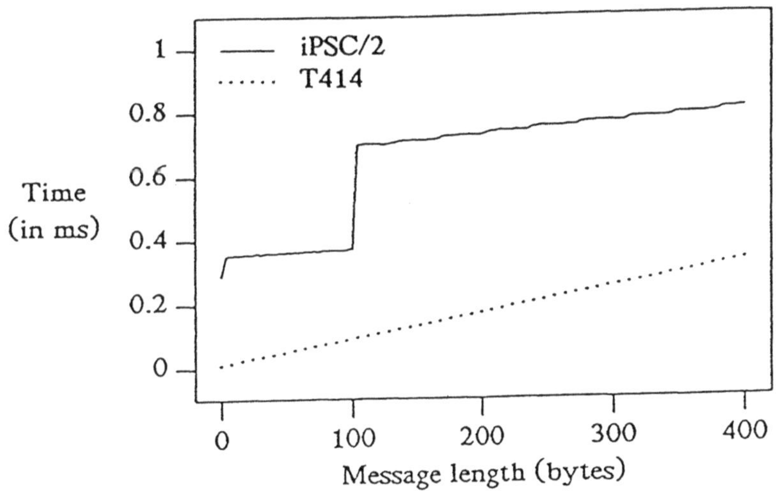

Fig. 1. Nearest neighbour communication speeds, time in milliseconds vs number of bytes: Transputer T414 (left, 10 Mbit/s link) and iPSC/2 node (right, 20 Mbit/s link).

b) On the **iPSC/2 hypercube** the nodes consist of Intel's 80386/387 processor/coprocessor combination with up to 16 Mbyte memory and up to eight 20 Mbit/sec communication channels per node. The communication is handled by a separate processor (Direct Connect Module) so that communication and computational work overlap and messages to non-neighbour nodes do not interfere with the computations in the intermediate nodes. The start-up time for sending a message to an arbitrary node is 350 μsec (unless its length exceeds 100 bytes), see Fig. 1. This large start-up time is mainly due to software overhead associated with the great functionality of the communication system. The (co)processor operates at 170 Kflops for 32-bit reals, thus communication of short messages is slow in comparison to computation.

3 Implementation of the algorithm

On both the grid of Transputers and the hypercube we implemented Gaussian Elimination for solving the system of linear equations

$$\mathbf{A}\,\mathbf{x} = \mathbf{b}\ ,$$

where $\mathbf{A} \in I\!R^{n \times n}$ is a band matrix with half bandwidth p and $\mathbf{b} \in I\!R^{n}$. We want to compare the efficiency of these implementations. Both use the same storage scheme. The major difference between them is the way in which data are communicated in the network.

a) Allocation scheme We consider a square $K \times K$ grid of processors as sketched in Fig. 2. The nodes $P(i,\ j)$ are numbered by row number i and column number j, $(i,\ j = 0\ ..\ K-1)$. If $K = 2^d$, it can be embedded in a hypercube of dimension $2d$. We allocate the matrix element a_{ij} to processor $P(i \text{ mod } K\ ,\ j \text{ mod } K)$ as proposed by Fox e.a. [3]; the matrix is wrapped around the grid in both directions.

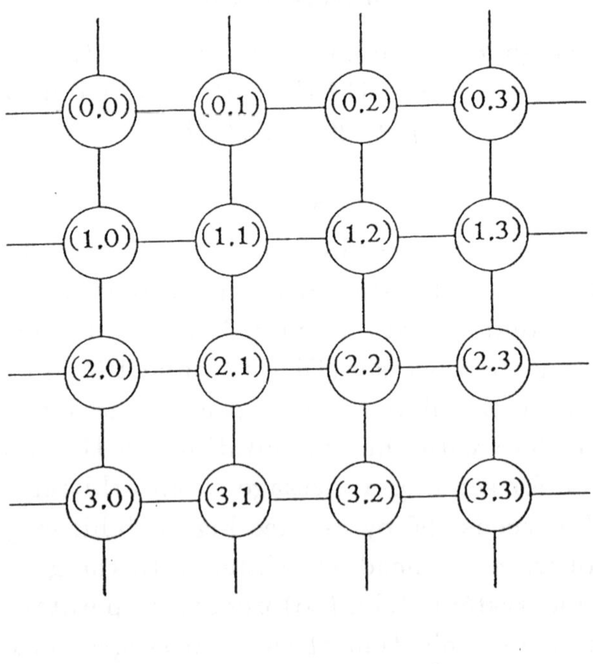

Fig. 2. A 4×4 network.

In [2] Bisseling & van de Vorst prove that this allocation scheme is optimal with respect to load balancing. The elements of the right-hand side **b** are placed in one processor column. Obviously each row or column of the matrix (and right-hand side) is contained in one processor row or column respectively. Other schemes, such as row- or column-allocation or the block-allocation scheme of Saad & Schultz [8], seem less efficient since they may imply a more unevenly distributed workload.

b) Communication scheme. During the execution of the algorithm data have to be passed in a column or in a row of the grid. In the hypercube such a row or column forms a subcube by itself. In such a subcube data are passed most efficiently via a balanced spanning tree of connections in the subcube, which takes d (= $^2\log K$) communication steps. In the Transputer based grid it is fastest to send the data around in the column or row (which forms a ring) in one direction, which takes $K-1$ communication steps.

c) LU-decomposition The LU-decomposition can roughly be described as follows :

FOR $k := 0$ TO $n-2$ DO
  find index *piv* of maximum in column $A[:, k]$
  interchange row $A[k, :]$ and row $A[piv, :]$ and distribute the latter one
  compute multipliers
  update subsequent rows of A
END

In the k-th step the diagonal processor $P(k \textbf{ mod } K, k \textbf{ mod } K)$ is the master of the elimination. The search for the pivot is done in the master-column. Each processor in this column determines the local maximum and forwards it to the master. The master then broadcasts the pivot information to all processors. The interchange of row k and the pivot row and the distribution of the latter row are combined and each column of processors works separately in parallel. In the Transputer based grid both rows are rolled in parallel through the grid in $K-1$ steps. The multipliers are determined in the master column and distributed rowwise. Upon reception of the multipliers each processor updates its own elements of the matrix. For more details see [1] and [9].

d) Solving the triangular systems These problems are handled in the same way as the LU-decomposition. For a single right-hand side however, this process takes place mainly in one column of processors and is much less efficient. For methods for dense systems see [5,7].

4 Numerical Experiments

The algorithm as described above has been carefully implemented and tested on the hypercube of dimension 4 by the first author and on a 4×4 Transputer network described in section 2 (K=4) by the second author. For a number of matrices, execution times have been measured. These are compared to execution times of the standard Gaussian elimination (without any communication overhead) executed on one processor. The execution time depends on the dimension n and the half bandwidth p (left and right bandwidth being equal), the speed-up however depends only on the bandwidth.

The program used on the iPSC/2 allows other allocation schemes than the one described in this paper. Therefore a rather complicated data structure is required. This leads to a substantial overhead, which could be removed by reducing the functionality of the program. Therefore we compare the parallel program with a sequential program that uses the same data structure, but in which all communication overhead has been removed. In this way it is possible to make a comparison between the speed-ups obtained on the Transputer network and on the iPSC/2.

		Decomposition			Triangular Solution		
n	p	Sequential	Parallel	Speed-Up	Sequential	Parallel	Speed-Up
100	10	1.66 s	0.21 s	7.9	0.20 s	0.09 s	2.2
400	20	27.00 s	1.97 s	13.7	1.70 s	0.67 s	2.5
900	30	137.00 s	9.35 s	14.6	5.81 s	2.65 s	2.2

Table 1. Execution times of Gaussian elimination on Transputer network.
n : dimension, p : half bandwidth.

		Decomposition			Triangular Solution		
n	p	Sequential	Parallel	Speed-Up	Sequential	Parallel	Speed-Up
100	10	0.98 s	0.59 s	1.7	0.36 s	0.45 s	0.7
400	20	13.25 s	3.17 s	4.2	2.33 s	2.33 s	1.0
900	30	63.70 s	10.25 s	6.2	7.54 s	6.34 s	1.2

Table 2. Execution times of Gaussian elimination on iPSC/2.
n : dimension, p : half bandwidth.

From these data we see that the speed-up is strongly influenced by the ratio $\frac{t_{comm}}{t_{calc}}$, where t_{comm} denotes the time to transmit a floating point number and t_{calc} denotes the time to perform a floating point operation. On the Transputer this ratio is small, such that for the LU-factorization a high speed-up can already be obtained with moderate matrix size. On the iPSC/2 hypercube however the communication overhead puts major restrictions on the attainable efficiency of both the LU-decomposition and the triangular solution.

References

[1] L. Beernaert and D. Roose, *Efficiency results for Gaussian elimination on the iPSC/2 hypercube* , preprint, 1988, submitted to J. Comp. Appl. Math.

[2] R. Bisseling and J. van de Vorst, *Parallel LU decomposition on a Transputer network*, 1988, Report Koninklijke/Shell-Laboratorium, Amsterdam

[3] G.C. Fox, M.A. Johnson, G.A. Lyzenga, S.W. Otto, J.K. Salmon and D.W. Walker, *Solving problems on concurrent processors* , Prentice Hall, 1988

[4] G.H. Golub and C.F. Van Loan, *Matrix Computations* , Johns Hopkins Univ. Press, Baltimore, Maryland, 1983.

[5] H.T. Heath and C.H. Romine, *Parallel solution of Triangular Systems on Distributed Memory Multiprocessors* , SIAM J. Sc. Stat. Comp. 9 (1988), pp. 558-588.

[6] Inmos ltd, *Occam Programming Manual* , Prentice Hall, 1987.

[7] G. Li and T.F. Coleman, *A parallel triangular solver for a distributed memory multiprocessor* , SIAM J. Sc. Stat. Comp. 9 (1988), pp. 485-502.

[8] Y. Saad and M.H. Schultz., *Parallel direct methods for solving banded linear systems* , Research Report YALEU/DCS/RR-387

[9] S. Van Praet and P.P.N. de Groen, *LU-decomposition on a parallel Transputer network* , preprint, 1988, submitted to J. Comp. Appl. Math.

Snapshots of Mobile Jacobi

Alan Edelman*

Department of Mathematics
Massachusetts Institute of Technology
Cambridge, MA 02139, USA
and
Thinking Machines Corporation
245 First St.
Cambridge, MA 02142, USA

Key words and phrases: eigenvalues, graphics, Jacobi

We wish to document the short but dramatic life of a special matrix undergoing the Jacobi diagonalization procedure to compute its eigenvalues. The combination of a massively parallel supercomputer with a real time graphics display allows us to understand algorithms in entirely new ways, giving insight that can be surprising as well as beautiful. In this note, we would like to give one illustration. We regret that the printed medium only allows for snapshots; the video is far more dramatic.

Jacobi's original method and its modification for parallel machines is carefully explained in many texts and recent papers on parallel algorithms (for example, see any of the references.) Here we provide snapshots to illustrate the underlying symmetries of a particular mobile scheme applied to a particular matrix.

The mobile scheme here is given by [5]. The key feature of this scheme is that the rotation parameters are computed only on the diagonal. The off diagonal elements march into the diagonal and are zeroed at the diagonal.

*Supported by Thinking Machines Corp. and a Hertz Foundation Fellowship

NATO ASI Series, Vol. F 70
Numerical Linear Algebra, Digital Signal Processing
and Parallel Algorithms
Edited by G. H. Golub and P. Van Dooren

Visually, this is quite striking. The elements make bishops' moves, some towards the main diagonal and some away. The latter will bounce off the border and return towards the main diagonal. Conceptually, in between each march step a rotation is computed and applied. In fact, the march step and the rotation application occur at the same time; it is a matter of storing the elements in permuted positions after applying the rotation.

The inner loop of the algorithm is as follows:

1. Compute rotations based on $\{a_{ii}, a_{i+1,i+1}, a_{i,i+1}\}$ for i odd.

2. Apply rotations and permute the pairs of rows and the pairs of columns i and $i+1$.

3. Repeat steps 1 and 2 for i even.

For more specific details, please consult the literature. This scheme allows for only local communication. The alternating blocks of white in figure (c) illustrate the zeros that were introduced on the super and sub-diagonals.

This scheme was originally implemented on the Connection Machine by Ewerbring, Luk, and Ruttenberg [3]. The Connection Machine is a 65,536 processor machine with processors arranged in a hypercube. Hypercubes conveniently unfold onto grids and thus this scheme is possible.

Here we document how Jacobi evolves. We found it convenient to represent the absolute value of each matrix element with one of five colors[1] depending on which of the intervals [0 .1], (.1 .5], (.5 .75], (.75 1], (1,∞) the value is found. The colors are white, purple, black, blue, and orange, respectively.

We take a very special matrix that is one on upper and lower triangles and zero on the interior diagonals. To be precise, we have a 128×128 matrix with fifty diagonals of ones in each of the triangles. Thus this is a fairly sparse, Toeplitz, zero-one matrix. Though arbitrary symmetric matrices lead to interesting symmetry patterns, we feel this special matrix leads to particularly beautiful patterns.

In (a) the matrix is born; the blue triangles indicate the ones while the inner band is zero. In (b) we notice the ones marching checkerboard

[1] In this volume, the colors are reproduced by a grey level proportional to the size of the elements

pattern toward the diagonal, but no rotations have occurred. Notice the holes that are left where the ones used to be. In (c) drama occurs, as the ones finally clash with the diagonal and non-trivial rotations emerge. Rotations continue with a later stage illustrated in (d). In (e), notice the off-diagonal elements dying out and finally in (f) we have convergence to diagonal.

The quilt patterns seen in (c) manifest themselves in arbitrary matrices though not as neatly. A related observation was made by [1]. We have also observed that often, though not always, the largest eigenvalue emerges more rapidly than the smaller ones. This phenomenon needs to be better understood, but it may point the way towards faster algorithms.

References

[1] M. Berry and A. Sameh, Parallel algorithms for the singular value and dense symmetric eigenvalue problems JCAM, to appear.

[2] R.P. Brent and F.T. Luk, The solution of singular-value and symmetric eigenvalue problems on multiprocessor arrays, *SIAM J. Sci. Stat. Comput.* *6* (1985), 69–84.

[3] L.M. Ewerbring, F.T. Luk, and A.H. Ruttenberg, Matrix computations on the Connection Machine, *Twenty First Asilomar Conference on Signals, Systems, and Computers.*

[4] G.H. Golub and C.F. Van Loan, *Matrix Computations*, The Johns Hopkins University Press, Baltimore, 1983.

[5] J. Modi and J.D. Price, Efficient implementation of Jacobi's diagonalization method on the DAP, *Numerische Mathematik 46* (1985), 443–454.

Figure a

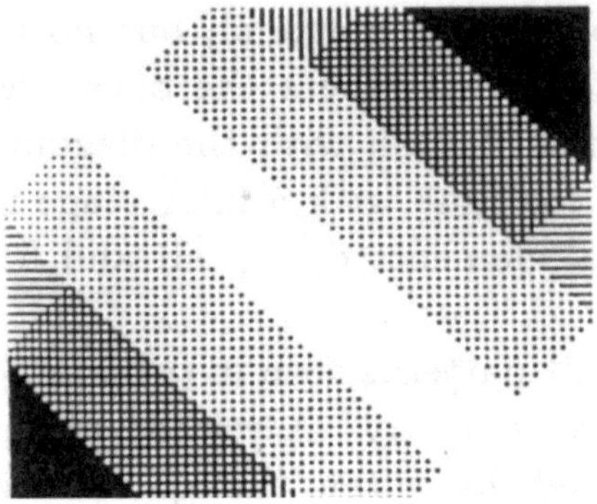

Figure b

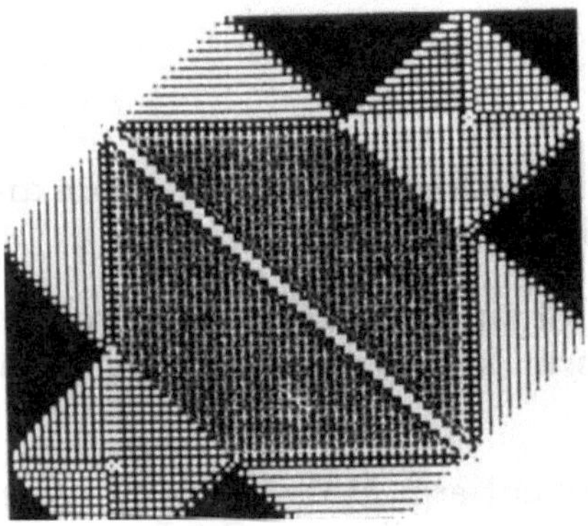

Figure c

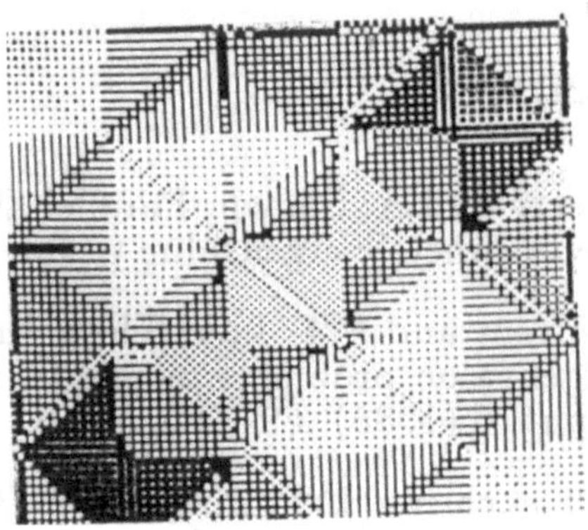

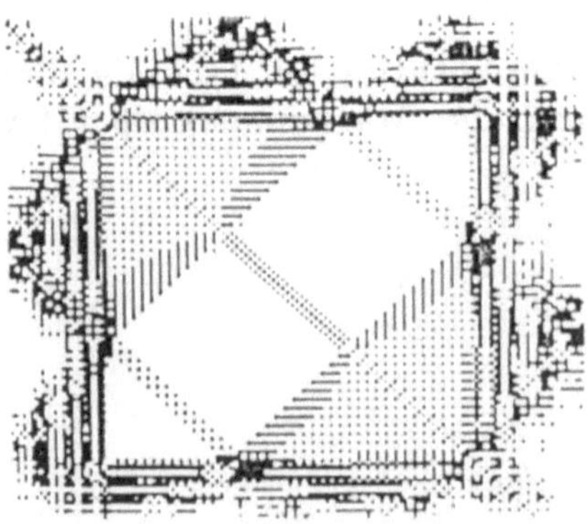

Figure e

Figure f

Computing Generalized Canonical Correlations

L. Magnus Ewerbring and Franklin T. Luk

School of Electrical Engineering
Phillips Hall
Cornell University
Ithaca, New York 14853, USA

Abstract

In this paper we consider the computation of generalized canonical correlations (gcc), a useful tool for linear prediction. The problem is reduced to a singular value decomposition of a product of three rank deficient matrices, for which a new algorithm is presented. It is then shown how the gcc may be computed directly from the data matrices.

1 Problem Definition

In canonical correlation analysis, we deal with m observations each of two random variables $\mathbf{x}$ and $\mathbf{y}$. Every observation of $\mathbf{x}$ contains n samples, and every of $\mathbf{y}$, p samples. We obtain two data matrices X and Y, which have respective dimensions $n \times m$ and $p \times m$, and we assume that $m > \max(n, p)$. Consider the $(n+p) \times (n+p)$ generalized joint covariance matrix Σ, given by

$$\Sigma = \begin{pmatrix} \Sigma_{xx} & \Sigma_{xy} \\ \Sigma_{yx} & \Lambda \end{pmatrix},$$

where Σ_{xx}, Σ_{xy}, Σ_{yx} denote correlation matrices of the two random variables, and Λ is a given symmetric and positive semi-definite matrix. Sup-

NATO ASI Series, Vol. F 70
Numerical Linear Algebra, Digital Signal Processing and Parallel Algorithms
Edited by G. H. Golub and P. Van Dooren

pose that

$$\begin{aligned} \text{rank}(X) &= r, \\ \text{rank}(\Lambda) &= s, \\ \text{rank}(Y) &= t, \end{aligned}$$

and

$$r \geq s \geq t.$$

Our problem is to find an $n \times r$ transformation J and a $p \times s$ transformation L such that

$$\begin{pmatrix} J & 0 \\ 0 & L \end{pmatrix}^T \begin{pmatrix} \Sigma_{xx} & \Sigma_{xy} \\ \Sigma_{yx} & \Lambda \end{pmatrix} \begin{pmatrix} J & 0 \\ 0 & L \end{pmatrix} = \begin{pmatrix} I_r & D \\ D^T & I_s \end{pmatrix}, \tag{1}$$

where D is a $r \times s$ diagonal matrix, and I_r and I_s denote identity matrices of order r and s, respectively.

Larimore [4] introduced the decomposition (1) as a valuable tool for some linear prediction problems, and discussed numerical procedures for the special case where both Σ_{xx} and Λ are nonsingular. In this paper we develop a new method for the general case where one or both matrices can be singular. Our algorithm can also be used to solve the rank deficient generalized total least squares problem described in Van Huffel [6].

2 Problem Reduction

To find the transformations, we begin by computing the square root factors $\Sigma_{xx}^{1/2}$ and $\Lambda^{1/2}$. They are, respectively, $r \times n$ and $s \times p$ matrices satisfying

$$\Sigma_{xx} = (\Sigma_{xx}^{1/2})^T \Sigma_{xx}^{1/2},$$

and

$$\Lambda = (\Lambda^{1/2})^T \Lambda^{1/2}.$$

As an example, consider the matrix Λ. We determine its square root via a triangular decomposition with pivoting on the diagonal elements, i.e.,

$$\Pi_\Lambda^T \Lambda \Pi_\Lambda = K_\Lambda^T K_\Lambda,$$

where Π_Λ denotes a $p \times p$ permutation matrix, and K_Λ an $s \times p$ upper trapezoidal matrix, so that

$$\Lambda^{1/2} = K_\Lambda \Pi_\Lambda^T . \tag{2}$$

A similar form is obtained for $\Sigma_{xx}^{1/2}$.

Denoting the Moore-Penrose generalized inverse of a matrix by the superscript $+$, we simplify equation (1) to

$$\begin{pmatrix} (\Sigma_{xx}^{1/2})^+ & 0 \\ 0 & (\Lambda^{1/2})^+ \end{pmatrix}^T \begin{pmatrix} \Sigma_{xx} & \Sigma_{xy} \\ \Sigma_{yx} & \Lambda \end{pmatrix} \begin{pmatrix} (\Sigma_{xx}^{1/2})^+ & 0 \\ 0 & (\Lambda^{1/2})^+ \end{pmatrix} = \begin{pmatrix} I_r & A \\ A^T & I_s \end{pmatrix}, \tag{3}$$

where the $r \times s$ matrix A is given by

$$A = ((\Sigma_{xx}^{1/2})^+)^T \Sigma_{xy} (\Lambda^{1/2})^+.$$

Our problem is reduced to finding a singular value decomposition (SVD) of A, say

$$U^T A V = D_A,$$

and the transformations for (1) are then given by

$$J = (\Sigma_{xx}^{1/2})^+ U , \tag{4}$$

and

$$L = (\Lambda^{1/2})^+ V . \tag{5}$$

3 SVD of a Matrix Product

For obvious numerical reasons, we do not form A explicitly. Instead, we want an implicit SVD algorithm for the matrix product

$$(E^+)^T F G^+ , \tag{6}$$

where the matrices E, F and G have dimensions $r \times n$, $n \times p$ and $s \times p$ respectively. We assume that $\text{rank}(E) = r$, $\text{rank}(G) = s$, and $r \geq s$. If $r < s$ we work with the transposed product

$$(G^+)^T F^T E^+.$$

In [2] Ewerbring and Luk presented an implicit SVD algorithm for the case where E and G are square and nonsingular, i.e., for a matrix product of the form

$$(E^{-1})^T F G^{-1}.$$

The purpose of this section is to show how (6) can be effectively reduced to this special case.

First, we compute RQ decompositions of E and G:

$$E = R_E Q_E^T \quad \text{and} \quad G = R_G Q_G^T,$$

where R_E is $r \times r$ upper triangular, Q_E is $n \times r$ with orthonormal columns, R_G is $s \times s$ upper triangular, and Q_G is $p \times s$ with orthonormal columns. Since R_E, Q_E, R_G, and Q_G all have full column rank, the generalized inverses of E and G are given by (cf. Greville [3])

$$E^+ = Q_E R_E^{-1} \quad \text{and} \quad G^+ = Q_G R_G^{-1}.$$

Hence, the matrix product (6) becomes

$$(R_E^{-1})^T Q_E^T F Q_G R_G^{-1}.$$

As the columns of both Q_E and Q_G are orthonormal, we may safely compute the $r \times s$ matrix

$$\hat{F} = Q_E^T F Q_G,$$

and apply the implicit SVD algorithm of [2] to the product

$$(R_E^{-1})^T \hat{F} R_G^{-1}\,. \tag{7}$$

In the above discussion, special care has been given to the computation of the generalized inverses; other approaches that require less computation may fail to provide the correct answer. For example, one may argue against computing a RQ decomposition of E, and choose to calculate instead a QR decomposition of the product FQ_G to obtain an $n \times s$ matrix Q_F with orthonormal columns. However, take the case where $r > s$. Since Q_F does not have full row rank and E does not have full column rank, it follows that

$$(E^+)^T Q_F = (Q_F^T E^+)^T \neq ((EQ_F)^+)^T.$$

So the extra RQ decomposition is necessary.

4 Direct Computation Using Data Matrices

For numerical considerations (e.g., not squaring the matrix condition numbers), we often wish to avoid forming explicitly the correlation matrices. Instead, we work directly with the data matrices X and Y. Provided that m is sufficiently large, we can estimate accurately the covariance matrices by

$$\begin{aligned} \Sigma_{xx} &\approx (1/m)XX^T, \\ \Sigma_{xy} &\approx (1/m)XY^T, \\ \Sigma_{yx} &\approx (1/m)YX^T. \end{aligned}$$

First, determine the QR decomposition with column pivoting of the $m \times n$ matrix X^T:

$$X^T = Q_X K_X \Pi_X^T,$$

where Q_X is an $m \times r$ matrix with orthonormal columns, K_X a $r \times n$ upper trapezoidal matrix, and Π_X an $n \times n$ permutation matrix. With

$$\Sigma_{xx}^{1/2} \approx (1/\sqrt{m}) K_X \Pi_X^T,$$

the matrix A can be approximated by the formula

$$A \approx (\sqrt{m}) Q_X^T Y^T (\Lambda^{1/2})^+ . \tag{8}$$

From equation (2) we get

$$A \approx (\sqrt{m}) Q_X^T Y^T \Pi_\Lambda K_\Lambda^+ .$$

As explained before, we can safely form the $r \times p$ matrix

$$\hat{Y} = Q_X^T Y^T \Pi_\Lambda .$$

So a simplified version of the SVD algorithm of Section 3, with $E = I_r$, can be applied to the matrix

$$\hat{Y} K_\Lambda^+ . \tag{9}$$

Note that this case is not the same as the generalized SVD (cf. Paige [5]). For ordinary canonical correlations where $\Lambda = \Sigma_{yy}$, we determine the QR decomposition with column pivoting of the matrix Y^T:

$$Y^T = Q_Y K_Y \Pi_Y^T,$$

and get

$$A \approx Q_X^T Q_Y,$$

a problem that has been well analyzed by Björck and Golub [1].

Acknowledgements

This work was supported in part by the Joint Services Electronics Program under contract F49620-87-C-0044. L.M. Ewerbring was also funded as a Graduate Fellow by the U.S. Army Research Office through the Mathematical Sciences Institute of Cornell University.

References

[1] Å. Björck and G.H. Golub, "Numerical methods for computing angles between linear subspaces," Math. Comput., vol. 27 (1973), pp. 579-594.

[2] L.M. Ewerbring and F.T. Luk, "Canonical correlations and generalized SVD: applications and new algorithms," Proc. SPIE, vol. 977, Real Time Signal Processing XI, Paper 23, 1988.

[3] T.N.E. Greville, "Note on the generalized inverse of a matrix product," SIAM Review, vol. 8 (1966), pp. 518-521.

[4] W.E. Larimore, "A unified view of reduced rank multivariate prediction using a generalized singular value decomposition," Technical Report, Scientific Systems Inc., Cambridge, Mass., 1987.

[5] C.C. Paige, "Computing the generalized singular value decomposition," SIAM J. Sci. Statist. Comput., 7 (1986), pp. 1126-1146.

[6] S. Van Huffel, "The generalized total least squares problem: formulation, algorithm and properties," Proceedings of the NATO Advanced Study Institute on Numerical Linear Algebra, Digital Signal Processing and Parallel Algorithms, Leuven, Belgium, August 1988.

Introduction to High Resolution Array Spectrum Estimation

D. R. Farrier

Department of Electrical Engineering
Southampton University
Southampton, S09 5NH, U.K.

Abstract

The basic principles of high resolution spectrum estimation using the SVD is described. The paper starts with a simple signal model for a linear array and explains why simpler signal processing tools are not powerful enough for many of the tasks which are required from systems. One particular SVD technique, the MUSIC algorithm, is described and some of the limitations on performance are described.

1 Introduction

Consider a signal from direction θ_1, incident on a linear equispaced array containing N receiving elements. It will be assumed that each sensor is able to receive signals from all directions with equal response for all directions. The separation between elements is a distance, d, so that the propagation distance between adjacent sensors for the signal from direction θ_1, is d.cosθ_1.

Assume all measurements have been passed through a bank of ideal narrow band filters, so that the signals may all be assumed to have the form $e^{j\omega k}$ where k is a time ordinate and ω is the frequency in radians/sec. Complex valued signals do not exist in practice, but they are a good representation to use for narrowband signals, i.e. signals which only have power within a very narrow range of frequencies. Let the signal $s_1(k)$ arrive at

NATO ASI Series, Vol. F 70
Numerical Linear Algebra, Digital Signal Processing
and Parallel Algorithms
Edited by G. H. Golub and P. Van Dooren

sensor number one at time k. Since the wave is travelling through space with velocity, v, the signal arriving at sensor number 2 at time k will have already been received by sensor 1 at an earlier time $t^{12} = \mathrm{d}\cos\theta_1/v$, so that using the relationship, $v = \omega\lambda/2\pi$, we have: $\omega t^{12} = 2\pi d\cos\theta_1/\lambda$. Similarly sensors 3 to N will have increasing delay, so that the received signals are

$$\begin{array}{rcl} X_1(k) & = & s_1(k) \\ X_2(k) & = & s_1(k)\, e^{-j2\pi \mathrm{d}\cos\theta_1/\lambda} \\ \cdots & & \cdots \\ X_N(k) & = & s_1(k)\, e^{-j2\pi(N-1)\mathrm{d}\cos\theta_1/\lambda} \end{array} \tag{1}$$

which may be written in matrix form as

$$\mathbf{X}(k) = \mathbf{h}_1\, s_1(k) \tag{2}$$

where $\mathbf{h}_1$ contains complex exponentials for a signal from direction θ_1:

$$\mathbf{h}_1^\dagger = [1, e^{j2\pi \mathrm{d}\cos\theta_1/\lambda}, \cdots, e^{j2\pi(N-1)\mathrm{d}\cos\theta_1/\lambda}]$$

where $\dagger$ indicates complex conjugate transpose.

If we have K incident signals, we may now generalize our model to:

$$\mathbf{X}(k) = [\mathbf{h}_1\, \mathbf{h}_2 \cdots \mathbf{h}_K] \begin{bmatrix} s_1(k) \\ \vdots \\ s_K(k) \end{bmatrix} = H\, \mathbf{s}(k) \tag{3}$$

where $\mathbf{h}_2 \cdots \mathbf{h}_K$ correspond to other incident direction vectors.

Arrays of receiving sensors are used in many situations such as radar, sonar and seismology and oil exploration. In all of these cases, the sensor array is used to interpret spatial properties about signals which are incident upon it. In many situations, linear arrays of sensors are used, although the same principles apply to all array geometries. When designing sensor arrays, it is common to try to choose the inter-element spacing, d, such that it equals half of a wavelength, but in many applications it is impossible to ensure that this occurs.

The array processing is often required to answer the following questions:
(1) How many signals are incident on the array?
(2) Which directions do they come from?

These questions are linked with questions regarding the detection of signals and tracking of moving targets.

In addition to the incident signals, there will always be some additive noise present, so that the received signal is

$$\mathbf{y}_k = H\mathbf{s}(k) + \mathbf{n}(k) \tag{4}$$

This noise can arise for many physical reasons, depending on the particular application. For example, for the sonar case, the ocean provides a continuum of very small signals from all possible directions. This makes it harder to distinguish between signal and noise. From a practical viewpoint, it is often useful to imagine this noise as providing a smooth distribution as a function of direction, whereas the "signals" provide discrete sources of power which are discontinuous as a function of direction. Other sources of noise arise from electrical noise within the sensors and quantisation noise caused by analogue to digital conversion.

The signal $\mathbf{y}_k$ is a data vector received at time k, and each measurement is made at the same time. For this reason, it is often called a "snapshot". At any one time, we can choose to use several snapshots which we may place in a data matrix, Y; given by

$$Y = [\mathbf{y}_1\, \mathbf{y}_2 \cdots \mathbf{y}_L] = H\,S + N \tag{5}$$

The number of snapshots will normally depend on whether the signal directions are changing with time. Ideally, it is useful if the number of snapshots, L, is large, since we will then obtain better performance, but we will often not be able to do this because of changes in the model which occur over extended time periods.

If the signals and noise are independent and normally distributed with zero mean, the covariance matrix is given by

$$R = H\,P_s\,H^\dagger + R_N \tag{6}$$

where P_s is the incident signal covariance matrix and R_N is the noise covariance matrix. To obtain this matrix it is necessary to average for infinite time, which also assumes stationarity. In practice, we only have a sample covariance matrix, which is obtained from the data matrix as

$$\hat{R} = YY^\dagger \tag{7}$$

The incident signal covariance matrix, P_s, may have full rank (k)or be rank deficient. A rank deficient situation can occur if there appears to be several signals, but these are the result of reflections off surfaces. The rank is physically the number of *distinct* signals.

We will now consider a number of special cases:

One Signal

For this situation the optimum is well known, and is easy to construct. A number of different optimality criteria all give the same solution. The Maximum Likelihood solution is the most rigorous, but engineers often prefer to consider Signal-to-Noise ratio (SNR). Fortunately, maximizing SNR is equivalent to maximizing the likelihood function, for the single signal case. If we construct:

$$\mathbf{h}(\phi) = \begin{bmatrix} 1 \\ e^{-j2\pi \mathrm{d}\cos\phi/\lambda} \\ e^{-j2\pi(N-1)\mathrm{d}\cos\phi/\lambda} \end{bmatrix} \tag{8}$$

We then form:

$$P(\phi) = \mathbf{h}^{\dagger}(\phi) Y Y^{\dagger} \mathbf{h}(\phi) \tag{9}$$

and plot P as a function of ϕ. This approach of using the signal model is only optimal if the noise is uncorrelated, and is widely used in many applications. It is often called *conventional beamforming.* In the signal direction $\phi = \theta$; $\mathbf{h}^{\dagger}(\phi)\, \mathbf{h}(\theta) = N$, and we obtain a large peak, and this peak will still be present when there is significant noise (i.e. we have signal-to-noise *Gain*).

K Signals

This situation is rather more complicated, since the maximum likelihood solution involves nonlinear equations which are difficult to solve. It is possible to solve them by iterative methods but convergence is difficult to prove. Suboptimal non-iterative methods, based on the use of SVD have therefore also been considered. We have:

$$R = E[Y Y^{\dagger}] = H\, P_s H^{\dagger} + R_N = H\, P_s H^{\dagger} + \sigma^2 I \tag{10}$$

if the noise is white (uncorrelated). If we perform an eigenvalue decomposition on the $(N \times N)$ matrix, R, we obtain:

$$R = U\,\Sigma\,U^{\dagger} \tag{11}$$

U contains the eigenvectors, and Σ is a diagonal matrix containing eigenvectors. This matrix may be written in two parts:

$$R = U_s\,\Sigma_s\,U_s^{\dagger} + U_n\,\Sigma_n\,U_n^{\dagger} \tag{12}$$

Σ_s: $(K \times K)$ is a diagonal matrix containing the K largest eigenvalues (which are presumed to be the signal eigenvalues); U_s: $(N \times K)$ contains the associated eigenvectors; and U_n: $(N \times (N-K))$ and Σ_n: $(N-K)\times(N-K)$ contain the remaining eigenvectors and eigenvalues (which are assumed to be the noise eigenvalues). U_s is called the signal subspace and U_n is called the noise subspace.

The above descriptions arise from the noise free case, since in the absence of noise, the matrix, R is only of rank K. It should be noted, however, that the signal eigenvectors do not usually have any physical significance, individually, and do not correspond to the signal models, $\mathbf{h}_k$. The noise eigenvectors obey the orthogonality relationship:

$$U_n^{\dagger}\mathbf{h}(\phi) = 0 \qquad \text{iff} \qquad \phi = \theta \tag{13}$$

This property is used by several methods. The most popular method called MUSIC [1,2] (**MU**ltiple **SI**gnal **C**lassification) makes use of this property by simply plotting:

$$F(\phi) = \mathbf{h}^{\dagger}(\phi)\,U_n\,U_n^{\dagger}\,\mathbf{h}(\phi) \tag{14}$$

as a function of ϕ. This is non-negative definite function and so nulls (or more generally, significant minima) of this function should correspond to signal directions. This method has two major advantages: (i) it applies to the many signal case and (ii) it has the beamforming structure which applied to the single signal case and is well understood. Sometimes the reciprocal of F is plotted, so that signal directions now correspond to maxima of $1/F$.

Eigenvector methods, such as Equation (13) are based on the noise free properties of R, and so are not strictly accurate when noise is present.

Noise tends to introduce "leakage" between the signal and noise subspaces, and this increases with increasing noise power or reduction of the number of snapshots, L. To predict the performance of eigenvector methods under these conditions it is necessary to perform a perturbation analysis [3]. This is especially important with small number of samples, or correlated noise [4].

Since R is a $N \times N$ matrix and Y is $N \times L$, it may be preferable to compute the SVD of Y rather than the eigendecomposition of R. This will certainly be more numerically stable, but if L is large it may require many more calculations.

Conclusions

A brief description of the basis for processing data from arrays of sensors has been given. In particular the use and limitations of SVD methods has been discusses.

References

[1] Schmidt, R., "Multiple emitter location and signal parameter estimation", Proc. RADC Spectral estimation Workshop (Rome, New York, 1979), pp. 243-258.

[2] Bienvenu, G. and Kopp, L., "Adaptivity to background noise spatial coherence for high resolution passive methods", Proc. IEEE ICASSP'80, Denver, CO, pp. 307-310.

[3] Kaveh, M. and Barrabell, A.J., "The statistical performance of the MUSIC and the minimum norm algorithm in resolving plane waves in noise", IEEE Trans. 1986, ASSP-34, pp. 331-341.

[4] Farrier, D.R., Jeffries, D.J. and Mardani R., "Theoretical Performance of the MUSIC algorithm", IEE Proc. Vol. 135, Part. F, 1988, pp. 216-224.

Modifications of the Normal Equations Method that are Numerically Stable

Leslie Foster*

Mathematics and Computer Science
San Jose State University
San Jose, California 95192, USA

Abstract

A method of solving min $\| b - Ax \|$ where A is an $m \times n$ matrix is presented which appears to be as stable as methods based on orthogonal factorization but in certain cases is nearly as efficient as methods based on factoring the normal equations. Iterative refinement is used in the new method. Also a class of methods is described that in terms of efficiency (for $m \geq 2n$) and, apparently, numerical stability is intermediate between orthogonalization methods and normal equations methods. The class can be considered to be a block implementation of the modified Gram-Schmidt algorithm.

1 Introduction

Recently, Lloyd Trefethen [1] proposed three unsolved mysteries in numerical linear algebra. The second of these is: "Why do we use non-orthogonal methods for solving (a square linear system) $Ax = b$, but orthogonal ones for many other linear algebra computations?". There are two themes in Trefethen's discussion: (a) non-orthogonal algorithms are twice as fast and (b) orthogonal algorithms are more stable. However, as Trefethen notes, in

*This work was supported in part by NSF grant DMS-8505783.

NATO ASI Series, Vol. F 70
Numerical Linear Algebra, Digital Signal Processing and Parallel Algorithms
Edited by G. H. Golub and P. Van Dooren

general there were no results available to pin to what extent either of these statements is valid.

Our interests will focus on the linear least squares problem

$$\min_x \| b - Ax \| \tag{1}$$

where A is an $m \times n$ matrix, x an n vector, b an m vector and the norm is the usual two norm. For this problem methods based on normal equations can be unstable but, for large m, are more efficient than methods based on orthogonal factorization. We will make some progress toward answering Trefethen's question by introducing a new method which appears to be just as stable as methods based on orthogonal factorization but in certain cases is as fast as the normal equations methods. We will also describe a class of methods that are intermediate between normal equations methods and the modified Gram-Schmidt orthogonalization method, both in terms of efficiency for $m \geq 2n$ and, apparently, numerical stability. The class of methods can be viewed as block implementation of the modified Gram-Schmidt technique.

As we will see iterative refinement will be an important tool in our development. In fixed precision arithmetic iterative refinement is of use in order to overcome numerical instabilities that are introduced by an algorithm but that are not inherent in a problem. For example such use appears elsewhere in this proceeding in relation to the downdating problem in signal processing [2] and to continuation methods [3]. We hope that our results will further indicate the power of iterative refinement for this purpose.

2 Existing Algorithms

The normal equations solution of (1) requires solving $A^T Ax = A^T b$ where ^{T} indicates transpose. The solution of this equation can be carried out using the Cholesky decomposition of $A^T A$: $R^T R = A^T A$ where R is right triangular. On the other hand (1) can be solved by calculating an orthogonal factorization of A using, for example, the QR algorithm based on Householder transformations [4] or the modified Gram-Schmidt (MGS) algorithm [5]. Our implementation of the Cholesky decomposition of $A^T A$ is somewhat non-standard and later we will need the MGS algorithm and so now we present both. In these algorithms a_k will refer to the kth column of the current A.

Cholesky Factorization of A^TA	Modified Gram-Schmidt Algorithm
For $k = 1, \ldots, n$	For $k = 1, \ldots, n$
$\quad r_{kk} = \sqrt{a_k^T a_k - \sum_{i=1}^{k-1} r_{ik} r_{ik}}$	$\quad r_{kk} = \sqrt{a_k^T a_k}$
$\quad$ For $j = k+1, \ldots, n$	$\quad$ For $j = k+1, \ldots, n$
$\qquad r_{kj} = (a_k^T a_j - \sum_{i=1}^{k-1} r_{ik} r_{ij})/r_{kk}$	$\qquad r_{kj} = (a_k^T a_j)/r_{kk}$
	$\qquad a_j \leftarrow a_j - a_k(r_{kj}/r_{kk})$

(2)

In this form both of the algorithms determine a new row of R as k increases. Also both algorithms can be implemented with the column interchanges described in Section 5.1 of [2] in this proceedings. This will produce a factorization of AP not A, where P is a permutation matrix and has the advantage that $\kappa(A)$, the condition number of the matrix A, can in practice be reliably estimated from the diagonal entries of R when A is not too ill- conditioned. We should also note that when solving (1) it is recommended that the MGS algorithm be applied to the matrix (A, b) formed by adding b as a new column of A. If R is the resulting triangular matrix, if R_n is its leading principal $n \times n$ submatrix and if $\hat{r}_{n+1}$ is the first n components of the last column of R then it follows [5] that the solution to (1) is the solution to the triangular system $R_n x = r_{n+1}$. The $(n+1)^{st}$ column should not be included in column interchanges if they are used.

The normal equations method requires $mn^2/2 + n^3/6$ multiplications whereas the MGS requires mn^2 multiplications and the solution of (1) by using Householder transformations requires $mn^2 - n^3/3$ multiplications. Thus for $m \gg n$ normal equations is approximately twice as fast as the other two methods and they will require approximately equal amounts of work.

To describe the numerical stability of the algorithms let x be the true solution to (1), let $\dot{x}$ be the solution calculated on a computer with relative machine precision ϵ, let $r = b - Ax$ and let κ be the condition number of

A. Then the condition number of the least squares problem (1) is [6]

$$\kappa(A,b) = \frac{\kappa\epsilon}{1-\kappa\epsilon}\left(1 + \frac{\kappa \| r \| + \| b \|}{\| A \| \| x \|}\right).$$

Note that here we are using $\kappa(A,b)$ to indicate the condition of the least squares problem and κ or $\kappa(A)$ for the condition of the matrix A. When (1) is solved by MGS or Householder based algorithms, if $\kappa\epsilon$ is somewhat less than 1, then [8] $\| x - \dot{x} \| \leq c_1\kappa(A,b)$ where in practice c_1 will not be large. On the other hand for normal equations, if $\kappa^2\epsilon$ is somewhat less than 1, then [8] $\| x - \dot{x} \| / \| x \| \leq c_2\kappa^2\epsilon$ where in practice c_2 will not be large. Thus the error in normal equations is always proportional to κ^2 whereas for the other two methods if the residual r is small then the error is only proportional to κ. We should also note that the Cholesky decomposition of A^TA may fail completely if $\kappa^2\epsilon \geq 1$.

3 Iterative Refinement

The iterative refinement technique that we will use is among the simplest techniques available for least squares problems. Let us assume that a method is available which can solve, approximately,

$$A^TAx = rhs \tag{3}$$

where rhs represents a right hand side and assume that an initial, perhaps incorrect, guess x_0 is available. Then we successively solve

$$A^TA(\delta x_p) = A^T(b - Ax_p) \quad \text{letting} \quad x_{p+1} = x_p + (\delta x_p) \tag{4}$$

This technique was originally proposed by Kahan [7] but did not receive much attention until the recent work by Bjork [8].

In order to solve Equation (3) Bjork considers solving $R^TR = rhs$ where R is produced by each of the three algorithms described in Section 2. We will term the iterations (4) normal equations plus iterative refinement (N + IR) when R comes from a Cholesky decomposition of A^TA and seminormal equations plus iterative refinement (SN + IR) when R comes from MGS or Householder transformations applied to A. Assuming the x_0 is the solution to $R^T Rx = A^T b$ Bjork does a careful error analysis for $p = 0$ and 1 and provides some discussion for $p > 1$. These results suggest the following convergence properties

$$\begin{gathered} \text{N + IR} \\ \| x - x_p \| / \| x \| = c_1 \kappa^2 \epsilon (c_2 \kappa^2 \epsilon)^p \\ \\ \text{SN + IR} \\ \| x - x_p \| / \| x \| = c_3 \kappa^2 \epsilon (c_4 \kappa \epsilon)^p \text{ for } p \geq 0 \end{gathered} \tag{5}$$

Bjork conjectures that convergence rates, like those in (5), will hold for SN + IR until the full accuracy permitted by the problem is achieved, that is, until

$$\| x - x_p \| / \| x \| \cong \kappa(A, b)\epsilon. \tag{6}$$

Our computational experiments support Bjork's conjecture and indicate that, if $\kappa^2\epsilon$ is somewhat less than one then N+IR also achieve this full accuracy.

The significant difference between SN+IR and N+IR is in the convergence rate factors $c_2\kappa\epsilon$ and $c_4\kappa^2\epsilon$. The SN + IR method converges faster and, furthermore, N + IR will not converge at all if $c_4\kappa^2\epsilon > 1$. However, it is interesting to ask for what matrix condition numbers κ will each method converge to full accuracy, i.e. satisfy (6), in p or less steps. To analyze this, for simplicity, we will assume that all the c's in (5) are approximately one. This is consistent with our numerical experiments. Equation (6) is certainly true if $\| x - x_p \| / \| x \| \cong \kappa\epsilon$ and assuming (5) it follows from simple algebra

$$\begin{gathered} \text{If } \kappa \leq \epsilon^{-p/(2p+1)} \text{ then} \\ \text{N + IR will reach full accuracy in } p \text{ steps} \\ \\ \text{If } \kappa \leq \epsilon^{-p/(p+1)} \text{ then} \\ \text{SN + IR will reach full accuracy in } p \text{ steps} \end{gathered} \tag{7}$$

For example for $\epsilon = 10^{-16}$, say, if $\kappa \leq 10^{-5.3}$ then N + IR will reach full accuracy in one step.

The amount of work per iterative refinement step of each of the above methods is $2mn + n^2$ multiplications which is an order of magnitude less work than required for the factorization. An important observation from these comments is that if A is not very ill-conditioned then N + IR will be just as accurate as orthogonalization methods while only requiring slightly more work than normal equations.

4 Block MGS and Improved Block MGS Algorithms

We will consider a block implementation of MGS. In developing our algorithm we will use $R_{\ell\ldots k,\ell\ldots k}$ to indicate the block submatrix of R in rows ℓ through k and columns ℓ through k, $r_{\ell\ldots k,j}$ will indicate the elements of R in rows ℓ through k and in column j, and $A_{\ell\ldots k}$ will indicate columns ℓ through k of A.

Our algorithm is a combination of the two algorithms in (2). To describe how this may be done, in the MGS algorithm let $\hat{a}_j$ refer to the current vectors stored in column j of the matrix A and let a_j represent the original column j. Then at the end of step k of the MGS algorithm in (2) $\hat{a}_j$ will be the portion of a_j perpendicular to the first k columns of the original A. Now consider the Cholesky factorization of the normal equations and suppose that for some k it was felt that any further continuation of this approach would produce excessive errors. One could, in principal, produce the same vectors $\hat{a}_j, j = k+1, \ldots, n$ that would occur at the end of step k of MGS by, for $j = k+1, \ldots, n$, solving

$$\min \| a_j - A_{1\ldots k} x_j \| \tag{8}$$

and assigning $\hat{a}_j = a_j - A_{1\ldots k} x_j$. Note that since the factorization of the first k rows of $A^T A$ is complete then, for each j, it follows that x_j can be calculated by solving $R_{1\ldots k,1\ldots k} x_j = r_{1\ldots k,j}$.

The idea of our block MGS (BMGS) algorithm then is to continue with a Cholesky decomposition of the normal equations for a while (we will discuss how long shortly) and then to form the residuals $\hat{a}_j$ as above. At this point the factorization could be finished with ordinary MGS; however in our implementation we choose not to do this. Rather note that $R_{k+1\ldots n,k+1\ldots n}$, the part of R remaining to be calculated, is the Cholesky factor of the matrix $\hat{A}_{k+1\ldots n}$ formed by $\hat{a}_j, j = k+1, \ldots, n$. Therefore we simply continue with the Cholesky factorization algorithm working with $\hat{A}_{k+1\ldots n}$; perhaps at some point again switching temporarily to an MGS step. This approach, we feel, will take maximum advantage of the increased efficiency of the Cholesky factorization algorithm.

A potential difficulty with the above technique is that any use of the Cholesky factorization algorithm could introduce errors beyond those inherent in the problem and consequently at times BMGS will be less accurate than simple MGS. However we can use iterative refinement to overcome this potential difficulty. In particular, after k steps of the normal equations algorithm we solve (8), for $j = k+1, \ldots, n$, using p steps of iterative refincment. As described earlier if $\kappa(A_{1\ldots k})$ satisfies the left hand condition in (7) then apparently the resulting x_j will have full accuracy and so the corresponding residual $\hat{a}_j$ will be just as accurate as the residual produced by MGS. We call the algorithm based on this technique improved block modified Gram-Schmidt (IBMGS).

Before presenting our algorithms we should note that since $r_{1\ldots k,j} = R_{1\ldots k,1\ldots k}x_j$ then IBMGS allows correction of a portion of the R matrix that was calculated by normal equations factorization. This is included in our algorithm. Also we should note that the BMGS algorithm requires an input parameter κ_1 to determine when to form residual vectors $\hat{a}_j$ and the IBMGS algorithm requires as input κ_1 and the number of iterations desired p. We discuss choice of these shortly. Note that, in principal, $\kappa(A_{1\ldots k}) = \kappa(R_{1\ldots k,1\ldots k})$ and, more generally, $\kappa(A_{\ell\ldots k}) = \kappa(R_{\ell\ldots k,\ell\ldots k})$ (see the algorithms). This is also used in the algorithms.

Our algorithms are:

```
ℓ = 1
while ℓ < n
   r_kk = sqrt(a_k^T a_k)
   while κ(R_{ℓ...k,ℓ...k}) ≤ κ_1 and k ≤ n
      For j = k + 1, ..., n
         r_kj = (a_k^T a_j − Σ_{i=ℓ}^{k−1} r_ik r_ij)/r_kk
      endfor
      k = k + 1
      r_kk = sqrt(a_k^T a_k − Σ_{i=ℓ}^{k−1} r_ik r_ik)
   endwhile
   k = k − 1
   For j = k + 1, ..., n
      Solve R_{ℓ...k,ℓ...k} x = r_{ℓ...k,j} for x
      a_j ← a_j − A_{ℓ...k} x
   endfor
   ℓ = k + 1
endwhile
```

The Block MGS Algorithm

```
ℓ = 1
while ℓ < n
   r_kk = sqrt(a_k^T a_k)
   while κ(R_{ℓ...k,ℓ...k}) ≤ κ_1 and k ≤ n
      For j = k + 1, ..., n
         r_kj = (a_k^T a_j − Σ_{i=ℓ}^{k−1} r_ik r_ij)/r_kk
      endfor
      k = k + 1
      r_kk = sqrt(a_k^T a_k − Σ_{i=ℓ}^{k−1} r_ik r_ik)
   endwhile
   k = k − 1
   For j = k + 1, ..., n
      Solve R_{ℓ...k,ℓ...k} x = r_{ℓ...k,j} for x
      a_j ← a_j − A_{ℓ...k} x
      For q = 1, ..., p
         Solve R^T_{ℓ...k,ℓ...k}(δr) = A^T_{ℓ...k} a_j for δr
         r_{ℓ...k,j} ← r_{ℓ...k,j} + (δr)
         Solve R_{ℓ...k,1...k}(δx) = r_{ℓ...k,j} for δx
         x ← x + (δx)
         a_j ← a_j − A_{ℓ...k} x
      endfor
   endfor
   ℓ = k + 1
endwhile
```

The Improved Block MGS Algorithm

(9)

If column interchanges are incorporated in the above algorithms then the condition numbers $\kappa(R_{\ell\ldots k,\ell\ldots k})$ required by the algorithm can be reliably estimated by $r_{\ell,\ell}/r_{kk}$. Inclusion of these interchanges is not difficult, but due to space considerations we do not describe this here.

In order to compare the efficiency of the algorithms in (2) with those in (9), for simplicity, let assume that in factoring A it was necessary to perform MGS steps (i.e. forming $a_j \leftarrow a_j - A_{\ell\ldots k}x$ for $j = k+1,\ldots,n$) once and that this occurred after row k of R was constructed. Then we can show that with p steps of iterative refinement IBMGS requires

$$mn^2/2 + n^3/6 + k(n-k)\left[m(2p+1) + k(p+1) - n/2\right] \text{ multiplications.}$$

The count for BMGS is obtained by letting $p = 0$. For $m \gg n$ this implies that, in this case, BMGS will require at most $3mn^2/4 + O(n^3)$ multiplications and therefore will be at least 25% faster than MGS. On the other hand, if we choose $p = 1$ and again assume that $m \gg n$ then IBMGS can require as many as $5mn^2/4 + O(n^3)$ multiplications and thus can be less efficient than MGS. However, it is significant to note that for k small or $k \cong n$ then both IBMGS and BMGS will require only slightly more work than normal equations and so will be faster than MGS or Householder based methods. Finally, we should note if $m \geq 2n$ then it can be shown that BMGS necessarily requires less work than MGS, no matter how frequently MGS steps are taken in the BMGS algorithm.

The accuracy of the above algorithms depends on the choice of κ_1. For IBMGS if one step of iterative refinement is used in the algorithm ($p = 1$) then as described earlier it would be appropriate to select $\kappa_1 = \epsilon^{-1/3}$. With such selection, although we have not formally proven it, we have found based on extensive experiments with many difficult test matrices, that IBMGS is just as accurate as methods based on orthogonal factorization. The BMGS algorithm, on the other hand, provides no guarantee that the calculated residuals $a_j \leftarrow a_j - A_{\ell \ldots k} x$ are as accurate as permitted by the data. Experimentally, it appears that the method is often much more accurate than direct factorization of the normal equations but at times not as accurate as methods based on orthogonalization. A choice κ_1 closer to one makes the accuracy of the methods closer to that of the MGS method and choice of κ_1 large make the accuracy similar to that of the normal equations method. In our experiments, while using a computer with machine $\epsilon \cong 10^{-16}$ we found $\kappa_1 = 10^{-3}$ to be a reasonable compromise.

Finally, in this section we should describe the use of the algorithms in (9) in solving (1). We recommend the same procedure described earlier for MGS. In particular b is added as a extra column of A and then the algorithm is applied to the larger matrix. Also we note that iterative refinement could follow any of our algorithms. Thus some of our experiments will describe results for with BMGS + IR, block modified Gram-Schmidt plus iterative refinement.

5 Numerical Experiments

In this section we will compare a number of the methods described above. All the test were done in double precision ($\epsilon \cong 10^{-16}$) using an AT compatible computer with an Intel 80386 processor and 80387 coprocessor. For the IBMGS method p was set to 1 and κ_1 was set to 5×10^{-6} and for the BMGS method κ_1 was set to 10^{-3}.

Problem V. This problem comes from [8]. The A matrix is a $21 \times n$ matrix given by $A = VD, v_{ij} = (i-1)^{(j-1)}$ and D is a diagonal matrix chosen so that $\| a_j \| = 1$. This matrix arises in fitting a nth degree polynomial to data. To examine matrices of varying conditions we selected $n = 6, 12, 14$ and 18. The corresponding condition numbers of A are 2.2×10^3, 1.0×10^8, 4.8×10^9 and 2.4×10^{13}. In the following table the right hand side

n	6		12				14			
iter. no.	0	1	0	1	...	11	0	1	2	...
Method										
Hh	13.4		9.5				8.6			
SN+IR	10.4	14.4	1.3	9.8			-2.0	6.5	8.7	
N+IR	9.9	14.4	1.4	2.2	...	9.8	0.6	0.6	0.6	...
BMGS+IR	10.9	14.4	10.2				5.9	5.9	8.7	
IBMGS	14.4		10.2				8.7			
MGS	14.4		10.2				8.7			

n	18				
iter. no.	0	1	2	...	
Method					
Hh	4.6				
SN+IR	-8.9	-2.9	-2.6	...	4.6
N+IR	-0.3	-0.5	-0.5	...	
BMGS+IR	2.9	2.9	4.1	...	4.5
IBMGS	4.5				
MGS	4.6				

Table 1: Significant Digits in Problem V

b was chosen so that the residual in (1) was zero. In these results the number of significant digits is defined as $\log(\| x - \dot{x} \| / \| x \|)$ where x is the true solution and $\dot{x}$ is the computed solution. The abbreviation Hh in the table stands for the Linpack implementation of Householder factorization.

In this table for $n = 18$, BMGS + IR required four iterations to converge and SN + IR required five iterations to converge. Entries in the table are discontinued after full accuracy is achieved.

We also did experiments with this example for right hand sides b so that (1) had large residuals. For these examples all the methods except N + IR provided accuracy equal to the Householder and MGS methods. Finally, we should note that we had excellent success with the IBMGS and BMGS methods when Problem V was rescaled so that a few rows of A had very large norms. For space reasons we do not include the details here.

Problems R. We consider the class of problems generated by multiplying a given diagonal matrix by lots (30 to 100) random elementary Householder transformations. The diagonal matrix can be chosen to prespecify the spectrum of the resulting matrix A. We intentionally generated spectrums that we thought would lead to difficult problems. For example, spectrums were selected with singular values whose logarithms (base 10) were chosen as follows: (A) uniformly from 0 to - 14; (B) uniformly from 0 to -8; and (C) half from a uniform distribution from 0 to -2 and half from a uniform distribution from -10 to -12. For each case we ran fifty different 80×20 random matrices. For each of Hh, SN + IR, BMGS (without iterative refinement), IBMGS and MGS the average of the number of correct significant digits over the fifty matrices was determined. For spectrums B and C, respectively, these averages were within .2 of 9.3 and 5.5 for every method. For case A, which is very ill-conditioned, all methods but BMGS averaged from 3.4 to 3.6 significant digits and BMGS averaged 1 significant digit. Note that on most of these examples directly factoring the normal equations will produce no correct significant digits.

Problems E. The matrices in these runs are 200×50 matrices with a s singular values between 10^{-10} and 10^{-12} and the remaining singular values between 1 and 10^{-4}. These runs are intended to test the efficiency of the algorithms. The Hh method required from 10 seconds (for s small) to 11.3 seconds (for $s \cong 50$) to solve (1). For $s = 0$, 2, 4 and 8, respectively, BMGS required 6.3, 7.0, 7.1 and 7.9 seconds whereas IBMGS required 6.1, 7.5, 8.7 and 10.6 seconds. For BMGS the maximum time required was 8.5 seconds (for $s = 25$) and for IBMGS the maximum time was 13.9 seconds. This supports our earlier discussion efficiency and indicates that for $m \gg n$

BMGS will be more efficient than orthogonalization based methods and for matrices with a few small singular values IBMGS will be more efficient than such methods. In these runs, except for $s = 0$, Hh and IBMGS had about five correct significant digits and BMGS had about 3 correct significant digits.

References

[1] L. N. Trefethen, *Three mysteries of gaussian elimination*, Signum Newsletter, pp. 2-5, Oct. 1985.

[2] A. Bjork, *Error analysis of least squares algorithms*, Proc. of the NATO Adv. Study Inst. on Numerical Linear Algebra, Digital Signal Processing and Parallel Algorithms, Leuven, Belgium, 1988.

[3] W. Govaerts and J. D. Pryce, *Block elimination with one iterative refinement solves bordered linear systems accurately*, Proc. of the NATO Adv. Study Inst. on Numerical Linear Algebra, Digital Signal Processing and Parallel Algorithms, Leuven, Belgium, 1988.

[4] G. H. Golub, *Matrix Computations*, Johns Hopkins University Press, Maryland, 1983.

[5] A. Bjork, *Solving linear least squares problems by Gram-Schmidt orthogonalization*, BIT 7: 1-21, 1967.

[6] P. A. Wedin, *Perturbation theory for pseudo-inverses*, BIT 13: 217-232, 1973.

[7] G. H. Golub and J. H. Wilkinson, *Note on iterative refinement of least squares solution*, Numer. Math 9: 139-148, 1966.

[8] A. Bjork, *Stability analysis of the method of seminormal equations for linear least squares problems*, Lin. Alg. and its Appl. 88/89: 31-48, 1987.

Block Elimination with Iterative Refinement for Bordered Linear Systems

W. Govaerts

Seminarie voor Hogere Analyse
Rijksuniversiteit-Gent, Belgium

J. D. Pryce

Computational Mathematics Group
R.M.C.S. Shrivenham, UK

Abstract

We describe (without proofs) some recent theory, confirmed by numerical experiments, which shows that Block Elimination with Iterative Refinement is an easily programmed and usually robust method for solving bordered linear systems with a border of width greater than 1. We compare it for accuracy, efficiency and convenience with the Generalized Deflated Block Elimination methods of Chan and Resasco.

1 Introduction

We consider a linear system

$$Mz = h$$

with a fairly well-conditioned matrix of bordered form

$$M = \begin{matrix} & \begin{matrix} n & m \end{matrix} \\ \begin{matrix} n \\ m \end{matrix} & \begin{pmatrix} A & b \\ c & d \end{pmatrix} \end{matrix} \tag{1}$$

NATO ASI Series, Vol. F 70
Numerical Linear Algebra, Digital Signal Processing
and Parallel Algorithms
Edited by G. H. Golub and P. Van Dooren

where n is large and m is small. (The notation means that A is n-by-n, b is n-by-m, and so on.) The case $m = 1$ is fairly common, especially in *continuation methods* [8] but in such applications as the computation of *symmetry-breaking bifurcations* and in *constrained optimization* calculations m is often larger than 1, e.g. [5,6]. The A in (1) may have special features so that software exists which solves systems like

$$Ax = f \tag{2}$$

especially cheaply (in terms of computer time, or memory, or both). We will call such software a *black-box solver*. For instance, with systems arising from discretizing differential equation problems, such software could be a banded or sparse matrix code, a preconditioned conjugate-gradient method, etc. It is then advisable to split the unknown vector z and the R.H.S. h in (1) into

$$z = \begin{matrix} n \\ m \end{matrix} \begin{pmatrix} x \\ y \end{pmatrix} \qquad \text{and} \qquad h = \begin{matrix} n \\ m \end{matrix} \begin{pmatrix} f \\ g \end{pmatrix}$$

and organize the solution of (1) so that most of the work consists in applying the black-box solver. One way to do this is the following *Block-Elimination* algorithm.

Algorithm BE
- BE1 Solve $Av = b$.
- BE2 Compute $\delta = d - cv$.
- BE3 Solve $Aw = f$.
- BE4 Solve $\delta y = g - cw$.
- BE5 Compute $x = w - vy$.

In [4] we discuss a singular-value inequality for block matrices that provides bounds for the condition of the m-by-m matrix δ. In particular, δ is well-conditioned if both M, A are. The algorithm fails spectacularly, however, if A is nearly singular, owing to fatal cancellation in step BE5.

A widely used remedy is the *Generalized Deflated Block Elimination* method (GDBE) of T. F. Chan and D. C. Resasco [1,2]. In [3], we proposed *Block Elimination with Iterative Refinement* (BEIR) as an alternative to GDBE in the case $m = 1$. We here present some results (without proofs) for BEIR in the general case and make a comparison with GDBE.

2 Block Elimination with Iterative Refinement

The algorithm of Block Elimination with k ($k = 0, 1, 2, \ldots$) steps of Iterative Refinement, or BE+k for short, consists of:

Step 1. Apply Algorithm BE, producing solution $\overline{x}$ and $\overline{y}$.
Step 2. For $i = 1, \ldots k$ apply Steps 3 to 5.
Step 3. Compute the residuals $\overline{f} = f - A\overline{x} - b\overline{y}$ and $\overline{g} = g - c\overline{x} - d\overline{y}$.
Step 4. Apply Algorithm BE with right-hand sides $\overline{f}, \overline{g}$, producing $\overline{\overline{x}}$ and $\overline{\overline{y}}$
Step 5. Compute $\overline{x} = \overline{x} + \overline{\overline{x}}, \overline{y} = \overline{y} + \overline{\overline{y}}$.

Remark that steps BE1, BE2 of Algorithm BE need not be repeated in Step 4. The essential cost of Step 4 is just one back-solve with A in BE3, where by the term *back-solve* we mean the process of solving a linear system with matrix A using an already computed factorization (LU, QR, or whatever) of A.

Remark also that Step 3 is performed in single-precision only.

At first sight, it seems unlikely that BE+k will converge when A is nearly singular so that BE (=BE+0) might well produce a solution with relative error greater than 1. In [3] however, we showed that for $m = 1$, BE+1 will in most practical cases produce x and y accurately even when A differs from a singular matrix only by roundoff error.

An important feature of BEIR is that it allows the use of a black-box solver S for a matrix $\overline{A}$ close to A instead of for A. This may lead to considerable time-saving in computations (like numerical continuation) which generate a fairly slowly changing sequence of matrices A.

Obviously, any method to solve systems (1) by using solvers for A must depend heavily on the behaviour of the solver S when applied to singular systems. Our error analysis in [3] assumes that there are matrices Q_1 and Q_2 close to the identity I (they depend on the RHS vector b but their closeness to I is independent of b) such that the computed solution $S(b)$ to $Ax = b$ satisfies

$$S(b) = Q_1 \overline{A}^{-1} Q_2 b$$

where $\overline{A} = A + E$ and $\|E\| \leq C_{LU} u \|A\|$. Here u is the machine roundoff unit.

In the case where $\overline{A}$ differs from A just by roundoff error, the classical Wilkinson error analysis for Gaussian Elimination shows that C_{LU} is a modest number, but in the continuation context mentioned above it could be much larger. The validity of this expression, involving the matrices Q_1 and Q_2, is attested by theory and by extensive numerical evidence for solvers that are based on Gaussian Elimination and for certain other cases. Essentially it is saying "the computed x is near to an x^* which is the exact solution for a b^* near to b". When $\overline{A}$ is well-conditioned, Q_1 and Q_2 can be absorbed into the classical Wilkinson bound but when $\overline{A}$ is nearly singular this is *not* the case.

Let now $\sigma_1 \geq \sigma_2 \geq \ldots \geq \sigma_n \geq 0$ be the singular values of $\overline{A}$ and define $\gamma = \sigma_{n-m+1}/\sigma_n$. (That $\overline{A}$, not A, is relevant here is a subtle-looking point but important to the detailed analysis of the method.) The main result of the error analysis for BE+1 is

Proposition 1 *The absolute error $\|\overline{z} - z\|$ is bounded by a modest multiple of $u\|z\|$ multiplied by the largest of the following numbers:*

$$\begin{aligned} &1 \\ n_1 &= C_{LU}^2 u \\ n_2 &= C_{LU}\sigma_n^{-1}\|M\|u \\ n_3 &= \sigma_n^{-1}\|M\|\gamma u. \end{aligned}$$

Essentially, n_1 is modest when C_{LU} is not unreasonably large; while n_2 is modest when the perturbation E of A is not too unfortunate, i.e. when there is a trade-off between how large C_{LU} is and how non-singular $\overline{A}$ is — for details see [3].

Assume for the moment that these conditions are satisfied. Then, for $m = 1$, BE+1 always produces an accurate solution as shown in [3]; while for $m > 1$ the relative error in z is of order $n_3 u$.

Further investigation in this case leads to a bound of the form

$$\text{(relative error)} \leq (D_1\|M\|u/\sigma_n).(D_2\gamma u)^k$$

for the solution by BE+k, where D_1 and D_2 are modest. Extensive numerical evidence confirms that this is also the size of the error that is to be expected in practice; the linear gain in accuracy as k increases is striking and so is the correlation between D_2 and the condition of M.

For well-conditioned M convergence fails only when γu is of order 1, which implies $\sigma_n \sim \|M\|u$ and $\sigma_{n-m+1} \sim \|M\|$. We are then in the remarkable case of a non-convergent algorithm that can be forced to fast convergence just by a random perturbation of appropriate size in the data matrix.

If n_1 is no longer modest and $C_{LU}u\,\text{cond}(M)$ is small relative to unity, then we expect $\sigma_n \sim \|M\|\sqrt{u}$.

If this is not the case, then the perturbation is unfortunate. If it is, the relative error by BE+k is of order $(C_{LU}u\,\text{cond}(M))^{k+1}$ and so is independent of the singular values of A.

At least in continuation applications, the case of an "unfortunate" perturbation is not merely something that arises now and again by chance, and just to be regarded as a nuisance event. The perturbation is related to states of the problem at previous continuation steps, and the appearance of "unfortunate" perturbations reflects that we are approaching a point of higher-order singularity, which must be studied separately anyway — cf. [3].

3 Comparison of BEIR with GDBE

Solvers Our requirements on the solver in the previous section are, in practice, satisfied by solvers based on LU and QR factorization. On the other hand, BEIR fails (in the case where A is nearly singular) when the solver consists in multiplication by an explicit computed inverse of A; but so does GDBE in this case. We conjecture that solvers that work for one of the methods will also work for the other.

Complexity GDBE as proposed in [2] requires $m + 1 + 4\mu$ backsolves where μ is at least the rank-deficiency of A (if this is not known, take $\mu = m$). This number can probably be decreased (one iteration of Algorithm SII in [2] is usually sufficient; also see [7]). On the other hand BE+k requires $m + 1 + k$ backsolves. For $m = 1$, BE+1 is certainly optimal. For $m > 1$ the choice may depend on the availability of solvers for systems near A, and on the distribution of the small singular values of the actual $\overline{A}$ that are used.

Reliability BEIR may fail to converge in exceptional cases (which have special significance, cf. previous section). If it converges, the accuracy obtained is always competitive with Gaussian Elimination performed on the whole matrix M. Since computation of the residual is part of the algorithm, the convergence or lack of it can easily be monitored.

GDBE was never found to fail for well-conditioned M. However, for rather ill-conditioned M it tends to produce results that are not competitive with GE on the whole of M. It is possible to extend the algorithm by adding iterative improvement steps (each step again requires one backsolve with A).

Flexibility BEIR is easier to program and, in typical applications, allows one to economize on matrix factorizations by using a (previously obtained) solver for a matrix near A.

References

[1] Chan T.F., *Deflation techniques and block-elimination algorithms for solving bordered singular systems*, SIAM J. Sci. Stat. Comput. 5 (1984) 121–134.

[2] Chan T.F. and Resasco D.C., *Generalized deflated block elimination*, SIAM J. Numer. Anal. 23 (1986) 913–924.

[3] Govaerts W. and Pryce J.D., *Block elimination with one iterative refinement solves bordered linear systems accurately*, University of Bristol School of Mathematics Report AM-88-01 (submitted to BIT).

[4] Govaerts W. and Pryce J.D., *A singular value inequality for block matrices*, Preprint (1988).

[5] Jepson A. and Spence A., *Singular points and their computations* in T. Kupper, H. Mittelmann and H. Weber (eds.) *"Numerical Methods for Bifurcation Problems"*, Int. Ser. of Numer. Math. vol. 70, Birkhäuser Verlag, Basel (1984) 195–209.

[6] Jepson A. and Spence A., *Folds in solutions of two-parameter systems and their calculation, Part I*, SIAM J. Numer. Anal. 22 (1985) 347–368.

[7] Moore G., *Some remarks on the deflated block elimination method*, Preprint, Imperial College Department of Mathematics (1987).

[8] Rheinboldt, Werner C., *"Numerical Analysis of Parametrized Nonlinear Equations"*, J. Wiley and sons, New York (1986).

The Efficient Calculation of the Eigenvalues, Eigenvectors and Inverses of a Special Class of Brownian Matrices

M. J. C. Gover*

Department of Mathematics
University of Bradford
Bradford, West Yorkshire, UK

Abstract

A Brownian matrix is defined by having all the elements in any row, to the right of the principle diagonal, equal, and all elements in any column, below the principal diagonal, equal. These matrices occur in signal processing problems, and it has been shown that the solution of Brownian systems of linear equations can be solved in $0(n)$ flops. The inverse of a nonsingular Brownian matrix can be found in $0(n^2)$ flops.

In this paper we consider a special class of Brownian matrices, which form a ring, and show that both the normal inverse, when nonsingularity applies, and some generalized inverses can be found in $0(n)$ flops. In addition, explicit formulae for the determinant, the eigenvalues and eigenvectors are given, also requiring $0(n)$ flops.

*Presented at the NATO Advanced Study Institute Interdisciplinary Meeting on Numerical Linear Algebra, Digital Signal Processing and Parallel Algorithms. Leuven, Belgium, August 1988

NATO ASI Series, Vol. F 70
Numerical Linear Algebra, Digital Signal Processing and Parallel Algorithms
Edited by G. H. Golub and P. Van Dooren

1 Introduction

The use of Toeplitz matrices in signal processing models is well known and occurs in problems concerning stationary signals. When a signal is not stationary a simple model which can be used involves the Brownian matrix which is defined as follows.

Definition 1.1 A matrix $B = [b_{ij}]$ of order n is *Brownian* if

$$b_{i,j+1} = b_{ij}, \quad j > i \text{ and } b_{i+1,j} = b_{ij}, \quad i > j$$

for all possible i and j.

In recent years several papers [1,4,5,6] have investigated properties of the Brownian matrix.

Remark 1.1 Brownian matrices are called *diagonal innovation* (DIM) matrices in [1] and [5].

In [1] and [6] Toeplitz-like recursive algorithms were used to solve a Brownian system of linear equations in $0(n^2)$ flops. However, it was shown, independently in [4] and [5] that this could be reduced to $0(n)$ flops. This is a consequence of the following result.

Theorem 1.1 A Brownian matrix A is congruent to a tridiagonal matrix.

Specifically, B is tridiagonal if $B = PAP^T$ where P is the lower triangular Toeplitz matrix defined by its first column $[1, -1, 0, \ldots, 0]^T$.

Remark 1.2 Some generalizations of the Brownian matrix are considered in [4] by either replacing the tridiagonal matrix B above by a band matrix and/or by replacing P by a more general Toeplitz matrix.

Remark 1.3 The tridiagonal matrix B is what occurs in the theory of random walks which is the discrete version of Brownian motion.

It is well known that the set of Toeplitz matrices is not closed under the operation of ordinary matrix multiplication and in [2] it is shown that in order to form a ring it is necessary to restrict Toeplitz matrices to be k-circulant.

It is easy to see that Brownian matrices are also not closed under multiplication and the purpose of this paper is to give the restrictions necessary in order that Brownian matrices form a ring $\mathcal{B}$. These restrictions and some of the properties of matrices in $\mathcal{B}$ are given in the next section. In Section 3 it is shown that Brownian matrices in $\mathcal{B}$ are similar to an almost diagonal matrix and so its eigenvalues and eigenvectors are easily found. Results on inverses are also given. Further details and proofs of these results can be found in [3].

2 A ring of Brownian matrices

Many of the results in this section depend on row and column sums, so that we begin by introducing a useful notation

Definition 2.1 If $F = [f_{ij}]$ is a matrix of order n, then

$$F_{Ri} = \sum_{k=1}^{n} f_{ik} \quad \text{and} \quad F_{Cj} = \sum_{k=1}^{n} f_{kj}\,. \tag{2.1}$$

If A and B are two Brownian matrices of order n, as given by Definition 1.1, then we have the following result.

Theorem 2.1 If $a_{12} \neq 0$ and $b_{21} \neq 0$ than AB is Brownian if and only if $A_{Ri} = \alpha$, $i = 2, 3, \ldots, n$ and $B_{Cj} = \beta$, $j = 2, 3, \ldots, n$ for some arbitrary constants α and β.

In fact, if $A_{Ri} = \alpha$ and $B_{Cj} = \beta$ then AB is Brownian without the extra restrictions, $a_{12} \neq 0$ and $b_{21} \neq 0$. Since similar restrictions on the columns of A and the rows of B are required for BA to be Brownian, we now consider the following class of matrices.

Definition 2.2 A matrix A of order n belongs to the class $\mathcal{B}$ if it is Brownian and for some constants α and β

$$A_{Ri} = \alpha, \quad i = 2, 3, \ldots, n \quad \text{and} \quad A_{Cj} = \beta, \quad j = 2, 3, \ldots, n\,. \tag{2.2}$$

It can now be shown that if $A \in \mathcal{B}$ and $B \in \mathcal{B}$ then $A \pm B$, AB and BA all belong to $\mathcal{B}$. Clearly the zero and unit matrices are also in $\mathcal{B}$ and so we obtain the central result of this paper.

Theorem 2.2 The class $\mathcal{B}$ given by Definition 2.2 is a ring.

When A is nonsingular its inverse can be obtained from its characteristic equation in terms of powers of A and thus from Theorem 2.2, $A^{-1} \in \mathcal{B}$.

Remark 2.1 A consequence of a later result is that the Moore-Penrose and group generalized inverses also belong to $\mathcal{B}$.

We next consider what effect (2.2) has on the elements of $A \in \mathcal{B}$. These row and column sum restrictions show that A is almost symmetric in the following sense.

Theorem 2.3 If $A \in \mathcal{B}$ then the submatrix formed by omitting its first row and column is symmetric.

As an immediate consequence of this the following result is obtained.

Corollary If $A \in \mathcal{B}$ then it is completely determined by the $n+2$ elements a_{11}, a_{12}, a_{21}, a_{22} and $a_{i,i+1}$, $i = 2, 3, \ldots, n-1$.

Remark 2.2 A general Brownian matrix is defined by $3n-2$ elements.

The result of Theorem 2.3 leads to an alternative definition of the ring $\mathcal{B}$ in terms of magic square matrices, details of which can be found in [2].

Definition 2.3 A matrix $M \in M(1{,}2)$ if $M_{Ri} = M_{Cj} = \alpha$ for some constant α.

If we now consider the *trailing* submatrices A_{n-r} of A, i.e. square matrices formed by omitting the first r rows and columns of A, $r = 1, 2, \ldots, n-1$ then A has the following characterisation.

Theorem 2.4 If A is a matrix of order n then $A \in \mathcal{B}$ if and only if $A_{n-r} \in M(1{,}2)$, $r = 1, 2, \ldots, n-1$ and

$$a_{1j} = a_{12}, \quad j = 3, 4, \ldots, n \text{ and } a_{i1} = a_{21}, \quad i = 3, 4, \ldots, n\,.$$

3 Efficient numerical methods for $A \in \mathcal{B}$

It has already been pointed out that a Brownian matrix is congruent to a tridiagonal matrix. If $A \in \mathcal{B}$ a much stronger result can be obtained.

Theorem 3.1 If $A \in \mathcal{B}$ then it is similar to B which is the direct sum of a 2×2 and a diagonal matrix of order $n-2$.

The elements of $\mathcal{B}$ can be written down simply from those of A and the similarity matrix R, below, is never used specifically.

If $A = [a_{ij}]$ and $A \in \mathcal{B}$ then

$$B = RAR^{-1} = \begin{bmatrix} a_{11} & a_{12} \\ (n-1)a_{21} & a_{22} + (n-2)a_{23} \end{bmatrix} \oplus \operatorname{diag}(a_{22} - a_{23}, \ldots, a_{n-1,n-1} - a_{n-1,n}) \tag{3.1}$$

where $\oplus$ denotes the direct sum and

$$R = \begin{bmatrix} 1 & 0 & 0 & 0 & \cdots & 0 & 0 \\ 0 & 1 & 1 & 1 & \cdots & 1 & 1 \\ 0 & -\frac{n-2}{n-1} & \frac{1}{n-1} & \frac{1}{n-1} & \cdots & \frac{1}{n-1} & \frac{1}{n-1} \\ 0 & 0 & -\frac{n-3}{n-2} & \frac{1}{n-2} & \cdots & \frac{1}{n-2} & \frac{1}{n-2} \\ \cdot & \cdot & \cdot & \cdot & & \cdot & \cdot \\ \cdot & \cdot & \cdot & \cdot & & \cdot & \cdot \\ \cdot & \cdot & \cdot & \cdot & & \cdot & \cdot \\ 0 & 0 & 0 & 0 & & -\frac{1}{2} & \frac{1}{2} \end{bmatrix} \tag{3.2}$$

The following consequences of Theorem 3.1 are easily obtained.

Theorem 3.2 The determinant of A, $|A|$ is given by

$$|A| = [a_{11}a_{22} + (n-2)a_{11}a_{23} - (n-1)a_{21}a_{12}] \prod_{k=3}^{n} (a_{k-1,k-1} - a_{k-1,k}).$$

Theorem 3.3 If λ_i are the eigenvalues of A then λ_1 and λ_2 are the roots of the equation

$$\lambda^2 - [a_{11} + a_{22} + (n-2)a_{23}]\lambda + a_{11}[a_{22} + (n-2)a_{23}] - (n-1)a_{21}a_{12} = 0$$

and $\lambda_i = a_{i-1,i-1}, -a_{i-1,i}, i = 3, 4, \ldots, n$. If $a_{12} \neq 0$ and $\lambda_1 \neq \lambda_2$ then the corresponding eigenvectors are

$$\underline{x}_i = \begin{cases} \begin{bmatrix} -a_{12} \\ \frac{1}{n-1}(a_{11} - \lambda_i)\underline{e}^{(n-1)} \end{bmatrix}, & i = 1, 2, \\ \begin{bmatrix} -\underline{e}_{i-1}^{(i-1)} \\ \frac{1}{n-i+1}\underline{e}^{(n-i+1)} \end{bmatrix}, & i = 3, 4, \ldots, n \end{cases}$$

where $\underline{e}_j^{(k)}$ is the jth column of the unit matrix of order k and $\underline{e}^{(k)} = [1,1,\ldots,1]^T$ is of the order k.

Remark 3.1 Minor modifications are required if $a_{12} = 0$ and/or $\lambda_1 = \lambda_2$.

Thus the determinant and the eigenvalues and eigenvectors of $A \in \mathcal{B}$ can be found explicitly from its elements in $0(n)$ flops.

We next consider the inverse of A when A is nonsingular. We first note that B^{-1} has the same form as B. If

$$B^{-1} = \begin{bmatrix} \beta_{11} & \beta_{12} \\ \beta_{21} & \beta_{22} \end{bmatrix} \oplus \operatorname{diag}(\beta_3, \beta_4, \ldots, \beta_n)$$

then it can be shown that $A^{-1} = [\alpha_{ij}]$, which also belongs to $\mathcal{B}$, can be written down using

$$\left.\begin{aligned} \alpha_{11} &= \beta_{11},\ \alpha_{12} = \beta_{12},\ \alpha_{21} = \beta_{21}/(n-1), \\ \alpha_{22} &= \tfrac{1}{n-1}[\beta_{22} + (n-2)\beta_3],\ \alpha_{23} = \tfrac{1}{n-1}[\beta_{22} - \beta_3], \\ \alpha_{i,i+1} &= \alpha_{i-1,i} + \tfrac{1}{n-i+1}(\beta_i - \beta_{i+1}),\ i = 3,4,\ldots,n-1, \\ \alpha_{ii} &= \beta_{i+1} + \alpha_{i,i+1},\ i = 3,4,\ldots,n-1,\ \alpha_{nn} = \alpha_{n-1,n-1}. \end{aligned}\right\} \tag{3.4}$$

Thus to find A^{-1}, we first write down B from A using (3.1), find B^{-1} and finally use (3.4). This requires just $0(n)$ flops as against $0(n^2)$ flops for a general Brownian matrix.

Remark 3.2 There seems to be no advantage using the similarity transformation of Theorem 3.1 over the existing algorithms [4,5] for solving $A\underline{x} = \underline{b}$, when $A \in \mathcal{B}$.

Finally if we replace R in (3.2) by $R_1 = DR$ where

$$D = \operatorname{diag}[1, \sqrt{\frac{1}{n-1}}, \sqrt{\frac{n-1}{n-2}}, \ldots, \sqrt{2}]$$

and consider $B_1 = R_1 A R_1^{-1}$ then B_1 has the same form as B in (3.1) and moreover R_1 is orthogonal.

The Moore-Penrose inverse $X = A^+$, of A, satisfies

$$AXA = A, \quad XAX = X \tag{3.5}$$

$$(AX)^* = AX, \quad (XA)^* = XA \tag{3.6}$$

and the group inverse $X = A^\#$ satisfies (3.5) and

$$AX = XA. \tag{3.7}$$

It is now easy to show that $A^+ \in \mathcal{B}$ and $A^\# \in \mathcal{B}$ when $A \in \mathcal{B}$. The determination of A^+ is therefore achieved by first finding $B_1 = R_1 A R_1^{-1}$, finding B_1^+ and then finding A^+. The process takes $0(n)$ flops and is similar to that for finding A^{-1} but with slight modifications due to the use of R_1 instead of R. The calculation of $A^\#$ is similar.

References

[1] G. Carayannis, N. Kalouptsidis and D.G. Manolakis, Fast recursive algorithms for a class of linear equations, IEEE Trans. Acoust. Speech, Signal Proc., ASSP-30: 227-239 (1982).

[2] P.J. Davis, *Circulant matrices*, Wiley, New York, 1979.

[3] M.J.C. Gover, A ring of Brownian matrices, Linear Algebra Appl., to appear.

[4] M.J.C. Gover and S. Barnett, Brownian matrices: properties and extensions, Int. J. Systems Sci., 17: 381-386 (1986).

[5] H. Krishna and S.D. Morgera, Fast $0(n)$ complexity algorithms for diagonal innovation matrices, IEEE Trans. Acoust. Speech, Signal Proc., ASSP-32: 1189-1193 (1984).

[6] B. Picinbono, Fast algorithms for Brownian matrices, IEEE Trans. Acoust. Speech, Signal Proc., ASSP-31: 512-514 (1983).

Improvements of Stepsize Control in Numerical Integration

Kjell Gustafsson

Department of Automatic Control
Lund Institute of Technology
S-221 00 Lund, Sweden

Abstract

Stepsize control in numerical integration of ODE is an essential tool to improve the precision of the produced solution. In addition, a well behaved stepsize sequence shortens the computing time needed to produce the solution. Normally an estimate of the solution error is fed back to adjust the stepsize, so that the error of the solution is kept below some user-specified level. In this paper stepsize adjustment is viewed as a feedback control problem. This gives insight as well as an improved algorithm.

Today's stepsize control is derived assuming a static relation between steplength and error. This relation is a consequence of an asymptotic reasoning where the steplength tends to zero. For certain combinations of integration method and differential equation the resulting stepsize controller yields an oscillatory and highly irregular stepsize sequence. Such a sequence may excite modes that the error estimator fails to recover, yielding an erroneous solution to the problem. A well known example of this behaviour is explicit Runge-Kutta methods applied to stiff differential equations.

To design a better stepsize controller a good model of the relation between the stepsize and the error of the solution is needed. The derivation of such a model is presented. Our analysis shows that for small stepsizes a static model is sufficient, while for the case of stepsize being limited by stability, a dynamic model has to be used.

NATO ASI Series, Vol. F 70
Numerical Linear Algebra, Digital Signal Processing and Parallel Algorithms
Edited by G. H. Golub and P. Van Dooren

The standard controller used to today does not account for this change in behaviour, which results in the anomalies described above.

The derived models are verified using numerical experiments. Data is collected when integrating nonlinear differential equations, and models are identified using standard system identification techniques. The identified models agree well with the theoretical results, and the change from a static to a dynamic relation is confirmed.

The new system model is used to construct an improved stepsize controller. It resolves the stepsize control anomalies and gives superior performance at little extra expense. Comparative numerical tests are presented.

The control theoretic view-point also gives a tool to address fundamental questions such as: What performance can be achieved? What trade-offs have to be done?

A Statistical Evaluation of Inverse Iteration

E. R. Jessup

Yale University
Department of Computer Science
New Haven, CT 06520, USA

Abstract

The eigenvalues of a symmetric tridiagonal matrix can be computed by bisection using Sturm sequences and their corresponding eigenvectors found by inverse iteration. The independent computing tasks comprising this combination are especially appropriate for implementation on a distributed memory multiprocessor, but the parallel efficiency of bisection and inverse iteration is not enough to recommend their use in general. Bisection produces eigenvalues to high accuracy. Inverse iteration results in eigenvectors with quality dependent on the spacing of the eigenvalues and on the starting vector. This paper examines the factors influencing the accuracy of inverse iteration. The effects of random starting vectors are discussed, and some uses of statistical analysis in the design of an inverse iteration algorithm are presented.

1 Introduction

The problem of computing the eigenvalues and eigenvectors of a real symmetric matrix arises in a variety of contexts including problems in quantum physics [8], structural analysis [12], and modelling acoustic and electromagnetic waveguides [11]. Symmetric tridiagonal matrices result directly from

NATO ASI Series, Vol. F 70
Numerical Linear Algebra, Digital Signal Processing and Parallel Algorithms
Edited by G. H. Golub and P. Van Dooren

the mathematical models describing some such applications and as a reduced form of dense or banded symmetric matrices from others. These matrices are often of very large order, and, in some cases, all eigenvalues and eigenvectors are required. Computing these complete eigensystems is a time-consuming process demanding extensive memory and thereby motivating a parallel approach. Distributed-memory multiprocessors, in particular, can provide both the storage and computing resources required for efficient solution of the large order problems.

Methods for computing eigenvalues and eigenvectors of a symmetric tridiagonal matrix include divide and conquer techniques [3,10], bisection with inverse iteration [15], and the QR method [2]. Of these, only bisection and inverse iteration consist of independent computing tasks for which straightforward and effective static load balancing of a distributed-memory multiprocessor is possible; reorthogonalization of eigenvectors computed by inverse iteration requires interprocessor communication but can be pipelined efficiently. Bisection and inverse iteration thus display both high speedup and fast execution time on a distributed-memory hypercube multiprocessor [7].

The speed and versatility of bisection and inverse iteration alone, however, are not enough to recommend their use in general. The residual error and the deviation from orthogonality of the computed eigenvector matrix must be considered as well. Bisection produces eigenvalues to high relative accuracy and often to high absolute accuracy [4]. Inverse iteration results in eigenvectors with quality dependent on the spacing of the eigenvalue spectrum and on the starting vector. Section 2 of this paper discusses the factors influencing the accuracy of inverse iteration. The effects on the accuracy of the solution of using random starting vectors are examined. In addition, some uses of statistical analysis in the design of an inverse iteration algorithm are presented. Section 3 includes a statistical analysis of the iterates resulting from random starting vectors. Conclusions follow in Section 4.

2 The Accuracy of Inverse Iteration

In [15,16], Wilkinson and Peters describe a procedure for computing the eigenvectors of an unreduced symmetric tridiagonal matrix T of order n given its eigenvalues. Although all eigenvalues of T are distinct in exact arithmetic, some may be computationally coincident in floating point arithmetic. Before starting inverse iteration, any such eigenvalues are perturbed to a spacing on the order of machine precision and dependent on the magnitudes of the matrix elements. These perturbed values are used in place of close eigenvalues as the shifts for inverse iteration in an attempt to produce linearly independent eigenvectors for coincident eigenvalues. Inverse iteration with a starting vector proportional to the vector of all ones scaled to prevent overflow is employed to compute the eigenvectors. Each iteration ends with a reorthogonalization by the Modified Gram-Schmidt procedure of all computed eigenvectors corresponding to eigenvalues separated by a distance of less than $O(10^{-3})$. Iteration continues in this way until the norm of the orthogonalized iterate is greater than one. Inverse iteration according to these criteria has been implemented as EISPACK's TINVIT [13].

The remainder of this section identifies the three main features of the above implementation influencing its accuracy, *i.e.*, the choice of starting vectors, the selection of vectors to be reorthogonalized, and the iteration stopping criterion. The details of the numerical experiments leading to the following observations and specifications of the resulting refined code will appear in a future paper. That paper will also include the proofs of the theorems presented in Section 3.

The starting vectors affect the accuracy of inverse iteration in several ways. To begin, when the same starting vector is employed for every eigenvector computed, perturbation of eigenvalues is insufficient to produce linearly independent eigenvectors when eigenvalues are strongly clustered. Experiments show, however, that use of a different random vector for each computed eigenvalue generally removes this linear dependence. While few definitive statements can be made about these or any other starting vectors without advance information about the eigenvectors, starting vectors with normally distributed random components lend themselves to statistical evaluation of their expected quality. In particular, it can be shown that the n different random starting vectors are likely to be linearly independent and, hence, to lead to linearly independent iterates. Furthermore, the probability that the starting vector is orthogonal to the target eigenvector is very small and, consequently, that rapid convergence of inverse iteration is expected. These results are formalized in Theorems 3.1 and 3.2.

In addition, the simple reorthogonalization criterion employed in TINVIT can produce groups of eigenvectors that, while orthogonal to each other, are not necessarily orthogonal to the other groups. A more stringent orthogonalization criterion based on the statistical analysis can lead to a complete set of eigenvectors orthogonal to near working precision in most cases. The relation of eigenvalue spacing, matrix order, and number of iterations to the expected deviation from orthogonality of the eigenvector matrix is quantified by Corollary 3.2 to Theorem 3.3.

The norm growth of the iterate and related speed of convergence of inverse iteration are also dependent on the quality of the starting vector. Comparing the norm to an arbitrarily fixed value can lead to too many or too few iterations for many problems. TINVIT usually suffers from the latter, and the accuracy of the resulting solution is worse than that achieved with additional iterations. Theorem 3.4 gives the probability that an iterate corresponding to a computationally distinct eigenvalue and, therefore, not requiring reorthogonalization will have greater than unit norm.

3 A Statistical Analysis

The following analysis proceeds in two parts. To begin, the likelihood that a chosen starting vector has appropriate direction is determined. The residual and orthogonality of subsequent iterates are then examined. In both cases, the quality of a vector is measured by its distance from the desired subspace. The proofs of the theorems below involve integrating distribution functions of quadratic forms in the vector components. Similar techniques have been used for matrix condition numbers [6]. The statistical fundamentals underlying these results may be found in [1,5,9,14].

The analysis of inverse iteration is based on the assumption that the starting vector x_0 has independent random components each having a normal distribution with mean 0 and variance 1 (normal (0,1)). This vector is then normalized so that $y_0 = x_0/\| x_0 \|_2$ has unit norm. Vectors of length n so generated are uniformly distributed on the unit n-sphere [5]. Unless otherwise specified, all vectors are represented in terms of the orthogonal basis of eigenvectors $\{u_1, u_2, \dots, u_n\}$ of the symmetric tridiagonal matrix T: $x_0 = (\xi_1, \xi_2, \dots, \xi_n)$ means $x_0 = \sum_{i=1}^n \xi_i u_i$. Use of the basis of eigenvectors is permitted because the distribution of the components of vectors uniformly distributed on the sphere is invariant under orthogonal transformations. (See [5], for example.)

3.1 The Quality of the Starting Vectors

Let $y_0 = \sum_{i=1}^n \eta_i u_i$ with unit norm. y_0 is considered a good approximation to u_i if η_i is much larger than any other component, *i.e.*, if $\eta_i^2 \geq 1 - \epsilon^2$ for some tiny error tolerance ϵ. Similarly, y_0 is nearly a linear combination of eigenvectors $u_1, \dots, u_d$ if $\sum_{i=1}^d \eta_i^2 \geq 1 - \epsilon^2$. Because the random vectors are uniformly distributed on the sphere, the probability that a random starting vector is a good approximation to such a linear combination is just the fraction of the surface area of the sphere defined by the good vectors. This fraction is at least as large as the bound given in Theorem 3.1. (Note that the result is independent of the set of eigenvectors chosen.)

Theorem 3.1. *Let $x_0 = (\xi_1, \dots, \xi_n)$, components each with a normal (0,1) distribution and $y_0 = x_0/\| x_0 \|_2 = (\eta_1, \dots, \eta_n)$. Given ϵ, the probability that $\sum_{i=1}^d \eta_i^2 \geq 1 - \epsilon^2$ is*

$$P(\sum_{i=1}^{d} \eta_i^2 \geq 1 - \epsilon^2) \geq 1 - \frac{1}{d\sqrt{\pi}}(n/d)^{d/2}(1-\epsilon^2)^{d/2},$$

for $1 \leq d \leq n-1$.

When $n = 3$, the probability that y_0 is a good approximation to u_3 (or $-u_3$) is at least as large as $P(\eta_3^2 \geq 1 - \epsilon^2) > 1 - \sqrt{3(1-\epsilon^2)/\pi}$. In this case, $d = 1$, and P is the probability that y_0 lies inside the double-sided cone around u_3 having vertex at the center of the unit sphere and making interior angle $\cos^{-1} \eta_3$ with eigenvector u_3. Equivalently, the cone has radius $\sqrt{1 - \eta_3^2}$. The probability that y_0 is a good approximation to a linear combination of u_1 and u_3 ($d = 2$) is $P(\eta_1^2 + \eta_3^2 \geq 1 - \epsilon^2) > 1 - \frac{3(1-\epsilon^2)}{4\sqrt{\pi}}$. Such a vector lies in a stripe around the sphere and makes an angle of at most $\sin^{-1}\eta_2 < \sin^{-1}\epsilon$ (for small ϵ) with u_1 or u_3. When $(1 - \epsilon^2) = 0.5$, the probability that y_0 approximates one eigenvector is at least 0.31, while the probability that it approximates a linear combination of two eigenvectors grows to 0.88.

The following theorem gives the probability that n randomly generated vectors are linearly dependent to working precision:

Theorem 3.2. *Let the vectors have independent random components each with a normal (0,1) distribution and unit norm $x_i^T x_i = 1$, and let that $\| \sum_{i=1}^n \alpha_i x_i \|_1 \leq \epsilon$ is*

$$P(\| \sum_{i=1}^{n} \alpha_i x_i \|_1 \leq \epsilon) \leq n(\epsilon/\sqrt{2\pi})$$

or the probability that the vectors $x_1, \ldots, x_n$ are linearly dependent to within a tolerance ϵ.

3.2 The Quality of the Iterates

The following results pertain to the outcome of inverse iteration and share a common terminology. The vector $x_0 = (\xi_1, \ldots, \xi_n), n \geq 2$, has independent random components each with a normal (0,1) distribution, and $y_0 = x_0/\|x_0\|_2$. The kth unnormalized iterate is denoted by $x_k = (T-\lambda)^{-k} y_0$ and the corresponding normalized iterate by $y_k = x_k/\|x_k\|_2 = (\eta_1, \ldots, \eta_n)$. Unless otherwise specified, λ most closely approximates λ_1, and $\lambda_1 = \ldots = \lambda_d$ for some d such that $1 \leq d \leq n$. Throughout, it is assumed that $|\lambda_d - \lambda|$ is small but not zero, and that $d << n$.

Note that because $(T - \lambda)^{-1}$ is not, in general, an orthogonal matrix, y_k is not uniformly distributed over the unit sphere, and the equivalence of surface area and probability that holds for the starting vector does not hold for the iterates. However, the iterate y_k may be expressed in terms of the initial vector x_0:

$$y_k = \frac{(T-\lambda)^{-k} y_0}{\| (T-\lambda)^{-k} y_0 \|_2} = \frac{(T-\lambda)^{-k} x_0/\| x_0 \|_2}{\| (T-\lambda)^{-k} x_0/\| x_0 \|_2 \|_2} = \frac{(T-\lambda)^{-k} x_0}{\| (T-\lambda)^{-k} x_0 \|_2}.$$

Thus, the squares of the components of y_k are

$$\eta_j^2 = \frac{\xi_j^2}{(\lambda_j - \lambda)^{2k}} \Big/ \sum_{i=1}^{n} \frac{\xi_i^2}{(\lambda_i - \lambda)^{2k}}$$

for $1 \leq j \leq n$. This observation leads to the following theorem.

Theorem 3.3. *Given $\epsilon^2 \leq 1$, the probability that $\sum_{i=1}^{d} \eta_i \geq 1 - \epsilon^2$ is*

$$P(\sum_{i=1}^{d} \eta_i \geq 1 - \epsilon^2) \geq 1 - \frac{1}{d\sqrt{\pi}} \left(\frac{n}{d}\right)^{d/2} (1 - \epsilon^2)^{d/2} \left|\frac{\lambda_d - \lambda}{\lambda_{d+1} - \lambda}\right|^{kd}.$$

This result differs from Theorem 3.1 in the inclusion of a term dependent on the separation of the cluster of eigenvalues from the closest distinct eigenvalue and for $k = 0$ reduces to Theorem 3.1. When $|\lambda_d - \lambda_{d+1}|$ is small, the iterate is more likely to have sizable components in the directions corresponding to eigenvalues outside of the cluster.

Generalization of this theorem to the case that $\lambda \approx \lambda_j = \ldots = \lambda_d$ is straightforward. The proof requires only substitution of the quantity $\min((\lambda_{j-1} - \lambda), (\lambda_{d+1} - \lambda))$ for the $(\lambda_d - \lambda)$ in the above theorem and its proof. This observation leads to the following corollaries.

Corollary 3.1. *(Residual) Suppose $\lambda \approx \lambda_j$ and $(\lambda_m - \lambda) = \min_{i \neq j}(\lambda_i - \lambda)$. Define the residual vector $r = (Ty_k - \lambda y_k)/\| T \|_2$. The probability that*

$$(\lambda_j - \lambda)^2 / \| T \|_2^2 \leq \| r \|_2^2 \leq [(\lambda_j - \lambda)^2 + \epsilon^2(\lambda_m - \lambda)^2] / \| T \|_2^2$$

is at least $1 - \sqrt{\frac{n}{\pi}}(1 - \epsilon^2)^{1/2} \left|\frac{\lambda_j - \lambda}{\lambda_m - \lambda}\right|^k$.

The lower bound holds with probability one and expresses the error inherent in using an approximate eigenvalue. As $|\lambda_j - \lambda_m|$ decreases, $|\lambda_m - \lambda|$ approaches $|\lambda_j - \lambda|$, and both the upper bound and the probability of attaining it decrease. When $|\lambda_j - \lambda_m|$ is very small, the probability of obtaining a small residual does not increase significantly with additional iterations, reflecting the difficulty of producing good approximations when the eigenvalues are clustered.

Corollary 3.2. *(Orthogonality) Suppose* $\lambda(j) \approx \lambda_j$ *and* $(\lambda_{m(j)} - \lambda(j)) = \min_{i \neq j}(\lambda_i - \lambda(j))$. *Let* $y_k(j)$ *be the result of* k *inverse iterations using the shift* $\lambda(j)$. *The probability that* $(e_j^T y_k(j))^2 \geq (1 - \epsilon^2)$ *for* $j = 1, \ldots, n$ *is at least* $1 - \sqrt{\frac{n}{\pi}}(1-\epsilon^2)^{1/2} \sum_{i=1}^{n} \left| \frac{\lambda_i - \lambda(i)}{\lambda_{m(i)} - \lambda(i)} \right|^k$.

Clustered eigenvalues lessen the probability that the n computed eigenvectors all approximate their appropriate eigenvectors. Moreover, when eigenvalues are very close, performing additional iterations does not markedly improve the probability of orthogonal results. Corollary 3.2 determines how many iterations are necessary to achieve orthogonality with a given probability, and, by quantifying when iteration will not lead to improvement, shows when reorthogonalization is necessary. Good orthogonality and residual are less likely when the problem size is large.

Theorem 3.4 *(Norm) Suppose* $\lambda \approx \lambda_j$, *where* λ_j *is distinct from* λ_{j-1} *and* λ_{j+1}. *The probability that* $\| (T - \lambda)^{-k} y_0 \|_2^2 \geq \alpha^2 \geq 1$ *is at least* $1 - \sqrt{n/\pi}|\lambda_j - \lambda|^k|\alpha|$.

When λ is a good approximation to λ_j, the probability of achieving at least unit norm is large and increases with iterations. The smaller the bound α, the more likely it is to be achieved. Because λ_j is well-separated from its neighbors, a large norm is a good indication of quality of the vector [15]. Developing a reliable stopping criterion based on norm for any spectrum requires an analysis of this sort for the results of reorthogonalization. A small norm after reorthogonalization indicates that successive iterates were linearly dependent.

4 Conclusions

The results of Section 3 provide the foundation for a refined implementation of inverse iteration. Simple experiments with random starting vectors show a marked increase in accuracy over those obtained for some more conventional choices: the statistical analysis provides a mechanism for improved efficiency. Corollary 3.2 shows the probability that reorthogonalization is needed, while Theorem 3.4 characterizes the behavior of the iterate norm when eigenvalues are well-separated. A good stopping criterion depends on a similar result for reorthogonalized vectors. Solving that problem and completing an experimental evaluation of the code based on the analysis are the subjects of ongoing work.

Acknowledgements

The author wishes to thank Jim Demmel, Leslie Fox, and Ilse Ipsen for several helpful discussions. The work presented in this paper was supported by the Office of Naval Research under grant N00014-86-K0310 and by an IBM Fellowship.

References

[1] M. Abramowitz and I.A. Stegun, *Handbook of Mathematical Functions,* Dover Publications, Inc., 1965.

[2] H. Bowdler, R.S. Martin, and J.H. Wilkinson, *The QR and QL Algorithms for Symmetric Matrices,* Num. Math., 11(1968), pp. 227–240.

[3] J.J.M. Cuppen, *A Divide and Conquer Method for the Symmetric Tridiagonal Eigenproblem,* Numer. Math., 36 (1981), pp. 177–95.

[4] J. Demmel and W. Kahan, *LAPACK Working Note#3: Computing Small Singular Values of Bidiagonal Matrices with Guaranteed Relative Accuracy,* Mathematics and Computer Science Division, Argonne National Laboratory, 1988.

[5] L. Devroye, *Non-Uniform Random Variate Generation,* Springer-Verlag, 1986.

[6] J.D. Dixon, *Estimating Extremal Eigenvalues and Condition Numbers of Matrices,* SIAM J. Numer. Anal., 20 (1983), pp. 812–814.

[7] I.C.F. Ipsen and E.R. Jessup, *Solving the Symmetric Tridiagonal Eigenvalue Problem on the Hypercube,* Research Report 548, Dept. Computer Science, Yale University, 1987.

[8] R.V. Jensen and R. Shankar, *Statistical Behavior in Deterministic Quantum Systems with Few Degrees of Freedom,* Phys. Rev. Lett., 54 (1985), pp. 1879–1882.

[9] N.L. Johnson and S. Kotz, *Continuous Univariate Distributions,* Houghton Mifflin Company, 1970.

[10] A.S. Krishnakumar and M. Morf, A Tree Architecture for the Symmetric Eigenproblem, *Proc. 27th Annual Symp. of SPIE,* 1983.

[11] J.R. Kuttler and V.G. Sigillito, *Eigenvalues of the Laplacian in Two Dimensions,* SIAM Review, 26 (1984), pp. 163–193.

[12] B.N. Parlett, *The Symmetric Eigenvalue Problem,* Prentice Hall, Englewood Cliffs, NJ, 1980.

[13] B.T. Smith, J.M. Boyle, J.J. Dongarra, B.S. Garbow, Y. Ikebe, V.C. Klema, and C.B. Moler, *Matrix Eigensystem Routines–EISPACK Guide, Lecture Notes in Computer Science, Vol. 6, 2nd edition* , Springer-Verlag, 1976.

[14] M.D. Springer, *The Algebra of Random Variables,* John Wiley and Sons, 1979.

[15] J.H. Wilkinson, *The Algebraic Eigenvalue Problem,* Clarendon Press, Oxford, 1965.

[16] J.H. Wilkinson and G. Peters, The Calculation of Specified Eigenvectors by Inverse Iteration, *Handbook for Automatic Computation, Vol.II: Linear Algebra,* Springer Verlag, 1971, pp. 418–439.

On Underdetermined Systems

W. R. Madych*

Department of Mathematics
University of Connecticut
Storrs, CT 06268, USA

1 Introduction

We summarize the results of a study [4] whose objective was to obtain some understanding of the nature of certain solutions of underdetermined systems of linear equations with a view to their role in the analysis of various practical inverse problems, specifically those associated with image restoration and computed tomography. To see the basic setup, consider the system of linear equations

$$\sum_{j=1}^{n} a_{ij}x_j = b_i\,, \qquad i = 1,\ldots,m\,. \tag{1}$$

Here, we take the field of scalars to be real. Using standard matrix notation equation (1) may be expressed as

$$Ax = b \tag{2}$$

where $A = (a_{ij})$ is an m by n and $b = (b_1,\ldots,b_m)^T$.

In what follows the matrix A and the data b are assumed to be known. We are interested in the set of solutions to (2), namely the set

$$S_{A,b} = \{x \;:\; Ax = b\}\,.$$

*Partially supported by a grant from the Air Force Office of Scientific Research, AFOSR-86-0145

NATO ASI Series, Vol. F 70
Numerical Linear Algebra, Digital Signal Processing
and Parallel Algorithms
Edited bv G. H. Golub and P. Van Dooren

To avoid complications which are not germain to the questions under consideration here, we always assume that $m < n$ and that A is of full rank. Thus $S_{A,b}$ is not empty; indeed, it is an $n - m$ dimensional affine manifold in the real Euclidean space R^n.

In the applications we have in mind, namely image restoration, computed tomography, and related inverse problems, A represents the mathematical model or a discrete analogue of the data aquisition scheme and b is the measured data. See [2], [5], [6]. Ideally, if one could constrain the feasible set of solutions appropriately, (2) should contain enough information to uniquely determine the x which gave rise to the data. Of course, the best that one can usually expect is to obtain a reasonable estimator.

Her we consider four methods for constraining the feasible set of solutions.

1.1 Minimum norm

Perhaps a more accurate description of this method would be the minimum quadratic norm. In any case the solution which arises by this method can be described by

$$x = A^T(AA^T)^{-1}b. \tag{3}$$

(We remind the reader that A is assumed to have rank m which implies that AA^T is invertible.) More complicated variants of this method are also used, see [1], but we do not consider them here.

1.2 Maximum entropy

The maximum entropy solution of (1) is an estimator whose components are of the form

$$x_j = p_j \exp\left(\sum_{i=1}^{n} \xi_i a_{ij}\right), \qquad j = 1, \ldots, n, \tag{4}$$

where $p = (p_1, \ldots, p_n)^T$ is a constant and the $\xi_i's$ are parameters chosen so that $x = (x_1, \ldots, x_n)^T$ satisfies (1). This form can be expressed more compactly as

$$x = F(\xi) = P \exp(A^T\xi), \tag{5}$$

where P is the constant diagonal matrix with diagonal p.

The parameter p is chosen to influence the behavior of the solution. in the discussion below, unless indicated otherwise, we always assume the components of p to be one.

It should be clear that any estimator of this form will have non-negative components. For more details see [2], [3].

1.3 Method one and method two

Method one generates solutions x of (2) which are of the form

$$x = \Phi(A^T\xi), \tag{6}$$

where if $y = (y_1, \ldots, y_n)^T$ then the j-th component of Φ is given by

$$\Phi_j(y) = \frac{1}{2}\{y_j + \sqrt{y_j + 4}\}, \qquad j = 1, \ldots, n.$$

Here $\xi = (\xi_1, \ldots, \xi_m)^T$ is a parameter as in (5).

Method two generates solutions x of (2) which are of the form

$$x = \Psi(A^T\xi), \tag{7}$$

where the j-th component of Ψ is given by

$$\Psi_j(y) = \psi(y_j), \qquad j = 1, \ldots, n,$$

and $t = \psi(s)$ is the function defined by

$$s = t - \frac{1}{t^3}. \tag{8}$$

Here also ξ is a parameter as in (6).

These methods generate unique estimators whenever $S_{A,b}$ has a non-empty intersection with the positive cone in R^n. For more details see [4].

2 Description of numerical experiments

To obtain a sense of the nature of the estimators generated by the methods described above we performed numerical experiments on relatively small linear systems. The systems considered were of the form

$$\frac{1}{k}\sum_{j=0}^{k-1} x_{i+j} = b_i\,, \qquad i = 1,\ldots,m\,, \tag{9}$$

where $k < n, m = n - k + 1$, and n is the number of variables x_j. System (9) is a typical example of a one dimensional blurring model. The sizes considered for our experiments ranged from $m = 15$, $n = 20$ to $m = 80$, $n = 100$ with the ratio m/n varying between 0.5 and 0.9. The experiments were performed as follows:

A non-negative pseudo-random vector $x = (x_1, \ldots, x_n)^T$ was generated via a canned subroutine and the data $b = Ax$ was computed. Then each of the methods outlined in section 3 was used to generate an estimator. We refer to these methods as minimum norm, maximum entropy, method one, and method two or, more briefly, MN, ME, M1 and M2, respectively. The resulting estimators $\hat{x}$ were plotted together with the true phantom x. In each case the error

$$\| x - \hat{x} \| = \left\{ \sum_{j=1}^{n} |x_j - \hat{x}_j| \right\}^{1/2}$$

was computed.

The results of these experiments can be loosely described as follows.

When the lower bound, zero in this case, was not a tight one for the phantom then all the methods generated estimates which were roughly equivalent. Namely, they differed from one another but the differences were judged not significant. The computed error varied but was roughly the same order of magnitude for all the methods.

On the other hand, when the lower bound for the phantom was reasonably tight then the methods which enforced the lower bound generated considerably better estimators. Indeed, the computed errors for estimators generated by ME, M1 and M2 were significantly smaller than the error for the estimator generated by MN. The estimators generated by ME, M1 and

M2 differed but the difference was judged not significant; the same was true of the corresponding computed error.

Similar experiments were performed on systems of the form

$$\sum_{j=1}^{n} x_j \cos \frac{2\pi ij}{n}, \qquad i = 0, \dots, k,$$

$$\sum_{j=1}^{n} x_j \sin \frac{2\pi ij}{n}, \qquad i = 1, \dots, k,$$

where $m = 2k + 1$ and m significantly less than n. Here the results were roughly the same as those reported above, although in the cases when the lower bound was tight on the phantom the differences seemed less dramatic.

References

[1] G. Golub and C. Van Loan, Matrix Computations, John Hopkins, Baltimore, 1983.

[2] S.F. Gull and J. Skilling, "Maximum entropy method in image processing", IEEE Proceedings, 131, (1984), 646-659.

[3] A.K. Livesey and J. Skilling, "Maximum Entropy Theory", Acta Cryst., A41, (1985), 113-122.

[4] W.R. Madych, "Solutions of Underdetermined systems of linear equations", CARC Technical Report, University of Connecticut, 1988.

[5] L.A. Shepp, and Y. Vardi, "Maximum likelihood reconstruction in positron emission tomography", IEEE Trans. on Medical Imaging, 1, (1982) 113-122.

[6] D.M. Titterington, "Regularisation procedures in signal processing and statistics", in Mathematics in Signal Processing, T.S. Durrani et *al.* eds., Clarendon, Oxford, 1987, 213-223.

A chart of numerical methods for structured eigenvalue problems

Angelika Bunse–Gerstner, Volker Mehrmann

Fakultät für Mathematik
Univ. Bielefeld
D–4800 Bielefeld 1, FRG

Ralph Byers

Department of Mathematics
University of Kansas
Lawrence, Kansas 66045, USA

Abstract

We consider eigenvalue problems for real and complex matrices with two of the following algebraic properties: symmetric, Hermitian, skew symmetric, skew Hermitian, symplectic, conjugate symplectic, J-symmetric, J-Hermitian, J-skew symmetric, J-skew Hermitian. In the complex case we found numerically stable algorithms that preserve and exploit both structures in 24 out of the 44 nontrivial cases with such a twofold structure. Of the remaining 20, we found algorithms that preserve part of the structure of 9 pairs. In the real case we found algorithms for all pairs studied. The algorithms are constructed from a small set of numerical tools, including orthogonal reduction to Hessenberg form, simultaneous diagonalization of commuting normal matrices, Francis' QR algorithm, the Quaternion QR-algorithm and structure revealing, symplectic, unitary similarity transformations.

NATO ASI Series, Vol. F 70
Numerical Linear Algebra, Digital Signal Processing and Parallel Algorithms
Edited by G. H. Golub and P. Van Dooren

Updating Choleski Factors in Parallel

J. J. Modi

University Engineering Department
Cambridge University
Trumpington Street
Cambridge, CB2 1PZ, UK

Abstract

In this paper we extend Bennett's algorithm for modifying the Choleski factors of a banded matrix which undergoes a rank m change, where m is small compared with n, the dimension of the matrix. We compare and contrast the implementation on the DAP and the T-RACK, respectively exhibiting features of SIMD and MIMD systems. The parallel update algorithm based on this technique is used for solving banded linear systems and has been incorporated in the finite element package for stress analysis problems, where it is found to be particularly effective for interactive analysis, when a sub-region of the mesh is repeatedly updated.

1 Introduction

Consider the system of linear equations $A\underline{x} = \underline{b}$, where A is an $n \times n$ matrix with bandwidth b', and $\underline{x}$ and $\underline{b}$ are vectors of order n. In a typical 3-dimensional finite element problem, with 3,000 nodes, n is approximately 10,000 and b' is 500. Once $\underline{x}$ has been computed it is often necessary to solve the modified system $\widetilde{A}\widehat{\underline{x}} = \widehat{\underline{b}}$, where A undergoes a rank m perturbation, producing a new matrix $\widetilde{A}$ and $\widehat{\underline{x}}$, $\widehat{\underline{b}}$ are of order n. In general $\widetilde{A}$ differs from A as follows

$$\widetilde{A} = A + XCY^T , \tag{1.1}$$

NATO ASI Series, Vol. F 70
Numerical Linear Algebra, Digital Signal Processing and Parallel Algorithms
Edited by G. H. Golub and P. Van Dooren

where X, Y are $n \times m$ matrices $(m < n)$ and C is $m \times m$. If the original system $A\underline{x} = \underline{b}$ is solved via the Choleski factorization $A = LDU^T$, where L, U^T have unit diagonal and are respectively lower and upper triangular, while D is a diagonal matrix, then the problem is one of computing the modified factors $\tilde{L}$, $\widetilde{D}$, and $\tilde{U}$ corresponding to $\tilde{A}$, in order to solve the modified system $\tilde{A}\underline{\hat{x}} = \underline{\hat{b}}$.

If we re-write (1.1) as

$$\tilde{A} = LDU^T + XCY^T = L(D + PCQ)U^T ,$$

then

$$\tilde{A} = L\overline{L}\,\overline{D}\,\overline{U}^T U^T = \tilde{L}\widetilde{D}\tilde{U}^T ,$$

giving the required factors of $\tilde{A}$; here P and Q are found by solving the equation $LP = X, UQ^T = Y; D + PCQ = \overline{L}\,\overline{D}\,\overline{U}^T$ is a Choleski factorization of a much simplified matrix; and $\tilde{L} = L\overline{L}, \widetilde{D} = \overline{D}, \tilde{U} = U\overline{U}$. (The product of two lower triangular matrices is itself lower triangular.)
A number of techniques for updating matrix factors in this way have been proposed which in general are based on Choleski factorization, Householder matrices or Givens rotations. The paper by Gill et al [1974] provides a review of some of these techniques.

A special case of this problem arises in constrained and unconstrained optimization calculations where a positive definite approximation to an $n \times n$ second derivative matrix is updated using the BFGS formula, based on rank one modification:

$$\tilde{A} = A + a\underline{y}\,\underline{z}^T ,$$

where a is a scalar, and $\underline{y}$, $\underline{z}$ are vectors of order n. (For further discussion of rank one modification in relation to the BFGS formula see Fletcher and Powell [1974] and Powell [1985].)

The purpose of the present paper is to discuss parallel implementation of the method based on Bennett's algorithm [1965], for the general case, which carries out these processes in a combined way, and arrives at the updated matrices $\tilde{L}, \widetilde{D}, \tilde{U}$ comparatively rapidly, under appropriate conditions. In particular we will consider its implementation on the DAP and on an array of transputers. The two architectures possess contrasting hardware features which influence the performance of the method in diverse ways.

2 The Method

The identity

$$(A + XCY^T)^i = A^i - A^i X (C^i + Y^T A^i X)^i Y^T A^i$$

where T, i indicates the transpose, and inverse respectively, shows how the inverse of a matrix can be modified when the term XCY^T is added to it. For a rank m update, approximately $3mn^2$ operations are required to modify the inverse, whereas in general inverse of a matrix requires approximately n^3 operations. Bennett's method requires $1mn^2$ operations, and proceeds by finding the first column of L and the first column of U and D_{11}, so that the problem is reduced to an $(n-1)\times(n-1)$ problem, which is then treated similarly to produce an $(n-2)\times(n-2)$ problem, and so on. Further details may be found in Bennett [1965]. We define the method, modified to handle banded matrices, in order to fix the notation, and to provide a setting for a discussion of its parallel implementation, in section 3.

Assuming we have a lower triangular matrix L with unit diagonal, an upper triangular matrix U^T also with unit diagonal, and a diagonal matrix D so that $A = LDU^T$, the update routine overwrites L, U, and D, with $\tilde{L}$, $\tilde{U}$, and $\widetilde{D}$ so that $\tilde{A} = \tilde{L}\widetilde{D}\tilde{U}^T$. The matrix $\tilde{A}$ is a rank $m << n$ modification to A so that

$$\tilde{A} = A + XCY^T ,$$

where X and Y are $n \times m$ matrices, and C is an $m \times m$ matrix . For a full $n \times n$ matrix A, with a rank m update, the number of operations on a serial machine is of order mn^2; for a banded matrix A with bandwidth b', it is of order mnb'.

Let $X_{i.}$ denote the i-th row of X.

Define $X^{(1)} = X$, $Y^{(1)} = Y$, $C^{(1)} = C$, $\underline{q}^{(0)} = \underline{p}^{(0)} = \underline{0}$

Set $i = 1$

restart: (i) **Compute** $(\underline{p}^{(i)})^T = X_{i.}^{(i)} C^{(i)}$

(ii) **Update** $D_{ii} = D_{ii} + Y_{i.}^{(i)} \underline{p}^{(i)}$

If $i = n$ **then goto** finish

(iii) **Compute** $(\underline{q}^{(i)})^T = Y_{i.}^{(i)} (C^{(i)})^T$

Set $j = 1$

flag: (iv) **Compute** $X_{(i+j).}^{(i+1)} = -L_{i+j,i} X_{i.}^{(i)} + X_{(i+j).}^{(i)}$

(v) **Compute** $Y_{(i+j).}^{(i+1)} = -U_{i,i+j} Y_{i.}^{(i)} + Y_{(i+j).}^{(i)}$

(vi) **Update** $L_{i+j,i} = L_{i+j,i} + \frac{1}{D_{ii}} X_{(i+j).}^{(i+1)} \underline{q}^{(i)}$

(vii) **Update** $U_{i,i+j} = U_{i,i+j} + \frac{1}{D_{ii}} Y_{(i+j).}^{(i+1)} \underline{p}^{(i)}$

Set $j = j + 1$

If $j < n - i + 1$ **and** $j \leq b'$ **then goto** flag

(viii) **Compute** $C^{(i+1)} = C^{(i)} - \frac{1}{D_{ii}} \underline{q}^{(i)} \underline{p}^{(i)T}$

Set $i = i + 1$

goto restart

finish print L, U, D.

The algorithm is numerically stable only when $L = U$, and $X = Y$ and both D and $\widetilde{D}$ are positive definite.

3 Parallel implementation on the DAP

In this section we discuss the implementation of the update algorithm on the Distributed Array Processor (DAP) which falls into the Single Instruction Multiple Data Stream (SIMD) category according to Flynn's machine taxonomy. It is an attached processor to the ICL 2980 mainframe, consisting of 64×64 processing units which can be thought of as arranged in a square grid with nearest neighbour connectivity. Each processing unit

has fast access to its own memory and that of its four neighbours, and the connection to the master control unit are by row and column 'highways'. A single instruction stream is broadcast to all units simultaneously, and the arithmetic is performed in hardware at bit level. DAPFORTRAN is the high level language provided on the DAP. It is similar to FORTRAN 77 although a number of additional features, such as ability to handle data sets of sizes 1, 64 and 4096, allow the programmer to take full advantage of the parallel processing capability of the DAP. With reference to the update algorithm defined in section 2, we consider the following parallel operations and implementation in DAPFORTRAN. (Upper case letters are used to conform to language specification.)
Line (i) involves multiplying the i-th row of X with the matrix C. In DAPFORTRAN this may be coded as:

$$P = SUMR(MATC(X(i,))^{*}C)$$

where the i-th row of X is expanded into a matrix whose columns are equal to the i-th row of X. The (*) indicates element-by-element multiplication. The SUMR function then sums all components in each column of the resultant matrix to produce a vector result which is assigned to P.

Line (ii) involves incrementing the $D(i,i)$ value by the inner product between the i-th row of Y and the vector P. In DAPFORTRAN the inner product is simply

$$SUM(Y(i,)^{*}P)$$

where the element-by-element multiplication between the i-th row of Y and P is summed using the SUM function which is based on recursive doubling technique.

Line (iii) involves matrix-vector product which is computed as in Line (i).

Lines (iv)-(vii) essentially involve element-by-element addition/subtraction and inner products.

Line (viii) involves outer product between P AND Q which may be coded as

$$MATC(P)^{*}MATR(Q)$$

where $MATC$ and $MATR$ take a vector argument and return a matrix whose columns (rows) are equal to the argument.

The global network consisting of a set of row and column highways allows the processing units to communicate with the master control unit. It permits rapid broadcasting of data to the array of processors, and also allows rapid global test operations. In particular, the functions *MATC*, *MATR*, *SUMC*, etc., take full advantage of this facility. In contrast, the *T-RACK* (section 4) does not possess global transmission lines, and thus communication of data between processing elements is more expensive. In such a system it is desirable to perform coarse grain computation within each processor, with fewer messages passing to others. The DAPFORTRAN operations associated with the update algorithm are basic matrix-vector operations, and there is minimum overhead in communication. For relatively low bandwidth b', matrices with size up to 4096×4096 can be updated by storing the bands as long vectors - by treating the DAP as a linear array with 4096 processors. This turns out to be particularly useful in practice. For even larger problems the standard slicing and crinkling techniques can be utilized in order to partition the problem as necessary (Modi [1988]).

4 Parallel implementation on the T-RACK

The T-RACK is designed for simulating novel parallel architectures. It consists of 64 32-bit floating-point transputers,which may be interconnected by a combination of fixed and program controlled reconfigurable links. This permits a variety of topologies to be configured in hardware, and thus renders the system adaptable to a wide range of applications.

The 16 cards, each containing four transputers, are placed in a rack. Each transputer holds 1Mbyte of Dynamic Random Access Memory (DRAM) and has processor speed of 10 MIPS, and four serial communication links operating at 20 Mbits/second. If these links are labelled 0,1,2 and 3, then the 0th link of the k-th transputer is connected to the 1st link of $(k+1)$th transputer. Thus there is a Hamiltonian path connecting all the 64 transputers. In addition the 2nd and 3rd links are connected to two 128×128 crossbar switchboards, which provides long distance communication between any two transputers. The system is front-ended with a SUN 3/110 colour workstation, and communication between the T-RACK and the SUN is via transputer link to/from a transputer based interface board which is accessed through the VME-bus in the SUN (see Fig. 1). The primary function of the control card is to set up the switch cards to the required configuration. The system may be run in any programming environment containing OCCAM 2 compiler on any network of transputers. Furthermore the system is modular, and multiple T-RACK can easily be configured.

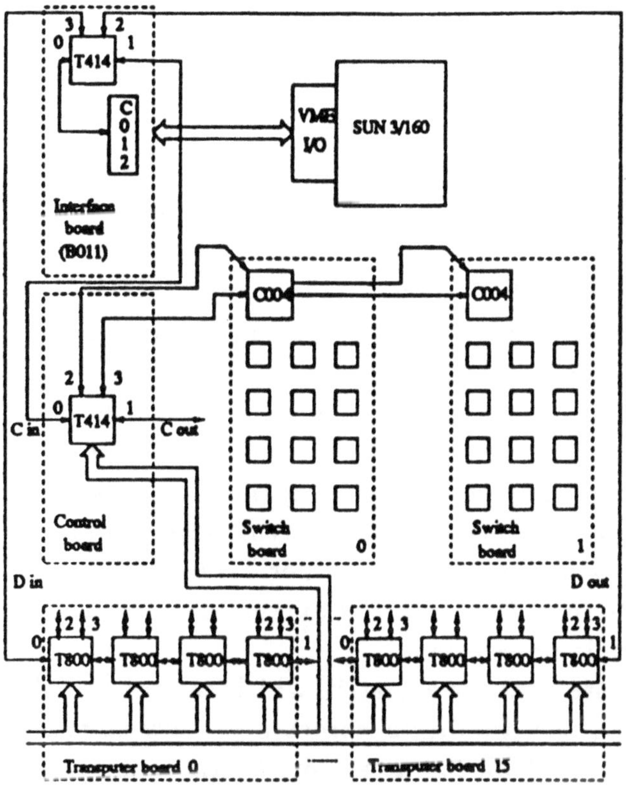

Fig. 1 T-RACK

Parallelism may be exploited at a number of different levels on the T-RACK. The following discussion identifies two levels. Furthermore there are several different strategies for specifying the parallelism, and again the discussion below illustrates one possible approach.

(a) Matrix-Vector Level

As seen in section 2, the core of the computation consists of vector-matrix operations. We illustrate the parallelism at this level by taking Line (i) which involves multiplying the i-th row of X with matrix C.

Allocate each processor to a column of C, so that each element of the i-th row of X is successively fed into the T-RACK, beginning at $p(1)$, where the computation of $X_{i1}C_{11}$ is carried out, and X_{i1} is broadcast to $p(2)$, using the nearest-neighbour connectivity of the processors. Thus, at the next stage, the computations $X_{i2}C_{21}$, in $p(1)$, and $X_{i1}C_{12}$, in $p(2)$, are performed simultaneously. The rest of the computation progresses similarly in a pipeline manner.

(b) **Overlapping Successive Updates**

In combination with (a), for a rank $m > 2$ update we may start the subsequent updates before completing the earlier ones. Thus we may treat a problem requiring a rank m modification as a series of low rank updates, the computation for which may be overlapped.
The ratio : computation / communication remains approximately the same in both cases when $m << n$.

It is interesting to find that in comparison with the DAP, which requires very little communication, here it becomes more important to take account of communication, incurred by passing the elements X_{i1} through the ring of processors, even though in both architectures only nearest-neighbour communication is utilised. On the other hand, because of the regularity of data movements in both cases, the extent to which the routing can affect the performance is not evident.

Finally we note that for a matrix which cannot be fully accommodated on the T-RACK, partitioning becomes necessary so that updates are performed in partitioned blocks at a time (with some overlap). The update algorithm can easily be extended to handle matrices consisting of partitioned blocks.

References

[1] Bennett J.M., 1965, Triangular factors of modified matrices, Numer. Math., 7, 217-221.

[2] Daniel J.W., Gragg W.B., Kaufman L. and Stewart G.W., 1976, Reorthogonalization and stable algorithms for updating the Gram-Schmidt QR factorization, Math. of Computation, 30, 136, 772-795.

[3] Fletcher R. and Powell M.J.D., 1974, On the modification of LDL^T factorizations, Math. of Computation, 28, 128, 1067-1087.

[4] Gentleman W.M., 1973, Least squares computations by Givens transformations without square roots, J. Inst. Math. Appl., 12, 329-336.

[5] Gill P.E., Golub G.H., Murray W. and Saunders M.A., 1974, Methods for modifying matrix factorizations, Math. of Computation, 28, 126, 505-535.

[6] Hammarling S., 1974, A note on modifications to the Givens plane rotation, J. Inst. Math. Applics., 13, 215-218.

[7] Livesley R.K. and Modi J.J., 1986, An investigation of the use of parallel computation to speed up finite element calculations, Int. Conf. 'Parallel Computing 85', eds. M. Feilmeier, et al., 149-156.

[8] Modi J.J., 1988, Parallel algorithms and matrix computation, Oxford University Press.

[9] Powell M.J.D., 1985, Updating conjugate directions by the BFGS formula, Internal report DAMPT 1985/NA11, Cambridge.

[10] Wilkinson J.H., 1965, The Algebraic Eigenvalue Problem, Oxford University Press.

Approximate Inversion of Partially Specified Positive Definite Matrices*

H. Nelis, E. Deprettere, and P. Dewilde

Department of Electrical Engineering
Delft University of Technology
2628 CD Delft, The Netherlands

Abstract

A fast algorithm is presented that can be used to compute an approximate inverse of a positive definite matrix that is specified only on a multiple band. The approximate inverse is the inverse of a matrix that closely matches the partially specified matrix. It has zeros in the positions that correspond to unspecified entries in the partially specified matrix. It is closely related to the inverse of the so-called maximum-entropy extension of the partially specified matrix. The algorithm is very well suited for implementation on an array processor.

1 Introduction

Suppose only some entries of a positive definite matrix are specified. Any Hermitian matrix that matches such a matrix is called an extension of that matrix. It was shown in [1] that if the diagonal entries of a partially specified positive definite matrix are specified, then there is a unique positive definite extension of that matrix that is such that its inverse has zeros in

*This work has been partially supported by the Dutch National Applied Science Foundation under grant FOM DEL 77.1260.

NATO ASI Series, Vol. F 70
Numerical Linear Algebra, Digital Signal Processing and Parallel Algorithms
Edited by G. H. Golub and P. Van Dooren

the positions that correspond to unspecified entries in the partially specified matrix. This extension is called the maximum-entropy extension of the partially specified matrix.

Our interest here is in a fast algorithm that can be used to compute an approximate inverse of a positive definite matrix that is specified only on a multiple band. We require that the approximate inverse has zeros in the positions that correspond to unspecified entries in the partially specified matrix. Furthermore, we assume that the maximum-entropy extension of the partially specified matrix is a good approximation of the completely specified matrix. For these reasons, we are interested in a fast algorithm that can be used to compute the inverse of the maximum-entropy extension of a positive definite matrix that is specified on a multiple band. Partially specified matrices of this type arise in applications with multi-dimensional geometries, such as multi-dimensional spectral estimation and finite element modelling.

It was shown in [2] how an algorithm known in interpolation theory as the Schur algorithm [3] can be used to compute the triangular factors of the inverse of the maximum-entropy extension of a positive definite matrix that is specified on a staircase band. The algorithm is fast and is very well suited for implementation on an array processor. Fast algorithms that can be used to compute the inverse of the maximum-entropy extension of a positive definite matrix that is specified on a multiple band are not known. Iterative algorithms for solving this problem in the context of multi-dimensional spectral estimation were proposed in [4] and [5]. However, these algorithms are computationally very intensive.

In this paper we present a fast algorithm that can be used to compute an approximate inverse of a positive definite matrix that is specified only on a multiple band. The approximate inverse is the inverse of a matrix that closely matches the partially specified matrix. It has zeros in the positions that correspond to unspecified entries in the partially specified matrix. The algorithm is a hierarchical extension of the Schur algorithm presented in [2], and is very well suited for implementation on an array processor.

2 Notation and Definitions

Matrices and entries of matrices will be denoted by italic uppercase letters. Block matrices and entries of block matrices will be denoted by bold uppercase letters. If a matrix A is partitioned as a block matrix, then this block matrix will be denoted by $\mathbf{A}$. Conversely, if a block matrix $\mathbf{A}$ is interpreted as a matrix with scalar entries, then this matrix will be denoted by A. For a partially specified positive definite block matrix $\mathbf{A}$, the symbol $\mathbf{A}_{ME}$ will denote the maximum-entropy extension of $\mathbf{A}$. For a completely specified positive definite block matrix $\mathbf{A}$, the symbols $\mathbf{L}_{\mathbf{A}}$ and $\mathbf{M}_{\mathbf{A}}$ will denote the upper triangular block matrices with upper triangular diagonal blocks with positive diagonal entries that are such that $\mathbf{A} = \mathbf{L}_{\mathbf{A}}\mathbf{L}_{\mathbf{A}}{}^* = \mathbf{M}_{\mathbf{A}}{}^*\mathbf{M}_{\mathbf{A}}$ ($*$ denotes conjugate transposition). For a block matrix $\mathbf{A}$, the symbol $\mathbf{A}(i,j)$ will denote the principal block matrix that lies in the rows and columns of $\mathbf{A}$ indexed by $i, \cdots, j$. Furthermore, the symbol $\nabla(\mathbf{A},(i,j))$ will denote the block matrix that is such that $\nabla(\mathbf{A},(i,j))(i,j) = \mathbf{A}$, and that is zero otherwise. The size of $\nabla(\mathbf{A},(i,j))$ will be defined by its context. The symbols I and 0 will be used to denote the identity matrix and the zero matrix respectively. The symbol J will be used to denote the matrix

$$J = \begin{bmatrix} I & 0 \\ 0 & -I \end{bmatrix}.$$

The size of I, 0, and J will be defined by their context. The support of a band $\mathcal{B}$ of a matrix A is the set $\{(i,j) \mid A_{ij} \in \mathcal{B}\}$. A band $\mathcal{B}$ will be said to be a staircase band if $\mathcal{B}$ includes the diagonal entries and if its support $\mathcal{S}$ is such that if $(i,j) \in \mathcal{S}$ with $i \leq j$ $(i \geq j)$, then $(i+r, j-s) \in \mathcal{S}$ $((i-r, j+s) \in \mathcal{S})$ for all $r, s \geq 0$ such that $i+r \leq j-s$ $(i-r \geq j+s)$.

3 The Schur Algorithm

If A is a positive definite matrix that is specified on a staircase band, then the Schur algorithm proposed in [2] can be used to compute the triangular factors $L_{A_{ME}}^{-*}$ and $M_{A_{ME}}^{-1}$ of the inverse of the maximum-entropy extension of A.

Theorem 1 ([2]) *Let A be a positive definite matrix that is specified on a staircase band $\mathcal{B}$. Let A be normalized to $A = V_A + I + V_A^*$, where V_A is a strictly upper triangular matrix, and let $U_A = V_A + I$. Then,*

1. *there exist elementary J-unitary matrices* [1] $\Theta_1, \ldots, \Theta_m$ *such that*

$$[U_A V_A]\Theta_1 \cdots \Theta_m = [U_A^{(m)} V_A^{(m)}],$$

where $U_A^{(m)}$ and $V_A^{(m)}$ are upper triangular matrices and $V_A^{(m)}$ has zeros on the upper triangular part of the staircase band $\mathcal{B}$.

2. *the product $\Theta_1 \cdots \Theta_m$ is such that*

$$\Theta_1 \cdots \Theta_m = \begin{bmatrix} U_{A_{ME}}^* & -V_{A_{ME}} \\ -V_{A_{ME}}^* & U_{A_{ME}} \end{bmatrix} \begin{bmatrix} L_{A_{ME}}^{-*} & 0 \\ 0 & M_{A_{ME}}^{-1} \end{bmatrix},$$

where $V_{A_{ME}}$ is a strictly upper triangular matrix, $U_{A_{ME}} = V_{A_{ME}} + I$, and $A_{ME} = V_{A_{ME}} + I + V_{A_{ME}}^* = L_{A_{ME}} L_{A_{ME}}^* = M_{A_{ME}}^* M_{A_{ME}}$.

3. *as a consequence of 2,*

$$[I\ I]\Theta_1 \cdots \Theta_m = [L_{A_{ME}}^{-*} M_{A_{ME}}^{-1}].$$

See [2] for more details. We suffice with stating that the algorithm proposed in [2] is fast and very well suited for implementation on an array processor.

---

[1] A matrix Θ is J-unitary if $\Theta^* J \Theta = J$.

4 The Hierarchical Schur Algorithm

If A is a positive definite matrix that is specified on a multiple band, that is, a band of the type

$$\begin{bmatrix}
x & x & & & x & x & & & & & & \\
x & x & x & & x & x & x & & & & & \\
& x & x & x & & x & x & x & & & & \\
& & x & x & & & x & x & & & & \\
x & x & & & x & x & & & x & x & & \\
x & x & x & & x & x & x & & x & x & x & \\
& x & x & x & & x & x & x & & x & x & x \\
& & x & x & & & x & x & & & x & x \\
& & & & x & x & & & x & x & & \\
& & & & x & x & x & & x & x & x & \\
& & & & & & x & x & & x & x & x \\
& & & & & & x & x & & & x & x
\end{bmatrix},$$

then the Schur algorithm proposed in [2] can not be used to compute the triangular factors of the inverse of the maximum-entropy extension of A. Indeed, this algorithm can only be used if a positive definite matrix is specified on a staircase band, and a multiple band is not a staircase band. Iterative algorithms for solving the above mentioned problem in the context of multi-dimensional spectral estimation were proposed in [4] and [5] . However, these algorithms are computationally very intensive. Our approach is as follows.

We start out with the following theorem. It describes the structure of the triangular factor $\mathbf{L}_{\mathbf{A}_{ME}}^{-*}$ of inverse of the maximum-entropy extension of a positive definite block matrix $\mathbf{A} = [\mathbf{A}_{ij}]$, $i, j = 1, \cdots, N$, that is specified on a band with support $\{ (i,j) \mid |i-j| \leq M \}$.

Theorem 2 (**[6]**) *Let* $\mathbf{A} = [\mathbf{A}_{ij}]$, $i, j = 1, \ldots, N$, *be a positive definite block matrix with blocks* $\mathbf{A}_{ij}$ *of size* n_i*-by-*n_j *that is specified on a band with support* $\{ (i,j) \mid |i-j| \leq M \}$. *Then, the columns of* $\mathbf{L}_{\mathbf{A}_{ME}}^{-*}$ *are such that* *for* $j = 1, \ldots, N - M - 1$

$$(\mathbf{L}^{-*}_{\mathbf{A}_{ME}})_{ij} = (\mathbf{L}^{-*}_{\mathbf{A}(j,j+M)})_{(i-j+1)1} \qquad i = j, \ldots, j + M$$
$$(\mathbf{L}^{-*}_{\mathbf{A}_{ME}})_{ij} = 0 \qquad i = j + M + 1, \ldots, N$$

and such that for $j = N - M, \ldots, N$

$$(\mathbf{L}^{-*}_{\mathbf{A}_{ME}})_{ij} = (\mathbf{L}^{-*}_{\mathbf{A}(N-M,N)})_{(i-(N-M)+1)(j-(N-M)+1)} \qquad i = j, \ldots, N.$$

Now, let A be a positive definite matrix that is specified on a multiple band. If it were possible to partition A as $\mathbf{A} = [\mathbf{A}_{ij}]$, $i, j = 1, \ldots, N$, such that, for some band with support $\{(i,j) \mid |i - j| \leq M\}$, the blocks in the band were completely specified and the blocks outside the band were unspecified, then, by Theorem 2, the triangular factor $\mathbf{L}^{-*}_{\mathbf{A}_{ME}}$ of the inverse of the maximum-entropy extension of $\mathbf{A}$ were built on the triangular factors of the inverses of the principle submatrices $\mathbf{A}(j, j + M)$, $j = 1, \ldots, N - M$. Clearly, it is impossible to partition A in this way. It is possible, however, to partition A as $\mathbf{A} = [\mathbf{A}_{ij}]$, $i, j = 1, \ldots, N$, such that, for some band with support $\{(i,j) \mid |i - j| \leq M\}$, the partially specified principle submatrices $\mathbf{A}(j, j+M)$, $j = 1, \ldots, N-M$, can be permuted to positive definite matrices that are specified on a staircase band (it was shown in [6] that the inverses of the maximum-entropy extensions of $\mathbf{A}(j, j + M)$, $j = 1, \ldots, N - M$, can be computed efficiently), and such that the blocks outside the band are unspecified.

Our first step is to approximate the triangular factor $\mathbf{L}^{-*}_{\mathbf{A}}$ of the inverse of the completely specified matrix $\mathbf{A}$ by a lower triangular matrix that is built on the triangular factors of the inverses of the maximum-entropy extensions of the partially specified principle submatrices $\mathbf{A}(j, j + M)$.

Definition 1 *Let A be a positive definite matrix that is specified on a multiple band. Let it be partitioned as $\mathbf{A} = [\mathbf{A}_{ij}]$, $i, j = 1, \ldots, N$, where the blocks $\mathbf{A}_{ij}$ are of size n_i-by-n_j, such that, for some band with support $\{(i,j) \mid |i - j| \leq M\}$, the partially specified principle submatrices $\mathbf{A}(j, j + M)$, $j = 1, \ldots, N - M$, can be permuted to positive definite matrices that are specified on a staircase band, and such that the blocks outside* *the band are unspecified. Then, the hierarchical approximation of A, denoted by A_H, is defined such that, if A_H is partitioned similarly as A, then the columns of $\mathbf{L}_{\mathbf{A}_H}^{-*}$ are such that for $j = 1, \ldots, N-M-1$*

$$\begin{aligned}
(\mathbf{L}_{\mathbf{A}_H}^{-*})_{ij} &= (\mathbf{L}_{\mathbf{A}(j,j+M)_{ME}}^{-*})_{(i-j+1)1} & i &= j, \ldots, j+M \\
(\mathbf{L}_{\mathbf{A}_H}^{-*})_{ij} &= 0 & i &= j+M+1, \ldots, N
\end{aligned}$$

and such that for $j = N-M, \ldots, N$

$$(\mathbf{L}_{\mathbf{A}_H}^{-*})_{ij} = (\mathbf{L}_{\mathbf{A}(N-M,N)_{ME}}^{-*})_{(i-(N-M)+1)(j-(N-M)+1)} \qquad i = j, \ldots, N.$$

It was shown in [7] that, under certain conditions, $||\mathbf{I} - \mathbf{L}_{\mathbf{A}_H}^{-1}\mathbf{L}_{\mathbf{A}}||_F$ is small. In the same paper it was shown that if $||\mathbf{I} - \mathbf{L}_{\mathbf{A}_H}^{-1}\mathbf{L}_{\mathbf{A}}||_F$ is small, then both the entries of A_H and A and the entries of A_H^{-1} and A^{-1} are close. However, the inverse of the hierarchical approximation of A does not have zeros in all positions that correspond to unspecified entries in the partially specified matrix. It has zeros in the positions that correspond to unspecified entries in A, except in those positions that correspond to unspecified entries in the blocks $\mathbf{A}_{ij}$ for which $|i-j| \leq M-1$, where $2 \leq i, j \leq N-1$.

Our last step is to search for a 'small' matrix Z that is such that $A_H^{-1} + Z$ has zeros in all positions that correspond to unspecified entries in A. To ensure that $A_H^{-1} + Z$ is positive definite, we require that Z is positive semi-definite. To find such a matrix Z, we first search for a matrix X that is such that $A_H^{-1} + X$ has zeros in all positions that correspond to unspecified entries in A. Next, we search for a matrix Y that is as close as possible to X (in a certain sense), that has zeros in the positions that correspond to unspecified entries in A, and that is such that $X - Y$ is positive semi-definite. A matrix X is easily found. Indeed, $A_H^{-1} + X$, with

$$X = \sum_{j=1}^{N-M-1} \nabla((\mathbf{A}(j, j+M)_{ME}(2, M+1))^{-1}, (j+1, j+M)),$$

has zeros in all positions that correspond to unspecified entries in A. This follows from the fact that

$$\mathbf{A}_H^{-1} + X = \sum_{j=1}^{N-M} \nabla(\mathbf{A}(j, j+M)_{ME}^{-1}, (j, j+M)),$$

and the latter matrix clearly has zeros in the positions that correspond to unspecified entries in A. The problem of finding Y can be formulated as a constrained nonlinear optimization problem [7], which can be solved by using standard techniques. These techniques, however, are computationally very intensive. It was shown in [7] that, under certain conditions,

$$Y = \sum_{j=1}^{N-M-1} \nabla(\mathbf{A}(j+1, j+M)_{ME}^{-1}, (j+1, j+M))$$

is a good approximation for X. Furthermore, since $\mathbf{A}(j+1, j+M)_{ME}^{-1}$ has zeros in the positions that correspond to unspecified entries in $\mathbf{A}(j+1, j+M)$, we have that Y has zeros in the positions that correspond to unspecified entries in A. The difference $Z = X - Y$, however, is not positive definite. In fact, it is indefinite.

Definition 2 *Let A be as defined in Definition 1. Then, the hierarchical banded-inverse approximation of A, denoted by A_{HBI}, is such that, if A_{HBI} is partitioned similarly as A, then*

$$\mathbf{A}_{HBI}^{-1} = \sum_{j=1}^{N-M} \nabla((\mathbf{A}(j, j+M)_{ME}^{-1}, (j, j+M)) -$$

$$\sum_{j=1}^{N-M-1} \nabla(\mathbf{A}(j+1, j+M)_{ME}^{-1}, (j+1, j+M)).$$

A_{HBI}^{-1} has zeros in all positions that correspond to unspecified entries in A. However, A_{HBI}^{-1} can not be guaranteed to be positive definite, since $Z = X - Y$ is indefinite. Since Z is 'small', we conjecture that A_{HBI}^{-1} may fail to be positive definite only if the completely specified matrix A is close to singular. This conjecture has been confirmed by numerical experiments [7].

References

[1] R. Grone, C. Johnson, E. Sá, and H. Wolkowicz. Positive definite completions of partial Hermitian matrices. *Linear Algebra and its Applications*, 58, 1984.

[2] P. Dewilde and E. Deprettere. Approximate inversion of positive matrices with applications to modelling. In *Modelling, Robustness and Sensitivity Reduction in Control Systems*, Springer-Verlag, 1987.

[3] I. Schur. Über Potentzreihen die im Innnern des Einheitskreises beschrankt sind. *Journal für die Reine und Angewandte Mathematik*, 147, 1917.

[4] S. Lang and J. McClellan. Multidimensional MEM spectral estimation. *IEEE Transactions on Acoustics, Speech, and Signal Processing*, 30, 1982.

[5] N. Rozario and A. Papoulis. Spectral estimation from nonconsecutive data. *IEEE Transactions on Information Theory*, 33, 1987.

[6] H. Nelis, E. Deprettere, and P. Dewilde. Approximate inversion of positive definite matrices, specified on a multiple band. In *Advanced Algorithms and Architectures for Signal Processing III, Proceedings SPIE*, 1988.

[7] H. Nelis, E. Deprettere, and P. Dewilde. *Maximum-entropy and minimum-norm matrix extensions.* Technical Report, Delft University of Technology, Department of Electrical Engineering, 1988.

Nearest "Unstable" Matrix Pencil to a Given Pencil

N.K. Nichols

Department of Mathematics
Reading University
Reading, UK

Abstract

We aim to generalize the concept of "distance to instability" for matrices to matrix pencils. Difficulties arise since infinitesimal perturbations can destroy the Kronecker canonical form of the pencil. If perturbations are constrained so that the structure of the pencil is retained, then there exist maximal finite perturbations such that the finite eigenvalues of the pencil remain in the left half plane.

A computable bound on the maximal perturbations is derived in terms of singular values of a particular matrix.

NATO ASI Series, Vol. F 70
Numerical Linear Algebra, Digital Signal Processing and Parallel Algorithms
Edited by G. H. Golub and P. Van Dooren

Parallel Recursive Least Squares on a Hypercube Multiprocessor

Charles S. Henkel*

Department of Nuclear Engineering
North Carolina State University
Raleigh, NC 27695, USA

Robert J. Plemmons†

Departments of Mathematics and Computer Science
North Carolina State University
Raleigh, NC 27695, USA

Abstract

We have developed an efficient parallel implementation of an algorithm for recursive least squares computations based upon the covariance updating method. The target architecture is a distributed-memory multiprocessor, and test results on an Intel iPSC/2 hypercube demonstrate the parallel efficiency of the algorithm. A 64-node system is measured to execute the algorithm over 48 times as fast as a single processor for the largest problem that fits on a single node (fixed size speedup). Moreover, the computation times increase only slightly with an increase in the number of processors when the problem size per processor remains constant. Applications include robust

*Research supported by the U.S. Department of Energy Nuclear Engineering Fellowship Program.

†Research supported by the Air Force Office of Scientific Research under grant no. AFOSR-88-0285 and the National Science Foundation under grant no. DMS-85-21154.

NATO ASI Series, Vol. F 70
Numerical Linear Algebra, Digital Signal Processing
and Parallel Algorithms
Edited by G. H. Golub and P. Van Dooren

regression in statistics and modification of the Hessian matrix in optimization, but the primary motivation for this work is the need for near real-time recursive least squares computations in signal processing.

1 Recursive Least Squares

The problem considered in this paper is that of designing an efficient parallel algorithm for recursive least squares computations based upon updating the covariance factorization. Our target architecture is the distributed-memory hypercube multiprocessor, and our tests are made on an iPSC/2 hypercube having an aggregate of 256 megabytes of memory and peak computational rate of 16 megaflops. The primary application of interest in this paper is adaptive signal processing. In particular, we develop a parallel implementation of a recursive least squares covariance updating scheme that may be used in conjunction with Kalman filtering applications.

The *linear least squares problem* can be posed as follows without reference to any specific application: Given a real $m \times n$ matrix X with full column rank n and a real m-vector s, find the n-vector w that solves

$$\min \|s - Xw\|_2, \tag{1}$$

where $\|\cdot\|_2$ denotes the usual Euclidean norm. The solution to (1) is given by $w = (X^T X)^{-1} X^T s$. If R denotes the upper triangular *Cholesky factor* of the cross product matrix $X^T X$, *i.e.*, $R^T R = X^T X$, then w can be obtained by solving the triangular systems $R^T v = X^T s, \quad Rw = v$, where v is an intermediate vector. However in many applications where accuracy and stability are important, R is computed directly from X by a sequence of orthogonal transformations; that is

$$QX = \begin{bmatrix} R \\ 0 \end{bmatrix}, \quad Q^T Q = I. \tag{2}$$

Then setting

$$Qs = \begin{bmatrix} s_1 \\ s_2 \end{bmatrix}, \; s_1 \quad \text{an } n\text{-vector}, \tag{3}$$

w is computed by solving the upper triangular system $Rw = s_1$. Such schemes are sometimes called "square root" methods in the signal processing literature. In addition, the matrix

$$P \equiv (X^T X)^{-1} \tag{4}$$

is called the *covariance matrix* for (1). It measures the expected errors in w. Its inverse $P^{-1} = X^T X$, often called the *normal equations matrix*, is known as the *information matrix* for (1) in the signal processing literature. It measures the information content in the experiment leading to (1).

In *recursive least squares* it is required to recalculate w when observations (*i.e.,* equations) are successively added to, or deleted from, the problem (1). For example, in many applications information arrives continuously and must be incorporated into the solution w. This is called *updating.* Alternatively, it is sometimes important to delete old observations and have their effect excised from w. This is called *downdating.* In some important applications the model itself is not stationary, requiring that an additional equation involving w be incorporated into the model [4]. For simplicity, *we consider only stationary models in this paper.*

There are two main approaches to solving recursive least squares problems; the *information matrix method* based upon modifying $P^{-1} = X^T X$ and the *covariance matrix method* based upon modifying P (called information filtering and Kalman filtering, respectively, in the literature [4], [5]). Instead of modifying P^{-1} or P directly, it is generally appropriate to modify their Cholesky factors [4]. Recursive least squares algorithms based upon modifying the Cholesky factor R of $X^T X$ on a hypercube multiprocessor were already studied in [3] by Henkel, Heath, and Plemmons. *Our attention in this paper is restricted to the covariance matrix method (Kalman filtering).*

Recursive least squares (RLS) updating computations using the information and covariance matrix methods can be described as follows. Consider the least squares problem (1) and, without loss of generality, assume that the additional data vector y and scalar σ for the equation $y^T w = y_1 w_1 + \cdots + y_n w_n = \sigma$ are appended to X and s forming

$$\tilde{X} = \begin{bmatrix} X \\ y^T \end{bmatrix}, \quad \tilde{s} = \begin{bmatrix} s \\ \sigma \end{bmatrix}. \tag{5}$$

One now seeks to solve the modified problem

$$\min \|\tilde{s} - \tilde{X}\tilde{w}\|_2 \tag{6}$$

for the updated least squares estimate vector $\tilde{w}$ at each recursive time step. We will not be concerned here with the use of exponential forgetting factors in the RLS computations.

The covariance matrix scheme for updating w to $\tilde{w}$ at each recursive step is now described. Here, the covariance matrix $P = (X^TX)^{-1}$ for w is updated to $\tilde{P}$, and $\tilde{w}$ is computed from w without the necessity of a triangular solution. Essentially, one begins with the inverse Cholesky factor R^{-1} of X^TX, and proceeds to update w to $\tilde{w}$. Observe that since $P^{-1} = X^TX = R^TR$, the covariance matrix P has the following factorization into a product of an upper triangular matrix R^{-1} with the lower triangular matrix R^{-T}: $P = (X^TX)^{-1} = R^{-1}R^{-T}$. Thus instead of updating R for P^{-1} we now update R^{-1} for P.

The process of modifying least squares computations by updating the covariance matrix P has been used in control and signal processing for some time in the context of linear sequential filtering [4]. One begins with estimates for $P = R^{-1}R^{-T}$ and w, and updates R^{-1} to $\tilde{R}^{-1}$ and w to $\tilde{w}$ at each recursive time step. Recently Pan and Plemmons [5] have described a parallel scheme for these computations which forms the basis for our hypercube implementation.

Observe first that with $\tilde{X}$ given in (5), $\tilde{P}^{-1}$ is given by $\tilde{P}^{-1} = \tilde{X}^T\tilde{X} = X^TX + yy^T = P^{-1} + yy^T$. Consequently, by the Sherman-Morrison formula, the updated w is given by $\tilde{w} = w + \tilde{P}y(\sigma - y^Tw)$. The vector $k \equiv \tilde{P}y$ is often called the Kalman gain vector (see, *e.g.*, [4], [5]). It weights the predicted residual $\sigma - y^Tw$.

The following algorithm for updating P to $\tilde{P}$ by updating R^{-T} to $\tilde{R}^{-T}$ and for computing the updated least squares estimate vector $\tilde{w}$ was established in Theorem 2 and Lemma 3 of [5]. Here we write $P = L^TL$, where $L \equiv R^{-T}$.

Algorithm (*Covariance Updating*) Given the current least squares estimate vector w, the current factor $L \equiv R^{-T}$ of $P = (X^TX)^{-1}$ and the observation $y^Tw = \sigma$ being added, the algorithm computes the updated factor $\tilde{L} \equiv \tilde{R}^{-1}$ of $\tilde{P}$ and the updated least squares estimate vector $\tilde{w}$.

1. Form the matrix vector product $a = Ly$.

2. Choose orthogonal plane rotations Q_i, to form

$$Q_m \cdots Q_1 \begin{bmatrix} -a \\ 1 \end{bmatrix} = \begin{bmatrix} 0 \\ \delta \end{bmatrix}, \quad \delta = \sqrt{1 + \|a\|_2^2}, \tag{7}$$

by reducing $-a$ to 0 from the top down, and apply

$$Q_m \cdots Q_1 \begin{bmatrix} L \\ 0^T \end{bmatrix} = \begin{bmatrix} \tilde{L} \\ u^T \end{bmatrix}. \tag{8}$$

3. Form $\tilde{w} = w - \frac{1}{\delta} u(\sigma - y^T w)$.

2 Test Results

In this section we report the actual performance of a parallel implementation of the algorithm on the iPSC/2 hypercube. Problems ranging in size from $n = 1024$ to size $n = 8192$ least squares estimates were considered. With 4 megabytes of local memory available on each processor, it was possible to run dense problems with values of n up to 1024 on a single processor in order to calculate speedups. We consider here both the standard fixed size speedup performance measurement as well as performance measurement concepts introduced recently in [1].

The fixed size speedups given in Table 2 are quite encouraging, in view of the difficulty reported in achieving high efficiency on the hypercube for linear algebra computations such as triangular solutions [2], and least squares downdating using the information matrix method [3].

By keeping our problem size per node constant as the number of processors increases (so the total $O(n^2)$ problem size increases proportionately) the resulting timing performance chart would ideally show constant time as a function of the number of processors (see [1]). The rationale here is that one is often interested in solving as large a problem as possible with a fixed number of processors. Notice that the *diagonal values in Table 1 remain almost constant*, increasing slightly as p increases when the problem size per node remains constant. In particular, time in seconds varies little with the hypercube dimension, as one would hope, except for the communication overhead in going from the serial version to the parallel version.

Table 1: *Times in seconds for covariance updating on the iPSC/2 hypercube.*

	Number of Processors p						
Dimension n	1	2	4	8	16	32	64
1024	19.100	9.593	4.886	2.494	1.268	.681	.393
1448		19.160	9.700	4.925	2.545	1.359	.775
2048			19.330	9.767	4.996	2.624	1.444
2896				19.450	9.884	5.118	2.749
4096					19.630	10.070	5.299
5793						19.870	10.320
8192							20.260

3 Observations

The parallel solution program could conceivably form the computational kernel for near real-time Kalman filtering applications on distributed memory MIMD architectures (and could possibly be implemented on related special purpose systolic array architectures). One might have the following scenario: A "black box" digital filter device is connected to the host processor for the parallel system. The host processor receives the digitized signals from the device and, after initialization of the process, maps the update signals in terms of y^T and σ to the processors, receiving from the processors the updated estimates in w, which in return are communicated to the filter device.

It is important to note that the results described in this paper show that the seemingly sequential least squares updating process can be effectively implemented on the hypercube. This work is strong evidence of the power of distributed-memory MIMD computing in recursive least squares filtering applications.

Table 2: *Fixed size speedups and efficiency of Algorithm 1 on the iPSC/2 for problem size* $n = 1024$.

Number of Processors p	Fixed Size Speedup	Efficiency in Percent
1	1	100
2	1.99	99
4	3.90	98
8	7.66	96
16	15.06	94
32	28.05	88
64	48.60	76

References

[1] J.L. Gustafson, G.R. Montry and R.E. Benner, *Development of Parallel Methods for a 1024 - processor Hypercube*, SIAM J. Sci. Stat. Comp., 9(1988), 609-638.

[2] M.T. Heath and C.R. Romine, *Parallel solution of triangular systems on distributed-memory multiprocessors*, SIAM J. Sci. Stat. Comp., 9(1988), 558-588.

[3] C.S. Henkel, M.T. Heath and R.J. Plemmons, *Cholesky downdating on a hypercube*, ACM Proc. Hypercube 3 Conf., Pasadena, CA, 1988.

[4] M. Morf and T. Kailath, *Square root algorithms and least squares estimation*, IEEE Trans. on Automatic Control, 20 (1975) 487-497.

[5] C.-T. Pan and R.J. Plemmons, *Parallel least squares modifications using inverse factorizations*, Computational and Applied Math., (1989), to appear.

A Divide and Conquer Method for the Unitary Eigenproblem and Applications*

Gregory Ammar†

Northern Illinois University
Dept. of Mathematical Sciences
DeKalb, IL 60115, USA

William Gragg‡

Naval Postgraduate School
Department of Mathematics
Monterey, CA 93943, USA

Lothar Reichel§

Bergen Scientific Centre IBM
Allégaten 36 N-5007 Bergen, Norway

Abstract

In divide and conquer methods for eigenproblems, the original eigenproblem is split into several smaller subproblems by using rank

*In memory of Peter Henrici

†This research was supported in part by the National Science Foundation under grant DMS-8704196, and by the Foundation Research Program of the Naval Postgraduate School

‡on leave from University of Kentucky, Department of Mathematics, Lexington, KY 40506, USA

§Norway on leave from University of Kentucky, Department of Mathematics, Lexington, KY 40506, USA

NATO ASI Series, Vol. F 70
Numerical Linear Algebra, Digital Signal Processing and Parallel Algorithms
Edited by G. H. Golub and P. Van Dooren

one modifications. Many of the subproblems can be solved in parallel, and from their solutions the solution of the original eigenproblem can be determined. This approach has for *symmetric* matrices been developed by Cuppen and refined by Dongarra and Sorensen. We describe a divide and conquer method for the computation of eigenvalues and eigenvectors of *unitary* matrices. The unitary matrices are represented in Schur parametric form. We also discuss the application of the unitary eigenproblem to the computation of Pisarenko frequency estimates in signal processing.

Utilization of the Matrix Pencil to Extract Poles of a Linear Time-Invariant System

Yingbo Hua, Tapan K. Sarkar*

Department of Electrical and Computer Engineering
Syracuse University, Syracuse, NY 13244, USA

P. Catalano, G. Casalegno and B. Audone

Aeritalia Corporation
Syracuse, NY 13244, USA

Abstract

For a (discrete) linear time-invariant system, its poles can be obtained by solving for rank reducing numbers of a matrix pencil. Three algorithms are presented for utilizing the matrix pencil to obtain the poles of the system.

1 Formulation of Matrix Pencil

It is known that the input sequence $x(k)$ and the output sequence $y(k)$ of a (causal) linear time-invariant system are, in general, related by

$$y(k) = \sum_{t=0,k} x(t)h(k-t) + h_0(k) \tag{1}$$

*This work has been supported in part by Aeritalia Corporation and the Office of Naval Research under Contract N00014-79-C-0598

NATO ASI Series, Vol. F 70
Numerical Linear Algebra, Digital Signal Processing and Parallel Algorithms
Edited by G. H. Golub and P. Van Dooren

where $h(k)$ is the unit response, and $h_0(k)$ is the zero input response. It is also true that for any practical systems,

$$y(k) + \sum_{t=1,M} a_t y(k-t) = \sum_{t=0,M} b_t x(k-t) \tag{2}$$

where M is the order of the system, and a_t's and b_t's are the coefficients determined by the system. The system poles are the roots of the characteristic polynomial

$$1 + \sum_{t=1,M} a_t z^{-t} = (1 - z_1 z^{-1})(1 - z_2 z^{-1}) \cdots (1 - z_M z^{-1}) \tag{3}$$

The poles, z_t's, may or may not be distinct. Now we assume that $x(k)$ and $y(k)$ are known for $k = 0, 1, \ldots, N-1$. An we define

$$\underline{x}_k = [x(k),\ x(k+1), \ldots, x(N-L+k-1)]^T \tag{4}$$

$$\underline{y}_k = [y(k),\ y(k+1), \ldots, y(N-L+k-1)]^T \tag{5}$$

where $k = 0, 1, \ldots, L$, and the superscript "T" denotes the transpose. Then it can be shown from (1), (2) and (3) that if $L \geq M$ and $N - L \geq M + \text{rank}[\underline{x}_1, \underline{x}_2, \ldots, \underline{x}_L]$ and if $x(k)$ meets some regular conditions (these assumptions are also made in the sequel), the matrix P_1 defined by

$$P_1 = [\underline{y}_1 - z\underline{y}_0,\ \underline{y}_2 - z\underline{y}_1, \ldots, \underline{y}_L - z\underline{y}_{L-1},\ \underline{x}_1, \underline{x}_2, \ldots, \underline{x}_L] \tag{6}$$

has rank equal to $M + \text{rank}[\underline{x}_1, \underline{x}_2, \ldots, \underline{x}_L]$ and decreases its rank by one iff $z = z_t$ for $1 \leq t \leq M$. For brevity we can write

$$P_1 = [Y_1 - zY_0,\ X] \tag{7}$$

where Y_1, Y_0 and X are defined according to (6). Note that P_1 is a generalized matrix pencil parametrized by z. However, it is clear that the rank reducing numbers of P_1 are the same as those of P_2 defined by

$$\begin{aligned} P_2 &= (I - XX^+)(Y_1 - zY_0) \\ &= Q_1 - zQ_0 \end{aligned} \tag{8}$$

where the superscript "+" denotes the Moore-Penrose inverse, XX^+ is the orthogonal projection onto the column space of X, and Q_1 and Q_0

are defined accordingly. It is not difficult to verify that Q_1 and Q_0 have the same column space and the same row space, and P_2 has rank M and decreases its rank by one iff $z = z_t$ for $1 \le t \le M$. Based on the above formulation, the following three algorithms can be easily proved for solving for the generalized eigenvalues (rank reducing numbers) of the matrix pencil P_2.

2 Algorithm 1

Following the approach in [2], we can decompose Q_0 into

$$Q_0 = Q_L Q_R \tag{9}$$

where Q_L is a $(N-L) \times M$ full rank matrix, and Q_R is an $M \times L$ full rank matrix. Then it can be shown [2] that the $M \times M$ matrix T_1 defined by

$$\begin{aligned} T_1 &= Q_L^+ Q_1 Q_R^+ \\ &= (Q_L^H Q_L)^{-1} Q_L^H Q_1 Q_R^H (Q_R Q_R^H)^{-1} \end{aligned} \tag{10}$$

has eigenvalues equal to $z_t' s$. Note that the geometric multiplicity of z_t is always one while the algebraic multiplicity of z_t is the number of the identical poles [1].

3 Algorithm 2

Similar to the ESPRIT algorithm in [4], the column space of $\begin{bmatrix} Q_1 \\ Q_0 \end{bmatrix}$ is spanned by some M independent vectors which we denote by the M columns of the $2(N-L)\times M$ matrix $\begin{bmatrix} Q_{1L} \\ Q_{0L} \end{bmatrix}$. It is clear that $Q_1 - zQ_0$ and $Q_{1L} - zQ_{0L}$ have the same generalized eigenvalues. Applying (9) and (10) to $Q_{1L} - zQ_{0L}$ immediately yields that the $M \times M$ matrix T_2 defined by

$$\begin{aligned} T_2 &= Q_{0L}^+ Q_{1L} \\ &= (Q_{0L}^H Q_{0L})^{-1} Q_{0L}^H Q_{1L} \end{aligned} \tag{11}$$

has eigenvalues equal to z_t's.

4 Algorithm 3

Since column spaces of Q_{1L} and Q_{0L} are same (since Q_1 and Q_0 have the same column space), we can find two $M \times M$ matrices Q_{1R} and Q_{0R} such that [5]

$$Q_{1L}Q_{1R} - Q_{0L}Q_{0R} = 0 \tag{12}$$

Applying (11) to (12) immediately yields that the $M \times M$ matrix T_3 defined by

$$T_3 = Q_{0R}Q_{1R}^{-1} \tag{13}$$

has eigenvalues equal to z_t's.

5 Discussion

Algorithms 1-3 requires different amount of computations. Algorithm 1 is the most costly while algorithm 3 is the least expensive. In practice, the data is always noisy so that the matrix reduction steps in the above algorithms can be carried out in least square sense. Namely, (9) is replaced by the rank M least square approximation (e.g., SVD) of Q_0, $\begin{bmatrix} Q_{1L} \\ Q_{0L} \end{bmatrix}$ is obtained by the M primary left singular vectors of $\begin{bmatrix} Q_1 \\ Q_0 \end{bmatrix}$, and $\begin{bmatrix} Q_{1R} \\ Q_{0R} \end{bmatrix}$ is obtained by the M right singular vectors, corresponding to the M smallest singular values, of $[Q_{1L}, -Q_{0L}]$. Since algorithms 1-3 apply three different ways of matrix reduction to Q_1 and Q_0 before the poles are obtained as the eigenvalues of T_1, T_2, and T_3, respectively, the three algorithms also differ in noise sensitivity. We have observed [1] that algorithm 3 tends to be the least sensitive to noise added to $y(k)$, and algorithm 2 tends to be the second best in noise sensitivity.

The parameter L is also very important in reducing the noise sensitivity. Both perturbation analysis and simulation results have shown [1-3] that the best choices of L are from $N/3$ to $2N/3$ (for $x(k) = 0$, $k \geq 1$).

The advantages both in computation and noise sensitivity of the matrix pencil method over the polynomial method [6-7] are discussed in detail in [2]. In particular, it has been shown that for $x(k) = 0$, $k \geq 1$, and $M = 1$, the first order variances (due to the white noise added on $y(k)$) of the extracted poles obtained by the matrix pencil method (algorithm 1) and the polynomial method [7] satisfy

$$Var(z_1)_{pencil} \leq Var(z_1)_{polynomial} \tag{14}$$

where the equality holds only when $L = 1$. For the polynomial method, L represents the polynomial degree.

References

[1] Y. Hua and T.K. Sarkar, "System identification by matrix pencil method", submitted for publication.

[2] Y. Hua and T.K. Sarkar, "Matrix pencil method for estimating parameters ...", to appear in IEEE Trans. ASSP.

[3] Y. Hua and T.K. Sarkar, "Matrix pencil method and its performance", in Proc. ICASSP'88.

[4] R. Roy, A. Paulraj, and T. Kailath, "ESPRIT - ...", IEEE Trans, ASSP, Vol. 34, No. 5, pp. 1340-1342, Oct. 1986.

[5] R. Roy and T. Kailath, "Total least square ESPRIT", in Proc. of 21th Asilomar Conf., Nov. 1987.

[6] D.W. Tufts and R. Kumaresan, "Estimation of ...", Proc. IEEE, Vol. 70, No. 9, pp. 975-989, Sept. 1982.

[7] R. Kumaresan and D. Tufts, "Estimating ...", IEEE Trans. ASSP, Vol. 30, No. 6, pp. 833-840, Dec. 1982.

Efficient Solution of Minimum Eigen-Problem of Hankel Systems by Conjugate Gradient Algorithm and FFT

Tapan K. Sarkar and Xiaopu Yang

Department of Electrical Engineering
Syracuse University
Syracuse, New York, USA

Abstract

The advantages of the conjugate gradient (CG) algorithm for the solution of the eigenvector corresponding to the minimum/maximum eigenvalue of a symmetric matrix over conventional methods, such as those found in IMSL, EISPACK or LINPACK library, are its efficiency and flexibility. Preliminary results for the solution of minimum eigen-problem of a Hankel system indicates the efficiency of the CG algorithm. The introduction of the FFT into computation demonstrates the flexibility of the CG algorithm, and further improves the efficiency of the algorithm.

1 Introduction

In this paper, we present a computationally efficient method for the solution of minimum eigen-problem of a Hankel system. The Hankel systems arise in many areas of digital signal processing. Both the covariance and autocorrelation matrices are special cases of Hankel matrix.

For the problem of interest, let us consider a data sequence

$$\{x\} = \{x_1, x_2, \ldots, x_{M+N-1}\}. \tag{1}$$

NATO ASI Series, Vol. F 70
Numerical Linear Algebra, Digital Signal Processing and Parallel Algorithms
Edited by G. H. Golub and P. Van Dooren

which is obtained from sampling the sum of ten sinusoidal signals with random phases plus zero mean white Gaussian noide, i.e.,

$$s(t) = \sum_{k=1}^{10} sin[2\pi(0.08 + 0.4k)t + 2\pi\, ran(k)] + w(t)\,, \tag{2}$$

where $ran(k)'s$ are random samples of a random variable uniformly distributed between -1 and 1, and $w(t)$ is the noise.

From the sequence $\{x\}$, we form an M by N Hankel matrix

$$X = \begin{bmatrix} x_1 & x_2 & \cdots & x_{N1} \\ x_2 & x_3 & \cdots & x_{N2} \\ . & . & \cdots & . \\ x_M & x_{M+1} & \cdots & x_{M+N-1} \end{bmatrix}, \tag{3}$$

To recover the sinusoids from the data samples x_i, it is required in Pisarenko's method [1] to obtain the minimum eigenvector of $X^T X$. We shall call this problem a minimum eigenproblem of a Hankel system for the Hankel matrix X involved.

In most cases, conventional methods, such as QR method, are invoked to solve the problem in spite of the fact that the methods solve for all the eigenvalues and eigenvectors, simultaneously, and hence are not efficient. As an alternative, conjugate gradient method was proposed to find the minimum eigenvector directly [2][3]. Recently, we have presented other versions of the conjugate gradient algorithm which have improved rate of convergence [4][5]. For the minimum eigen-problem of a Hankel system, introduction of the FFT into computation further improves the efficiency of the CG algorithm.

2 Conjugate Gradient Algorithm

For the symmetric matrix A, finding its minimum eigenvalue corresponds to obtaining the least value of the functional $F(x)$ defined by the Rayleigh quotient:

$$F(x) = < Ax, x > / < x, x > = x^T Ax/(x^T x)\,, \tag{4}$$

where $x \in R$, and is assumed to be never identically zero.

The conjugate gradient algorithm starts with an initial guess $x(0)$ for the minimum eigenvector and update it as

$$x(k+1) = x(k) + \alpha(k)\, p(k)\,. \tag{5}$$

In (5), $\alpha(k)$ is obtained by optimal line search, that is, by minimizing the functional $F(x)$ along the direction $p(k)$ passing through the point $x(k)$.

It can be shown that $\alpha(k)$ is always real and given by

$$\alpha(k) = -B + \sqrt{B^2 - 4CD}/(2D)\,, \tag{6}$$

where

$$D = p_b(k)\, p_c(k) - p_a(k)\, p_d(k)\,, \tag{7}$$

$$B = p_b(k) - \lambda(k)\, p_d(k)\,, \tag{8}$$

$$C = p_a(k) - \lambda(k)\, p_c(k)\,, \tag{9}$$

$$p_a =< p(k), Ax(k) > / < x(k), x(k) >\,, \tag{10}$$

$$p_b =< p(k), Ap(k) > / < x(k), x(k) >\,, \tag{11}$$

$$p_c =< p(k), x(k) > / < x(k), x(k) >\,, \tag{12}$$

$$p_d =< p(k), p(k) > / < x(k), x(k) >\,, \tag{13}$$

$$\lambda(k) = F[x(k)] =< x(k), Ax(k) > / < x(k), x(k) > \,. \tag{14}$$

At $x(k+1)$, a new search direction is selected according to the rule

$$p(k+1) = r(k+1) + \beta(k)\, p(k)\,, \tag{15}$$

with $\beta(-1) = 0$, and

$$r(k+1) = [\lambda(k+1)x(k+1) - Ax(k+1)]/ < x(k+1),\, x(k+1) > \tag{16}$$

In (15), $\beta(k)$ is selected such that $p(k+1)$ is conjugate to $p(k)$ with respect to the Hessian of $F(x)$ around the point $x(k+1)$. One of the formulas for $\beta(k)$ is

$$\beta(k) = - < r(k+1),\, r(k) - r(k+1) > / < p(k),\, r(k) - r(k+1) >\,, \tag{17}$$

which is equivalent to a couple of others in generating Hessian conjugated search directions.

The equations (5)-(17) constitute the conjugate gradient algorithm for the minimum eigen-problem of a symmetric matrix.

3 Numerical Result

A sample result for the problem posed in the introduction (M=50, N=21) is presented here to demonstrate the efficiency of the CG algorithm. The CG algorithm and the EIGRS routine of IMSL program package are both applied to the problem of interesting given in the introduction. The EIGRS routine is written based on the QR algorithm. The eigenvalues obtained and the cpu time used by both algorithms are given below along with the true eigenvalues obtained with the EIGRS routine utilizing double precision. It is seen that the CG algorithms is more efficient and accurate than the QR algorithm in this example.

Table I

Comparison of the CG and QR algorithms

CG algorithm		EIGRS (QR) algorithm		True
λ_{min}	CPU time (ms)	λ_{min}	CPU time (ms)	λ_{min}
0.205749	5	0.205907	27	0.205763

4 Improved Efficiency in Computation using FFT

Due to special structure of the Hankel matrix, we never explicitly form X^TX in the conjugate gradient algorithm. That is, we compute $p_a(k)$, $p_b(k)$, and $r(k+1)$ according to

$$p_a(k) = < Xp(k), Xx(k) > / < x(k), x(k) > , \tag{18}$$

$$p_b(k) = < Xp(k), Xp(k) > / < x(k), x(k) > , \tag{19}$$

and

$$r(k+1) = [\lambda(k+1)\, x(k+1) - X^TXx(k+1)] / < x(k+1), x(k+1) > \,. \tag{20}$$

The advantages of not forming X^TX are

1. we are dealing with the raw data, and not the processed data which contains round-off and truncation error, and can not be rectified.

2. we avoid multiplication of matrix by matrix, $X^T X$, and instead two more multiplications of matrix by vector are needed at each iteration which are computed using FFT (NlogN operations as opposed to N^2). The gain in efficiency depends on the number of iterations required. The efficiency is improved by introducing the FFT into the computation.

In the CG algorithm, most of the computation time is used in the evaluation of Xp, Xx and $X^T(Xx)$. These are multiplications of matrix by vector required in each iteration. We propose to minimize the operation count of the multiplication from MN to approximately $2(M+N-1)\log(M+N-1)$ operations by employing FFT.

By observing the Hankel structure of matrix X, the product

$$Xp = \begin{bmatrix} x_0 & x_1 & x_2 \\ x_1 & x_2 & x_3 \\ x_2 & x_3 & x_4 \\ x_3 & x_4 & x_5 \\ x_4 & x_5 & x_6 \end{bmatrix} \begin{bmatrix} p_2 \\ p_1 \\ p_0 \end{bmatrix} \tag{21}$$

can be treated as the convolution of two sequences $\{x\}$ and $\{y\}$, where

$$\{x\} = \{x_0, x_1, x_2, x_3, x_4, x_5, x_6\}, \tag{22}$$

and

$$\{y\} = \{p_0, p_1, p_2, 0, 0, 0, 0\}. \tag{23}$$

The result of the product is embedded in the convolution of the two sequences $\{x\}$ and $\{y\}$. This is clear as one observes the 3rd, 4th, 5th, 6th and 7th elements of the convolution of $\{x\}$ and $\{y\}$. The convolution can be efficiently carries out by FFT. So if one performs the FFT of the sequences $\{x\}$ and $\{y\}$ and multiply them and then take the inverse FFT to obtain the sequence $\{c\}$ as

$$\{z\} = (FFT)^{-1}(FFT\{x\} \cdot FFT\{y\}), \tag{24}$$

then the desired result is the 3rd to the 7th elements of $\{z\}$. The total operation count are MN for direct multiplication and $2(M+N-1)\log(M+N-1)$ using FFT. If $M+N-1$ is not a number of type 2^n, zero padding could be used. It is seen that the application of the FFT becomes more efficient as the matrix size increases. It is interesting to point out that the FFT of the sequence of $M+N-1$ elements of $\{x\}$ needs to be computed only once, and it could be used to compute Xp, Xx, and $X^T(Xx)$ as well in each iteration. There is also a saving from storage point of view. Instead of requiring the storage of MN elements of matrix X, one needs to store only $M+N-1$ elements in an array $\{x\}$. This idea of utilizing FFT to evaluate matrix products has been used before [6-9].

5 Conclusion

For the minimum/maximum eigen-problem of a symmetric matrix, the conjugate gradient is often more efficient and accurate than the conventional eigen-problem algorithms. Moreover, it is more convenient to be used in connection with signal processing techniques such as the adaptation and the FFT. Hence, it offers a good alternative to conventional algorithms for the solution of the minimum/maximum eigen-problem.

References

[1] Pisarenko V.F., "The retrieval of harmonics from a covariance function", *Geophys. J. R. Astron. Soc.*, Vol. 33, pp. 347-366, 1973.

[2] Chen H., Sarkar T.K., Brule J. and Dianat S.A., "Adaptive spectral estimation by the conjugate gradient method", *IEEE Trans. ASSP.*, Vol. ASSP-34, pp. 272-284, April 1986.

[3] Sarkar T.K. and Yang X., "Application of the conjugate gradient and steepest descent for computing the eigenvalues of an operator", Accepted for publication, *Signal Processing*.

[4] Yang X., *Conjugate Gradient Algorithms for the Solution of Extreme Eigen-problem of a Hermitian Matrix and its Applications in Signal*

Processing, Ph. D. Dissertation, Syracuse University, Chapter 3, pp. 29-64.

[5] Yang X. and Sarkar T.K., "A survey of conjugate gradient algorithms for solution of extreme eigen-problems of a symmetric matrix", Accepted for publication, *IEEE Trans. ASSP.*

[6] Sarkar T.K., Arvas E. and Rao S.M., "Application of FFT and the conjugate gradient method for the solution of electromagnetic radiation from electrically large and small conjuction bodies", *IEEE Trans. Antenna and Propagation*, pp. 635-640, May 1986.

[7] Sarkar T.K., "On the application of the generalized biconjugate gradient method", *Journal of Electromagnetic Waves and Applications*, Vol. 1, No 3, pp. 223-242, 1987.

[8] Sarkar T.K. and Yang X., "Accurate and efficient solution of Hankel matrix systems by FFT and the conjugate gradient methods", *Proc, ICASSP 84*, pp. 1835-1838, 1984.

[9] Sarkar T.K. and Yang X., "A limited survey of various conjugate gradient methods for solving complex matrix equations arising in electromagnetic wave interactions", *Journal of Wave Motion*, Dec., 1988.

Progress Towards a Systolic SVD Array Implementation

David E. Schimmel

School of Electrical Engineering
Cornell University
Ithaca, NY 14850, USA

Abstract

The past five years have seen a great deal of interest in parallel architectures for the computation of the singular value decomposition. We present a linear array approach which is unique in that there is no angle computation within the array, I/O time is balanced with computation time, communication is nearest neighbor, and simple multiply and accumulate processors are utilized. In addition, special properties of the SVD algorithm are exploited in order to obtain a VLSI design with reduced area. We are currently designing a chip using two micron CMOS design rules. When completed we estimate that our array will compute the SVD at a rate of between 5n and 10n MFlops/second, where n is the dimension of the array.

NATO ASI Series, Vol. F 70
Numerical Linear Algebra, Digital Signal Processing and Parallel Algorithms
Edited by G. H. Golub and P. Van Dooren

On the Convergence of Cyclic Jacobi Methods

Gautam Shroff

Computer Science Dept.
Rensselaer Polytechnic Institute
Troy, New York, USA

Robert Schreiber

Computer Science Dept.
Rensselaer Polytechnic Institute
Troy, New York, USA

1 Jacobi Methods

Jacobi's method is an iterative algorithm for diagonalizing a symmetric matrix $A = A^{(1)}$, in which at iteration k a plane rotation $J_k = J_k(i_k, j_k, \theta_k)$ is chosen so as to annihilate a pair of off diagonal elements in the matrix $A^{(k)}$

$$\begin{pmatrix} c & -s \\ s & c \end{pmatrix} \begin{pmatrix} a_{ii}^{(k)} & a_{ij}^{(k)} \\ a_{ji}^{(k)} & a_{jj}^{(k)} \end{pmatrix} \begin{pmatrix} c & s \\ -s & c \end{pmatrix} = \begin{pmatrix} a_{ii}^{(k+1)} & 0 \\ 0 & a_{jj}^{(k+1)} \end{pmatrix}$$

where $c = \cos\theta_k$ and $s = \sin\theta_k$.

The off diagonal mass of the matrix is thereby reduced at each iteration. If certain conditions are satisfied, the $A^{(k)}$ converge to a diagonal matrix and the product of the J_k converges to a matrix of orthogonal eigenvectors [6].

Jacobi's method was the standard algorithm for symmetric eigenvalue problems until the invention of the more efficient QR algorithm. There has

NATO ASI Series, Vol. F 70
Numerical Linear Algebra, Digital Signal Processing
and Parallel Algorithms
Edited by G. H. Golub and P. Van Dooren

been renewed interest Jacobi's algorithm in recent years because of its potential for parallel implementation, as well as its generalization to a number of other problems like the singular value decomposition [10], the generalized singular value decomposition [14] and the unsymmetric eigenvalue problem [4,5]. A number of parallel implementations of Jacobi methods have been proposed, systolic array implementations [2,11] as well as for coarse grained parallel machines [1]. A parallel Jacobi method involves annihilating the off diagonal elements in some special order so that more than one element can be annihilated at once.

2 Jacobi Methods using Parallel Orderings

The ordering used to annihilate the off diagonal elements is an important consideration in proving global convergence of a Jacobi method. Forsythe and Henrici [6] proved convergence for cyclic by rows ordering. Examples of orderings that may not converge were first shown by Hansen [7]. The issue of convergence of the Jacobi method for different orderings has never been conclusively settled. However Luk and Park [12] proved convergence for some parallel orderings by showing that they can be obtained from the cyclic by rows ordering through combinatorial transformations that preserve the global convergence property of an ordering.

We have generalized and extended Luk's ideas regarding convergent parallel orderings by completely characterizing the class of orderings that are related to the cyclic by rows ordering via any known convergence preserving transformations. We show that this class is characterized by simple properties which are easy to test [16] (i.e., in polynomial time). The approach is outlined below.

First, the class of Wavefront Orderings is introduced which consists of exactly those orderings that are *equivalent* to the cyclic by rows ordering of Forsythe and Henrici. Equivalence of orderings was introduced by Hansen [7]. Two off diagonal elements (p, q) and (r, s) are said to commute if p, q, r, s are distinct integers. In any ordering of the off diagonal elements, adjacent commuting elements can be transposed to obtain a new ordering that is equivalent to the original one. Equivalent orderings are related to each other through a sequence of such transpositions. In Fig.1, ordering O_1

is equivalent to O_2 since one is obtained from the other by the transposition of the adjacent commuting elements $(1,5)$ and $(2,4)$. Jacobi methods using two equivalent orderings produce the same matrices $A^{(k)}$ after every sweep, i.e. at $k = 1, N, 2N \ldots$ where N is the number of off diagonal elements in the matrix. It follows that *equivalence* is a convergence preserving transformation.

Next the class of Weakly Wavefront Orderings is introduced using the notion of *weakly equivalent* orderings. Weakly equivalent orderings are related to each other by a sequence of transformations that can be of two types. First, adjacent commuting elements can be transposed as before. In addition, cyclic shifts of the orderings are also allowed. In Fig.1, O_3 is obtained from O_2 by a cyclic shift of 3 and the transposition of adjacent commuting elements $(1,2)$ and $(4,5)$. Jacobi methods using weakly equivalent orderings produce the same matrices $A^{(k)}$ as k tends to infinity. So global convergence is still preserved under the weak equivalence transformation. The Weakly Wavefront orderings are defined as those weakly equivalent to cyclic by rows. (The idea of weak equivalence of orderings was implicitly used by Luk [12] but not formally developed and generalized.)

Finally the most general class of P-Wavefront orderings is introduced, which consists of orderings that are related to the weakly wavefront orderings via permutations of rows and columns of the matrix. In Fig.2, O_2 is obtained from O_1 by a permutation π of rows and columns. Convergence is preserved because if we start with a convergent ordering O_1, then the permuted ordering O_2 is nothing but O_1 applied to the permuted data matrix $\Pi^T A \Pi$ (where $A\Pi$ performs permutation π on the columns of A). The class of P-Wavefront orderings consists of all the orderings that can be obtained from the cyclic by rows ordering through convergence preserving transformations.

The Wavefront Orderings are exactly those orderings that have the following simple property [16]. Each off diagonal element (i,j) is annihilated after the elements $(i-1,j)$ and $(i,j-1)$, and before the elements $(i+1,j)$ and $(i,j+1)$.

The Weakly Wavefront Orderings are characterized by the property of having a *good splitting*. A good splitting of an ordering is a partition of the elements of the ordering into two sets X and Y that satisfy the following properties [16].

$$\begin{pmatrix} x & 1 & 2 & 3 & 6 \\ & x & 4 & 5 & 7 \\ & & x & 8 & 9 \\ & & & x & 10 \\ & & & & x \end{pmatrix} \quad \begin{pmatrix} x & 1 & 2 & 3 & 5 \\ & x & 4 & 6 & 7 \\ & & x & 8 & 9 \\ & & & x & 10 \\ & & & & x \end{pmatrix} \quad \begin{pmatrix} x & 2 & 4 & 5 & 7 \\ & x & 6 & 8 & 9 \\ & & x & 10 & 1 \\ & & & x & 3 \\ & & & & x \end{pmatrix}$$

$$O_1 \qquad\qquad O_2 \qquad\qquad O_3$$

Figure 1: Equivalence, Weak Equivalence

1. If $(p,q) \in X$, then all $(i,j), i \leq p$ and $j \leq q$ are also in X. Similarly, if $(p,q) \in Y$, then all $(i,j), i \geq p$ and $j \geq q$ are also in Y.
2. Each set X and Y is a Wavefront ordering by itself.
3. Every pair of elements $(r,s) \in Y$ and $(i,j) \in X$ such that (i,j) precedes (r,s) in O is commuting pair.
4. The pair $(1,2) \in X$.

These properties are illustrated in Fig.3. The splitting is shown by enclosing the set Y in a box. The first splitting shown (s_1) is a good splitting but the second (s_2) is not because it violates the third property above.

It can be shown that the property of possessing a good splitting may be verified in $O(n^4)$ time (n being the size of the matrix). In addition an algorithm for testing if an ordering is a P-Wavefront ordering can be constructed by testing various permutations of the ordering, and those weakly equivalent to it, for the good splitting property. It can be shown that of the $n!$ permutations of an ordering only $O(n^2)$ need to be tested [16]. Therefore, membership in the class of Provably convergence P-Wavefront orderings can be decided in polynomial time.

Does the class of P-Wavefront orderings cover all known provably convergent orderings? Interestingly the answer is no. Nazareth [13] proved convergence for a class of orderings. Some Nazareth orderings are not P-Wavefront and vice versa. To characterize all known provably convergent orderings, we would have to include all Nazareth orderings as well as those that can be obtained from them through convergence preserving transformations. This is an area of ongoing research.

$$
\begin{pmatrix}
 & 1 & 3 & 5 & 8 & 10 \\
 & & 7 & 9 & 11 & 13 \\
 & & & 12 & 14 & 15 \\
 & & & & 2 & 4 \\
 & & & & & 6 \\
 & & & & &
\end{pmatrix}
\;\overset{\Pi : \begin{pmatrix} 12345 \\ 12543 \end{pmatrix}}{\longrightarrow}\;
\begin{pmatrix}
 & 1 & 8 & 5 & 3 & 10 \\
 & & 11 & 9 & 7 & 13 \\
 & & & 2 & 14 & 6 \\
 & & & & 12 & 4 \\
 & & & & & 15 \\
 & & & & &
\end{pmatrix}
$$

Figure 2: Permutation Equivalence

$$
\underset{s_1}{\begin{pmatrix}
 & 1 & 3 & 5 & 8 & 10 \\
 & & 7 & 9 & 11 & 13 \\
 & & & 12 & 14 & 15 \\
 & & & & \boxed{\begin{matrix} 2 & 4 \\ & 6 \end{matrix}} \\
 & & & & &
\end{pmatrix}}
\quad
\underset{s_2}{\begin{pmatrix}
 & 1 & 3 & 5 & 8 & 9 \\
 & & 6 & 7 & 10 & 11 \\
 & & & \boxed{\begin{matrix} 2 & 4 & 12 \\ & 13 & 14 \\ & & 15 \end{matrix}} \\
 & & & & &
\end{pmatrix}}
$$

Figure 3: Good Splitting Property

3 Parallel Block Jacobi Methods

The class of P-Wavefront orderings contains a number of parallel orderings including those considered by Luk and others. Also included are some completely new orderings that were never investigated before.

In addition it is important to note that these results regarding convergence of various Jacobi orderings for the symmetric eigenvalue problem can be extended in a straight forward fashion to develop convergent parallel orderings for all those generalizations of the Jacobi method where convergence results for the row ordering are known. These include the Kogbetliantz algorithm for the SVD problem [10,6], the Eberlein method for the unsymmetric eigenproblem [4,8] and the generalized singular value problem [14].

Regarding the SVD problem, it has recently been conjectured that preserving triangular form of the matrix A during the Jacobi iterations can improve the rate of convergence in certain situations [3,9]. It is also shown that the well known rows ordering preserves triangular form [9]. By our results on Jacobi orderings, it follows that all Weakly Wavefront orderings, including some parallel orderings, also preserve triangular form and have the same good convergence properties.

Among the many convergent orderings discovered in the class of Weakly Wavefront orderings there are a particular a class of block Jacobi orderings which we call Block Wavefront Jacobi methods. Block algorithms access data from memory in large chunks or blocks. In modern computer architectures including parallel architectures, input/output is usually as expensive or more compared to computation, making block algorithms attractive for these machines. In Block Wavefront Jacobi methods the annihilation of the off diagonal elements proceeds in a blocked fashion, wherein elements a single submatrix are all annihilated before proceeding to another submatrix. This variety of Block Jacobi methods were first proposed by Schreiber [15] and have been used by Bischof [1] for the SVD problem with good results.

4 Conclusions

We have introduced the class of P-Wavefront orderings. Cyclic Jacobi methods for the symmetric eigenvalue, SVD and other problems using orderings from this class converge without the need of thresholds. This class is characterized by properties which are easy to test (polynomial time). We have also shown that there are some parallel block Jacobi methods that fall in this class. However, there are provably convergent orderings that are not P-Wavefront orderings.

5 Acknowledgements

This research was supported by the Office of Naval Research under Contract N00014-86-K-0610.

References

[1] Christian Bischof. *Computing the Singular Value Decomposition on a Distributed System of Vector Processors.* Technical Report 87-869, Department of Computer Science, Cornell University, 1987.

[2] R.P. Brent and F.T. Luk. The solution of singular-value and symmetric eigenvalue problems on multiprocessor arrays. *SIAM J. Sci. Stat. Comp.*, 6(1):69–84, 1985.

[3] J.P. Charlier and P. Van Dooren. On Kogbetliantz's SVD algorithm in the presence of clusters. *Lin. Alg. Appl.*, 95:135–160, 1987.

[4] P.J. Eberlein. A Jacobi method for the automatic computation of eigenvalues and eigenvectors of an arbitrary matrix. *J. SIAM*, 10:74–88, 1962.

[5] P.J. Eberlein. On the Schur decomposition of a matrix for parallel computation. *IEEE Trans. Comp.*, 36:167–174, 1987.

[6] G.E. Forsythe and P. Henrici. The cyclic Jacobi method for computing the principal values of a complex matrix. *Trans. Amer. Math. Soc.*, 94:1–23, 1960.

[7] E.R. Hansen. On cyclic Jacobi methods. *J. SIAM*, 11:448–459, 1963.

[8] V. Hari. On the global convergence of the Eberlein method for real matrices. *Numer. Math.*, 39:361–369, 1982.

[9] V. Hari and K. Veselic. On Jacobi methods for singular value decompositions. *SIAM J. Sci. Stat. Comp.*, 8(5):741–754, 1987.

[10] E. Kogbetliantz. Diagonalization of general complex matrices as a new method for solution of linear systems. In *International Congress on Mathematics, Amsterdam*, pages 356–357, 1954.

[11] F.T. Luk and H.T. Park. *On Parallel Jacobi Orderings.* Technical Report EE-CEG-86-5, School of Electrical Engineering, Cornell University, 1986.

[12] F.T. Luk and H.T. Park. *A Proof of Convergence for two Parallel Jacobi SVD Algorithms.* Technical Report EE-CEG-86-12, School of Electrical Engineering, Cornell University, 1986.

[13] L. Nazareth. On the convergence of the Cyclic Jacobi method. *Lin. Alg. Appl.*, 12:151–164, 1975.

[14] C.C. Paige and P. Van Dooren. On the quadratic convergence of Kogbetliantz's algorithm for computing the singular value decomposition. *Lin. Alg. Appl.*, 77:301–303, 1986.

[15] R. Schreiber. Solving eigenvalue and singular value problems on an undersized systolic array. *SIAM J. Sci. Stat. Comp.*, 7(2):441–451, 1986.

[16] G. Shroff and R. Schreiber. *On the Convergence of the Cyclic Jacobi Method for Parallel Block Orderings.* Technical Report RPI-CS-88-11, Computer Science Dept., Rensselaer Polytechnic Institute, May 1988. Revised July 1988, To appear in SIAM J. Mat. Anal. Appl.

Numerically Stable Fast Transversal Filters for Recursive Least-Squares Adaptive Filtering *

Dirk T.M. Slock and Thomas Kailath

Department of Electrical Engineering
Information Systems Laboratory
Stanford University
Stanford, CA 94305, USA

Abstract

In this paper, a solution is proposed to the long-standing problem of the numerical instability of Fast Recursive Least-Squares Transversal Filter (FTF) algorithms with exponential weighting, which is an important class of algorithms for adaptive filtering. A framework for the analysis of the error propagation in FTF algorithms is first developed; within this framework, we show that the computationally most efficient $7N$ form is exponentially unstable. However, by introducing redundancy into this algorithm, feedback of numerical errors becomes possible; a judicious choice of the feedback gains then leads to a numerically stable FTF algorithm with complexity $9N$. The results are presented for the complex multichannel joint-process filtering problem.

*This work was supported in part by the Joint Services Program at Stanford University (US Army, US Navy, US Air Force) under Contract DAAG29-85-K-0048 and the SDI/IST Program managed by the Office of Naval Research, under Contract N00014-85-K-0550.

NATO ASI Series, Vol. F 70
Numerical Linear Algebra, Digital Signal Processing and Parallel Algorithms
Edited by G. H. Golub and P. Van Dooren

1 Introduction

The joint-process adaptive filtering problem has been described extensively in the literature. We will adhere to the notation used in [1]. The advantages of RLS algorithms over the LMS algorithm in aspects such as tracking behaviour are well-known. However, in practice the use of the LMS algorithm is widespread due to its computational simplicity. More recently fast RLS algorithms have been introduced and esp. the FTF algorithm appears to be a computationally attractive alternative to LMS. However, its bad numerical properties have given rise to as rescue devices and have discouraged potential users. The results presented in this paper change this picture drastically. We will present the complex multichannel algorithm, but give the analysis for the single-channel case.

2 General Approaches for Analysis of Error Propagation

Discrete-time adaptive algorithms can be viewed as discrete-time non-linear dynamical systems. Let $\Theta(T)$ denote the (row) vector of algorithmic quantities that are recursively computed. Then the adaptive algorithms considered can be written in state-space form as

$$\Theta(T) = f\left(\Theta(T-1), d(T), X_N(T)\right) \quad . \tag{1}$$

Now, any numerical implementation of the algorithm will generate inaccuracies due to round-off and representation errors in the processor. Therefore, the algorithm will run with perturbed quantities $\widehat{\Theta}(T)$ and can be described by

$$\widehat{\Theta}(T) = f\left(\widehat{\Theta}(T-1), d(T), X_N(T)\right) + V(T) \tag{2}$$

where the driving term $V(T)$ represents the round-off noise. If the algorithm turns out to be robust w.r.t. numerical errors and the wordlength used is sufficiently long, then the perturbations $\Delta\Theta(T) \triangleq \widehat{\Theta}(T) - \Theta(T)$ will be small w.r.t. the infinite precision quantities $\Theta(T)$. Therefore we will consider the linearization of the system (2) around the unperturbed trajectory $\Theta(.)$. This yields

$$\Delta\Theta(T) = \Delta\Theta(T-1)F(T) + V(T) \tag{3}$$

where $F(T) \triangleq \nabla_\Theta f(\Theta, d(T), X_N(T))\Big|_{\Theta=\Theta(T)}$. If the linearized error system (3) is exponentially stable, then the implemented algorithm (2) is locally exponentially stable around the unperturbed trajectory. If the system (3) is unstable however, then the algorithm behaviour is unacceptable.

We will concentrate on the *error propagation*: how do errors generated at time T propagate in subsequent recursions. The main problem of interest is thus the analysis of the stability properties of the autonomous system

$$\Delta\Theta(T) = \Delta\Theta(T-1)F(T) \tag{4}$$

Note that this is a time-varying system and that $F(T)$ depends on $\{x(t),\ t = 0,\dots,T\}$. There are two approaches which can be taken now, depending on whether $x(.)$ is considered a stochastic process or a deterministic signal.

Stochastic approach

We have in general that (5)

$$\Delta\Theta(T_2) = \Delta\Theta(T_1)\Phi(T_1,T_2) \triangleq \Delta\Theta(T_1)F(T_1+1)\cdots F(T_2)$$

and assume that $x(.)$ is a stationary process. Then $F(.)$ will be asymptotically stationary for $\lambda < 1$. It can be shown (see [2]) that even though the system (4) is time-varying and stochastic, the stationarity of the process $x(.)$ induces a constancy in the asymptotic stability properties of the system. Now, this only tells us something about the existence of a constant underlying eigendecomposition. Little information is given as to how to compute it. In order to get an idea of these time-invariant system dynamics, a so-called *averaging technique* has frequently been used in the literature [3],,[4]. The averaging principle leads to studying the stability properties of the average $F(T)$. It can be formulated as follows.

Theorem 1 (Averaging Principle) *Assume* $F(T) = I_n - (1-\lambda)G(T)$ *where* $G(.)$ *is a sequence of bounded (as* $\lambda \to 1$*) matrices. Then the dynamics of* $\Phi(T_1,T_2)$ *can be approximated up to first order in* $(1-\lambda)$ *by the dynamics of* $\overline{F}(T_1,T_2)^{T_2-T_1}$ *where* $\overline{F}(T_1,T_2) \triangleq \frac{1}{T_2-T_1}\sum_{T=T_1+1}^{T_2} F(T)$.

If we assume $x(.)$ stationary and ergodic, then for sufficiently large $T_2 - T_1$, we can approximate $\overline{F}(T_1, T_2)$ by $EF(T)$. This might prove to be a tractable approach, leading to the study of the eigenvalue decomposition of a constant matrix.

Deterministic approach

If $x(T)$ is a known deterministic signal, then the sequence $F(.)$ can be computed in principle and the stability properties of system (4) can be investigated for the given signal. If $x(T)$ is periodic with period T_1, it readily follows that the dynamics of system (4) will be governed by the magnitudes of the T_1^{th} root of the eigenvalues of $\Phi(T, T+T_1)$.

3 The FTF algorithm

The FTF algorithm is given in Table I. Consider the following redundant ways for computing certain quantities. Apart from the possibility for computing $r_N^p(T)$ as $B_{N,T-1}X_{N+1}(T)$, which we will denote by $r_N^{pf}(T)$ (superscript f denotes filtering), there is another possibility $r_N^{ps}(T)$ that follows from the last entry in the order downdate of $\tilde{C}_{N+1,T}$, see I-(8) (superscript s denotes computation by manipulation of scalars). The two ways of computing $\tilde{C}_{N+1,T}^N$ follow immediately from the corresponding computations of $r_N^p(T)$. Also $\gamma_N^{-1}(T)$ can be computed in two ways, once as the output of the filter $\tilde{C}_{N,T}$ as in I-(13), and also by using its interpretation as inverse of the energy of the residual in estimating σ, see I-(4,12). The most computationally efficient form of the FTF algorithm (with complexity $7N$) [1],[5] uses $r_N^{ps}(T), \gamma_N^{-s}(T)$ and avoids the inner products of the filtering operations.

When infinite precision is used, the two ways of computing quantities yield identical answers. However, in any practical implementation they will differ and their difference is purely due to numerical error. Hence, these difference signals are output signals from the error propagation system, and one can now start thinking of using them in a feedback mechanism in order to influence the error propagation. However, an important question that arises here is: which feedback structure to use? Indeed, these difference signals can be fed back to any point in the algorithm, since their values would be zero if infinite precision were available. The specific structure we propose can be described as follows: feed the difference signals back into the computation of the quantities they are associated with. Or in other words: take as final value for those quantities a convex combination of their two ways of computing, e.g.

$$r_N^p(T) = r_N^{ps}(T) + K\left(r_N^{pf}(T) - r_N^{ps}(T)\right) = K r_N^{pf}(T) + (1-K)\, r_N^{ps}(T) \quad (6)$$

and similarly for $\tilde{C}_{N+1,T}^N$ and $\gamma_N^{-1}(T)$. Now, the quantity $r_N^p(T)$ is used in several instances in the algorithm, so if we use different values for the feedback constant K at those different places, then we have more freedom in affecting the error system dynamics. The motivation behind this particular choice of feedback structure is that it is intuitively appealing and it will allow us to stabilize all unstable modes in the error propagation, which is the ultimate goal. The algorithm presented in Table I reflects all the error feedback thus introduced. Note that the original $7N$ algorithm is recovered by choosing all $K_i = 0$.

The joint-process extension is clearly decoupled from the prediction problem and its error propagation dynamics are therefore completely identical to those (for $W_{N,T}$) in the RLS algorithm, Hence the results of the previous section are immediately applicable here also. The prediction problem on the other hand needs to be studied separately and we will now concentrate on this part only. This brings us to the issue of defining a state vector for the FTF algorithm:

$$\Theta(T) \triangleq \left[\overline{A}_{N,T}\ \alpha_N(T)\ \overline{B}_{N,T}\ \beta_N(T)\ \tilde{C}_{N,T}\ \gamma_N^{-1}(T)\right] \quad (7)$$

where $\overline{A}_{N,T}$ and $\overline{B}_{N,T}$ are defined by $A = \left[I \overline{A}\right]$ and $B = \left[\overline{B} I\right]$ respectively. The error propagation system can be described by

$$\Delta\Theta(T) = \Delta\Theta(T-1)\ F(T) \quad (8)$$

and $F(T)$ is $(3N+3) \times (3N+3)$. The general expression for $F(T)$[2] is prohibitive enough to make impossible any inference about its dynamics. Therefore we will consider the following steps.

Limiting case: $\lambda \to 1$.

In this case, an exact analysis is possible. We get asymptotically

$$F(T) = \begin{bmatrix} I & * & 0 & * & 0 & 0 \\ 0 & 1 & 0 & 0 & 0 & 0 \\ 0 & 0 & I & * & 0 & 0 \\ 0 & 0 & 0 & 1 & 0 & 0 \\ * & 0 & * & * & S^H-(1-K_4)u_N\overline{B}_{N,T-1} & * \\ 0 & * & 0 & * & 0 & 1-K_3 \end{bmatrix} \tag{9}$$

where u_N is the Nth unit vector and S is a $(N \times N)$ shift matrix (ones on the first subdiagonal). In what follows, we will concentrate on the block entries of $F(T)$ and consider $F(T)$ as a (6×6) block matrix. $F_{ij}(T)$ will denote block entry (i,j). Note that the eigenvalues of a product of consecutive $F(T)$ are the products of the diagonal elements. Also, $B_{N,T} \rightarrow B_N$.

For the first four components of $\Delta\Theta(T)$, we get a random walk phenomenon. For $\Delta\tilde{C}_{N,T}$ and $K_4 = 0$, the eigenvalues of the companion matrix F_{55} are the zeros of the backward prediction filter B_N. Now, it is known that the backward filter in the prewindowing method has its roots inside the unit-circle. Hence, for prewindowing with exponential weighting, the roots of the backward predictor are in a circle with radius $\frac{1}{\sqrt{\lambda}}$. For $\lambda < 1$, this means that the backward prediction filter may not be minimum-phase. However, $B_{N,T}$ is an unbiased estimate of B_N, and hence, on the average, the roots of $B_{N,T}$ will be inside the unit-circle. Nevertheless, one may want to choose $K_4 \neq 0$ in order to influence the eigenvalues of F_{55}. It is obvious that the optimal values are $K_4 = K_3 = 1$. This makes the eigenvalues associated with $\Delta\tilde{C}_{N,T}$ and $\Delta\gamma_N^{-1}(T)$ zero.

Averaging analysis

We consider an averaging analysis for the general case. From the previous analysis, it is clear that $F(T)$ is $(1-\lambda)$-close to identity only for part of the system. Hence, the averaging analysis will not yield quantitative results but only qualitative insight in the dynamics of the error system. The averaging yields approximately that the eigenvalues are given by the average diagonal entries $\overline{F_{11}} \cdots \overline{F_{66}}$, which are

$$\lambda I,\quad \lambda,\quad \left\{\tfrac{1}{\lambda} - [K_1 + (2-K_5)(1-\lambda)](1-\lambda)\right\} I,$$
$$\tfrac{1}{\lambda} - [2K_2 - (1-K_5)(1-\lambda)](1-\lambda)\,,\quad S^H - (1-K_4)\,u_N B_N,\quad 1-K_3 \tag{10}$$

This shows that the original algorithm ($K_i = 0$) is exponentially unstable, due to the $\frac{1}{\lambda}$ entries for $\Delta\overline{B}_{N,T}$ and $\Delta\beta_N(T)$. It indicates that the algorithm can be stabilized by proper choice of the feedback parameters K_i. However, it is not true that the eigenvalues corresponding to the last four entries in $F(T)$ can be given arbitrary values, as follows from the above approximate analysis. An exact analysis can be done for simple deterministic cases. We consider two examples with $N = 1$.

Deterministic example 1

Let $x(T) \equiv 1$ (dc signal). This is the extreme case of a perfectly predictable signal. Let ν_i denote the eigenvalue associated with F_{ii}. We find that $\nu_1 = \nu_2 = \nu_4 = \lambda$. Furthermore, if we choose $K_3 = 1$, then $\nu_6 = 0$, and a quadratic equation remains for ν_3, ν_5. The coefficients of this equation depend on K_1, K_4, which can be chosen to make $\nu_3 = \nu_5 = 0$.

Deterministic example 2

Let $x(T) = 0.5 + 0.5(-1)^T$. This is the extreme case of a perfectly unpredictable signal. Here we investigate $F(T-1)F(T)$. Again we find that $K_3 = 1$ is highly preferable and also $K_5 = 1$ (though only of secondary importance). With these choices, we find $\nu_2 = \lambda^2$, $\nu_5 = \nu_6 = 0$, and $\nu_4 = \lambda^2 + 2(1-\lambda^2)(1-K_2)$. A quadratic equation remains for ν_1, ν_3 with K_1 as only free parameter. Minimization of $\max\{|\nu_1|, |\nu_3|\}$ w.r.t. K_1, as a function of λ yields an optimal $\nu_1 = \nu_3$. This reveals that the system cannot be stabilized for small λ ($< .4$). This seems to be due to an increasing discrepancy between $\alpha_N(T)$ and $\beta_N(T)$.

Optimal choice of K_i

The influence of the different feedback parameters can be summarized as follows. K_5 is only of secondary importance. K_4 is also not necessary for stabilization, but only adds additional freedom in the shaping of the error system dynamics. $K_3 = 0$ leads to an $8N$ algorithm, and to a random walk for $\Delta\gamma_N^{-1}(T)$. This has been confirmed by simulations, which show that the errors increase perfectly linearly with time, over millions of samples. Eventual blow-up may take a very long time, but will occur eventually. It can be shown that the solution proposed in [6] approximately corresponds to this $8N$ algorithm and their parameter ρ is related to K_1 ($= K_2$ for this comparison). However, this solution is clearly insufficient. Furthermore, as is clear from above framework, their interpretation of the influence of the parameter ρ is quite misleading. Finally, $K_1 = K_2 = 0$ leads to the original $7N$ algorithm with exponential divergence.

In general, the optimal K_i can be a function of λ, N, and the signal characteristics (x). The optimal choice for K_4 appears to be a tricky matter. Any value for K_4 which is significantly different from zero appears to make the optimal values for the other K_i depend heavily on the signal characteristics. Also, $K_4 = 1$ leads to much poorer performance than $K_4 = 0$, or even instability for any signal for which $B_N \neq 0$. A very small positive value for K_4 appears to work best. Furthermore, for $K_4 \approx 0$, the optimal values for K_1, K_2 and K_3 are virtually signal independent.

A computer program has been written to search by simulation for the optimal K_i for a white noise signal. The optimality criterion used is

$$\sum_{T=0}^{T_1} \left\{ \left(r_N^{ps}(T) - r_N^{pf}(T) \right)^2 + \left(\gamma_N^{-s}(T) - \gamma_N^{-f}(T) \right)^2 \right\} \tag{11}$$

where T_1 is of the order of 10^5, depending on N. $K_4 = 0$ and $K_5 = 1$ were kept fixed, and the search was done over the remaining 3D space. Searches were performed for $N = 2^i$, $i = 0, 1, \ldots, 7$ and $\lambda = 1 - \frac{j}{10N}$, $j = 0, 1, \ldots, 10$. The results indicate that stability can be achieved for all N and for $\lambda \in \left[1 - \frac{1}{2N}, 1\right]$. This is a useful range for λ for tracking purposes in all practical applications.

As far as the sensitivity for the optimal values of the K_i is concerned, we found that K_3 and K_2 are not sensitive at all. For K_3, any value around 1 will do, so $K_3 = 1$ will do fine in all cases. For $\lambda = .98$ and $N = 20$, any value of K_2 in the range [.6, 10.] will do and $K_2 = 2.5$ will probably do in all cases. For K_1, the criterion is a bit more sensitive but $K_1 = 1.6$ should work fine. For the case of $\lambda = .98$ and $N = 20$, a search for the optimal K_i was performed with an extremely colored noise, obtained by passing white noise through the filter $H(z) = 1/(1 - .999z^{-1})$. It was found that the optimal K_i obtained with white noise are also optimal for this colored noise.

To compare the thus stabilized FTF algorithm with the RLS algorithm, it appears that by proper fine-tuning of the K_i, one can make $\nu_{\max}$ comparable to λ for a certain range of λ values so that the dynamics of the error propagation are comparable for both algorithms. The error generation situation seems to be in favor of the FTF algorithm since this algorithm requires an order of magnitude less operations than RLS, and thus will generate less round-off error. This intuition appears to agree with the known *high accuracy* property of the FTF algorithm. Simulations indicate that both algorithms have comparable error levels. As far as the influence of the data conditioning is concerned, F_{11} corresponds to the behaviour of $\Delta W_{N,T}$, while F_{22} can be likened to the behaviour of the Riccati equation in RLS. Indeed, F_{22} shows exponential instability with rate $\frac{1}{\lambda}$ in a certain subspace if the data matrix is rank deficient. In general, the error level increases with the condition number of the data matrix. One can conclude that any condition number can be handled, given that the wordlength used is long enough. Conversely, for a given wordlength it is possible to create sufficiently unexciting signals that will make the algorithm unstable. This is true for both RLS and FTF.

References

[1] J.M. Cioffi and T. Kailath. "Fast, recursive least squares transversal filters for adaptive filtering". *IEEE Trans. on ASSP*, ASSP-32(2):304–337, April 1984.

[2] D.T.M. Slock and T. Kailath. "Numerically Stable Fast Transversal Filters for Recursive Least-Squares Adaptive Filtering". Submitted, IEEE Trans. ASSP.

[3] C.G. Samson and V.U. Reddy. "Fixed Point Error Analysis of the Normalized Ladder Algorithm". *IEEE Trans. Acoust. Speech Sig. Proc.*, ASSP-31(5):1177–1191, Oct. 1983.

[4] B.D.O. Anderson *et al. Stability of Adaptive Systems: Passivity and Averaging Analysis.* The MIT Press, Cambridge, Massachusetts, 1986.

[5] G. Carayannis, D. Manolakis, and N. Kalouptsidis. "A fast sequential algorithm for least squares filtering and prediction". *IEEE Trans. ASSP*, ASSP-31(6):1394–1402, 1983.

[6] J-L Botto and G.V. Moustakides. "Stabilization of Fast Recursive Least-Squares Transversal Filters for Adaptive Filtering". In *Proc. ICASSP 87 Conf.*, volume 1, pages 403–407, Dallas, Texas, April 1987.

Table I
Multichannel Complex FTF Algorithm

#	Computation	Cost
Prediction Problem		
1	$e_N^p(T) = A_{N,T-1} X_{N+1}(T)$	Np^2
2	$\tilde{C}^{0H}_{N+1,T} = -\lambda^{-1}\alpha_N^{-1}(T-1)\, e_N^p(T)$	$1.5p^2 + 0.5p$
3	$\tilde{C}_{N+1,T} = \left[0 \;\; \tilde{C}_{N,T-1}\right] + \tilde{C}^0_{N+1,T} A_{N,T-1}$ except $\tilde{C}^N_{N+1,T}$	$(N-1)p^2$
4	$\gamma^{-1}_{N+1}(T) = \gamma_N^{-1}(T-1) - \tilde{C}^0_{N+1,T}\, e^p_N(T)$	p
5	$r_N^{pf}(T) = B_{N,T-1} X_{N+1}(T)$	Np^2
6	$\tilde{C}^{NfH}_{N+1,T} = -\lambda^{-1}\beta_N^{-1}(T-1)\, r_N^{pf}(T)$	$1.5p^2 + 0.5p$
7	$\tilde{C}^{Ns}_{N+1,T} = \tilde{C}^{N-1}_{N,T-1} + \tilde{C}^0_{N+1,T} A^N_{N,T-1}$	p^2
8	$r_N^{ps}(T) = -\lambda\beta_N(T-1)\, \tilde{C}^{NsH}_{N+1,T}$	$1.5p^2 + 0.5p$
9	$r_N^{p(i)}(T) = K_i\, r_N^{pf}(T) + (1-K_i) r_N^{ps}(T) \quad i = 1,2,5$	$3p$
10	$\tilde{C}^N_{N+1,T} = K_4\, \tilde{C}^{Nf}_{N+1,T} + (1-K_4)\tilde{C}^{Ns}_{N+1,T}$	p
11	$\left[\tilde{C}_{N,T} \;\; 0\right] = \tilde{C}_{N+1,T} - \tilde{C}^N_{N+1,T} B_{N,T-1}$	Np^2
12	$\gamma_N^{-s}(T) = \gamma^{-1}_{N+1}(T) + \tilde{C}^{Ns}_{N+1,T} r_N^{p(5)}(T)$	p
13	$\gamma_N^{-f}(T) = 1 - \tilde{C}_{N,T} X_N(T)$	Np
14	$\gamma_N^{-1}(T) = K_3\gamma_N^{-f}(T) + (1-K_3)\gamma_N^{-s}(T)$	1
15	$e_N(T) = e_N^p(T)\,\gamma_N(T-1)$	p
16	$A_{N,T} = A_{N,T-1} + e_N(T)\left[0 \;\; \tilde{C}_{N,T-1}\right]$	Np^2
17	$\alpha_N^{-1}(T) = \lambda^{-1}\alpha_N^{-1}(T-1) - \tilde{C}^{0H}_{N+1,T}\, \gamma_{N+1}(T)\, \tilde{C}^0_{N+1,T}$	$p^2 + p$
18	$r_N^{(i)}(T) = r_N^{p(i)}(T)\, \gamma_N^s(T) \quad i = 1,2$	$2p$
19	$B_{N,T} = B_{N,T-1} + r_N^{(1)}(T)\left[\tilde{C}_{N,T} \;\; 0\right]$	Np^2
20	$\beta_N(T) = \lambda\beta_N(T-1) + r_N^{(2)}(T)\, r_N^{p(2)H}(T)$	p^2
	for multichannel also:	
21	$\tilde{C}^{N(2)}_{N+1,T} = K_2\, \tilde{C}^{Nf}_{N+1,T} + (1-K_2)\tilde{C}^{Ns}_{N+1,T}$	p
22	$\gamma^{-(2)}_{N+1}(T) = \gamma_N^{-s}(T) - \tilde{C}^{N(2)}_{N+1,T}\, r_N^{p(2)}(T)$	p
23	$\beta_N^{-1}(T) = \lambda^{-1}\beta_N^{-1}(T-1) - \tilde{C}^{N(2)H}_{N+1,T}\, \gamma^{(2)}_{N+1}(T)\, \tilde{C}^{N(2)}_{N+1,T}$	$p^2 + p$
Joint-Process Extension		
24	$\epsilon_N^p(T) = d(T) + W_{N,T-1} X_N(T)$	Npq
25	$\epsilon_N(T) = \epsilon_N^p(T)\,\gamma_N(T)$	q
26	$W_{N,T} = W_{N,T-1} + \epsilon_N(T)\, \tilde{C}_{N,T}$	Npq
Single real channel total (3 divides)		$9N + 20$
Multichannel total (2 divides)		$N\left(6p^2 + 2pq + p\right) + 7.5p^2 + 14.5p + q + 1$

Applications of Analytic Centers

György Sonnevend*

Institut für Angewandte Mathematik und Statistik
Universität Würzburg
8700 Würzburg, West-Germany

1 Introduction

In this paper we present some applications of a recently introduced concept, the "analytical center" of a convex inequality system. Let $F = \{f_\alpha(x,p)\}$ denote a family of functions depending on parameters $p \in P$ and elements α of an index set A, which are concave in $x \in X$, on the sets $S_{\alpha,p}$ where they are nonnegative. Here X is a Hilbert space fixed together with the sets A and P. Below we shall consider the specific case when for each (α,p) the function $f_\alpha(x,p)$ is linear or concave quadratic in x or when it is the determinant of a symmetric, positive semidefinite matrix, whose elements depend linearly on x. Thus we consider inequality systems (A,p) with feasible sets

$$S(A,p) := \{x \mid f_\alpha(x,p) \geq 0, \alpha \in A\} = \bigcap_{\alpha \in A} S_{\alpha,p}.$$

We propose a special "solution" (i.e. a feasible point) $x = x(A,p)$ of such systems, called the analytic center of (A,p) (in fact $x(A,p)$ will depend on (A,p) and not only on $S(A,p)$) and show that it can be applied for the

*on leave from Dept. Numerical Analysis, Inst. of Mathematics, Eötvös University, 1088, Budapest, Muzeum K. 6-8, Föep.

NATO ASI Series, Vol. F 70
Numerical Linear Algebra, Digital Signal Processing
and Parallel Algorithms
Edited by G. H. Golub and P. Van Dooren

solution of many problems associated with $S(A,p)$, especially to the four problems listed below.
1.) Decide by a (low complexity and stable) numerical algorithm whether $S(A,p)$ is nonvoid for a particular value of the parameter p (and if so, find one feasible point in it).
2.) Having defined a particular solution $x(A,p)$ we would like to be able to "update" or to "downdate" this solution when the parameter p is changed in P, or when we enlarge or diminish the index set (by adding or deleting elements).
3.) The approximation problem. In order to estimate the feasible set $S(A,p)$ we would like to approximate it both from inside and from outside by concentrical and similar ellipsoids around $x(A,p)$, whose homothecity (similarity) ratio is possibly small and which are as easily constructed as $x(A,p)$ is.
4.) In order to solve an optimization problem of the form

$$\sup\{f_0(x,p) | \; x \in S(A,p)\} =: \lambda^*$$

where $f_0(x,p)$ is supposed to be concave in x (at least on $S(A,p)$), we are interested to study the path of analytic centers of the one parameter family of inequality systems

$$S(\lambda, A, p) = \{x | \; \lambda \leq f_0(x,p), \; x \in S(A,p)\}, \lambda \leq \lambda^*.$$

Obviously points of this "central path" converge to the optimal set if $\lambda \uparrow \lambda^*$. we are interested to propose the "center" solution by assuring an invariance property for it of the following form: whenever G is a group acting both on the parameter space and on the space X, it should be true that – for each $g \in G$ and $p \in P$

$$T_g(p)x(A,p) = x(A,gp)$$

here T_g denotes the linear transformation corresponding to g (i.e. a representation of G; we shall be interested in the affine group, in its subgroup of volume preserving transformations and in the real orthogonal group).
In section 2 we present some precise theorems proved in [14], see also [10] and [12] which provide partial solutions to problems 1. – 4. for the case when the index set A is finite and the functions f_i are linear or quadratic.

We show that one can construct effective algorithms for following the central path by a predictor corrector method. The corrector is the well known Newton's method for solving the simple algebraic equation characterizing the center $x(\lambda)$, it turns out that its "large" domain of quadratic convergence is approximately equal to the interior ellipsoid we will be able to construct. A lot of nice internal structure in this special class of problems is exploited to construct efficient extrapolation methods, i.e. predictors.
In section 4 we show the generalization to the case of an infinite set A (endowed with a measure invariant under G) is straightforward. In fact in a number of interesting cases the arising solution concept i.e the analytic centre coincides with the "maximum entropy" solution of moment or extension problems and it can be computed surprisingly simply, in a finite small number of operations. The main examples for this are connected to the Nevanlinna - Pick moment problems and to the problems of minimum norm (and maximum entropy) extensions of operators see [16], [7], [6], [3], [8], [2]. As an application of analytic centers we present a new look on the following two problems, see [4] - [9].
Problem ME. Suppose the elements of a matrix are fixed on a subset I of indexes (i,j). Find, if possible that extension of this matrix (i.e. specify its undefined elements in a way) which makes it positive semidefinite and its determinant maximal (when not stated otherwise we shall assume that in this problem all extensions yield symmetric matrices).
Problem MN. A problem (in fact very much related to the previous one) we study is that of finding a maximal spectral norm extension of a partially given matrix. An important special case of this problem (with its applications) has been studied in [4], note that the problem studied in [5] also belongs to this class of problems modulo a change of norms.
In section 3 we present the application to problems of (feedback) control and estimation in linear discrete and continuous time systems under pointwise bounded disturbance and control inputs and pointwise bounded measurement errors, these are the problems which the classical linear-quadratic optimal filtering and control theory solves under the assumption of integral bounds. In contrast to the methods of "nondifferentiable" convex analysis proposed hitherto analytic centers allow us to use "smooth", in fact analytic objects.
Finally in section 4 we show, that some difficult, essentially nonconvex

feasibility problems, like that of constructing low rank approximations of special type (Hankel, Toeplitz or positive definite) to partially given matrices and minimum degree stable (or positive real) rational interpolants to discrete or continuous time transfer functions (of the same type) can be approached using analytic centers.

2 Basic properties of analytic centers (for concave quadratic inequalities)

In this section we suppose that A is the finite set $\{1,..,m\}$ and will drop the symbols A and p (whenever not needed) in the notations. Supposing $f_1, ..., f_m$ to be concave on the sets $S_1, ..., S_m \subset I\!R^n$, where they are positive, we assume that $S(m) = \cap_{k=1}^m S_k$ is bounded and has a nonempty interior in $I\!R^n$. Consider the problem

$$\sup\{\Psi(x)|\ f_k(x) \geq 0,\ k = 1, ..., m\}, \quad \text{where} \quad \Psi(x) = \prod_{k=1}^{m} f_k(x)^{1/m} \quad (2.1)$$

It is easy to show, that $\Psi(x)$ is a concave function and that it is strongly concave if the following assumption A is made

$$-\sum_{k=1}^{m} D^2 f_k(x) + \nabla f_k(x)(\nabla f_k(x))^\star \quad \text{is positive definite}$$

for all x in $S(m)$. The latter condition holds for example if at least one of the f_k-s is strongly concave on $S(m)$ or if all f_k are linear, see [14] and [15] for this and the following statements.
The unique solution of (2.1) $x = x(f_1, ..., f_m)$ satisfies the equation

$$\sum_{k=1}^{m} \frac{\nabla f_k(x)}{f_k(x)} = 0 \quad (2.2)$$

and is its unique solution under assumption A. The Hessian of the function $-m \log \Psi(x)$ has the simple form

$$\sum \frac{D^2 f_k(x)}{f_k(x)} - \frac{\nabla f_k(x)(\nabla f_k(x))^\star}{f_k^2(x)} =: H(x) \quad (2.3)$$

<u>Theorem 1.</u> Suppose that the functions $f_1, ..., f_m$ are concave quadratic and that assumption A holds, then for the feasible set $S(m)$ of the inequality system defined by the requirement of the positivity of the functions f_k, $k = 1, ..., m$, the following two-sided ellipsoidal approximation is valid

$$x(f_1, ..., f_m) + \frac{1}{2\sqrt{m}}\tilde{E} \subseteq S(m) \subseteq x(f_1, ..., f_m) + \sqrt{2m}\tilde{E} \qquad (2.4)$$

where :
$\tilde{E} = \tilde{E}(f_1, ..., f_m) = \{x| < D^2\Psi(x(f_1, ..., f_m))x, x > \leq \Psi(x(f_1, ..., f_m))\}$ is an ellipsoid (constructed in a natural way from the second order taylor expansion of Ψ around $x(f_1, ..., f_m)$).
This theorem is proved in [15] using the following interpretation of the optimality condition (2.2): if $x = x(f_1, ..., f_m)$ is taken as the center of duality in $\mathbb{R}^n$ with respect to the unit ball, then the arithmetical mean of the centers of the dual ellipsoids $E_k^D, k = 1, ..., m$ is zero,i.e. the origin (here, of course, we used the observation that the dual E^D of an ellipsoid E containing the origin is again an ellipsoid even if E is not centered at the origin). An other proof (using the same ellipsoid E) but yielding a similarity ratio $25m$ (instead of $2\sqrt{2}$) has been given by Jarre, see [10], his method of proof can be extended (to yield similar approximations) to a larger class of problems, where the functions $f = f_k$ satisfy a relative Lipschitz condition of the following form

$$| < z(D^2f(x) - D^2f(y))z > | \leq M\|x - y\|_{H(y)} < z, D^2f(y)z >$$

for all x, y with $\|x - y\|_{H(y)} \leq 1/2$ and all z, where - as usual $\|z\|_C$ denotes the C-norm $< z, Cz >^{1/2}$ and M is a positive constant. We suspect that the problems, where the det function - over the class of positive definite, symmetric matrices, under linear conditions is to be maximized (see below for interesting examples) - belong to this class.
For the solution of problem 4, i. e. for the extrapolation of the central curve $x(\lambda)$, $\lambda \geq \lambda^*$ several algorithms have been proposed and analysed already. Since this curve is analytic in λ, high degree extrapolations seem to be more effective than lower order ones, in fact preliminary computational experiments indicate that one gains in raising the order of a simple, polynomial predictor at least till order 5. In [15] we proposed an extrapolation algorithm using rational functions based on the analysis of the problem with

$m = 1$, f_0, f_1 quadratic. This algorithm is a generalisation of the method of conjugate gradients and of an underlying observation that multipoint Pade approximants to Stieltjes functions are again Stieltjes functions (with the 'right' support for their poles) and converge geometrically to the original function. Of course, knowing the connections between analytic centers and moment problems of the Nevanlinna -Pick type (see below) it is not very surprising that functions from the same class (namely Stieltjes functions) and orthogonal polynomials occur also in the finite dimensional case.
In the case of linear programs, where f_k, $k = 1, ..., m$ are linear in x one can prove, see [15] for a future reference to this recent result obtained in cooperation with Gongyun Zhao, that for a simple, first order (i.e. linear) extrapolation algorithm, with adaptive stepsize selection the total number of iterations needed to reach the parameter value $\lambda = -\delta + \lambda^*$ from λ_0 can be estimated by *const.* $m^{1/4} \log((\lambda^* - \lambda^0))/\delta)$. For an 'optimal' zero order method $m^{1/4}$ here should be replaced by $m^{1/2}$, note however that the latter estimation has been proved also for general quadratic problems in [10].
In problems where we have (in addition) a finite set of linear equality constraints on the variable x, the analytic center is defined as the maximizer of the same function under the additional constraints. For example, a linear inequality system

$$S(p) = S(b^m, a^m) := \{x| \ b_i - < a_i, x > \geq 0, \quad 1 \leq i \leq m\}$$

is 'equivalent'- by introducing the variables $\mu_i = b_i - < a_i, x >, \ \ i = 1, ..., m$ and eliminating x, to the constrained system

$$P(k^N, c^N) = \{\mu| < k_j, \mu >= c_j, j = 1, ..., N = m - n,$$
$$\mu_i \geq 0, \quad i = 1, ..., m\} \tag{2.7}$$

The analytic center $\mu(k^m, c^m)$ is the solution of the problem

$$\sup\{\sum \log \mu i| \ \ \mu \in P(k^N, c^N)\} \tag{2.8}$$

A geometric characterisation of the optimal $\mu = \bar{\mu}(k^N, c^N)$ which shows that $P(k^N, c^N)$ can be represented as the intersection of an $(m-1)$-dimensional simplex S, is given in [12]. Centers and central paths for linear programs have been studied also in [1], where - as in [12] and [15] - further references are given to a rapidly growing field of research started with the contribution of Karmarkar to the earlier theory of 'logarithmic' potential functions.

3 Application to control and estimation problems

We will denote the vectors of state, disturbance input, control input observed output, observation error and output to be controlled by x, v, u, y, w and z respectively. The relations between these are described by

$$x(k+1) = Fx(k) + Gu(k) + Dv(k), \quad x(0) = 0 \tag{3.1}$$

$$y(k) = Hx(k) + Jw(k), \quad z(k) = Cx(k), k = 0, 1, ... \tag{3.2}$$

The bounds for the values of the variables u, v and w will be described by the requirement that they belong to unit balls (in the respective spaces) either in the Euclidean or in a 'box' norm at each moment of time. This will imply quadratic or linear constraints on the linear forms in the variables $(x, ..., z)$:

$$x(j+1) - Fx(j) \in G_1, \quad y(j) - Hx(j) \in G_2, \quad u(j) \in G_3, j \leq k \tag{3.3}$$

(we can regard the $2r$ linear constraints specifying a box constraint,say $a_i \leq u_i \leq b_i$, $i = 1, ..., r$ as equivalent to r quadratic constraints $(b - u_i)(u_i - a_i) \geq 0$, $i = 1, ..., r$. A stationary feedback controller is a function $U(\xi(k))$, where $\xi(k)$ is the value of an auxiliary variable ξ at time k evolving according to a recursion of the following form

$$\xi(k+1) = F(\xi(k), y(k)) \tag{3.4}$$

for some function F, (3.4) is called a dynamic observer.Let us first consider the problem of observation, i.e. of state reconstruction.We see that (3.3) is a quadratic inequality system for the unknowns $x(0), ..., x(k)$ given $y(0), ..., y(k)$. Using Theorem 1 we can find a two-sided ellipsoidal approximation for the set of possible values of the unknown vector $x^k = (x(0), ..., x(k))$, thus specially for the set of localisation for the vector $x(k)$ (note that projections of ellipsoids are again ellipsoids). In order to be able to update these approximations (i.e. the centers $\bar{x}_k(k)$ and the corresponding ellipsoids $E_k(k)$ we select a window length d ,i. e. delete equations and variables corresponding to time moments $j \leq k - d$. If k is raised to $(k+1)$ then two constraints are added, resp. deleted and we can use the central

values $[x_k(k-d), ..., x_k(k)]$ as starting points for (the Newton iteration for) computing the new center $[\bar{x}_{k+1}(k-d+1), ..., \bar{x}_{k+1}(k+1)]$. An alternative is to compute first a starting value for $x_{k+1}(k+1)$ by solving the 'mini' problem (i.e. to compute its central solution $\bar{x}^0(k+1)$) :
$x(k+1) - Fx(k) \in G_1, \;\; y(k+1) - Hx(k+1) \in G_2, x(k) \in E_k(k)$
and use $[\bar{x}_k(k-d+1), ..., \bar{x}_k(k), \bar{x}^0(k+1)]$ as the starting value (for these corrector steps we may need eventually to apply a short path following method, in [18] a special algorithm is given for computing the centers of linear inequality systems with an estimation of its complexity).
The (dual) control problem - assuming first complete observability and absence of disturbances) is that of finding - under some state constraints $z_k \in G_4$ which we neglect for brevity - the set of states accesible from the origin, we propose an approximation to it by computing an ellipsoidal approximation for a finite (Minkowski) sum of ellipsoids (or boxes):

$$x(k) \in \sum_{j=0}^{k-1} F^j G_3 = \{x| \;\; x = \sum_{j=1}^{k} F^j u_{k-j}, \;\; f(u_{k-j}, G_3) \geq 0\} \tag{3.5}$$

where f is a quadratic function or the r-th root of the product of the factors $(u_i - a_i)(b_i - u_i)$ corresponding to G_3 according to our assumption. Obviously the equality constraints in (3.5) are linear. In order to improve the above ellipsoidal approximations (by a more expensive algorithm) we propose to solve - at each step k - the 'convex' problem of finding (a good approximation of) the ellipsoid of maximal volume in the current set of localisation, note that the logarithm (or n-th root) of the determinant is a concave function and $AG \subseteq S(p)$, G being the unit ball, is also a 'convex' condition, when we restrict A to the set of symmetric, positive semidefinite matrices.Thus we can apply the homotopy approach proposed above for the problem 4.
For continuous time systems the following observer can be proposed (its recursive implementation being under current investigation)- say , for quadratic constraints: solve the Euler - Lagrange variational problem

$$\inf_x \int_{t_0}^{T} (\log(\rho^2 - \|\dot{x}(t) - Fx(t)\|^2) + \log(\sigma^2 - \|y(t) - Hx(t)\|^2))dt$$

whose extremals satisfy a smooth (analytic), second order differential equation and the transversality condition $\dot{x}(t) - Fx(t) = 0$ at $t = t_0$ and $t = T$.

For the more difficult problems of feedback control the tools presented here are also applicable, we refer to [5] and [14]. One can prove that the optimal observers can be chosen to be linear functions in the observed variables, but optimal feedback controls, in the class of 'games' (3.1)-(3.2) for achieving a maximal (minimal) value for the minimal (maximal) value of the norm of $z(k)$, say for $N_0 \leq k \leq N_1$ in evaison- pursuer games (arising also in robotics) must be nonlinear, so that one has to look to alternatives to the classical or even to the more recent (see [5]) linear compensators.

4 Analytical centers as maximum entropy extensions in moment problems

Historically the first examples of easily computable analytic centers arose (in the infinite dimensional setting) in the theory of moment problem of Nevanlinna - Pick type. As an example we briefly formulate some of the main results for the class of positive - real transfer functions on the unit disk D of the complex plane, for a comprehensive treatment see [7], the references therein and [13], [14] for some of the results only briefly mentioned below. Let $z_1, ..., z_n$ be n distinct points of D and $c-1, ..., c_n$ be given complex numbers. The interpolation (moment) problem: find a nonnegative measure μ on the unit circle T such that

$$c_i = \Omega^\mu(z_i) = \frac{1}{2\pi} \int_{-\pi}^{\pi} \frac{e^{i\theta} - t}{e^{i\theta} + t} \mathrm{d}\mu(\theta) \quad i = 1, ..., N \tag{4.1}$$

is solvable iff the matrix

$$\frac{c_i + \bar{c}_j}{1 - z_i \bar{z}_j} \tag{4.2}$$

is positive semidefinite, moreover in the case of positive definiteness of this matrix there is a unique mass distribution $d\mu(\theta) = \mu'(\theta)\mathrm{d}\theta$ which maximizes the "entropy" integral (an "obvious" generalization to (2.7))

$$d(c^N, z^N) = \sup \int_{-\pi}^{\pi} \log \mu'(\theta) \mathrm{d}\theta \tag{4.3}$$

under the "moment conditions (4.1). Moreover the corresponding function $\Omega^\mu(z) = a_N(z)/q_N(z)$, $\deg a_N \leq \deg q_N = N$ and the optimal density is easily computable by a stable $O(N^2)$ operation algorithm, in fact

$$\mu'(\theta) = |\frac{p_N(z)}{q_N(z)}|^2, \quad \text{where } p_N(z) = (z - z_1) \cdot \ldots \cdot (z - z_N)$$

and $q_N(z)$ is the N-th orthogonal polynomial with respect to the weight $|p_n(z)|^{-2} d\mu(\theta), e^{i\theta} = z$, where μ is any of the measures in (4.1); concerning this maximum entropy solution see also [11] and the original contribution in [2].

In [13] and [14] we have proposed and analyzed an algorithm for the sequential i.e. adaptive choice of the interpolation points $z_1, \ldots, z_n$, to get – by $p_N(z)/q_N(z)$ a better recovery of the measure than what is achievable by any passive, i.e. simultaneous choice of the points. For a precise statement about the superiority of sequential algorithms see [13]. Note that the choice $z_i = 0, \quad i = 1, \ldots, N$ corresponds to measuring the correlation coefficients (to estimate the spectral measure $d\mu$ corresponding to a stationary Gaussian process). In the latter application a finite length sample $[x_1, \ldots, x_M]$ is given and we argue that the values of $\Omega(z_k)$ – for not too small values of $\delta(z_k) = 1 - |z_k|$ – can be much more accurately recovered from $[x_1, \ldots, x_M]$ - using the isometry $x_k \leftrightarrow e^{ik\theta}$ (as an element of $L_2(d\mu, T)$ - than the high order correlation coefficients. Having computed approximate values $\tilde{c}_1, \ldots, \tilde{c}_N$ within accuracy $\epsilon_1, \ldots, \epsilon_N$, we propose first to make a "regularization" by solving – again by the method proposed for the solution of problem 4

$$\max\{\det \frac{c_k + \bar{c}_j}{1 - z_k \bar{z}_j} | \ |c_k - \tilde{c}_k| \leq \epsilon_k, \quad k = 1, \ldots, N\}. \tag{4.4}$$

In order to find out whether, for some fixed k, there exists a positive real interpolant $a_k(z)/q_k(z)$

$$\frac{a_k(z_i)}{q_k(z_i)} = c_i, \ \deg a_k \leq \deg q_k \leq k, \ \text{where } \frac{N}{2} \leq k < N, \ i = 1, \ldots, N, \tag{4.5}$$

for the regularized data $c_1, \ldots, c_n$ at $z_1, \ldots, z_n$ we propose the following algorithm. First we reduce the order of the above interpolant p_N/q_N by any one of the nowadays standard techniques of "stochastic model reduction" using singular value analysis, balanced realizations or, say, by a Hankel norm optimal reduction of the model filter corresponding to p_N/q_N. This yields usually a good k-th order approximation a_k^0/q_k^0 for a_N/q_N which however does not satisfy the prescribed interpolation conditions. In a second step we consider the convex feasibility problem

$$a(z_i) = c_i q(z_i) \quad i = 1, ..., N, \quad \deg a \ \leq \deg q \leq k, \tag{4.6}$$

$$|q(z) - q_k^0(z)| \leq (1-\alpha)|q_k^0(z)| \quad \text{for all } z \in D \tag{4.7}$$

$$|a(z) + q(z) - (a_k^0(z) + q_k^0(z))| \leq (1-\gamma)|a_k^0(z) + q_k^0(z)|, \tag{4.8}$$

$$|a(z) - q(z) - (a_k^0(z) - q_k^0(z))| \leq \delta |a_k^0(z) + q_k^0(z)|, \quad \text{for all } |z| = 1 \tag{4.9}$$

Here suitable values for γ, δ and α should be selected (by trial) such that

$$\delta + \beta < \gamma \tag{4.10}$$

where β is any number satisfying

$$|a_k^0 - q_k^0| \leq \beta |a_k^0 + q_k^0| \tag{4.11}$$

In the last inequality β can in general be selected to be smaller than 1, since the positive real condition for a fraction a/q is that

$$|q(z) - a(z)| \leq |q(z) + a(z)| \text{ for } |z| = 1$$

$$\text{and } |q(z)| > 0, \quad \text{for } |z| < 1 \tag{4.12}$$

This shows that if the numbers c_i are near to $c_i^0 := a_k^0(z_i)/q_k^0(z_i)$ (4.11) can be always satisfied by suitable constants. If in addition (4.10) can be satisfied, then any solution a/q of (4.6) - (4.9) is a positive real interpolant: indeed (4.8) implies that $|a(z) + q(z)| \geq \gamma |a_k^0(z) + q_k^0(z)|$ and $|a(z) - q(z)|$ can be estimated by the sum of $|a(z) - q(z) - [a_k^0(z) - q_k^0(z)]|$ and $|a_k^0(z) - q_k^0(z)|$.

We propose now the following homotopy method to check the feasibility of system (4.6) – (4.9). If c_i is replaced by c_i^0 then the latter will be the analytic center of this system which we get – by definition – maximizing the sum of the integrals of the logarithmic residuals, for example the first term of the objective function will be

$$\int_D \log((1-\alpha)^2 |q_k^0(z)|^2 - |q_k^0(z) - q(z)|^2) dz$$

while in the second and third term the integration over T, we used squared modules to have an analytic objectiv function in the $2k+1$ free variables. Now we propose to deform c_i^0 to c_i by the usual linear homotopy and follow the path of centers in the $(2k+1)$ dimensional space of the coefficients.
Note that if we are only interested in the stability requirement (4.6), a stable Pade-type approximant (stable minimal "realization") is obtained (when selecting the minimal possible value of k) whenever we use a similar (simpler extension - reduction step), an interesting alternative being Caratheodory - Fejér extension and Hankel norm optimal reduction, to find the right value of α we have to use a further homotopy.

5 On Maximum entropy, minimum norm and low rank extensions of symmetric matrices

Suppose, that A is a symmetric $(m \times m)$ partially known matrix: let J be a subset of $[1, ..., m] \times [1, ..., m]$ and

$$S(A_J) = \{B|\ B_{ij} = A_{ij}, (ij) \in I,\ B^{\star} = B\} \tag{5.1}$$

We consider the problem ME:

$$\sup\{\det(B)|\ \ B \in S(A_J),\ \ B > 0\} \tag{5.2}$$

requiring thus in addition to (5.1), that B must be positive definite.
For the case of index sets A forming a staircase i.e. satisfying the condition

$$(i, j) \in J\ \ i \leq j (\text{resp. } i \geq j) \text{ implies } (i+r, j-s) \in J, (\text{resp. } (i-r, j+s) \in I)$$

for all $r, s \geq 0$, the optimal solution of problem (5.2) – which exists if all the completely specified main submatrices are positive definite – can be computed surprisingly simply: in a finite (small) number of arithmetical operations (square rooting included), (if the cardinality of J is small) see [7] and [6] for these solutions (which originated from the maximum entropy extension of Toeplitz matrices, a special case of the extension, interpolation problem considered above for $z_1 = z_2 = ...z_n = 0, N = m$). A closely related problem is the minimum norm extension problem, i.e. problem MN

$$\min\{\mu |\ B + \mu I \geq 0, \quad I\mu - B \geq 0, B \in S(A_J)\} =: \mu^\star \tag{5.3}$$

where the index sets J of interest are in general different from those used in problem (5.2). Of special interest is the case when $I = (1, ..., n_1) \times (1, ..., m) \cup (1, ..., m) \times (1, ..., n_1)$, for some $n_1 < m$, i.e. A is unspecified in the "southwest corner". It turns out see below that the solution of the problem (5.3) can be obtained (following our method to solve the problems 4) from the solutions of the problem

$$\max\{\det(B + \mu I) \cdot \det(\mu I - B)| \quad -\mu I \leq B \leq \mu I, \quad B \in S(A_J)\}, \tag{5.4}$$

where $\mu \geq \mu^\star$ is fixed (see below). In fact the solution of (5.4) is the central solution of the corresponding feasibility problem, discovered in operator theory, see [4], [16], [8]. It has the following closed form expression, assuming

$$B = \begin{pmatrix} A_0 & A_1^\star \\ A_1 & X \end{pmatrix}$$

$$X_{opt} = -A_1(\mu^2 I - A_0^2)^{-1} A_0 A_1^\star. \tag{5.5}$$

This can be easily proved: using the notion and the main formula for the Schur complement one obtains that the feasible set of problem (5.4) is precisely given by the two inequalities for X

$$-\mu I + A_1(\mu I + A_0)^{-1} A_1^\star \leq X \leq \mu I - A_1(\mu I - A_0)^{-1} A_1^\star \tag{5.6}$$

Since the matrices on both sides have the same eigenvectors the maximization of the product of the determinants is trivial: $(\alpha - x)(x - \beta)$ is maximal for $x = (\alpha + \beta)/2$. In fact problem (5.3) can be even solved – in a similar way – for nonsymetric data, see[4].

Note first that the det function over the feasible set (5.1) is very similar to the function Ψ above – one has just one factor, therefore we can expect, that the solution (5.2) will have the nice properties what the analytic centers are known to share for "algebraically simple" functions (linear, quadratic, product of linear etc.), esp. centering properties connected to ellipsoidal approximations, good predictibility of the central path, in fact some of these

can be proved using the results in [4], [6]-[8].
We can interpret now the matrix extension problem (5.2) as a moment problem noting that B has a representation in the form of an integral of elementary (nondecomposable) symmetric positive definite matrices $bb^\star\mu(b)$, $|b| = 1$.
As another "central" point in the feasible set of (5.2) we propose the solution of the problem (where G is the unit ball)

$$\sup\{\int_G \log\mu(b)\mathrm{d}b|\ \ B = \int_G bb^\star\mu(b)\mathrm{d}b, B\in S(A_J)\} \tag{5.7}$$

It is easy to prove that $\bar{\mu}(b) = (\sum_{i,j\in J}\alpha_{ij}b_ib_j)^{-1}$, where $\alpha_{ij}, (i,j)\in J$ are uniquely determined by the data i.e. the values of $A_{ij}, (ij)\in J$. Since the positive definiteness of B can be expressed by requiring that $< Bx, x >\geq 0$, for all $x\in G$, an other center can be obtained by solving the problem

$$\sup\{\int \log < Bb, b > \mathrm{d}b|\ \ B_{ij} = A_{ij},\ (ij)\in J\} \tag{5.8}$$

the optimal solution is characterized by the condition

$$[\int \frac{bb^\star}{< Bb, b >}\mathrm{d}b]_{k,l} = 0 \ \text{ for } \ (k,l)\notin J \tag{5.9}$$

Note that the objective functions (5.7), (5.8) are only invariant under orthogonal transformations $B\to Q^\star BQ$ (while the determinant function is invariant under area preserving linear transformations $T^\star AT, \det T = 1$). Of course for the class of problems (5.2) by the peculiarity of the data set – the "value" of A, $< Ae_i, e_j >$ is fixed over specific pairs of orthogonal vectors – the former smaller group seems to be relevant. It is tempting to interpret the matrix in (5.8) as an inverse of B, then we have the same optimality conditions as in (5.2): the inverse of the optimal extension must be supported on the index set J, see [9].
Finally we note that the solution of the problem of deciding whether – for a partially given matrix of the same class as in (5.2) there is an extension B_k which is symmetric, positive semidefinite and of rank less than a given number $k < m$, (this is interesting , e.g. for the class of Toeplitz matrices.) The following algorithm (analoguous to the one proposed is section 4) can be applied: Compute first the solution B of (5.2), with its structure preserving "optimal" approximant of rank k (having the additional structure,

Toeplitz or whatever), $\bar{B}_k$. Consider the convex feasibility problem (its analytic center). $\bar{B}_k - B \geq 0, \quad B_{ij} = A_{ij}, (i,j) \in J$, maximizing $\det(\bar{B}_k - B)$ or one of the integrals used in (5.7), (5.8).

References

[1] D.A. Bayer and Lagarias, The nonlinear geometry of linear programming, I, II,III, preprints, AT&T Bell Laboratories, Murray Hill, New Jersey, 1986-7.

[2] J.P. Burg, Maximum entropy spectral analysis,in Modern Spectrum Analysis (D.G. Childiers ed.), IEEE Press, New York, 1978, 34-39.

[3] T. Constantinescu et al., Schur analysis of some completion problems, Preprint N.62, INCREST, Bucharest, 1986.

[4] Ch. Davies et al.,Norm preserving dilatations and their applications to optimal error bounds, SIAM J. Num. Anal., 19 (1983) 445-469.

[5] M. A. Dahleh, J.B. Pearson, Optimal Rejection of Bounded Disturbances, IEEE Trans. Aut. Contr., 33 (1988) 722-731.

[6] P. Dewilde, E.F.A. Deprettere, The generalized Schur algorithm: Approximation and Hierarchy, in Operator Theory .Advances and Applications, Birkhäuser (1988) in press.

[7] H. Dym, J- contractive Matrix Functions, Reproducing Kernel Hilbert Spaces and Interpolation, preprint 1988, Dept. Theor. Math., The Weizman Inst., Rehovot.

[8] B. Fritzsche, B. Kirstein, A matrix extension problem with entropy optimization, Optimization 19 (1988) 85-99.

[9] R. Grone et al., Positive definite completions of partial Hermitian matrices, Linear Algebra and its Applications 58 (1984) 109-124.

[10] F. Jarre: On the Convergence of the Method of Analytic Centers when applied to Convex Quadratic Programs, report No. 35, 1987, Schwerpunktprogramm der DFG für anwendungsbezogene Optimierung und

Steuerung, submitted to Math. Programming in revised form, March 1988.
F. Jarre: Convergence of the Method of Analytic Centers for Generalized Convex Programs, DFG - report No. 67, May 1988.

[11] H.J. Landau, Maximum enzropy and the moment problem, Bull. Amer. Math. Soc. 16 (1987) 47-77.

[12] G. Sonnevend: An "Analytical Centre" for Polyhedrons and New Classes of Global Algorithms for Linear (Smooth, Convex) Programming, Proc. 12th IFIP Conf. on System Modelling and Optimization, Budapest 1985, Lecture Notes in Control and Inf. Sciences 84, 866-876, Springer Verlag 1986

[13] G. Sonnevend, Sequential and stable methods for recovering the reflectivity function of a layered medium, in Model Optimization in Exploration Geophysics, 2, (ed. A. Vogel),1987, Fr. Vieveg & Sohn, 60-72.

[14] G. Sonnevend, Existence and numerical computation of extremal invariant sets in linear differential games, Lect.Notes in Contr. and Inf. Sci. 22 (1981) 251-260.

[15] G. Sonnevend and J. Stoer: Global Ellipsoidal Approximations and Homotopy Methods for Solving Convex Analytic Programs, Report No. 40, Jan. 1988, Schwerpunktprogramm der Deutschen Forschungsgemeinschaft – Anwendungsbezogene Optimierung und Steuerung, Inst. f. Ang. Mathematik und Statistik, Universität Würzburg, to appear in Applied Mathematics and Optimization

[16] B.Szökefalvi-Nagy, C. Foias, Harmonic Analysis of Operators on Hilbert Space, North Holland, Amsterdam, 1970.

[17] L.N. Trefethen, M. Gutknecht, Pade, Stable Pade and Chebyshev approximation, in Numerical Analysis (D.F. Griffiths and G.A. Watson eds.) 1986.

[18] P. Vaidya, A locally well behaved potential function and a simple Newton method to find the center of a polytope, preprint AT&T Bell Laboratories, Murray Hill, New Jersey, 1986.

Solving Triangular System in Parallel is Accurate

Nai-kuan Tsao*

Wayne State University
Detroit, MI 48202, USA

Abstract

An error complexity analysis of two algorithms for solving a unit diagonal triangular system is given. The results show that the usual sequential algorithm is optimal in terms of having the minimal maximum and cumulative error complexity measures. The parallel algorithm described by Sameh and Brent is shown to be essentially equivalent to the optimal sequential one.

1 Introduction

In [1] Sameh and Brent have shown that, given $n^3/68 + O(n^2)$ processors, a triangular system of n equations Ax=b may be solved in O(log^2n) steps. They have also shown that if $\tilde{x}$ is the computed solution then $(A + \delta A_p)\,\tilde{x} = b$, where δA_p is bounded by, $||\delta A_p|| \leq \alpha(n)\epsilon||A||$. Here, $||.||$ stands for the ∞-norm, $\alpha(n) = O(n^2 logn), \epsilon$ is the unit roundoff, $\kappa(A)$ is the condition number of A. This bound can be very large compared to that of Wilkinson [2], where $||\delta A_s|| \leq n\epsilon||A\epsilon$.

In this paper we show that the proposed parallel algorithm is essentially equivalent to the usual sequential one in terms of our error complexity measures.

*This work was supported in part by an ASEE-NASA 1988 Summer Faculty Fellowship and in part by NASA while the author was visiting ICOMP, NASA Research Center, Cleveland, Ohio.

NATO ASI Series, Vol. F 70
Numerical Linear Algebra, Digital Signal Processing and Parallel Algorithms
Edited by G. H. Golub and P. Van Dooren

2 Solving the Triangular System

Given A, B and it is desired to find x such that

$$Ax = b, A = \begin{pmatrix} 1 & & & \\ a_{21} & 1 & & \\ \vdots & \vdots & \ddots & \\ a_{n1} & a_{n2} & \cdots & 1 \end{pmatrix}, b = \begin{pmatrix} b_1 \\ b_2 \\ \vdots \\ b_n \end{pmatrix}.$$

Now the exact solution x can be expressed as

$$x = M_{n-1} \ldots M_1 b, \quad M_i = I - a_i e_i^T,$$

$$a_i = [0(1:i), a_{i+1,i}, \ldots, a_{ni}]^T.$$

The usual sequential algorithm is given as

$$x = M_{n-1}(\ldots(M_2(M_1 b))\ldots)$$

or

Algorithm 1.
$x_1 = b_1$
for i=2 to n do
  $t_0 = b_i$
  for j=1 to i-1 do
    $t_j = t_{j-1} + a_{ij} x_j$
  $x_i = t_{i-1}$

Algorithm 1 requires $O(n^2)$ steps for a single processor and $O(n)$ steps if n processors are available.

On the other hand, the product forming x can also be evaluated in parallel by divide-and-conquer. For n=8 we can evaluate x as follows:

$$x = ((M_7 M_6)(M_5 M_4))((M_3 M_2)(M_1 b))$$

or in general for $n = 2^\nu$:

Algorithm 2.

for i=1 to n-1 do
  $M_i^{(0)} = M_i$
$b^{(0)} = b$
for j=1 to ν do
  for k=$n/2^j - 1$ down to 1 do
    $M_k^{(j)} = M_{2k+1}^{(j-1)} M_{2k}^{(j-1)}$
  $b^{(j)} = M_1^{(j-1)} b^{(j-1)}$
$x = b^{(\nu)}$

Algorithm 2 requires $O(n^3)$ each in multiplications and additions. However, with $n^3/68 + O(n^2)$ processors available, it requires only $O(log^2 n)$ steps[1].

The backward error analysis of the above two algorithms gives us the following theorem[1,2]:

Theorem 1. Let $\tilde{x}$ and $\bar{x}$ be the computed solutions using Algorithm 1 and Algorithm 2, respectively, then they satisfy

$$(A + A_s)\tilde{x} = b, \quad (A + A_p)\bar{x} = b$$

where

$$\|A_s\|_\infty \leq n\epsilon \|A\|_\infty,$$

$$\|A_p\|_\infty \leq O(n^2 logn)\epsilon\kappa^2(A)\|A\|_\infty, \quad \kappa(A) = \|A\|_\infty \|A^{-1}\|_\infty.$$

Thus the upper bound for $\|A_p\|_\infty$ can be quite large compared to that for $\|A_s\|_\infty$.

In this paper we take an alternative approach using the fact that in either algorithm the basic type of computation is

$$z = fl(\sum_{i=1}^{\lambda(z)} z_i)$$

where each z_i is a product of original data. Using the basic round-off model given by Wilkinson[2]:

$$fl(x * y) = xy\delta, \; fl(x \pm y) = (x \pm y)\Delta = x\Delta \pm y\Delta, \quad |\delta|, |\Delta| \leq 1 + \epsilon$$

the computed z can be expressed as

$$z = \sum_{i=1}^{\lambda(z)} z_i \delta^{\tilde{\sigma}_i} \Delta^{\sigma_i}$$

Two measures can be defined:

$$\sigma_a(z) = \max_{1 \le i \le \lambda(z)} (\sigma_i), \quad s_a(z) = \sum_{i=1}^{\lambda(z)} \sigma_i$$

$$\sigma_m(z) = \max_{1 \le i \le \lambda(z)} (\tilde{\sigma}_i), \quad s_m(z) = \sum_{i=1}^{\lambda(z)} \tilde{\sigma}_i$$

It turns out that $\sigma_m(z)$ and $s_m(z)$ are independent of the algorithm chosen to evaluate z under some mild restrictions[3]. Thus $\sigma_a(z)$ and $s_a(z)$ only will be used to compare Algorithm 1 and Algorithm 2. In such case we will assume that exact multiplications are possible for our computation. We have the following theorem[3]:

Theorem 2. If Algorithm 1 is used to find x, then

$$\sigma_a(x_i) = i - 1, \quad s_a(x_i) = 2^{i-1}(i-1).$$

If Algorithm 2 is used to find x, then

$$\sigma_a(x_i) \left\{ \begin{array}{l} = i - 1 \textit{ if inner products are evaluated from right to left} \\ \le 1.5(i-1) \textit{ if inner products are evaluated in parallel} \\ \qquad \textit{by divide and conquer} \\ \le \lceil \log i \rceil (2^{\lceil \log i \rceil} - 1) \textit{ if inner products are evaluated} \\ \qquad \textit{from left to right} \end{array} \right\}$$

$$s_a(x_i) \left\{ \begin{array}{l} = (i-1)2^i - 1 \textit{ if inner products are evaluated from right to left} \\ \le 1.5(i-1)2^{i-1} \textit{ if inner products are evaluated in parallel} \\ \qquad \textit{by divide and conquer} \\ \le 2^{i-1} \lceil \log i \rceil (2^{\lceil \log i \rceil} - 1) \textit{ if inner products are evaluated} \\ \qquad \textit{from left to right} \end{array} \right\}$$

We conclude from Theorem 2 that Algorithm 2 is essentially equivalent to the usual sequential Algorithm 1 in terms of our error complexity measures.

References

[1] A. H. Sameh and R. P. Brent, Solving triangular systems on a parallel computer, SIAM J. Numer. Anal., 14(1977), pp.1101-1113.

[2] J. H. Wilkinson, Rounding Errors in Algebraic Processes, Prentice-Hall, Englewood Cliffs, NJ, 1963.

[3] N. K. Tsao, Solving triangular systems in parallel is not bad, submitted for publication.

Storage Schemes for Parallel Eigenvalue Algorithms

Robert A. van de Geijn

Department of Computer Sciences
The University of Texas at Austin
Austin, Texas 78712, USA

Abstract

In this paper, we examine storage schemes for parallel implementation on distributed memory MIMD multiprocessors of algorithms for solving the algebraic eigenvalue problem. We show that a novel storage scheme, which we call block Hankel-wrapped storage, allows better utilization of the processors than column-wrapped storage when implementing Jacobi's method or the nonsymmetric QR algorithm.

1 Introduction

With the advent of parallel processors, much research has concentrated on the parallel implementation of algorithms in numerical linear algebra. Recent papers on parallel eigenvalue algorithms include [4,10] for symmetric, tridiagonal matrices. For full symmetric matrices, Jacobi methods are again of interest [2,3,8,11,12,16]. Papers examining the implementation of QR methods for nonsymmetric matrices include [1,5,14,15].

In this paper, we examine storage schemes for implementation of eigenvalue algorithms on distributed memory multiprocessors. The *column-wrapped* storage scheme, which is useful for parallel implementation of algorithms for solving linear systems [9], does not seem to allow efficient parallel implementation of eigenvalue algorithms. An alternate storage scheme,

NATO ASI Series, Vol. F 70
Numerical Linear Algebra, Digital Signal Processing and Parallel Algorithms
Edited by G. H. Golub and P. Van Dooren

which we call *block Hankel-wrapped* storage, shows promise for parallel implementation of a wide range of algorithms for solving the eigenvalue and singular value problems [16]. In §2, the column-wrapped and block Hankel-wrapped storage schemes are defined. Next, we examine the usefulness of these schemes for efficient parallel implementation of Jacobi's method in §3 and the nonsymmetric QR algorithm in §4. Concluding remarks are given in §5.

Throughout this note, we assume all matrices, vectors, scalars, and arithmetic to be real. We further assume that the parallel processor has p processors, $\mathbf{P}_0, \ldots, \mathbf{P}_{p-1}$, and $\mathbf{P}_i$ is adjacent to $\mathbf{P}_j$ if $|i-j|\%p = 1$, where .%. indicates the modulo operator. In other words, we assume it is possible to embed a ring within the network of processors.

2 Parallel Storage Schemes

A popular storage scheme for distributed memory parallel computers is the column-wrapped storage scheme. It assigns the ith column of given matrix A to processor $\mathbf{P}_{(i-1)\%p}$.

As an alternative, we define the class of *block Hankel-wrapped* storage schemes. Given m, for simplicity we assume the dimension of matrix A is $n = mhp$ for some integer h. Partition the matrix so that $A = (A_{ij})$, $1 \leq i, j \leq n$, where $A_{ij} \in \mathbf{R}^{h \times h}$. For general m, the block Hankel-wrapped storage scheme (BHW$_m$) assigns A_{ij} to $\mathbf{P}_{\lfloor (i+j-2)/m \rfloor \% p}$. The super-scripts in the following figure indicate the processor to which the blocks are assigned when $m = 1$:

$$\begin{array}{ccccc}
A_{11}^{(0)} & A_{12}^{(1)} & A_{13}^{(2)} & \cdots & A_{1p}^{(p-1)} \\
A_{21}^{(1)} & A_{22}^{(2)} & A_{23}^{(3)} & \cdots & A_{2p}^{(0)} \\
A_{31}^{(2)} & A_{32}^{(3)} & A_{33}^{(4)} & \cdots & A_{3p}^{(1)} \\
\vdots & \vdots & \vdots & & \vdots \\
A_{p1}^{(p-1)} & A_{p2}^{(0)} & A_{p3}^{(1)} & \cdots & A_{pp}^{(p-2)}
\end{array}$$

It should be noted that this storage scheme is an example of a *skewed storage scheme* [7].

For reasons explained in the next section, we prefer BHW$_2$, which assigns submatrices to processors as illustrated by

$$\begin{matrix} A_{11}^{(0)} & A_{12}^{(0)} & A_{13}^{(1)} & \cdots & A_{1(2p)}^{(p-1)} \\ A_{21}^{(0)} & A_{22}^{(1)} & A_{23}^{(1)} & \cdots & A_{2(2p)}^{(0)} \\ A_{31}^{(1)} & A_{32}^{(1)} & A_{33}^{(2)} & \cdots & A_{3(2p)}^{(0)} \\ \vdots & \vdots & \vdots & & \vdots \\ A_{(2p)1}^{(p-1)} & A_{(2p)2}^{(0)} & A_{(2p)3}^{(0)} & \cdots & A_{(2p)(2p)}^{(p-1)} \end{matrix}$$

3 Jacobi's Method

Define a plane rotation in the (i, j)-plane by

$$P_{ij} = \begin{pmatrix} I_{i-1} & & & & \\ & c & & s & \\ & & I_{j-i-1} & & \\ & -s & & c & \\ & & & & I_{n-j} \end{pmatrix}.$$

Given symmetric matrix A, one can choose plane rotation P_{ij} so that $P_{ij}AP_{ij}^{\mathrm{T}}$ has a zero (i, j) element. Jacobi methods for finding the eigenvalues of A successively compute rotations to annihilate off-diagonal elements of A. The following sequential algorithm is known as the *column-serial* Jacobi method [6]:

Algorithm 1

```
do until convergence
   for i=1,...,n-1
      for j=i+1,...,n
         compute P_ij
         update A = P_ij A P_ij^T
```

One iteration of the outer loop annihilates each off-diagonal element exactly once and is called a *sweep*.

To enhance data locality, Algorithm 1 can be reformulated as follows [11,13]:

Algorithm 2

```
do until convergence
   for j=1,...,n/2
```

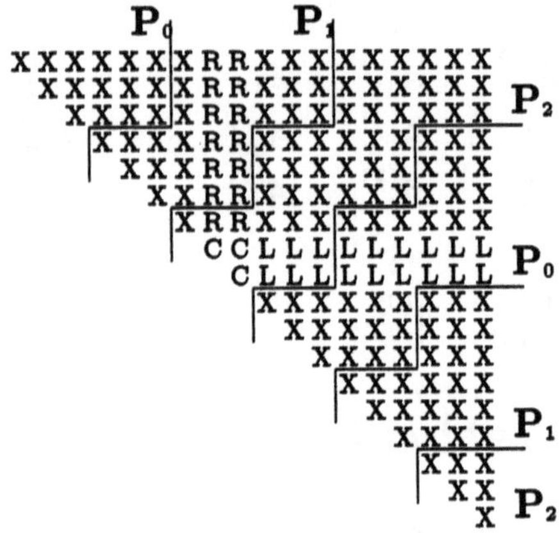

Figure 1: Jacobi's method: annihilating $(i, i+1)$

```
for i=1,...,n-1
    compute P_{i(i+1)}
    update A = P_{i(i+1)} A P^T_{i(i+1)}
    exchange rows and cols i and i+1
```

Notice that only elements of the first super-diagonal are annihilated and as a result only adjacent rows and columns are involved in the application of a plane rotation.

If A is distributed using column-wrapped storage, the application of $P_{i(i+1)}$ from the left results in the work being distributed among the processors. However, applying $P^T_{i(i+1)}$ from the right requires columns i and $i+1$ to be brought together, e.g. on the processor that holds column i. Next, the computation is performed by one processor, while the other processors are idle and hence the work is not balanced among the processors. Moreover, the time complexities for communication and computation are of the same order. The former problem can be overcome by annihilating many super-diagonal elements simultaneously [11]. In this case, communication overhead still stands in the way of efficient utilization of the processors.

Assume the upper triangular part of A is instead distributed using the BHW_2 storage scheme, as illustrated in Figure 1 for $p = 3$ and $h = 3$. This figure also indicates the computation required to annihilate a typical superdiagonal element $(i, i+1)$. Here, Xs are nonzero elements; Cs indicate the elements from which $P_{i(i+1)}$ is computed; and Ls and Rs indicate elements affected by applying $P_{i(i+1)}$ from the left and right, respectively. Once the appropriate processor has computed the rotation and distributed it to all processors, the computation required to update the matrix is distributed among the processors. Whenever a boundary is encountered, a limited amount of additional communication is required.

We chose BHW_2 over BHW_1 since on distributed memory multiprocessors it is important to store data that is involved in a given computation on neighboring processors. The Jacobi method computes plane rotations from submatrices

$$\begin{pmatrix} a_{ii} & a_{i(i+1)} \\ a_{(i+1)i} & a_{(i+1)(i+1)} \end{pmatrix}.$$

Notice that $m = 1$ assigns elements a_{ii} and $a_{(i+1)(i+1)}$ to processors that are not neighbors. If $m = 2$, elements involved in the computation or application of a rotation are always on neighboring processors.

The storage scheme allows further parallelism, details of which can be found in [16].

4 The QR Algorithm

The QR decomposition of an $n \times n$ matrix A is given by $A = QR$, where Q is unitary and R is upper triangular. A typical sequential QR algorithm can be described by

Algorithm 3

$$A^{(0)} = \hat{Q}^{\mathrm{T}} A \hat{Q}$$

`for k=0,1,...`

$$A^{(k)} - s_k I \rightarrow Q^{(k)} R^{(k)}$$
$$A^{(k+1)} \leftarrow R^{(k)} Q^{(k)} + s_k I,$$

where $\{s_k\}$ is called the shifting sequence, chosen to speed up convergence. The first statement of the algorithm reduces A to upper Hessenberg form, that is $A^{(0)}$ has zeroes below the first subdiagonal. For a detailed description of QR algorithms, see [6].

Each iteration of the loop indexed by k can be implemented by successively computing plane rotations in the $(i+1, i)$-plane that annihilate the $(i+1, i)$ element of the matrix, as follows:

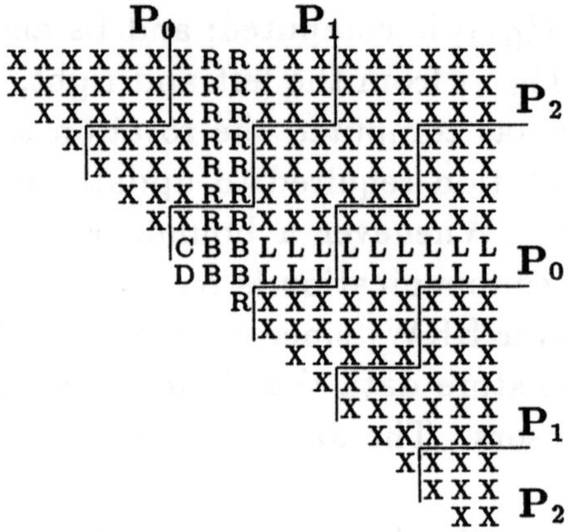

Figure 2: The Implicit QR Algorithm: annihilating $(i+2,i)$

Algorithm 4

$A^{(k)} = A^{(k)} - s_k I$
for i=1,...,n-1
  compute $P_{(i+1)i}$ **so that**
    $P_{(i+1)i}A^{(k)}$ **has zero** $(i+1,i)$ **entry**
  update $A^{(k)} = P_{(i+1)i}A^{(k)}$
for i=1,...,n-1
  update $A^{(k)} = A^{(k)}P^{\mathrm{T}}_{(i+1)i}$
$A^{(k+1)} = A^{(k)} + s_k I$

The shift is subtracted explicitly from $A^{(k)}$ and added back to $A^{(k+1)}$. Notice that if $A^{(k)}$ is upper Hessenberg, so is $A^{(k+1)}$.

An implicitly shifted alternative computes $A^{(k+1)}$ from $A^{(k)}$ as follows:

Algorithm 5

compute P_{21} **so that**
  $P_{21}(A^{(k)} - s_k I)$ **has zero** $(2,1)$ **entry**
$A^{(k)} = P_{21}A^{(k)}P^{\mathrm{T}}_{21}$
for i=1,...,n-2
  compute $P_{(i+2)(i+1)}$ **so that**
    $P_{(i+2)(i+1)}A^{(k)}$ **has zero** $(i+2,i)$ **entry**
  update $A^{(k)} = P_{(i+2)(i+1)}A^{(k)}P^{\mathrm{T}}_{(i+2)(i+1)}$
$A^{(k+1)} = A^{(k)}$

Column-wrapped storage again leads to difficulties in balancing the work during the application of the rotation from the right, as well as excessive communication overhead [5].

If $A^{(k)}$ is stored using BHW_2, the computation required for a given i of the inner loop is illustrated by Figure 2 for $p = 3$ and $h = 3$. In this figure, D is the element to be annihilated; the rotation is computed from C and D; Ls and Rs are elements affected by applying the rotation from the left and right, respectively; and Bs are elements which are affected by applying the rotation from both sides. Once the appropriate processor has computed and distributed the rotation, all processors are involved in updating the matrix. In general, each processor must apply the rotation to $2h$ pairs of elements, except for the processor that computes the rotation, which performs slightly more work. Some additional communication is necessary when a boundary is encountered. Communication overhead can be reduced by pipelining the computation and application of rotations.

5 Conclusion

In this paper, we have introduced a new storage scheme that appears to be more appropriate than column-wrapped storage for parallel implementation of the Jacobi method and the nonsymmetric QR algorithm on distributed memory parallel computers. We are currently evaluating parallel implementations of these eigenvalue algorithms.

The emphasis of this paper is on the merits of the storage scheme rather than the algorithms themselves. It should be noted that the Jacobi method for symmetric matrices presented in §3 can be easily changed to a Jacobi method for finding the singular value decomposition of an upper triangular matrix. The techniques for implementing the implicitly shifted QR algorithm can also be used for parallel implementations of the explicitly and double shifted QR algorithm. Efficient ways for adding a deflation scheme to the QR algorithm are being studied, as well as parallel implementation of the reduction to upper Hessenberg form.

We believe the Hankel-wrapped storage scheme is appropriate for a wide range of algorithms for finding the eigenvalues and singular values of matrices. Its usefulness for solving dense linear systems is also under investigation.

References

[1] Boley D., "Solving the Generalized Eigenvalue Problem on a Synchronous Linear Processor Array," *Parallel Computing*, 3, 123-166, 1986

[2] Berry M. and Sameh A., "Parallel Algorithms for the Singular Value and Dense Symmetric Eigenvalue Problems," CSRD report No. 761, 1988

[3] Brent R. and Luk F., "The Solution of Singular Value and Symmetric Eigenvalue Problems on Multi-processor Arrays," *SIAM J. Sci. Stat. Comput.*, 6, 69-84, 1985.

[4] Dongarra J.J. and Sorensen D.C., "A Fully Parallel Algorithm for the Symmetric Eigenvalue Problem," *SIAM J. Sci. Stat. Comput.*, Vol. 8, No. 2, March 1987

[5] Geist G.A., Ward R.C., Davis G.J. and Funderlic R.E., "Finding Eigenvalues and Eigenvectors of Unsymmetric Matrices using a Hypercube Multiprocessor," unpublished manuscript

[6] Golub G.H. and Van Loan C.F., *Matrix Computations*, Johns Hopkins Press, 1983

[7] Hockney R.W. and Jesshope C.R. *Parallel Computers*, Adam Hilger Ltd., Bristol, 1981

[8] Ipsen I.C.F and Saad Y. "The Impact of Parallel Architectures on the Solution of Eigenvalue Problems," *Large Scale Eigenvalue Problems*, J. Cullum and R.A. Willoughby (Ed.), Elsevier Science Publishers, 1986

[9] Ipsen I.C.F., Saad Y. and Schultz M. H., "Complexity of Dense-Linear-System Solution on a Multiprocessor Ring," *Linear Algebra and its Applications*, 77:205-239, 1986

[10] Lo S.S., Philippe B. and Sameh A., " Multiprocessor Algorithm for the Symmetric Tridiagonal Eigenvalue Problem," *SIAM J. Sci. Stat. Comput.*, Vol. 8, No. 2, March 1987

[11] Luk F.T. and Park H., "On the Equivalence and Convergence of Parallel Jacobi SVD Algorithms," *Proceedings of SPIE, Advanced Algorithms and Architectures for Signal Processing II*, Vol. 826, 152-159, 1988

[12] Sameh A., "On Jacobi and Jacobi-Like Algorithms for a Parallel Computer," *Math. Comp.*, vol. 25, pp. 579-590, 1971

[13] Stewart G.W., "A Jacobi-like Algorithm for Computing the Schur Decomposition of a Non-Hermitian Matrix," *SIAM J. Sci. Stat. Comp.*, B 6, pp. 853-64, 1985

[14] Stewart G.W., "A Parallel Implementation of the QR Algorithm," Dept. of Comp. Sci., Univ. of MD, TR-1662, May 1986

[15] Van de Geijn R.A., "Implementing the QR-Algorithm on an Array of Processors," Ph.D. thesis, Dept. of Comp. Sci., Univ. of MD, TR-1897, 1987

[16] Van de Geijn R. A., "A Novel Storage Scheme for Parallel Jacobi Methods," The University of Texas at Austin, Dept. of Computer Sciences, TR-88-26

Iterative Solution Methods for Large, Sparse Systems of Linear Equations Arising from Tomographic Image Reconstruction

M.C.A. van Dijke, H.A. van der Vorst and M.A. Viergever

Delft University of Technology
Faculty of Mathematics and Informatics
2600 AJ Delft, The Netherlands

The problem of reconstructing an object from its projections can be solved using two different approaches: analytic methods and algebraic methods. Analytic methods, based on integral transforms such as the inverse Radon transform, are considerably more efficient than algebraic methods. In many practical cases, however, such as limited angle tomography or problems where refraction plays an important role, algebraic techniques have to be used.

The reconstruction problem then consists of solving a large sparse system of linear equations. The sparsity of the system matrix is unstructured and the system is often both underdetermined and overdetermined. The size of the system matrix usually prohibits the use of direct solution methods based on the calculation of the generalized inverse of the system matrix. Consequently one has to settle for iterative solution methods. There are two classes of iterative methods in use for systems of equations derived from tomographic problems:

1: **projection methods** (Lanczos, Conjugate Gradients, LSQR etc.)
2: **Gauss methods** (Gauss-Seidel (ART), Gauss-Jacobi (SIRT) etc.)

NATO ASI Series, Vol. F 70
Numerical Linear Algebra, Digital Signal Processing and Parallel Algorithms
Edited by G. H. Golub and P. Van Dooren

Members of these two families are compared with respect to their eigenvalue spectra (rate of convergence) as well as to the quality of the resulting images.

Gauss methods implicitly scale the system they are solving. This influences the rate of convergence by the change of the eigenvalue spectrum. The scaling also implies that the iteration process no longer converges to the Minimum Norm (Least Squares) solution of the original problem but to that of another (scaled) problem (1). Our recent work concentrates on this phenomenon.

Projection methods do not have this implicit scaling. We have found that application of CG to the system may be inferior to iteration by members of the Gauss family, for instance ART, even through CG is considered optimal. Application of CG to the scaled system, however, appeared to be superior to ART and to applying CG to the original system.

Besides the comparison between the two different classes of iterative solution methods, an attempt is made to speed up the rate of convergence by subdividing the system into blocks, as was done successfully on ART by Eggermont et. al. (2). The effect of this partitioning on the eigenvalue spectrum is investigated. Also some theoretical insight in the effect of changing the ordering of the rows (blocks) of the system matrix on the rate of convergence is obtained. Furthermore some new convergence proofs are given.

References

[1] van der Sluis A. and van der Vorst H.A. (1987). Numerical solution of large sparse linear algebraic systems arising from tomographic problems. In: Seismic Tomography, Nolet, G. (ed.), Reidel, Dordrecht.

[2] Eggermont P.P.B., Herman G.T. and Lent A. (1981). Iterative algorithms for large partitioned systems, with application to image reconstruction. Linear Algebra Appl. 40, 37-67.

The Generalized Total Least Squares Problem : Formulation, Algorithm and Properties

Sabine Van Huffel*
ESAT Laboratory
Department of Electrical Engineering
K.U.Leuven
B – 3030 Heverlee, Belgium

Abstract

The Total Least Squares (TLS) method has been devised as a more global fitting technique than the ordinary least squares technique for solving overdetermined sets of linear equations $AX \approx B$ when errors occur in all data. If the errors on the measurements A and B are uncorrelated with zero mean and equal variance, TLS is able to compute a strongly consistent estimate of the true solution of the corresponding unperturbed set $A_0X = B_0$. In this paper the TLS computations are generalized in order to maintain consistency of the solution in the following cases : first of all, some columns of A may be error-free and secondly, the errors on the remaining data may be correlated and not equally sized. Hereto, a numerically reliable Generalized TLS algorithm GTLS, based on the Generalized Singular Value Decomposition (GSVD), is developed. Additionally, the equivalence between the GTLS solution and alternative expressions of consistent estimators, described in literature, is proven. These relations allow to deduce the main statistical properties of the GTLS solution.

*Senior Research Assistant of the Belgian N.F.W.O. (National Fund of Scientific Research)

NATO ASI Series, Vol. F 70
Numerical Linear Algebra, Digital Signal Processing and Parallel Algorithms
Edited by G. H. Golub and P. Van Dooren

1 Introduction

Every linear parameter estimation problem gives rise to an overdetermined set of linear equations $AX \approx B$. Whenever both the data matrix A and observation matrix B are subject to errors, the Total Least Squares (TLS) method can be used for solving this set. Much of the literature concerns the **classical** TLS problem $AX \approx B$ in which all columns of A are subject to errors, and several algorithms have been suggested in order to compute the classical TLS solution $\widehat{X}$, see e.g. [8-9],[19-20-21]. If the errors on the measurements A and B are uncorrelated with zero mean and equal variance, then under mild conditions this TLS solution $\widehat{X}$ is a strongly consistent estimate of the true solution X of the corresponding unperturbed set $A_0X = B_0$, i.e. $\widehat{X}$ converges to X with probability one as the number of equations tends to infinity. However, in many linear parameter estimation problems **some** columns of A may be **error-free**. Moreover, the errors on the remaining data may be **correlated** and **not** equally sized. In order to maintain consistency of the result when solving these problems, the classical TLS formulation can be generalized as follows (M^{-T} denotes the transposed inverse of matrix M) :

<u>**Generalized TLS formulation**</u> :

Given a set of m linear equations in $n \times d$ unknowns X :

$$AX \approx B \qquad A \in R^{m\times n}, B \in R^{m\times d} \text{ and } X \in R^{n\times d} \tag{1}$$

$$\text{Partition} \quad A = [A_1; A_2] \quad A_1 \in R^{m\times n_1}, A_2 \in R^{m\times n_2} \text{ and } n = n_1 + n_2 \tag{2}$$

$$X = [X_1^T; X_2^T]^T \quad X_1 \in R^{n_1\times d} \text{ and } X_2 \in R^{n_2\times d} \tag{3}$$

and assume that the columns of A_1 are error-free and that nonsingular error equilibration matrices $R_D \in R^{m\times m}$ and $R_C \in R^{(n_2+d)\times(n_2+d)}$ are given such that the errors on $R_D^{-T}[A_2; B]R_C^{-1}$ are equilibrated, i.e. uncorrelated with zero mean and same variance.

Then, a GTLS solution of (1) is any solution of the set

$$\widehat{A}X = A_1X_1 + \widehat{A}_2X_2 = \widehat{B} \tag{4}$$

where $\widehat{A} = [A_1; \widehat{A}_2]$ and $\widehat{B}$ are determined such that (R denotes the range)

$$R(\widehat{B}) \subseteq R(\widehat{A}) \tag{5}$$

$$\|R_D^{-T}[\Delta\widehat{A}_2;\Delta\widehat{B}]R_C^{-1}\|_F = \|R_D^{-T}[A_2 - \widehat{A}_2; B - \widehat{B}]R_C^{-1}\|_F \quad \text{is minimal} \quad (6)$$

The problem of finding $[\Delta\widehat{A}_2;\Delta\widehat{B}]$ such that (5-6) are satisfied, is referred to as the **GTLS problem.** Whenever the solution is not unique, GTLS singles out the **minimum norm** solution, denoted by $\widehat{X} = [\widehat{X}_1^T;\widehat{X}_2^T]^T$.

The error equilibration matrices R_D and R_C are the square root of the error covariance matrices $C = E(\Delta^T\Delta)$ and $D = E(\Delta\Delta^T)$ respectively where E denotes the expected value operator and $\Delta_{m\times(n_2+d)}$ represents the errors on the noisy data $[A_2;B]$. Often, only C and D are known : in these cases, the matrices R_C and R_D are simply obtained from their Cholesky decomposition, i.e. $C = R_C^TR_C$ and $D = R_D^TR_D$, R_C and R_D upper triangular. Although the TLS problem for solving the univariate problem ($n = 1$), i.e. linefitting, is quite old [1], [17], it has only been recently extended to the multivariate problem. In the field of **numerical analysis**, this problem was first studied by Golub and Van Loan [8].We generalized the algorithm of Golub and Van Loan [8] to all cases in which their algorithm fails to produce a solution and described the properties of these so-called nongeneric TLS problems [19],[21]. In the field of **statistics**, Gleser [6] studied the same problem. His estimate, called **multivariate errors-in-variables regression** estimate, coincides with the TLS solution given by Golub and Van Loan [8] whenever the TLS problem has a unique minimizer. Also in the field of **experimental modal analysis**, the TLS technique (more commonly known as the H_v technique), was studied recently [13]. And finally in the field of **system identification**, Levin [14] first studied the same problem. His method, called the **eigenvector method** or **Koopmans-Levin method** [4], computes the same estimate as the TLS algorithm whenever the TLS problem has a unique solution. If **some columns** of the data matrix A in the set $AX \approx B$ are **error-free**, the classical TLS algorithms can be generalized in order to compute the more general TLS estimate $\widehat{X} = [\widehat{X}_1^T;\widehat{X}_2^T]^T$ satisfying the TLS criteria (5-6) with $R_C \sim I$ and $R_D \sim I$ [19],[7] (see section 2). In particular, this algorithm is able to compute the **Compensated Least Squares** (CLS) estimate as derived by Guidorzi [10] and Stoica and Söderström [18]. When the only disturbance of the input-output sequences is given by white noise, the CLS, GTLS and eigenvector methods all give the same estimate. Observe that our GTLS algorithm is computationally more efficient than the computation procedure presented in [18]. For a detailed appraisal of the TLS method and its generalizations, see [19], [22].

2 The generalized TLS algorithm GTLS

As outlined below, the GTLS algorithm is based on an **implicit GSVD** method [3], which computes an SVD of a triple matrix product $E^{-1}FG^{-1}$ without explicitly forming the products and without inverting E or G. This guarantees its better numerical performance. Moreover, by first performing a QR factorization [9], only the GSVD of a smaller submatrix is required which makes the GTLS algorithm computationally more efficient than methods described in [4], [18].

Given : An $m \times d$ matrix B and an $m \times n$ matrix $A = [A_1; A_2]$ whose first n_1 columns A_1 have full column rank and are error-free, $n = n_1 + n_2$ and $m \geq n + d$.
The error equilibration matrices $(R_D)_{m\times m}$ and $(R_C)_{(n_2+d)\times(n_2+d)}$, as defined in the GTLS formulation.

Step 1 : QR and QL factorizations

1.a. Compute the QR factorization of $[A_1; A_2; B]$:

$$[A_1; A_2; B] = Q_{AB} \begin{bmatrix} R_{AB} \\ 0 \end{bmatrix} \quad \text{with } R_{AB} = \begin{array}{c} \begin{bmatrix} R_{11} & R_{12} \\ 0 & R_{22} \end{bmatrix} \begin{array}{l} n_1 \\ n_2+d \end{array} \\ \begin{array}{cc} n_1 & n_2+d \end{array} \end{array}$$

where R_{AB} is upper triangular.

1.b. If $R_D \sim I$ *then* $E_{22} \leftarrow I$ *else* compute the QL factorization of $R_D Q_{AB}$:

$$R_D Q_{AB} = Q_{\widehat{D}} L_{\widehat{D}} \text{ with } L_{\widehat{D}}^T = \begin{array}{c} \begin{bmatrix} E_{11} & E_{12} & E_{13} \\ 0 & E_{22} & E_{23} \\ 0 & 0 & E_{33} \end{bmatrix} \begin{array}{l} n_1 \\ n_2+d \\ m-n-d \end{array} \\ \begin{array}{ccc} n_1 & n_2+d & m-n-d \end{array} \end{array}$$

where $L_{\widehat{D}}^T$ is upper triangular.
If $n_1 = n$ then begin $Z_2 \leftarrow -I_d$; go to step 2.d. end.

1.c. If R_C upper triangular then $R_{\widehat{C}} \leftarrow R_C$ else compute the QR factorization of R_C :

$$R_C = Q_{\widehat{C}} R_{\widehat{C}} \qquad (R_{\widehat{C}})_{(n_2+d)\times(n_2+d)} \text{ upper triangular}$$

Step 2 : GSVD

2.a. Compute the implicit GSVD

$$U^T E_{22}^{-1} R_{22} R_{\widehat{C}}^{-1} V = \text{diag}(\sigma_1, \ldots, \sigma_{n_2+d}) \qquad \sigma_{i-1} \geq \sigma_i \quad i = 2, \ldots, n_2+d \tag{7}$$

with U and $V = [v_1, \ldots, v_{n_2+d}]$ orthonormal and σ_i the generalized singular values.

2.b. If not user determined, compute the rank $r(\leq n_2)$ by means of a user-defined rank determinator R_0 :

$$\sigma_1 \geq \cdots \geq \sigma_r > R_0 \geq \sigma_{r+1} \geq \cdots \geq \sigma_{n_2+d}$$

2.c. If $R_C \sim I$ then $Z_2 \leftarrow [v_{r+1}, \ldots, v_{n_2+d}]$ else solve$R_{\widehat{C}} Z_2 = [v_{r+1}, \ldots, v_{n_2+d}]$ by back substitution .

2.d. If $R_D \sim I$ then $\widehat{Z} = 0$ else solve $E_{22}\widehat{Z} = R_{22} Z_2$ by back substitution.

2.e. Solve $R_{11} Z_1 = E_{12}\widehat{Z} - R_{12} Z_2$ by back substitut ion.
If $n_1 = n$ then begin $\widehat{X} \leftarrow Z_1$; stop end.

Step 3 : GTLS solution $\widehat{X} = [\widehat{X}_1^T; \widehat{X}_2^T]^T$

3.a. If $R_C \not\sim I_{n+d}$, $d > 1$ and $r < n_2$, orthonormalize $\begin{bmatrix} Z_1 \\ Z_2 \end{bmatrix}$ using a QR factorization :

$$\begin{bmatrix} Z_1 \\ Z_2 \end{bmatrix} = Q_z R_z; \quad \begin{bmatrix} Z_1 \\ Z_2 \end{bmatrix} \leftarrow Q_z$$

3.b. Perform Householder transformations Q such that

$$\begin{bmatrix} Z_1 \\ Z_2 \end{bmatrix} Q = \begin{bmatrix} W & Y \\ 0 & \Gamma \end{bmatrix} \begin{matrix} n \\ d \end{matrix} \quad \text{and } \Gamma \text{ upper triangular} \tag{8}$$
$$\qquad\qquad\qquad n_2 - r \quad\;\; d$$

If Γ nonsingular then solve $\widehat{X}\Gamma = -Y$
else lower the rank r with the multiplicity of σ_r
go back to step 2.c.

END

The following comments are in order :

- Step 2 of the GTLS algorithm, based on the canonical correlation computation procedure of [3], reduces all 3 matrices involved in the GSVD to upper triangular form of equal dimension. In step 2.a, the algorithm PSVD-2 of [3] can readily be applied to find the **implicit GSVD**. For more details and an analysis of the computational complexity, see [3]. The special case where $R_D \sim I_m$ reduces PSVD-2 to the well-known GSVD algorithms [15-16] for computing the SVD of the product FG^{-1} implicitly. These algorithms are all based on an implicit Kogbetliantz approach and are suitable for parallel implementation.

- If $R_D \sim I_m$, the GTLS algorithm reduces to the algorithm described in [22]. Of course, one can always perform the product $R_D^{-T}[A;B]$ explicitly and then apply the GTLS algorithm outlined in [22] in order to increase the computational efficiency. This approach is however not recommended in general since computing $R_D^{-T}[A;B]$ would usually lead to unnecessarily large numerical errors if R_D is ill-conditioned with respect to the solution of equations.

- If R_C or R_D are singular, the GSVD algorithm PSVD-2 can be adapted by using matrix adjoints (cf. [16]). Note that in this case the GSVD is not necessarily given by the SVD of $E_{22}^{\dagger} R_{22} R_{\widehat{C}}^{\dagger}$ († denotes the pseudo-inverse). Similarly to [16], the GSVD algorithm could be adapted for the case that $m < n + d$ by adding zero rows and columns, if necessary, to give square matrices of equal dimension. These extensions are not yet fully analyzed.

- If $R_C \sim I_{n+d}$ and $R_D \sim I_m$, the GSVD in step 2 is simply the **ordinary SVD** of R_{22} so that in this case the GTLS algorithm reduces to the **classical TLS** algorithms [8-9], [19, sec.1.8.1], [21] for the case that $n_1 = 0$ and [19, sec.1.8.2] for the case that $n_1 \neq 0$. Observe also that the GTLS algorithm solves the **ordinary LS** problem if all columns of A are error-free ($n_1 = n$).

- If Γ in (8) is nonsingular (resp., singular), the GTLS solution is called generic (resp., **nongeneric**). For more details, see [19], [21].

3 Properties of the generalized TLS solution

The following important theorem allows to derive the main statistical properties of the GTLS solution :

Theorem 1 *Consider the equations (1-2) and let A_1 and A_2 have full column rank. Denote by σ' (resp., σ) the minimal generalized singular value of the matrix pair $(R_D^{-T}A, R_{Ca}^*)$ (resp., $(R_D^{-T}[A;B], R_C^*)$) where* $R_C^* = \begin{matrix} \begin{bmatrix} 0 & 0 \\ 0 & R_C \end{bmatrix} & \begin{matrix} n_1 \\ n_2+d \end{matrix} \\ \begin{matrix} n_1 & n_2+d \end{matrix} & \end{matrix} = \begin{matrix} \begin{bmatrix} R_{Ca}^* & R_{Cab}^* \\ 0 & R_{Cb}^* \end{bmatrix} & \begin{matrix} n \\ d \end{matrix} \\ \begin{matrix} n & d \end{matrix} & \end{matrix}$ *is upper triangular. Let σ have multiplicity d and denote $D = R_D^T R_D, C_a^* = R_{Ca}^{*T} R_{Ca}^*$ and $C_{ab}^* = R_{Ca}^{*T} R_{Cab}^*$.*
If $\sigma' > \sigma$, the GTLS solution is given by :

$$\widehat{X} = (A^T DA - \sigma^2 C_a^*)^{-1}(A^T DB - \sigma^2 C_{ab}^*) \tag{9}$$

Proof : see [22, Theorem 4].

If $R_D \sim I$, (9) is a well-known expression in **linear regression** analysis and **statistics**. Its consistency and other statistical properties have been investigated by Gallo [5]. Gleser [6] studied the special case that $R_C \sim I$ and $n_1 = 0$, corresponding to the classical TLS problem. Using their results, it can be concluded that under mild conditions the GTLS solution is a **consistent** estimate of the true but unknown parameters X of the **general errors-in-variables model**, defined as :

$$B_0 = A_0 X = A_1 X_1 + (A_2)_0 X_2; \quad A_2 = (A_2)_0 + \Delta A_2 \text{ and } B = B_0 + \Delta B \tag{10}$$

X_1 and X_2 are the true but unknown parameters to be estimated, A_1 and $(A_2)_0$ are of full column rank. They consist of constants as well as B_0. A_1 is known but $(A_2)_0$ and B_0 not. The observations A_2 and B of the unknown values $(A_2)_0$ and B_0 contain measurement errors ΔA_2 and ΔB such that the rows of $[\Delta A_2; \Delta B]$ are independently and identically distributed (i.i.d.) with zero mean and known positive definite covariance matrix $C = R_C^T R_C$, up to a factor of proportionality.
Statistical properties of the GTLS solution for the case that the true values A_0 and B_0 in model (10) are random variables, have been proven by Kelly [11]. And finally, Aoki and Yue [2], Kotta [12] and Stoica and Söderström [18] studied the statistical properties of the GTLS solution for Toeplitz-like sets of equations arising in **ARMA modelling** and **system identification**.These models are given by :

$$y(t) + a_1 y(t-1) + \cdots + a_{n_a} y(t-n_a) = b_1 u(t-1) + \cdots + b_{n_b} u(t-n_b) \quad (11)$$

where the $\{u_j\}$ and $\{y_j\}$ are the input and output sequences respectively and $\{a_j\}$ and $\{b_j\}$ are the unknown constant parameters of the system. Assuming enough observations, (11) gives rise to an overdetermined set of equations. If the only disturbances of the outputs and the inputs (if they can't be measured exactly) are given by stationary, zero mean, white noise of equal variance, i.e. $R_C \sim I$ and $n_1 = 0$ or n_b, strong consistency of the GTLS solution of this set can be proven [2], [18]. Kotta [12] extended these results to prove consistency for the case that the disturbances are not necessarily white and the covariance matrix $C = R_C^T R_C$ of the correlated noise on the input-output data is known, up to a factor of proportionality.

Acknowledgements

The author would like to thank Gene Golub for valuable discussions and pointing out [3].

References

[1] R.J. Adcock. *The Analyst* **5** (1878), pp. 53-54.

[2] M. Aoki and P.C. Yue. *IEEE Trans. Autom. Contr.* **15** (1970), pp. 541-548.

[3] L.M. Ewerbring and F.T. Luk, Canonical Correlations and Generalized SVD : Applications and New Algorithms. *Techn. Rep. EE-CEG-88-11*, Cornell Univ.,Ithaca, NY , July 1988 (to appear in J. Comput. Appl. Math.).

[4] K.V. Fernando and H. Nicholson. *IEE Proc. D* **132** (1985), pp. 30-36.

[5] P.P. Gallo. *Commun. Statist.-Theor. Meth.* **11** (1982), pp. 973- 983.

[6] L.J. Gleser. *Ann. Statist.* **9** (1981), pp. 24-44.

[7] G.H. Golub, A. Hoffman and G.W Stewart. *Lin.Alg. and its Appl.* **88/89** (1987), pp. 317-327.

[8] G.H. Golub and C.F. Van Loan. *SIAM J. Numer. Anal.* **17** (1980), pp. 883-893.

[9] G.H. Golub and C.F. Van Loan, *Matrix Computations.* The Johns Hopkins Univ. Press, Baltimore, Maryland, 1983.

[10] R.P. Guidorzi. *Automatica* **11** (1975), pp.361-374.

[11] G. Kelly. *Ann. Statist.* **12** (1984), pp. 87-100.

[12] U. Kotta. *Proc. 5th IFAC Symp. on Identification and System Parameter Estimation*, Darmstadt, FRG (1979), pp.453-458.

[13] J. Leuridan, D. De Vis, H. Van der Auweraer and F. Lembregts. *Proc. 4th Int. Modal Analysis Conf.*, Los Angeles, Feb. 3-6 (1986), pp. 908-918.

[14] M.J. Levin. *IEEE Trans. Autom. Contr.* **9** (1964), pp. 229-235.

[15] F.T. Luk. *J. Parallel Distrib. Comput.* **2** (1985), pp. 250-260.

[16] C.C. Paige. *SIAM J. Sci. Stat. Comput.* **7** (1986), pp. 1126- 1146.

[17] K. Pearson. *Philos.Mag.* **2** (1901), pp. 559-572.

[18] P. Stoica and T. Söderström. *Int. J. Control* **35** (1982), pp. 449-457.

[19] S. Van Huffel, Analysis of the total least squares problem and its use in parameter estimation. *Doct. Dissertation*, Dept. Electr. Eng., K.U.Leuven, Belgium, June 1987.

[20] S. Van Huffel and J. Vandewalle. *J. Comput. and Applied Math.* **21** (1988), pp. 333-341.

[21] S. Van Huffel and J. Vandewalle. *SIAM J. Matrix Anal. Appl.* **9** (1988), pp. 360-372.

[22] S. Van Huffel and J. Vandewalle, Analysis and properties of the generalized total least squares problem $AX \approx B$ when some or all columns in A are subject to error. *Int. Rep.*, K.U.Leuven, Dept. of Electr. Eng., Mar. 1988 (submitted to SIAM J. Matrix Anal. Appl.).

Understanding Old Deficiencies in the Conventional Recursive Least-Squares (CRLS) Scheme

M.H. Verhaegen

Delft University of Technology
Department of Electrical Engineering
Control Laboratory
NL-2628 CD Delft, The Netherlands

Abstract

This paper *extends* the well-known fact, that the normal equations approach in solving least-squares problems operates "well" when these problems are "well" conditioned, to the recursive formulation of this approach. The latter scheme is indicated in this paper as the conventional RLS scheme. In particular this paper highlights that for those well-conditioned problems the conventional RLS scheme *does not* suffer from the divergence phenomenon, the loss of symmetry and positive definiteness of the parameter covariance matrix.

1 Introduction

Recursive least-squares (RLS) schemes are used in a broad class of practical applications, such as in the self-tuning regulator schemes developed by [1].

Still a number of deficiencies of what is referred in this paper as the conventional recursive least-squares (CRLS) occur when using this implementation in practice. These deficiencies are known as divergence phenomenon. However, classifying these phenomenon (a common practice in science in

NATO ASI Series, Vol. F 70
Numerical Linear Algebra, Digital Signal Processing and Parallel Algorithms
Edited by G. H. Golub and P. Van Dooren

an attempt to grasp their unpredictable nature) has not been sufficient to explain their nature.

Recently, in [3], a theoretical explanation has been given of the cause of both the loss of symmetry and the loss of positive definiteness in the CRLS. The results in [3] related the loss of symmetry phenomena to a "wrong" way of implementing the CRLS scheme. A straightforward means of overcoming this deficiency was also derived from these insights, which has been clearly demonstrated by an experimental simulation study.

In this paper we concentrate on the experimental validation of the insights revealed in [3] about the nature of the loss of positive definiteness of the parameter covariance matrix. These insights are evaluated for two implementation of the CRLS scheme that preserve the symmetry of the parameter covariance matrix. The first scheme was presented in [3] and is indicated by the CLS2 (conformal with the notation used in [3]) or "correct" implementation of the CRLS algorithm. And the second scheme, which was not studied in [3], is the widely used implementation (see e.g . [2]) which only updates the upper (or lower) triangular part of the parameter covariance matrix. This particular scheme is indicated as the CLS0 implementation.

This paper is organized as follows. In Section 2, we define precisely the two above-mentioned RLS schemes. Then in Section 3 the theoretical result explaining the loss of positive definiteness is recalled from [3] and the results of the experimental analysis are summarized in Section 4. Finally, some concluding remarks are made in Section 5.

2 The Conventional Recursive Least-Squares (CRLS) Scheme

Let the linear scalar regression model be denoted as:

$$y_k = \vartheta' \varphi_k + e_k \tag{1}$$

where the regressor vector $\varphi_k \in R^n$, e_k is a zero mean discrete white noise sequence with variance σ_k^2 and $(.)'$ denotes the transpose. For this regression model, we now state the recursive relationships, known as the (conventional) RLS scheme:

$$\hat{\vartheta}_k = \hat{\vartheta}_{k-1} + K_k(y_k - \hat{\vartheta}'_{k-1}\varphi_k) \tag{2}$$

$$K_k = \frac{P_{k-1}\varphi_k}{\lambda\sigma_k^2 + \varphi'_k P_{k-1}\varphi_k} \tag{3}$$

$$P_k = \frac{1}{\lambda}\left(P_{k-1} - \frac{P_{k-1}\varphi_k\varphi'_k P_{k-1}}{\lambda\sigma_k^2 + \varphi'_k P_{k-1}\varphi_k}\right) \tag{4}$$

where λ is the exponential forgetting factor ≤ 1. In order to use this RLS scheme we still have to specify the initial conditions $(\hat{\vartheta}_{-1}, P_{-1})$.

The "correct" implementation of the CRLS scheme, indicated in [3], is represented in Table 1 as the CLS2. Further, we also consider the CLS0 implementation, which only computes the upper (or lower) triangular part of the parameter covariance matrix P_k. This is scheme indicated in [2] as the conventional Kalman filter scheme. This CLS0 implementation does not have to compute $P_{k-1} \times \varphi_k$ but can simply make use of the transpose of fiP in Table 1.

Label	mathematical expression	Execution in CLS2
fiP	$\varphi'_k P_{k-1}$	$\varphi'_k \times P_{k-1}$
r_k^e	$\lambda\sigma_k^2 + \varphi'_k P_{k-1}\varphi_k$	$\lambda \times \sigma_k^2 + fiP \times \varphi_k$
K_k	$\dfrac{P_{k-1}\varphi_k}{\lambda\sigma_k^2 + \varphi'_k P_{k-1}\varphi_k}$	$\dfrac{P_{k-1} \times \varphi_k}{r_k^e}$
$P_{k-1}^\star$	$\dfrac{P_{k-1}\varphi_k\varphi'_k P_{k-1}}{\lambda\sigma_k^2 + \varphi'_k P_{k-1}\varphi_k}$	$K_k \times fiP$
P_k	$\frac{1}{\lambda}\left(P_{k-1} - \dfrac{P_{k-1}\varphi_k\varphi'_k P_{k-1}}{\lambda\sigma_k^2 + \varphi'_k P_{k-1}\varphi_k}\right)$	$\frac{1}{\lambda}\left(P_{k-1} - P_{k-1}^\star\right)$

Table 1: The CLS2 implementation of the conventional RLS.

3 Theoretical Explanation of the Loss of Positive Definiteness

In this section we summarize the theoretical results explaining the loss of positive definiteness in the following theorem. For a proof of the theorem we refer to [3].

Theorem: *When* implementing the set of equations (2-4) on a digital computer with machine precision ϵ, and the following conditions are satisfied: (1) $\|\overline{P}_k - \overline{P}'_k\| = 0$, (2) $\Lambda(\overline{P}_k) > 0$ and (3) $\kappa(\overline{P}_k) > \epsilon^{-1}$, *then* $\Lambda(\overline{P}_{k+1}) > 0$. □

Here $\overline{P}_k$ denotes the quantity stored in the computer and $\Lambda(\overline{P}_k)$, $\kappa(\overline{P}_k)$ respectively denote the eigenvalues and the condition number of $\overline{P}_k$.

4 Numerical Simulation Tests

In this section, we examine experimentally the round-off error propagation in the CLS0 and CLS2 implementation. The experiments have been conducted on a Micro Vax II.

The regression model that was used to generate data was taken from [3], and has the following form:

$$y_k = \vartheta_1 \varphi_k(1) + \cdots + \vartheta_4 \varphi_k(4) \tag{5}$$

In the following tests we considered only the measurement sequence 1 of $[y_k, \varphi_k]$ in [3], which guarantees that for the regression model (5) the regressor vector φ_k is *persistently excited.*

In the first test, we performed mixed precision computations in order to study the propagation of round-off errors. Based on the assumption that the single precision updated quantities (denoted by $\overline{(.)}$) are the erroneous quantities and the double precision updated quantities are error-free (denoted by $(.)$), the "total" round-off error on P_k is given as:

$$\|\delta_{tot} P_k\| = \|\overline{P}_k - P_k\| \tag{6}$$

In the subsequent tests, $\|.\|$ was taken the Frobenius norm.

Using persistently excited data, quantity (6) is computed for both the CLS0 and CLS2 implementations for $\lambda = 0.75$. This represents a rather severe circumstance, since the estimates now (approximately) only rely on the last 4 measurements. Still under this extreme condition, Fig. 1 clearly demonstrates that *no divergence* of the round-off errors occurs for either implementation.

In the second test, the same measurement sequence as in the previous test is taken. Now we assumed the following regression model inside the CRLS scheme:

$$y_k = \vartheta_1\varphi_k(1) + \cdots + \vartheta_4\varphi_k(4) + \vartheta_5\varphi_k(4) \tag{7}$$

Taking $P_{-1} = diag(100, 100, 100, 100, 100)$ and an exponential forgetting factor $\lambda = 0.95$, the evolution of $\kappa(P_k)$ is plotted in Fig. 2a for the CLS2 implementation. Clearly, the matrix P_k now becomes singular. Shortly after the recursion instant when P_k remains singular, its diagonal elements turn *negative*. This is clarified in Fig. 2b, where we plotted only the first 4 diagonal elements of P_k.
The results corresponding to the CLS0 implementation are very similar and are therefore not shown in this paper.

5 Concluding Remarks

This paper clarifies the theoretical explanation given in [3] about the loss of positive definiteness phenomenon in the conventional RLS scheme by means of an experimental evaluation. This analysis demonstrated that the two critical requirements to preserve this property of the parameter covariance matrix P_k are: (1) the preservance of the symmetry of P_k (two implementations of the CRLS scheme with this property are outlined in this paper) and (2) strictly ensuring that P_k does not become numerically singular.

The latter requirement clearly reveals that when the underlying regression problem is *well conditioned*, the recursive schemes that preserve symmetry of P_k, guarantee bounded errors on the quantities updated in these schemes. This is a well-known result in numerical analysis for the normal equations approach, which is now extended to the RLS method.

In practice this latter requirement means that we have to excite our system under consideration properly during estimation, *plus* we have to know correctly the order of the regression model under consideration. By order we refer to the number of components in the regressor vector.

Since the CLS0 (which only updates the upper (or lower) part of the P_k) is the most efficient (generally applicable) recursive least-squares scheme available, the analysis of this paper demonstrates that this scheme might be of interest in applications making use of vector processors or systolic arrays.

References

[1] Åström, K.J., U. Borisson, L. Ljung, and B. Wittenmark "Theory and applications of self-tuning regulators" *Automatica*, **13**, pp. 457-476, 1977

[2] Bierman, G.J. "Factorization methods for discrete sequential estimation" *Academic Press*, New York, 1976

[3] Verhaegen, M.H. "Round-off error propagation in four generally applicable, recursive, least-squares-estimation schemes" Internal Report R88.066, Delft University of Technology, July 1988 (accepted for publication in Automatica)

Algorithms for Optical Computing and their Sensitivity

Erik I. Verriest

Georgia Institute of Technology
Atlanta, GA 30332, USA

1 Introduction

Fiber optics and guided wave optics have played an increasing role in communications, in recent years, and are expected to continue to grow. This activity, which was largely oriented to switching networks, will have direct impact on signal and data processing as well.

The use of optics in computing offers great potential for large-scale, high speed computing power. However, the favorable and unfavorable attributes of optics must be well understood in relation to those of electronics, so that hybrid opto-electronic systems can be designed for maximal utilization of the desirable features in both. This will provide the advantage needed for large scale complex processing problems. The favorable characteristics in optics are: large bandwidth (hence high speed), parallelism, interconnectivity, and the availability of readily generated special functions (such as sine and cosine). The main disadvantage of optical processors are their low accuracy. This makes optical systems appropriate for fast, first pass processors, where high accuracy is not necessary. This paper zooms in on a recently designed coherent electro-optical Givens Rotation Device [1]. Enabling LU-decompositions, it possesses great potential in fast parallel processing. A study of the accuracy is primordial, in order to predict the behaviour and applicability of many proposed optical computation and signal processing schemes. Besides the electro-optical matrix triangularization, also hyper-

NATO ASI Series, Vol. F 70
Numerical Linear Algebra, Digital Signal Processing and Parallel Algorithms
Edited by G. H. Golub and P. Van Dooren

bolic rotators, and a polynomial equation solver, all based on the coherent optical Givens rotator will be described.

2 The Integrated Optical Givens Rotation Device

As already mentioned, the signal medium is coherent light. The (complex) phasor notation is used to describe the electric fields. The device uses electro-optical grating diffraction and phase shifting. The required sine/cosine-evaluation, multiplication and addition operations are achieved by the device physics (Bragg diffraction) itself [3].

At the interface of two transparent media with different refraction index, an incident wave splits into a reflected and a transmitted beam. In a medium of periodically varying refraction-index material, multiple reflections and transmissions occur. Interference of these beams leads to several transmitted and reflected beams. The directions are such that the path difference is a multiple of the wavelength in the medium.

A sinusoidal modulation in the index of refraction can be induced by the electro-optic effect using interdigitated electrodes. If an optical phasor a is incident at the first Bragg angle on a thick grating, a transmitted phasor $\cos\psi$, and a diffracted phasor $\exp[j(\phi_n - \pi/2)]\sin\psi$ result. The ψ is the grating strength parameter. It is proportional to the applied potential V, while also depending on parameters as the spatial period of the electrodes, the thickness of the grating, the optical wavelength, and the electro-optic parameters of the material and configuration. With another beam, b, incident at the symmetrical Bragg angle, superposed on the first, the total effect is

$$\begin{aligned} \bar{a} &= a\exp[j(\phi_n - \pi/2)]\sin\psi + b\cos\psi \\ \bar{b} &= a\cos\psi + b\exp[j(-\phi_n - \pi/2)]\sin\psi \end{aligned}$$

ϕ_n is a constant relating to the phase reference of the grating. An elementary rotation can thus be obtained by adding external phaseshifters, respectively with phaseshift $\Gamma_1 = \phi_n + \pi/2$ and $\Gamma_2 = -\phi_n + \pi/2$. This is also possible with the electro-optic effect.

An integrated optical implementation can be fabricated as in figure 1., where the two channel waveguides ($8\mu m$ wide) are cut in a lithium niobate crystal. The light is guided in a TM mode. The details of the implementation of the device are given in [1]. The intervening physical principles are standard, and have found applications in many other devices. For detection of the output phasors, homodyning techniques are required.

3 Quaternion Representation

As the signal in each channel is represented by a phasor, the signals in the two channels can be represented simultaneously in a two-dimensional right vector space over C. With a basis $\{1, i\}$ in C, this space is isomorphic to Q, the set of quaternions. A two-channel signal can be represented by a "quasor" (in analogy to phasor) $\underline{q} = (a + jb) + i(c + jd) = a + jb + ic + kd$, following the rules of quaternion multiplication. Either component of the quaternion can be picked up (physically) by homodyne detection. All two-channel operations can now be represented in this quasor-notation. The Givens Rotation (over angle θ) gives $\underline{q}_{\text{out}} = e^{i\theta}\underline{q}_{\text{in}}$. One only need to caution that the Quaternion algebra is non-commutative. The real parts (under proper calibration, all input and output signals are "in phase") can graphically be displayed in the $\{1, i\}$-plane, the "Q-plane". In this Q-plane, circles centered at O are the orbits under Givens rotation, while the straight line segments, connecting antipodal points, $\underline{q}$ and $-\underline{q}$, are the orbits under phase shifts. A general 2×2 unitary operation transforms the quasor $\underline{q}$ to $e^{j\alpha}e^{i\theta}e^{j\beta}\underline{q}e^{j\gamma}$.

4 Matrix Triangularization

Elementary Givens rotations can be used to triangularize a matrix. The "numerical" signal flow lends itself directly to optical architectures using the integrated optical rotation devices. In [2], a parallel and a pipelined architecture are discussed. The pipelined version is reproduced below (not in its most time-compressed form). A rotation device is located at each intersection of the channel waveguides. The squares represent photodetectors. Because of the inversion of the top and the bottom, with the device configuration as shown, the triangularized matrix appears "upside down".

5 Accuracy Analysis

The previous triangularization scheme looks nice, and natural, but does it work? Several sources of potential danger can be identified. First there are the modelled (or known) deviations from the ideal behaviour, they stem from simplifications in the physical model. Bends in the waveguides are necessary to allow the interconnection, and these are not loss-free. There exist also higher order diffracted waves, which were here neglected. As the total signal power in and out of the crystal must remain constant, the sum of the first order transmitted and diffracted signal does not add to the input signal power. Secondly, there are the unmodelled deviations, due to misadjustments and tolerances. Their effects can be nihilized by proper calibration [2]. A third source of randomness will be neglected as well: the quantum nature of the light. This is acceptable, since the device is envisioned to operate well above the quantum limit.

The higher diffraction orders cause the rotation angle (grating strength) to be nonlinearly related to the grating potential. Hence, for a unit input at the top, the two outputs of the (phase-compensated) device are $Bs(V) = \sin[K_s(V)V]$, and $Bc(V) = \cos[K_c(V)V]$, for a grating potential V. Empirically, it was found that if $K(V) = (K + \Delta K(V))$, over the entire 360° range, then $\max|\Delta K/K| \cong 0.024$. Because of the symmetry in the two channels, the Bragg-diffraction cell has the transfer characteristic represented by the left quasor multiplier $\sigma(V)\exp(i\theta(V))$, where $\sigma^2(V) = Bs^2(V) + Bc^2(V)$, and $\tan\theta(V) = Bs(V)/Bc(V)$. For the intended applications, the nonlinearity of the $\theta(V)$ is not a problem, as long as the relation is monotone. Figure 3 displays the total output power $\sigma^2(V)$ versus grating potential.

The corresponding digital accuracy for this analog device may be found by assuming a k-bit table look-up function evaluation. If the exact function is $y = f(x)$, the k-bit rounded approximation is $r_k(f(r_k(x)))$. Dividing both the domain D, and the range R of f in 2^k bins, the digital accuracy is $\Delta_k(y) = r_k(f(r_k(x))) - f(x) \leq 2^{-(k+1)}(R + |f'(x)|D)$. For $f(x) = \sin x$ one finds $\Delta_k \leq 2^{-k}(1+\pi)$, and $|\sigma - 1| \leq (1+\pi)2^{-k+\frac{1}{2}}$. By matching the bounds for analog and digital system, an accuracy of less than 10 bits over the $(-2\pi, 2\pi)$ range was found. Higher accuracy is achievable if the domain is restricted to the potential corresponding to $[-\pi, \pi]$, or even smaller, but then cascades of equal devices have to be used to produce a full range of rotations.

For a device operating as "constructor", a null needs to be detected in an output beam. Zero-detection can be performed by an incoherent detector, thus simplifying the circuit considerably. The threshold D_o of the detector induces another error. A zero will be decided for a signal ϵD_o, for $|e| \leq 1$. It induces a relative error in the computed norm of the input quasor of approximately (up to second order) $\epsilon(1 + \epsilon(D_o^2 - 1)/2)$, and an absolute error in angle less than $\epsilon(1 + D_o/(1 - e))$. A statistical analysis of the accuracy in matrix triangularization was shown in [2].

Adders and beamsplitters can be derived by using the rotation device with a fixed grating potential corresponding to an angle of $\pi/4$. The undesirable factor $1/\sqrt{2}$ appearing at the outputs, can be removed at the expense of some slightly more complex circuitry, using two more rotators and an additional zero-detection. An inverter is obtained by rotation over $\pi/2$.

6 Hyperbolations

The role of hyperbolic rotations is ubiquitous in many signal processing problems, as illustrated elsewhere in this book. Unfortunately no physical principles are now known that would lead in a straightforward way to the necessary hyperbolic functions. However, one can exploit a one-to-one relation $\theta \to \phi_G(\theta)$, between Hyperbolations (over θ) and Rotations (over ϕ), determined via the identities

$$\tan\phi = \sinh\theta \Longleftrightarrow \sin\phi = \tanh\phi \Longleftrightarrow \cos\phi = 1/\cosh\theta$$

This rotation angle $\phi_G(\theta)$ corresponding to the hyperbolation θ is called the Angle of Gudermann. These identities are easily verified graphically. If H is the point on the unit hyperbola, with coordinates $[\cosh\theta, \sinh\theta]$, then a horizontal through H intersects the tangent to the top of unit hyperbola (and unit circle) in a point P. The vector OP makes an angle $\phi_G(\theta)$ with the horizontal axis Ox. The identities then follow by writing the vertical coordinate of P as $\sinh\theta = \tan\phi$.

Alternatively, by writing

$$\begin{pmatrix} x \\ y \end{pmatrix} = \begin{pmatrix} \cos\phi & -\sin\phi \\ \sin\phi & \cos\phi \end{pmatrix} \begin{pmatrix} a \\ b \end{pmatrix} \text{ as } \begin{pmatrix} a \\ y \end{pmatrix} = \begin{pmatrix} \cosh\theta & \sinh\theta \\ \sinh\theta & \cosh\theta \end{pmatrix} \begin{pmatrix} x \\ b \end{pmatrix}$$

the identities are obvious from an algebraic point of view. This relation leads directly to the Hyperbolation device depicted in figure 4.

It consists of a rotator, set at the required Gudermann-angle ϕ, a beam splitter, and a summer. A signal with tunable (also via a rotation device) amplitude K enters the beam splitter. The amplitude is ramped, until a zero is detected at the output of the adder. In summary, hyperbolic scalings are obtained via rotation over the corresponding angle of Gudermann. The $\theta - \phi$ correspondence is not needed explicitly. If the device is used as an "applicator", its potential settings are given directly as voltages by a "master", and not via the theoretical hyperbolation angle θ.

Finally, let us remark that the necessary detections are all zero-detections, so that incoherent detectors may be used. It may be possible to integrate the detector and its electronics on the same chip. The Q-plane plot is shown in figure 5.

The Constructor (or Master) is obtained as a slight variation of the previous. The auxiliary signal K and the beamsplitter are removed, and the rotation angle ϕ is now tuned by the detector.

7 Coherent Polynomial Equation Solver

Consider a linear array of $n+1$ Givens rotation devices, with top-output connected to the bottom-input of the next device. Let the input to the bottom of the first be x_0, and the remaining (top) input at the $k-th$ device be a_{k-1}, for $k = 1$ to $n+1$. The grating potential on all rotation devices in the linear array are set equal.

Denoting the interconnecting signal between the $k-th$ and $(k+1)-st$ device by x_k, and the top-output after the $(n+1)-st$ stage by x_{n+1}, and iterating the $x_{k+1} = cx_k + sa_k$, where c and s are respectively the cosine and the sine corresponding to the set grating potential, one gets

$x_{n+1} = c^{n+1}x_0 + sP_n(a,c)$. $P_n(a,x)$ denotes the $n-th$ order polynomial $a_0x^n + a_1x^{n-1} + \ldots + a_n$. With $x_0 = 0$, the signal exiting the last top is proportional to the polynomial evaluated at c. A sweep of the grating potential, and zero-detection of x_{n+1}, determines the real zeros c^*, in the interval [-1,1]. This solution is available, up to the factor a_0, as the optical signal a_0c^* from the bottom-output of the first device. There is also always a "phantom" root, $c = 1$, attributable to the sine factor.

For a root at c^*, the gradient with respect to the grating potential is $-(s^*)2P_n'(a,c^*)$. This quantifies the accuracy. It is possible to boost the accuracy of the root determination by using feedback. An $(n+2)-th$ device, whose potential can be set independently of the others in the array (corresponding to $\underline{c}$, and $\underline{s}$ say), is added to the linear array, one output of this stage is fed back to x_0. The top-input is kept at zero. This gives the top- and bottom outputs x_0 and t respectively as $x_0 = \underline{c}x_{n+1}$ and $t = -\underline{s}x_{n+1}$. This test-signal t is needed for the zero-detection. After some algebra, $t = -\underline{s}s/(1-\underline{c}c^{n+1})P_n(a,c)$. The gradient, evaluated at a solution c^* is a factor $\underline{s}/(1-\underline{c}c^{n+1})$ steeper than for the corresponding linear open loop array. By proper tuning of the auxiliary rotation, this factor can be made larger than 1. Since $P_n(a,x) = x^nP_n(b,x^{-1})$, where b is the vector of components of a, but in reversed order, the zeros of $P_n(a,x)$ outside (-1,1) are found by inverting the zeros of $P_n(b,z)$ inside (-1,1).

Finally, with four devices (two parallel in a vertical plane connected to two parallel horizontal ones) at each stage, all roots inside the unit circle of a complex polynomial can be found. The real and imaginary parts of the coefficients are fed to the two top inputs at each stage. If all "horizontal" planes rotate over θ, and all vertical ones over τ, the polynomial $P_n(a_R + ja_I, z)$ is evaluated at $e^{j\tau}\cos\theta$. Zero-detection yields the roots. The reverse polynomial determines the roots outside the unit circle.

Acknowledgements

The author acknowledges support from the U.S. Air Force (AFOSR-87-0308) and the Joint Services Electronics Program (DAAL-03-87-K-0059). Figure 3 is by courtesy of Dr. E. Glytsis of the Georgia Institute of Technology.

References

[1] M.M. Mirsalehi, T.K. Gaylord, and E.I. Verriest, "Integrated-Optical Givens Rotation Device," *Applied Optics*, May 15, 1986, pp. 1608-1614.

[2] T.K. Gaylord and E.I. Verriest, "Matrix Triangularization Using Arrays of Integrated Optical Givens Rotation Devices," *Computer*, December 1987, pp. 59-66.

[3] H. Kogelnik, "Coupled Wave Theory for Thick Hologram Gratings," *Bell Syst. Tech. J.*, 48, 2909 (1969).

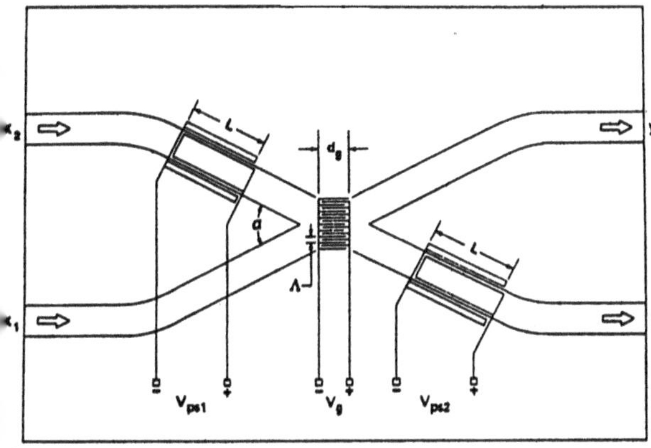

Figure 1: The Optical Givens Rotator

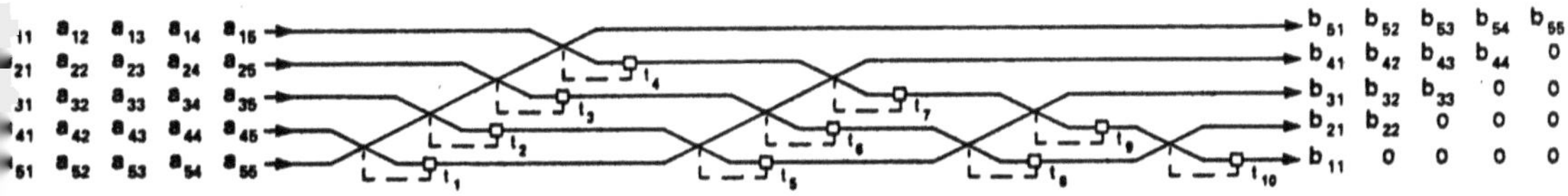

Figure 2: The Triangularization Pipeline

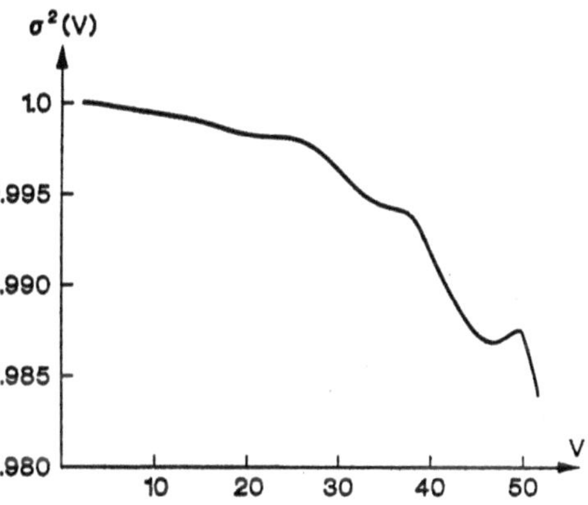

Figure 3: $\sigma^2(V)$

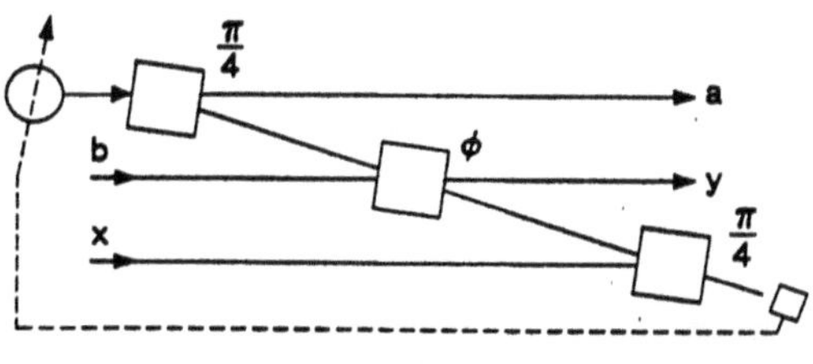

Figure 4: The Optical Hyperbolator

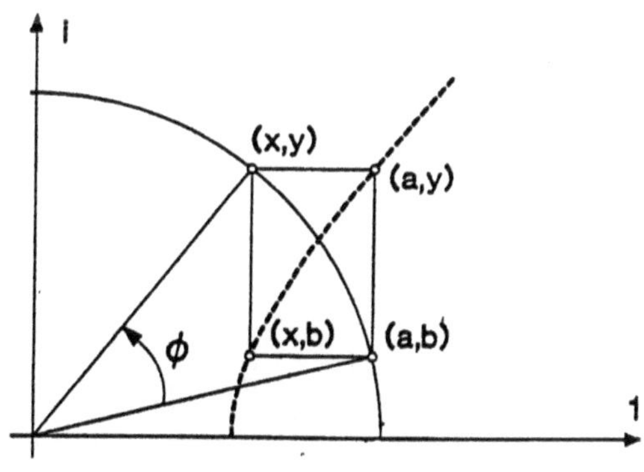

Figure 5: Q-Plane for the Hyperbolator

A Constrained Eigenvalue Problem*

Walter Gander

Institut für Informatik
ETH Zentrum
CH-8092 Zürich, Switzerland

Gene H. Golub

Department of Computer Science
Stanford University
Stanford, CA 94305, USA

Urs von Matt

Institut für Informatik
ETH Zentrum
CH-8092 Zürich, Switzerland

Abstract

In this paper we consider the following mathematical and computational problem. Given the quantitics

A: $(n+m)$-by-$(n+m)$ matrix, symmetric, $n > 0$

N: $(n+m)$-by-m matrix with full rank

$\mathbf{t}$: vector of dimension m with $\|(\mathrm{N}^{\mathrm{T}})^{+}\mathbf{t}\| < 1$

Determine an $\mathbf{x}$ such that

$$\mathbf{x}^{\mathrm{T}}\mathrm{A}\mathbf{x} = \min$$

*This work was in part supported by the National Science Foundation under Grant NSF CCR-8412314 and by the US Army (DAAL03-87-K-0095).

NATO ASI Series, Vol. F 70
Numerical Linear Algebra, Digital Signal Processing and Parallel Algorithms
Edited bv G. H. Golub and P. Van Dooren

subject to the constraints

$$(i)\ \mathbf{N}^T\mathbf{x} = \mathbf{t}$$
$$(ii)\ \mathbf{x}^T\mathbf{x} = 1.$$

Variants of this problem occur in many applications [1,5,7,8,11]. The problem has been studied previously when $\mathbf{t} = \mathbf{0}$, the null vector, (cf. [4,6]). When $\mathbf{t} \neq \mathbf{0}$, then the problem becomes more complicated.

We show how to eliminate the linear constraint (i). Then three different methods are presented for the solution of the resulting Lagrange equations.

1 Introduction

In this paper we consider the following mathematical and computational problem. Given the quantities

A: $(n+m)$-by-$(n+m)$ matrix, symmetric, $n > 0$
N: $(n+m)$-by-m matrix with full rank
t : vector of dimension m with $\|(\mathbf{N}^T)^+\mathbf{t}\| < 1$

Determine an $\mathbf{x}$ such that

$$\mathbf{x}^T\mathbf{A}\mathbf{x} = \min \tag{1}$$

subject to the constraints

$$\mathbf{N}^T\mathbf{x} = \mathbf{t} \tag{2}$$
$$\mathbf{x}^T\mathbf{x} = 1. \tag{3}$$

Variants of this problem occur in many applications [1,5,7,8,11]. The problem has been studied previously when $\mathbf{t} = \mathbf{0}$, the null vector, (cf. [4,6]). When $\mathbf{t} \neq \mathbf{0}$, then the problem becomes more complicated.

2 Problem Simplification

In this section we will normalize the problem in order to study its solvability. These considerations are not only of theoretical interest, but they also serve as a basis for the numerical calculations.

2.1 Elimination of the Linear Constraint

To simplify the first constraint (2), we use the QR-decomposition of N

$$P^T N = \begin{bmatrix} R \\ 0 \end{bmatrix} \tag{4}$$

where P denotes an orthogonal matrix, and R is a m-by-m upper triangular matrix. Now the problem can be formulated as

$$\begin{aligned} \mathbf{x}^T A\mathbf{x} &= \mathbf{x}^T PP^T APP^T\mathbf{x} = \min \\ N^T\mathbf{x} &= \begin{bmatrix} R^T & 0 \end{bmatrix} P^T\mathbf{x} = \mathbf{t} \\ \mathbf{x}^T\mathbf{x} &= \mathbf{x}^T PP^T\mathbf{x} = 1. \end{aligned}$$

We now make the definitions,

$$P^T AP =: \begin{matrix} & m & n & \\ & \left[\begin{matrix} B \\ \Gamma \end{matrix}\right. & \left.\begin{matrix} \Gamma^T \\ C \end{matrix}\right] & \begin{matrix} m \\ n \end{matrix} \end{matrix} \quad ; \tag{5}$$

$$P^T\mathbf{x} =: \begin{bmatrix} \mathbf{y} \\ \mathbf{z} \end{bmatrix} \begin{matrix} m \\ n \end{matrix} \quad . \tag{6}$$

Note that $C^T = C$. Now

$$\begin{aligned} \mathbf{x}^T A\mathbf{x} &= \begin{bmatrix} \mathbf{y}^T & \mathbf{z}^T \end{bmatrix} \begin{bmatrix} B & \Gamma^T \\ \Gamma & C \end{bmatrix} \begin{bmatrix} \mathbf{y} \\ \mathbf{z} \end{bmatrix} \\ &= \mathbf{y}^T B\mathbf{y} + \mathbf{y}^T\Gamma^T\mathbf{z} + \mathbf{z}^T\Gamma\mathbf{y} + \mathbf{z}^T C\mathbf{z} \\ &= \mathbf{y}^T B\mathbf{y} + 2\mathbf{z}^T\Gamma\mathbf{y} + \mathbf{z}^T C\mathbf{z}, \\ N^T\mathbf{x} &= \begin{bmatrix} R^T & 0 \end{bmatrix} \begin{bmatrix} \mathbf{y} \\ \mathbf{z} \end{bmatrix} = R^T\mathbf{y} = \mathbf{t}, \\ \mathbf{x}^T\mathbf{x} &= \begin{bmatrix} \mathbf{y}^T & \mathbf{z}^T \end{bmatrix} \begin{bmatrix} \mathbf{y} \\ \mathbf{z} \end{bmatrix} = \mathbf{y}^T\mathbf{y} + \mathbf{z}^T\mathbf{z} = 1. \end{aligned}$$

So we have reduced the constraint (2) to an ordinary linear system with an upper triangular matrix. Now

$$\mathbf{y} = \mathrm{R}^{-\mathrm{T}}\mathbf{t} \tag{7}$$

and with the help of the definitions

$$s^2 := 1 - \mathbf{y}^{\mathrm{T}}\mathbf{y} > 0 \tag{8}$$
$$\mathbf{b} := -\Gamma\mathbf{y}, \tag{9}$$

we get the simplified problem

$$\begin{aligned} \mathbf{z}^{\mathrm{T}}\mathrm{C}\mathbf{z} - 2\mathbf{b}^{\mathrm{T}}\mathbf{z} &= \min \\ \mathbf{z}^{\mathrm{T}}\mathbf{z} &= s^2. \end{aligned} \tag{10}$$

This problem has been extensively studied in the literature (cf. [2,7,8,10]).

2.2 Stationary Points

In order to calculate the stationary points, we set up the so-called *Lagrange principal function*:

$$\Phi(\mathbf{z}, \lambda) := \mathbf{z}^{\mathrm{T}}\mathrm{C}\mathbf{z} - 2\mathbf{b}^{\mathrm{T}}\mathbf{z} - \lambda(\mathbf{z}^{\mathrm{T}}\mathbf{z} - s^2) \tag{11}$$

Differentiating Φ by $\mathbf{z}$ and λ yields the equations

$$\begin{aligned} 2\mathrm{C}\mathbf{z} - 2\mathbf{b} - 2\lambda\mathbf{z} &= \mathbf{0} \\ \mathbf{z}^{\mathrm{T}}\mathbf{z} - s^2 &= 0 \end{aligned}$$

or normalized

$$\begin{aligned} \mathrm{C}\mathbf{z} &= \lambda\mathbf{z} + \mathbf{b} \\ \mathbf{z}^{\mathrm{T}}\mathbf{z} &= s^2. \end{aligned} \tag{12}$$

Now let us compare the values $\mathbf{z}^{\mathrm{T}}\mathrm{C}\mathbf{z} - 2\mathbf{b}^{\mathrm{T}}\mathbf{z}$ of different tuples $(\lambda, \mathbf{z})$. Following the proof given in [3,12], a short calculation shows that the smallest λ is needed in order to minimize the value $\mathbf{z}^{\mathrm{T}}\mathrm{C}\mathbf{z} - 2\mathbf{b}^{\mathrm{T}}\mathbf{z}$. So in place of the original minimization we can solve the *Lagrange equations*

$$\begin{aligned} \mathrm{C}\mathbf{z} &= \lambda\mathbf{z} + \mathbf{b} \\ \mathbf{z}^{\mathrm{T}}\mathbf{z} &= s^2 \\ \lambda &= \min. \end{aligned} \tag{13}$$

3 Solvability

We will now investigate the solvability of the Lagrange equations (13). Simultaneously this analysis will point out a first method to solve the problem.

3.1 Explicit Secular Equation

For our discussion, we need the eigenvalue decomposition

$$\mathrm{C} = \mathrm{QDQ}^{\mathrm{T}} \tag{14}$$

where $\mathrm{D} = \mathrm{diag}(\delta_1, \dots, \delta_n)$, $\delta_1 \leq \delta_2 \leq \dots \leq \delta_n$ and $\mathrm{Q}^{\mathrm{T}}\mathrm{Q} = \mathrm{I}$. Thus the Lagrange equations (13) are transformed as follows:

$$\begin{aligned} \mathrm{QDQ}^{\mathrm{T}}\mathbf{z} &= \lambda \mathrm{QQ}^{\mathrm{T}}\mathbf{z} + \mathbf{b} \\ \mathbf{z}^{\mathrm{T}}\mathbf{z} &= \mathbf{z}^{\mathrm{T}}\mathrm{QQ}^{\mathrm{T}}\mathbf{z} = s^2. \end{aligned}$$

With the definitions

$$\mathbf{u} := \mathrm{Q}^{\mathrm{T}}\mathbf{z} \tag{15}$$

$$\mathbf{d} := \mathrm{Q}^{\mathrm{T}}\mathbf{b}, \tag{16}$$

this can be simplified to

$$\begin{aligned} \mathrm{D}\mathbf{u} &= \lambda \mathbf{u} + \mathbf{d} \\ \mathbf{u}^{\mathrm{T}}\mathbf{u} &= s^2 \\ \lambda &= \min . \end{aligned} \tag{17}$$

Let us suppose $\lambda \notin \lambda(\mathrm{D})$. Then the inverse $(\mathrm{D} - \lambda \mathrm{I})^{-1}$ exists and $\mathbf{u}$ has the representation

$$\mathbf{u} = (\mathrm{D} - \lambda \mathrm{I})^{-1}\, \mathbf{d}. \tag{18}$$

For $\mathbf{u}$ to solve the normalization condition $\mathbf{u}^{\mathrm{T}}\mathbf{u} = s^2$ it must hold that

$$\mathbf{u}^{\mathrm{T}}\mathbf{u} = \mathbf{d}^{\mathrm{T}}\,(\mathrm{D} - \lambda \mathrm{I})^{-2}\,\mathbf{d} = \sum_{i=1}^{n} \left(\frac{d_i}{\delta_i - \lambda} \right)^2 = s^2.$$

We define

$$f(\lambda) := \sum_{i=1}^{n} \left(\frac{d_i}{\delta_i - \lambda} \right)^2 - s^2 \tag{19}$$

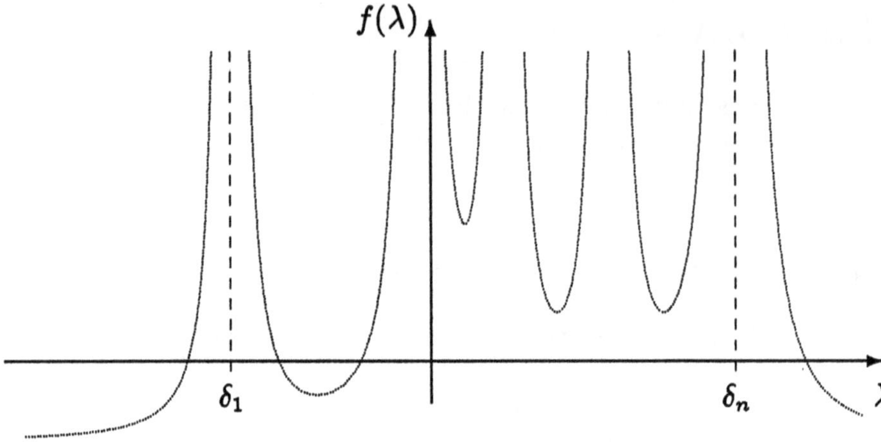

Figure 1: Graph of the secular function

as the so-called *explicit secular function* (see Figure 1). Thereby the smallest zero of the *explicit secular equation*

$$f(\lambda) := \sum_{i=1}^{n} \left(\frac{d_i}{\delta_i - \lambda} \right)^2 - s^2 = 0 \tag{20}$$

implies a unique solution $\mathbf{u}$ of the Lagrange equations (13).

3.2 Implicit Secular Equation

The above computation of the smallest λ, that solves the Lagrange equations (13), can be carried out even without the calculation of the eigenvalue decomposition (14) of C. This is useful, when we want to avoid this factorization numerically. In this case the *implicit secular equation*

$$f(\lambda) = \mathbf{b}^{\mathrm{T}}(\mathrm{C} - \lambda \mathrm{I})^{-2}\mathbf{b} - s^2 = 0 \tag{21}$$

must be solved.

4 Zero Finder

Now, we want to calculate the smallest zero of the secular equation (20). We will solve it by using an iterative method. Suppose we know an approximation $\lambda^{(i)}$. Then we can approximate $f(\lambda)$ with the replacement function

$$g(\lambda) = \frac{a}{(b-\lambda)^2} - s^2 \tag{22}$$

in such a way that

$$g(\lambda^{(i)}) = f(\lambda^{(i)})$$
$$g'(\lambda^{(i)}) = f'(\lambda^{(i)}).$$

The zero of $g(\lambda)$ will determine the next approximation $\lambda^{(i+1)}$.

A short calculation yields the values

$$a = 4\frac{\left(f(\lambda^{(i)}) + s^2\right)^3}{f'^2(\lambda^{(i)})}$$
$$b = \lambda^{(i)} + 2\frac{f(\lambda^{(i)}) + s^2}{f'(\lambda^{(i)})}$$

and for the zero $\lambda^{(i+1)} = b - \sqrt{a}/s$ of the replacement function $g(\lambda)$ we get

$$\lambda^{(i+1)} = \lambda^{(i)} - 2\frac{f(\lambda^{(i)}) + s^2}{f'(\lambda^{(i)})}\left(\frac{\sqrt{f(\lambda^{(i)}) + s^2}}{s} - 1\right).$$

It can be shown that this iteration process will yield a strictly decreasing sequence of approximations $\lambda^{(i)}$. The reader is referred to [9,12].

4.1 Implicit Secular Equation

If we do not want to compute the eigenvalue decomposition (14) of C, we have to consider the evaluation of the *implicit secular function*

$$f(\lambda) = \mathbf{b}^T(\mathrm{C} - \lambda\mathrm{I})^{-2}\mathbf{b} - s^2 \tag{23}$$

and its derivative

$$f'(\lambda) = 2\mathbf{b}^T(\mathrm{C} - \lambda\mathrm{I})^{-3}\mathbf{b}.$$

With the definitions

$$\mathbf{u} := (\mathrm{C} - \lambda\mathrm{I})^{-1}\mathbf{b}$$
$$\mathbf{u}' := (\mathrm{C} - \lambda\mathrm{I})^{-1}\mathbf{u},$$

these values can be expressed as

$$f(\lambda) = \mathbf{u}^{\mathrm{T}}\mathbf{u} - s^2$$
$$f'(\lambda) = 2\mathbf{u}^{\mathrm{T}}\mathbf{u}'.$$

Therefore each iteration step requires the solution of two linear systems with the matrix $\mathrm{C} - \lambda\mathrm{I}$.

5 Quadratic Eigenvalue Problem

The two previously mentioned methods have the property that they reduce the problem to finding the solution of a one-dimensional secular equation. Considering the problem from another point of view the Lagrange equations (13) can be reduced to a quadratic eigenvalue problem. For the derivation let us assume $\lambda \notin \lambda(\mathrm{C})$. In this case $\mathbf{z}$ can be written as $\mathbf{z} = (\mathrm{C} - \lambda\mathrm{I})^{-1}\mathbf{b}$. Taking into account the normalization condition for $\mathbf{z}$ we get the secular function

$$f(\lambda) = \mathbf{b}^{\mathrm{T}}(\mathrm{C} - \lambda\mathrm{I})^{-2}\mathbf{b} - s^2,$$

of which the zeroes are to be computed. The task looks different if we make the definition

$$\gamma := (\mathrm{C} - \lambda\mathrm{I})^{-2}\mathbf{b}$$

so that $(\mathrm{C} - \lambda\mathrm{I})^2\gamma = \mathbf{b}$. Instead of the secular equation, we have to solve the system

$$\mathbf{b}^{\mathrm{T}}\gamma - s^2 = 0 \tag{24}$$
$$(\mathrm{C} - \lambda\mathrm{I})^2\gamma = \mathbf{b}. \tag{25}$$

The first condition (24) can also be formulated as

$$1 = \frac{1}{s^2}\mathbf{b}^{\mathrm{T}}\gamma.$$

Using this factor 1 as a coefficient of $\mathbf{b}$ in (25) we get the *quadratic eigenvalue problem*

$$(\mathrm{C} - \lambda\mathrm{I})^2\gamma = \frac{1}{s^2}\mathbf{b}\mathbf{b}^{\mathrm{T}}\gamma. \tag{26}$$

Note that the restriction $\lambda \notin \lambda(\mathrm{C})$ is no longer necessary. Of course, we must face the fact that the set of solutions for λ has been extended by these manipulations, for two equations cannot be formulated as a single one without consequences. But it can be shown [12], that again we have to compute the smallest eigenvalue of (26).

5.1 Solving the Quadratic Eigenvalue Problem

The quadratic eigenvalue problem (26) can be reduced to an ordinary eigenvalue problem by properly chosen transformations. With the definition

$$\eta := (\mathrm{C} - \lambda \mathrm{I})\gamma \tag{27}$$

the following equations can be established:

$$\begin{aligned} \mathrm{C}\gamma - \eta &= \lambda\gamma \\ \mathrm{C}\eta - \frac{1}{s^2}\mathbf{b}\mathbf{b}^{\mathrm{T}}\gamma &= \lambda\eta. \end{aligned}$$

In matrix terms this leads to:

$$\begin{bmatrix} \mathrm{C} & -\mathrm{I} \\ -\frac{1}{s^2}\mathbf{b}\mathbf{b}^{\mathrm{T}} & \mathrm{C} \end{bmatrix} \begin{bmatrix} \gamma \\ \eta \end{bmatrix} = \lambda \begin{bmatrix} \gamma \\ \eta \end{bmatrix}. \tag{28}$$

Thus we have transformed the original quadratic eigenvalue problem into an equivalent linear one, that can be solved with traditional methods.

Acknowledgement

We wish to thank Professor A. M. Lesk who stimulated the research described in this paper through a personal communication.

References

[1] N. R. DRAPER, *"Ridge Analysis" of Response Surfaces*, Technometrics, 5 (1963), pp. 469–479.

[2] G. E. FORSYTHE AND G. H. GOLUB, *On the stationary values of a second-degree polynomial on the unit sphere*, SIAM J. Appl. Math., 13 (1965), pp. 1050–1068.

[3] W. GANDER, *Least Squares with a Quadratic Constraint*, Numer. Math., 36 (1981), pp. 291–307.

[4] G. H. GOLUB, *Some Modified Matrix Eigenvalue Problems*, SIAM Review, 15 (1973), pp. 318–334.

[5] G. H. GOLUB, M. HEATH AND G. WAHBA, *Generalized Cross-Validation as a Method for Choosing a Good Ridge Parameter*, Technometrics, 21 (1979), pp. 215–223.

[6] G. H. GOLUB AND C. F. VAN LOAN, *Matrix Computations*, The Johns Hopkins University Press, Baltimore, 1983.

[7] J. J. MORÉ, *The Levenberg-Marquardt Algorithm: Implementation and Theory*, in Proc. of the Biennial Conference held at Dundee, ed. A. Dold and B. Eckmann, Springer-Verlag, 1978, pp. 105–116.

[8] J. J. MORÉ AND D. C. SORENSEN, *Computing a Trust Region Step*, SIAM J. Sci. Stat. Comput., 4 (1983), pp. 553–572.

[9] CHR. H. REINSCH, *Smoothing by Spline Functions. II*, Numer. Math., 16 (1971), pp. 451–454.

[10] E. SPJØTVOLL, *A Note on a Theorem of Forsythe and Golub*, SIAM J. Appl. Math., 23 (1972), pp. 307–311.

[11] E. SPJØTVOLL, *Multiple Comparison of Regression Functions*, Ann. Math. Statist., 43 (1972), pp. 1076–1088.

[12] U. VON MATT, *A Constrained Eigenvalue Problem*, Diploma Thesis, Abteilung für Informatik, ETH Zürich, 1988.

On GR Algorithms for the Eigenvalue Problem

David S. Watkins

Department of Pure and Applied Mathematics
Washington State University
Pullman, WA 99164, USA

Abstract

This paper outlines a general theory of GR algorithms which subsumes the theory of QR, LR, and similar algorithms. A generic GR algorithm is described, and its convergence properties are discussed. A generic chasing algorithm, which performs a generic GR step implicitly, is also described. The more general point of view introduced here gives new insights which may lead to more efficient algorithms for the general eigenvalue problem.

Key words. eigenvalue, QR algorithm, GR algorithm, subspace iteration, chasing the bulge

AMS(MOS) subject classifications. 65F15, 15A18

Running head: GR Algorithms

1 Introduction

This paper outlines a general theory of GR algorithms which subsumes the theory of QR, LR, SR, HR and similar algorithms [5,2]. Details will be given in [3] and [4]. Our object of study, a generic GR algorithm, is introduced in §2. Our formulation is very general. It allows multiple GR

NATO ASI Series, Vol. F 70
Numerical Linear Algebra, Digital Signal Processing and Parallel Algorithms
Edited by G. H. Golub and P. Van Dooren

steps of any multiplicity; single and double-step algorithms are just special cases. Every GR algorithm is a form of subspace iteration in which a change of coordinate system is made at each step. This insight is the key to a clear understanding of why the algorithms converge. It is interesting that the convergence of all GR algorithms is based on the convergence of the same sequence of subspaces (modulo choice of shifts). What sets the various algorithms apart from one another is the varying quality of the transforming matrices used to perform the change of coordinates at each step. Convergence cannot be guaranteed unless the condition numbers of the accumulated transforming matrices remain bounded as the iteration proceeds. In §3 we describe a generic chasing algorithm and show that it implicitly performs a generic GR step. The general theory suggests new approaches to the eigenvalue problem which may not have been thought of before. For example, it might turn out that the best algorithm for the general eigenvalue problem is a combination of the QR and LR algorithms which uses the very stable reflectors when necessary and the inexpensive but less stable Gaussian elimination transformations otherwise.

2 The Generic *GR* Algorithm

The generic GR algorithm is based on a generic GR decomposition. A GR *decomposition* is any well-defined rule by which every matrix C in some large class of matrices $\mathcal{C}$ can be expressed as a product

$$C = GR,$$

where G is nonsingular and R is upper triangular. Given a GR decomposition, we can define a corresponding GR algorithm. Let A be a matrix whose eigenvalues are desired. The GR algorithm generates a sequence (A_i) of similar matrices as follows. A_0 is taken to be A or some convenient matrix similar to A, say $A_0 = G_0^{-1}AG_0$. Given A_{i-1}, let p_i be some polynomial of low degree such that $p_i(A_{i-1}) \in \mathcal{C}$. The step from A_{i-1} to A_i is governed by the two equations

$$p_i(A_{i-1}) = G_iR_i, \tag{1}$$

$$A_i = G_i^{-1}A_{i-1}G_i. \tag{2}$$

Under suitable conditions the sequence (A_i) will tend to upper triangular, or at least block triangular, form, yielding information about the eigenvalues. Information about eigenvectors and invariant subspaces is obtained by accumulating the transforming matrices G_i. The choice of the p_i has much to do with the rate of convergence. The degree of p_i is called the *multiplicity* of the ith step. If p_i has degree 1, it is a *single* step. If the degree is 2, it is a *double* step, and so on.

The easiest way to understand the GR algorithm is to view it as subspace iteration with a change of coordinates at each step. The ith step begins with the decomposition $p_i(A_{i-1}) = G_i R_i$. Since R_i is upper triangular, this shows that, for all k, $p_i(A_{i-1})\langle e_1, \ldots, e_k\rangle = \langle g_1, \ldots, g_k\rangle$, where $g_1, \ldots, g_n$ are the columns of G_i. The step is concluded with the similarity transformation $A_i = G_i^{-1} A_{i-1} G_i$, which is a change of coordinate system which maps $\langle g_1, \ldots, g_k\rangle$ back to $\langle e_1, \ldots, e_k\rangle$. Thus the entire step consists of a step of subspace iteration applied to $\langle e_1, \ldots, e_k\rangle$, followed immediately by a coordinate transformation which maps the resulting space back to $\langle e_1, \ldots, e_k\rangle$. To see the combined effect of k steps, define the accumulated transforming matrices

$$\hat{G}_i = G_1 G_2 \cdots G_i \qquad \hat{R}_i = R_i \cdots R_2 R_1,$$

and let $\hat{p}_i = p_i \cdots p_2 p_1$. Then one proves easily by induction that

$$\hat{p}_i(A_0) = \hat{G}_i \hat{R}_i.$$

Thus $\hat{p}_i(A_0)\langle e_1, \ldots, e_k\rangle = \langle \hat{g}_1, \ldots, \hat{g}_k\rangle$, so $\langle \hat{g}_1, \ldots, \hat{g}_k\rangle$ approaches an invariant subspace as $i \to \infty$, if the p_i are chosen in a reasonable manner. Since also obviously

$$A_i = \hat{G}_i^{-1} A_0 \hat{G}_i, \tag{3}$$

we expect A_i to tend to block triangular form as $i \to \infty$. Whether or not this actually happens depends on the answers to two questions: (1) Do the subspaces $\hat{p}_i(A_0)\langle e_1, \ldots, e_k\rangle$ really converge to an invariant subspace of A_0? (2) Do the condition numbers of the accumulated transforming matrices $\hat{G}_i$ remain bounded? Precise convergence theorems are stated and proved in [3]. In practice the shift polynomial p_i is chosen just before the ith step, based on A_{i-1}, but let's suppose just for a moment that the p_i have all been chosen in advance. Then all GR algorithms, QR, LR, SR, etc., will produce the same subspace sequences $\hat{p}_i(A_0)\langle e_1, \ldots, e_k\rangle$. The difference lies in the varying quality of the transforming matrices $\hat{G}_i$.

3 The Generic Chasing Algorithm

In practice *GR* algorithms are usually implemented implicitly by chasing algorithms. Let's consider a generic chasing algorithm. We are going to look at a single step in detail, so I will change the notation. Let $A \in C^{n\times n}$ be in irreducible upper Hessenberg form, and let p be a polynomial of degree m. Our chasing algorithm will transform A to $B = G^{-1}AG$, where G is related to $p(A)$ by $p(A) = GR$ for some R, and B is upper Hessenberg.

The algorithm starts by computing $x = p(A)e_1$. This vector satisfies $x_{m+1} \neq 0$, and $x_i = 0$ for $i > m + 1$. The next step is to determine a nonsingular matrix $G_0 = \operatorname{diag}\{\tilde{G}_0, I_{n-m-1}\}$, with $\tilde{G}_0 \in C^{(m+1)\times(m+1)}$, whose first column is proportional to x, and transform A to $A_0 = G_0^{-1}AG_0$. It is a simple matter to check that A_0 has the form shown in the diagram on the left.

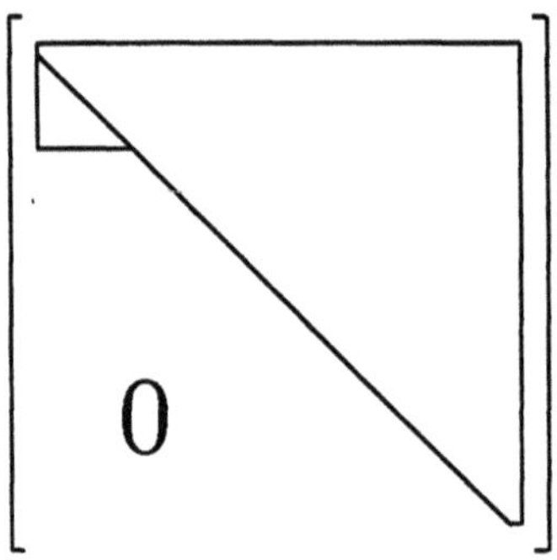

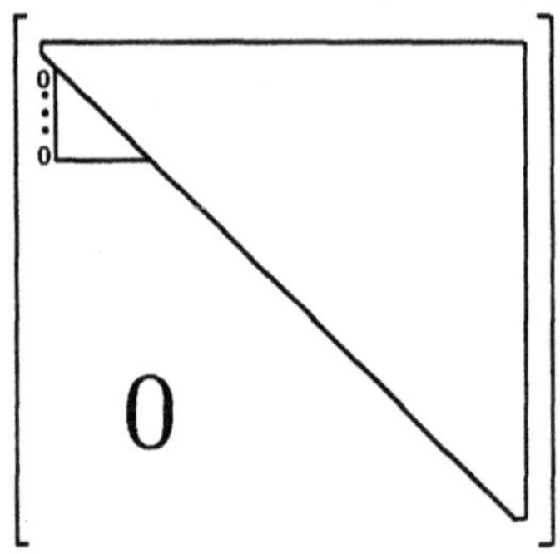

That is, it is almost in upper Hessenberg form, except that it has a triangular *bulge* with height m rows, base m columns, and vertex at position $(m + 2, 1)$. The rest of the algorithm consists of returning A_0 to upper Hessenberg form. The first step in this direction is to transform A_0 to $A_1 = G_1^{-1}A_0G_1$, where $G_1 = \operatorname{diag}\{1, \tilde{G}_1, I_{n-m-2}\}$, and $\tilde{G}_1 \in C^{(m+1)\times(m+1)}$ has the property $\tilde{G}_1^{-1}y_1 = ce_1$, where $y_1 \in C^{m+1}$ is the vector consisting of the $(2,1)$ through $(m + 2, 1)$ entries of A_0. It is a simple matter to verify that A_1 has the form shown in the diagram on the right. The bulge has been "chased" one position down and to the right, so that its vertex now lies at position $(m+3,2)$. The next step produces $A_2 = G_2^{-1}A_1G_2$, whose bulge has been moved down and over one more position, in analogous fashion. After $n-2$ such steps the bulge will have been chased off the bottom of the matrix. This completes the step. We have $G = G_0G_1G_2\cdots G_{n-2}$ and $B = A_{n-2} = G^{-1}AG$. It is also possible to devise algorithms which chase several columns at once, using tools such as the WY representation of reflectors [1], but we will not discuss that possibility here. To see that the chasing algorithm effectively performs a GR step of the form (1), (2), we will introduce some terminology and state two elementary lemmas. This material is nothing new, except that it now appears in a more general context.

Let $S \in C^{n\times n}$ be a nonsingular matrix with columns $s_1,\ldots,s_n$. Following the geometers we define the *flag* of S, denoted flag(S), to be the nested sequence of $n-1$ subspaces $\{\langle s_1\rangle, \langle s_1,s_2\rangle, \langle s_1,s_2,s_3\rangle,\ldots,\langle s_1,s_2,\ldots,s_{n-1}\rangle\}$ determined by the columns of S. It is a simple matter to prove the following lemma.

Lemma 1: Two nonsingular matrices $S, G \in C^{n\times n}$ have the same flag if and only if there is a nonsingular upper triangular matrix R such that $S = GR$.

Thus, for nonsingular $p(A)$, a GR decomposition $p(A) = GR$ produces a G for which flag$(G) =$ flag$(p(A))$.

Given $z \in C^n$ and $C \in C^{n\times n}$, the *Krylov matrix* $K(C,z) \in C^{n\times n}$ is defined by $K(C,z) = [z, Cz, C^2z,\ldots,C^{n-1}z]$. The following lemma is proved in [2, pp. 208-9].

Lemma 2: Let $z \in C^n$, $S, C, H \in C^{n\times n}$, suppose S is nonsingular, its first column is proportional to z, and $H = S^{-1}CS$. Then H is in irreducible upper Hessenberg form if and only if flag$(S) =$ flag$(K(C,z))$.

Now let's apply these results to the study of the chasing algorithm. We restrict our attention to the usual case, in which $p(A)$ is nonsingular. This implies that B has irreducible upper Hessenberg form. The singular case does not cause any problems; we will not discuss it here. Since $x = p(A)e_1$, and A commutes with $p(A)$, we see immediately that $p(A)K(A,e_1) = K(A,x)$. Since A has irreducible upper Hessenberg form, $K(A,e_1)$ is upper triangular and nonsingular. Therefore, by Lemma 1, flag($p(A)$) = flag($K(A,x)$). Because each of $G_1, \ldots, G_{n-2}$ has e_1 as its first column, the first column of G is the same as that of G_1, whence it is proportional to $x = p(A)e_1$. We have $B = G^{-1}AG$, and B is in irreducible upper Hessenberg form, so by Lemma 2, flag(G) = flag($K(A,x)$). Thus flag(G) = flag($p(A)$), which implies that $p(A) = GR$ for some upper triangular R. This proves that the generic chasing algorithm performs a step of the generic GR algorithm. The initial description of the GR algorithm specified that there must be a well-defined rule for performing GR decompositions. In the case of the chasing algorithm, the chasing algorithm is itself the rule.

In the chasing algorithm there are numerous ways of constructing the G_i. If we take them to be reflectors, we get the QR algorithm. If we use Gauss transformations with partial pivoting, we get the LR algorithm with partial pivoting. In some cases the choice of transformation is dictated by the structure of the matrix. For example, if A_0 is normal, and we wish to preserve that property, we should stick to the QR algorithm. If A_0 is Hamiltonian, and we wish to preserve that property, we take the G_i to be symplectic. This gives the SR algorithm. But let us consider the case that A_0 has no special structure to exploit. No matter how we choose the G_i, we are doing subspace iteration. The convergence theory suggests that our only consideration is to make the transforming matrices as well-conditioned as possible. Stability considerations would seem to argue for the same. We might then conclude that we should use only reflectors, which are optimally conditioned; that is, we should use the QR algorithm. This conclusion ignores the question of cost, but it is, in fact, the choice which has been made. It is an interesting choice. Consider for the moment a related problem, that of solving the linear system $Ax = b$. Two classes of direct methods for this problem are Gaussian elimination and the QR decomposition using reflectors. Here Gaussian elimination with partial pivoting has won the day, in spite of the fact that it is not unconditionally stable. The argument usually given is that Gaussian elimination costs half as much and works well in practice. In the case of the eigenvalue problem the opposite argument is made. The QR algorithm is unconditionally stable, so it should be used instead of LR, in spite of the fact that each LR step costs half as much. Of course the decision was based on practical experience; sometimes the LR algorithm just does not perform adequately. Nevertheless, I cannot help but believe that the LR algorithm still has a role to play. It "usually" does the job, and when it does, it is about twice as fast as QR. Perhaps what is needed is a hybrid algorithm, one which uses reflectors to get past the tough spots and Gauss transformations otherwise. After all, there is no reason why the two types of transformation cannot be mixed, even within a step. All that matters is that the condition numbers of the transforming matrices be kept under control.

Acknowledgment: This work was supported in part by the United States National Science Foundation under grant DMS-8800437.

References

[1] C. Bischof and C. Van Loan, *The WY representation for products of Householder matrices*, SIAM J. Sci. Stat. Comput., 8 (1987), pp. s2-s13.

[2] W. Bunse and A. Bunse-Gerstner, *Numerische lineare Algebra*, Teubner, Stuttgart, 1985.

[3] D. S. Watkins and L. Elsner, *Convergence of algorithms of decomposition type for the eigenvalue problem*, in preparation.

[4] D. S. Watkins and L. Elsner, *Chasing algorithms for the eigenvalue problem*, in preparation.

[5] J. H. Wilkinson, *The Algebraic Eigenvalue Problem*, Oxford, 1965.

Numerical Analysis of Nonlinear Equations in Computer Vision and Robotics

Layne T. Watson*

Spatial Data Analysis Laboratory
Virginia Polytechnic Institute & State University
Blacksburg, VA 24061, USA

Abstract

The need to apply sophisticated numerical analysis algorithms to computer vision and robotics problems is urgent, and these fields provide unique challenges different from the engineering problems normally encountered by numerical analysts. The opportunities for cross-fertilization are vast; within computer vision, facet modelling, surface approximation, three-dimensional object recognition, and range data analysis require splines, generalized polynomials, classical approximation theory, approximation in various norms, robust statistics, and fixed point theory. Fundamental vision problems such as shape from shading, structure from motion, consistent labelling, and surface segmentation involve nonlinear equations, nonlinear optimization, quasi-Newton and homotopy algorithms. Robot control, kinematics, and planning problems involve modern differential and algebraic geometry, modern control theory, computational geometry, and homotopy theory.

*This work was supported in part by Air Force Office of Scientific Research Grant 85–0250.

NATO ASI Series, Vol. F 70
Numerical Linear Algebra, Digital Signal Processing and Parallel Algorithms
Edited by G. H. Golub and P. Van Dooren

1 Introduction

This note selects four representative problems from computer vision and robotics, whose solution will require sophisticated mathematical and numerical techniques. These particular problems, while perhaps not unsolved, are closely related to major open questions in computer vision and robotics. The four problems, with the general area indicated in parentheses, are: 1) constrained transform coding (signal data compression), 2) line-based structure from motion (computer vision), 3) inverse manipulator kinematics (robotics), 4) surface facet modeling (image processing).

2 Constrained transform coding

Transform coding is a technique for (single or multiband) image data compression, qualitatively different from another technique known as differential pulse code modulation (DPCM). Roughly speaking, transform coding achieves compression by doing a factor analysis and then transmitting only the dominant factors. DPCM achieves compression by transmitting only variations in the data. Image data compression (where information is purposely lost) is to be distinguished from data encoding, where the goal is to transmit the data exactly but with fewer bits (no information lost).

Block transform techniques break the image up into blocks, and transform code each block. The reason why errors occur at block boundaries in the highly compressed transform coded images is apparent: there has never been any constraint in transform coding that the surfaces must match up at the block boundaries.

If a picture is transform coded block by block in a left-to-right, top-to-bottom scan, such a surface matching constraint requires that when the current block's surface is extended to the block on its left and to the block above, then these surfaces must match. Of course the matching concept can be extended to making the surfaces match up to an nth derivative. In any case, requiring that the left and top edges match leaves us in a situation where, knowing nothing more than the left and top boundaries of the current block's surface, we may interpolate or estimate the remainder of the block's surface. As in DPCM, the differences between the actual surface and the estimated surface can be transmitted. However, these differences do not have to be transmitted in a pixel-by-pixel manner as in DPCM. Rather they are transmitted parametrically as surface differences. The parameters are exactly analogous to the coefficients transmitted in a transform coding procedure. Thus we have defined a constrained transform coding procedure which by its nature is a combination of transform coding with differential pulse code modulation. The new procedure has the property that there will be no block boundary mismatch errors and it retains the coding efficiency advantage of transform coding.

Let T be a matrix whose columns are the desired basis vectors of some transform coding scheme. The columns of T need not be orthonormal. Let y be the vector representing the pixels in the block currently being transform coded and let α be the vector representing the transformed coefficients for the block y.

In accordance with the identity between discrete least squares function fitting and orthogonal projections, each column of T is a function sampled at the same specified values, which for the case of transform coding is just the spatial coordinates of the pixels in the block of y, and $T\alpha$ is the sampled values of the fitted surface at the pixel locations on the block. Now, these functions which are sampled to form the columns of T can be sampled at each of the spatial coordinates of the exterior top and left borders of the block y, thereby defining a matrix S having the same number of columns as T and having one row for each exterior top or left border pixel. The extrapolation of the estimated surface of the block y to the coordinates at the exterior top and left border pixels of y is then given as $S\alpha$.

Let z be the vector specifying the values of the surface at the exterior top and left border pixels of block y. Constraining the fitted surface to match at the top and left border of the block amounts to requiring that α minimize $\|z - S\alpha\|$. Therefore, the constrained transform coding problem is determining the coefficient vector α which minimizes $\|y - T\alpha\|$, subject to the constraint that α minimizes $\|z - S\alpha\|$ first.

It is shown in [1] that the α which achieves this constrained minimization can be represented as $\alpha_p + \alpha_h$ where α_p is any vector minimizing $\|z - S\alpha_p\|$ and $T\alpha_h$ is the orthogonal projection of $y - T\alpha_p$ onto T ker S. Calculating this α is a challenging numerical linear algebra problem.

Problems in surface approximation, image data compression, scene analysis, and other areas of computer vision frequently reduce to "nonstandard" linear algebra problems, of which constrained transform coding is just one example.

3 Line-based structure from motion

The recovery of 3-dimensional structural information from 2-dimensional image data is an inherently nonlinear process. The recovery problem may also be ill-posed, as optical illusions and ambiguous line drawings demonstrate. Two fundamentally different ways of recovering 3-dimensional structure are 1) based on light intensity variations recorded in the image (*shape from shading*), and 2) based on multiple views in space or time of the same scene (*structure from motion*). An example of the latter is considered here [2].

Given multiple views of a scene, the same objects must be found in different views. This is the infamous *correspondence problem*, and is highly nontrivial (for other than special situations, it remains an open question). Arguing that lines are easier to match between views than individual points, and assuming that lines have been matched correctly, one gets the problem described here for recovering 3-dimensional depth information (the "structure") from only 2-dimensional image screen coordinates. (Surfaces would be even easier to match between views, but correctly extracting surfaces (another open problem known as *segmentation*) is much harder than extracting lines.)

Consider then looking at a scene containing a certain number of *lines* from a certain number of *views*. Each viewing screen has its own coordinates, and we choose two points on the screen image of each line. These points are represented in the coordinates of the viewing screen, for each view. The x values give these coordinates. We assume we can match the lines in one view with their corresponding lines in the other views, but the points chosen on the lines in one view are not correlated to the points chosen on the lines in any other view. We wish to extract from this 2-dimensional view information 3-dimensional information about the lines in the scene.

The convention for the subscripts is:

subscript name:	view	line	coordinate	point
subscript symbol:	p	q	r	s
subscript values:	1,2,3	1,2,3,4	1,2	1,2

The direction vector for the qth line in the pth view is

$$\begin{aligned} L_{p,q} &= (L_{p,q,1}, L_{p,q,2}, L_{p,q,3}) \\ &= (\lambda_{p,q} x_{p,q,1,2} - x_{p,q,1,1}, \lambda_{p,q} x_{p,q,2,2} - x_{p,q,2,1}, (1-\lambda_{p,q})c). \end{aligned}$$

The cosine of the angle between the ith and jth lines using the measurements taken from the kth view is then

$$\Lambda_{k,(i,j)} = \frac{L_{k,i} \cdot L_{k,j}}{|L_{k,i}|\,|L_{k,j}|}.$$

Expressing the fact that these angle cosines are invariant yields

$$\Lambda^2_{1,(i,j)} = \Lambda^2_{2,(i,j)} = \Lambda^2_{3,(i,j)}, \quad 1 \le i < j \le 4.$$

This leads to 12 equations

$$\Lambda^2_{1,(i,j)} = \Lambda^2_{2,(i,j)}, \quad 1 \le i < j \le 4$$

and

$$\Lambda^2_{2,(i,j)} = \Lambda^2_{3,(i,j)}, \quad 1 \le i < j \le 4$$

in the 12 unknowns $\lambda_{p,q}$, $p = 1,2,3$; $q = 1,\ldots,4$.

$$x_{p,q,r,s}, \quad p = 1,2,3, \quad q = 1,\ldots,4, \quad r = 1,2, \quad s = 1,2,$$

and c (the focal length of the lens) are constants. $x_{p,q,r,s}$ is the measured image x ($r = 1$) or y ($r = 2$) coordinate for the first ($s = 1$) or second ($s = 2$) point on the qth line in the pth image.

The resulting polynomial system of equations, after clearing fractions and square roots, has total degree 8^{12}. Quasi-Newton methods applied to such polynomial systems fail totally, and a naive approach based on homotopy methods would require tracking 8^{12} paths, clearly impractical! Very sophisticated algebraic geometry theory and numerical analysis algorithms are going to be required to crack such problems.

4 Inverse manipulator kinematics

The inverse position problem (IPP) for six-revolute-joint manipulators is a basic problem in mechanisms. Given parameters describing a manipulator and a desired hand position and orientation, the problem is to find all the sets of joint angles that put the hand in this position and orientation. The IPP has the form $f(z) = 0$ where

$$\begin{aligned} f_l &= a_{l,1}z_1z_3 + a_{l,2}z_1z_4 \\ &+ a_{l,3}z_2z_3 + a_{l,4}z_2z_4 \\ &+ a_{l,5}z_5z_7 + a_{l,6}z_5z_8 \\ &+ a_{l,7}z_6z_7 + a_{l,8}z_6z_8 \\ &+ a_{l,9}z_1 + a_{l,10}z_2 \\ &+ a_{l,11}z_3 + a_{l,12}z_4 \\ &+ a_{l,13}z_5 + a_{l,14}z_6 \\ &+ a_{l,15}z_7 + a_{l,16}z_8 \\ &+ a_{l,17} \qquad (l = 1, \ldots, 4) \\ f_l &= z_{2l-9}^2 + z_{2l-8}^2 - 1 \qquad (l = 5, \ldots, 8) \end{aligned}$$

The values of the coefficients for various manipulator problems are given in [3]. We 2-homogenize via $z_j \leftarrow z_j/z_9$ for $j = 1,2,5,6$ and $z_j \leftarrow z_j/z_{10}$ for $j = 3,4,7,8$, letting z_9 and z_{10} be the 2-homogeneous variables. There are four equations of type (1,1), two of type (2,0), and two of type (0,2). The Bezout number is the coefficient of $\alpha_1^4\alpha_2^4$ in

$$D = (1\alpha_1 + 1\alpha_2)^4(2\alpha_1 + 0\alpha_2)^2(0\alpha_1 + 2\alpha_2)^2,$$

which in this case is 96.

The IPP might have an infinite number of physical solutions, in cases where the hand position and orientation can be attained by the free rotation of a joint and in other degenerate configurations. We focus on finding the geometrically isolated solutions. Usually these are nonsingular, but not always. Let P^k denote k-dimensional complex projective space. In P^8 the 1-homogeneous form of the system has total degree 256, an infinite number of solutions at infinity, and (we can prove) at most 64 finite solutions (unless there are an infinite number), while we observe at most 32 finite solutions (unless there are an infinite number). (The complicated formulas by which manipulator data generates polynomial coefficients make exact analysis here difficult.) In $P^4 \times P^4$ the 2-homogeneous form of the system has Bezout number 96, 8 solutions at infinity (each of multiplicity 4), and (we observe) 32 finite solutions and 32 (additional) solutions at infinity. Numerical results confirm that the problem should be solved in its 2-homogeneous form rather than the 1-homogeneous.

The IPP is solved by a globally convergent homotopy algorithm [4, 5] using a homotopy map of the form $\rho(\lambda, z) = (1 - \lambda)g(z) + \lambda f(z)$. There is some choice for g. We could simply choose a 2-homogeneous g with 96 nonsingular solutions. We might do this if we wanted to solve the IPP only once. However, if we wish to solve this problem many times, we can create a much more efficient homotopy. It turns out that $f(z) = 0$ has exactly the same set of 8 multiplicity-four solutions at infinity in $P^4 \times P^4$, independent of the choice of coefficients. Therefore we can choose g to have the same form as f and ignore these solutions as start points. This leaves just 64 paths to track. We obviate the issue of finding a g that is easily solvable by simply choosing one and solving it using another homotopy. This becomes the start system for all other IPP systems.

Note what was required in the above discussion: globally convergent homotopy theory (neither quasi-Newton nor simple continuation methods suffice for polynomial systems), advanced algebraic geometry to reformulate the problem in $P^4 \times P^4$ (instead of the natural C^8), and a synthesis of these two areas to prove that the Bezout count of 96 paths can be effectively reduced to 64 paths.

5 Surface facet modelling

Digital images can be thought of as noisy samples of some underlying intensity surface, which is composed of *facets*. A fundamental problem is how to compute this underlying intensity surface, its facets, and how to group these facets into *regions* comprising *objects* [6]. To go from facets to regions, we must make explicit use of the sensing physics. We assume that the camera's optical axis is in the y-direction and that the origin is at the camera lens. The camera observes a Lambertian surface $y = s(x, z)$ having reflectivity coefficient $r(x, z)$.

We assume that the illumination is part distant point source and part diffuse. The diffuse part can illuminate all parts of the Lambertian surface $s(x, z)$ equally well. The point source can illuminate only those parts of the surface $s(x, z)$ that are not occluded. We let $I_1(x, z)$ be a 0-1 valued function such that $I_1(x, z) = 1$ when the surface point $(x, s(x, z), z)$ is not occluded and $I_1(x, z) = 0$ when the point is occluded. We let $I_0(x, z)$ be a constant.

Each surface point $(x, s(x, z), z)$ has perspective projection (x', z') on the image where $x' = Fx/s(x, z)$, $z' = Fz/s(x, z)$, and F is the distance that the image plane lies in front of the camera. (The analysis is easier to understand using the dual plane to the plane that the film actually lies in.) The received irradiance at (x', z') is $I_1(x, z)r(x, z)\cos\theta + I_0 r(x, z)$ where θ is the angle between the surface normal and the illumination direction.

This equation gives us some information about the surfaces and reflectances involved if we know what kind of boundary is causing the observed differences. Consider computing the value $(lighter - darker)/darker$ across some boundary produced by the facet model. Then if the boundary is due to reflectance change alone, we have:

$$\begin{aligned}\frac{lighter - darker}{darker} &= \frac{(I_1 r_1 \cos\theta + I_0 r_1) - (I_1 r_2 \cos\theta + I_0 r_2)}{I_1 r_2 \cos\theta + I_0 r_2} \\ &= \frac{(I_1 \cos\theta + I_0)(r_1 - r_2)}{(I_1 \cos\theta + I_0) r_2} = \frac{r_1 - r_2}{r_2}\end{aligned}$$

If the boundary is due to shadow alone, we have:

$$\frac{lighter - darker}{darker} = \frac{(I_1 r \cos\theta + I_0 r) - I_0 r}{I_0 r} = \frac{I_1}{I_0}\cos\theta.$$

If the boundary is due to shape curvature alone, we have:

$$\begin{aligned}\frac{lighter - darker}{darker} &= \frac{(I_1 r \cos\theta_1 + I_0 r) - (I_1 r \cos\theta_2 + I_0 r)}{I_1 r \cos\theta_2 + I_0 r} \\ &= \frac{I_1 r(\cos\theta_1 - \cos\theta_2)}{I_1 r \cos\theta_2 + I_0 r} = \frac{\cos\theta_1 - \cos\theta_2}{\cos\theta_2 + I_0/I_1}\end{aligned}$$

The facet model permits regions to be identified which are not constant in irradiance, but which may be slowly changing. If these slow changes can be assumed to be due to curvature alone (no reflectance or illumination changes), then it is possible to learn something about the shape of the surfaces from the irradiance changes. However, the knowledge we seek from this kind of relationship is not as total as the shape from shading reconstruction which Horn attempted. We only seek qualitative relationships about how the shape causes the observed irradiance change and, therefore, examine the partial derivatives on the image assuming there is no illumination or reflectance change on the object. We expect that the direction of highest number in irradiance on the image at a particular pixel corresponds to the position on the object having a surface normal that is turning toward the illumination direction.

Letting the received irradiance at the point (u,v) be $J(u,v)$, we obtain that the partial derivative of J with respect to u is

$$J_u = -I_1 r \sin\theta\,(\theta_x x_u) - I_1 r \sin\theta\,(\theta_z z_u) \quad \text{where} \quad \cos\theta = \frac{a_x s_x - a_y + a_z s_z}{\sqrt{s_x^2 + s_z^2 + 1}};$$

the point illumination source comes from direction (a_x, a_y, a_z); θ_x, θ_z, x_u, z_u, s_x, and s_z denote respective partial derivatives; and $u = Fx/s(x,z)$ and $v = Fz/s(x,z)$. Expanding the resulting expression yields:

$$\begin{aligned} J_u \;=\; & \frac{I_1 r(a_x, a_y, a_z)}{F(s - z s_z - x s_x)} \left\{ s(s - z s_z)\frac{\partial}{\partial x}\left[\begin{pmatrix} s_x \\ -1 \\ s_z \end{pmatrix} \frac{1}{\sqrt{s_x^2 + s_z^2 + 1}}\right] \right. \\ & \left. + s z s_x \frac{\partial}{\partial z}\left[\begin{pmatrix} s_x \\ -1 \\ s_z \end{pmatrix} \frac{1}{\sqrt{s_x^2 + s_z^2 + 1}}\right]\right\}. \end{aligned}$$

There is a similar expression for J_v. Thus, the observed irradiance partial derivative in the u-direction on the image is proportional to a linear combination of the surface normal velocities times the cosine of the angle between the point source illumination and the direction of the linear combination of surface normal velocities.

These partial differential equations are a complete and accurate description of the surface $s(x,z)$, but their solution remains an open problem.

References

[1] L. T. Watson, R. M. Haralick, and O. A. Zuniga, Constrained transform coding and surface fitting, IEEE Trans. Commun., 31 (1983), 717–726.

[2] A. Mitiche, S. Seida, and J. K. Aggarwal, Line-based computation of structure and motion using angular invariance, in Proc. Workshop on Motion: Representation and Analysis, Kiawah Island, SC, May, 1986, IEEE Computer Society Press.

[3] A. P. Morgan, A. Sommese, and L. T. Watson, Finding all solutions to polynomial systems using HOMPACK, General Motors Technical Report GMR-6109, GM Research Labs, Warren, MI 48090, 1988.

[4] L. T. Watson, Numerical linear algebra aspects of globally convergent homotopy methods, SIAM Review, 28 (1986), 529–545.

[5] L. T. Watson, S. C. Billups, and A. P. Morgan, HOMPACK: A suite of codes for globally convergent homotopy algorithms, ACM Trans. Math. Software, 13 (1987), 281–310.

[6] R. M. Haralick, From facets to regions, manuscript, private communication, 1985.

A Linear Systolic Array for the Adaptive MVDR Beamformer

B. Yang and J.F. Böhme

Lehrstuhl für Signaltheorie
Ruhr-Universität Bochum
4630 Bochum, FRG

Abstract

We propose an efficient algorithm and a linear systolic array for the adaptive MVDR (Minimum Variance Distortionless Response) beamformer. We show that the beamformer output can be computed in a fully pipelined fashion without determining the optimum weight vector. For N antenna elements and M look directions to which the beamformer should be steered, we require $O(N^2/2+MN)$ operations[1] per sample time.

1 Introduction

The MVDR beamformer [1]–[4] is an optimum beamformer in the sense of minimum variance distortionless response. It minimizes the beamformer output power subject to a linear constraint of unity response for a specific look direction. For time varying signal environments, adaptive beamformers are required. This involves the solution of a linearly constrained adaptive least squares problem.

In a previous paper, Schreiber [1] proposed an adaptive algorithm. It is based on updating the Cholesky factorization of the spectral density

[1]For easy comparison of the computational amounts of different algorithms, we do not count the scaling operations due to the forgetting factors for adaptive beamforming throughout this paper.

NATO ASI Series, Vol. F 70
Numerical Linear Algebra, Digital Signal Processing and Parallel Algorithms
Edited by G. H. Golub and P. Van Dooren

matrix. However, he needed solving a triangular equation system and computing vector inner products, in addition. These operations can not be implemented by a single systolic array.

In a recent paper, Bojanczyk and Luk [2] derived an alternative algorithm to overcome this problem. Their approach requires $O(N^2/2 + 3MN^2/2)$ operations per sample time for steering the beamformer with N antenna elements to M different look directions.

In this paper, we show a more efficient method [3]. Based on a modification and generalization of the Schreiber's algorithm, our approach eliminates the needs of solving linear equation systems and computing vector inner products. It requires $O(N^2/2 + MN)$ operations per sample time. The algorithm can be implemented either by a trapezoidal systolic array, such as proposed by McWhirter and Shepherd [4], or by a linear one [3].

2 The Adaptive MVDR Beamformer

We assume an antenna array of N sensor elements and of known but arbitrary geometry. We use the frequency domain formulation. Let $x(\omega)$ be the N-element complex column vector of the sensor outputs for a particular frequency ω. The beamformer output is a weighted sum of the sensor outputs

$$y(\omega, \vartheta) = w(\omega, \vartheta)^H x(\omega) \quad ,$$

where $w(\omega, \vartheta)$ is the complex weight vector to be suitably chosen for steering to the bearing angle ϑ at frequency ω. The superscript H denotes the complex conjugate transposition. The expected beamformer output power is given by

$$E|y(\omega, \vartheta)|^2 = w(\omega, \vartheta)^H E[x(\omega)x(\omega)^H] w(\omega, \vartheta) = w(\omega, \vartheta)^H R(\omega) w(\omega, \vartheta) \quad ,$$

where $R(\omega)$ denotes the spectral density matrix of the array data. The MVDR beamformer is defined as follows,

$$\begin{aligned} \text{minimize} \quad & E|y(\omega, \vartheta)|^2 \quad \text{over} \quad w(\omega, \vartheta) \\ \text{subject to} \quad & w(\omega, \vartheta)^H d(\omega, \vartheta) = 1 \quad . \end{aligned}$$

The steering vector $d(\omega, \vartheta)$ corresponds to the ideal sensor output vector $x(\omega)$ which we obtain for a single source at angle ϑ. Thus, by maintaining the beamformer response at a proper level for signals arriving from ϑ, all contributions to the beamformer output from noise and interferences are minimized.

Solving the above linearly constrained least squares problem, we find the beamformer output

$$y(\omega,\vartheta) = \frac{d(\omega,\vartheta)^H R(\omega)^{-1} x(\omega)}{d(\omega,\vartheta)^H R(\omega)^{-1} d(\omega,\vartheta)} \quad . \tag{1}$$

$R(\omega)$ is assumed to be positively definite. In the following, we omit the frequency ω and bearing ϑ arguments unless they are needed explicitly.

For adaptive beamforming purposes, R is usually estimated by the rank-one update

$$R(t) = \mu_1 R(t-1) + \mu_2 x(t) x(t)^H \quad , \tag{2}$$

where t is a time index and $0 < \mu_1, \mu_2 < 1$ are two forgetting factors to be suitably chosen. Equation (2) has the nice property that $R(t)$ is positively definite if $R(t-1)$ is such, too.

3 Update Algorithm

In most applications, we are only interested in the resulting beamformer output rather than the optimum weight vector. For this case, Schreiber [1] proposed an adaptive algorithm based on the Cholesky factorization of $R(t)$

$$R(t) = L(t) L(t)^H \quad .$$

$L(t)$ is a complex, lower triangular matrix with positive diagonal elements. Defining the new variables,

$$\tilde{d}(t) = L(t)^{-1} d \, , \quad \tilde{x}(t) = L(t)^{-1} x(t) \, , \quad D(t) = |\tilde{d}(t)|^2 \, ,$$

the beamformer output (1) can be expressed by

$$y(t) = \tilde{d}(t)^H \tilde{x}(t) / D(t) \quad . \tag{3}$$

Suppose that we know $L(t-1)$ and $x(t)$. The new Cholesky factor $L(t)$ can be obtained by applying an unitary transformation to a suitably constructed matrix

$$U(t) \begin{bmatrix} \sqrt{\mu_1} L(t-1)^H \\ \sqrt{\mu_2} x(t)^H \end{bmatrix} = \begin{bmatrix} L(t)^H \\ 0 \end{bmatrix} \quad . \tag{4}$$

The unitary matrix $U(t)$ ($U(t)^H U(t) = I_{N+1}$) is designed to eliminate the row vector $\sqrt{\mu_2}x(t)^H$ while preserving the triangular structure of $L(t-1)^H$. This can be accomplished, for example, by using a sequence of N complex Givens rotations

$$U(t) = U_N(t) \cdots U_2(t) U_1(t) \quad .$$

Each of them consists of a scaling and a "true" plane rotation

$$U_i(t) = \begin{bmatrix} I_{i-1} & & & \\ & \cos\theta_i & & \sin\theta_i \\ & & I_{N-i} & \\ & -\sin\theta_i & & \cos\theta_i \end{bmatrix} \begin{bmatrix} I_N & 0 \\ 0 & e^{-j\varphi_i} \end{bmatrix} \qquad (i = 1, \ldots, N) \quad . \tag{5}$$

We first scale the last row of the left hand side matrix in (4) with $e^{-j\varphi_i}$ to make its i-th element real. This corresponds to a real Givens rotation applied to the real and the imaginary parts. Then, we apply a real Givens rotation to the i-th and the last rows for eliminating the element which has been previously made real. Note that $L(t-1)^H$ is upper triangular, and all its diagonal elements are real and positive.

Schreiber also showed that the same unitary transformation $U(t)$ can be used to update $\tilde{d}(t)$ and $D(t)$,

$$U(t) \begin{bmatrix} \tilde{d}(t-1)/\sqrt{\mu_1} \\ 0 \end{bmatrix} = \begin{bmatrix} \tilde{d}(t) \\ z(t) \end{bmatrix} \quad ,$$

$$D(t) = D(t-1)/\mu_1 - |z(t)|^2 \quad .$$

These equations follow from the orthogonal property of $U(t)$. That is, for

$$U(t)[v_1 \; v_2] = [\tilde{v}_1 \; \tilde{v}_2] \quad ,$$

we have

$$v_i^H v_j = \tilde{v}_i^H \tilde{v}_j \qquad (i, j = 1, 2) \quad . \tag{6}$$

However, for computing the beamformer output (3), solving a triangular equation system $\tilde{x}(t) = L(t)^{-1}x(t)$ and computing the inner product $\tilde{d}(t)^H \tilde{x}(t)$ are still required.

These operations can be avoided if we apply $U(t)$ to an additional column vector

$$U(t) \begin{bmatrix} 0 \\ 1/\sqrt{\mu_2} \end{bmatrix} = \begin{bmatrix} \tilde{x}(t) \\ \alpha(t) \end{bmatrix} \quad .$$

By exploiting (6), we can show that $\tilde{x}(t)$ computed above equals $L(t)^{-1}x(t)$, and the vector inner product $\tilde{d}(t)^H\tilde{x}(t)$ can be replaced by a simple scalar multiplication $-\alpha(t)z(t)^H$. The beamformer output (3) can thus be computed by

$$y(t) = -\alpha(t)z(t)^H/D(t) \quad .$$

Now, we assume that the beamformer should be steered to M different look directions ϑ_i corresponding to M steering vectors $d_i = d(\vartheta_i)$ $(i = 1,\ldots,M)$. By attaching the bearing index i to each direction dependent quantity, the adaptive MVDR beamforming algorithm can be summarized as follows.

Initial step : initialize $L(0)^H$, $\tilde{d}_i(0)$, $D_i(0)$ $\quad (i=1,\ldots,M)$
For $t = 1,2,\ldots$ do
1. Scaling operations :

$$\sqrt{\mu_1}\cdot L(t-1)^H, \quad \sqrt{\mu_2}\cdot x(t)^H, \quad (1/\sqrt{\mu_1})\cdot\tilde{d}_i(t) \qquad (i=1,\ldots,M) \tag{7}$$

2. Unitary transformation :

$$U(t)\begin{bmatrix} \sqrt{\mu_1}L(t-1)^H & 0 & \tilde{d}_1(t-1)/\sqrt{\mu_1} & \cdots & \tilde{d}_M(t-1)/\sqrt{\mu_1} \\ \sqrt{\mu_2}x(t)^H & 1/\sqrt{\mu_2} & 0 & \cdots & 0 \end{bmatrix}$$
$$= \begin{bmatrix} L(t)^H & \tilde{x}(t) & \tilde{d}_1(t) & \cdots & \tilde{d}_M(t) \\ 0 & \alpha(t) & z_1(t) & \cdots & z_M(t) \end{bmatrix} \tag{8}$$

3. Beamformer output computations :

$$D_i(t) = D_{i-1}(t)/\mu_1 - |z_i(t)|^2 \ , \tag{9}$$

$$y_i(t) = -\alpha(t)z_i(t)^H/D_i(t) \qquad (i=1,\ldots,M) \tag{10}$$

End

We remark that by computing the quantities $D_i(t) = d_i^H R(t)^{-1} d_i$, the above algorithm also implements the Capon's beamformer [5] calculating the inverse $1/D_i(t)$ of the estimated beamformer output power. The order of the numerical operations per sample time is $O(N^2/2 + MN)$ if we do not include the scaling operations (7) into account.

4 A Linear Systolic Array

McWhirter and Shepherd [4] derived independently a similar algorithm as above for the adaptive MVDR beamformer. They proposed a trapezoidal systolic array whose array size increases with the number of the look directions. An alternative approach is the use of a linear systolic array with a fixed array size.

Figure 1 shows a linear systolic array implementing the adaptive MVDR beamformer. It consists of two rows of processing elements (PEs). The upper one performs the scaling operations (7). The lower one contains two different types of PEs, P2 and P3, which implement the unitary transformation (8) and the scalar operations (9) and (10). Each of the N PEs P2 carries out one of the N complex Givens rotations (5). At first, the two rotation angles φ_i and θ_i are calculated and stored in the i-th ($i = 1, \ldots, N$) PE P2 from left. Then, the same complex Givens rotation with the stored angles is applied to successive input data pairs for eliminating the i-th element of x. Note that two computation modes for the PEs P2, computing the rotation angles and performing rotation operations, are required. This is controlled by a control bit VR which flows with the most left input data column through the PEs. This control bit is not drawn in Figure 1. By setting and delaying it suitably, we can show that all N elements of x are successively eliminated in the N PEs P2.

We remark that the PE P2 can be realized by a CORDIC (COordinate Rotation DIgital Computer) processor [3,6] which is especially suitable for performing rotation-like operations. Based on a iterative shift-add operations, it can perform a variety of functions of the initial values x_0, y_0 and z_0 (table 1). For $m = 1$, we see that the CORDIC processor performs just the desired operations. In this case, the control bit VR for the systolic array is identical to the programming bit VR for the CORDIC processor.

Finally, we mention that the above systolic array can also be used for beamforming at multiple frequencies. For this purpose, the input data are arranged one after another for distinct frequencies and pass successively through the systolic array.

	$m = 1$	$m = 0$	$m = -1$
VR=1	$x = \sqrt{x_0^2 + y_0^2}$ $z = z_0 + \epsilon \arctan(y_0/x_0)$	$x = x_0$ $z = z_0 + \epsilon y_0/x_0$	$x = \sqrt{x_0^2 - y_0^2}$ $z = z_0 + \epsilon \text{arctan}(y_0/x_0)$
VR=0	$x = x_0 \cos z_0 - \epsilon y_0 \sin z_0$ $y = y_0 \cos z_0 + \epsilon x_0 \sin z_0$	$x = x_0$ $y = y_0 + \epsilon x_0 z_0$	$x = x_0 \cosh z_0 + \epsilon y_0 \sinh z_0$ $y = y_0 \cosh z_0 + \epsilon x_0 \sinh z_0$

Table 1: CORDIC functions $(\epsilon = \pm 1)$

References

[1] Schreiber, R., "Implementation of Adaptive Array Algorithms", IEEE Trans. ASSP, vol.34, no.5, pp.1038–1045, Oct. 1986

[2] Bojanczyk, A.W. and Luk, F.T., "A Novel MVDR Beamforming Algorithm", Proc. SPIE, Advanced Algorithms and Architectures for Signal Processing II, vol.826, pp.12–16, San Diego, 1987

[3] Yang, B. and Böhme, J.F., "Systolic Implementation of a General Adaptive Array Processing Algorithm", Proc. IEEE ICASSP, pp.2785–2789, New York, April 1988

[4] McWhirter, J.G. and Shepherd, T.J., "An Efficient Systolic Array for MVDR Beamforming", Proc. IEEE Int. Conf. on Systolic Arrays, May 1988

[5] Capon, J., "High Resolution Frequency-Wavenumber Spectrum Analysis", Proc. IEEE, vol.57, pp.1408–1418, 1969

[6] Walther, J.S., "A Unified Algorithm for Elementary Functions", IRE Trans., vol.EC-8, no.3, pp.330–334, 1959

Solving Band Systems of Linear Equations by Iterative Methods on Vector Machines

Zahari Zlatev

Air Pollution Laboratory
Danish Agency of Environmental Protection
Ris National Laboratory
DK-4000 Roskilde, Denmark

Abstract

The use of iterative methods in the solution of systems of linear algebraic equations whose coefficient matrices are band is an important problem, because such systems do appear in many fields of engineering and science. Band matrices, symmetric and positive definite as well as unsymmetric, can efficiently be treated on many vector machines. The possibility of exploiting zero diagonals within the band in order to reduce the storage and the computing time will be discussed (also such matrices appear very often when large-scale problems are to be handled). The treatment of some special matrices, as tri-diagonal and five-diagonal matrices, will be sketched. It will be shown that on vector machines iterative methods may perform better than the direct methods even in the case where the direct methods are better than the iterative methods on scalar machines. The reason for this phenomenon is the fact that many iterative methods can be designed so that one works with long and contiguous arrays during the whole computational process. This can not be achieved when systems of linear algebraic equations with band matrices are solved by direct methods.

NATO ASI Series, Vol. F 70
Numerical Linear Algebra, Digital Signal Processing and Parallel Algorithms
Edited by G. H. Golub and P. Van Dooren

1 Band Systems of Linear Algebraic Equations

Band systems of linear algebraic equations appear very often when partial differential equations are discretized. These systems are as a rule very large; in some air pollution models for studying long range transport phenomena the number of equations being greater than 250000. Therefore it is important to be able to solve such systems efficiently on vector processors. Efficiency can be achieved by reorganizing the computations, so that the calculations are carried out by diagonals. If this is done, then one is working with both **long** and **contiguous** arrays when such systems are solved by iterative methods. Computations with both long and contiguous arrays can not be carried out when direct methods are to be used. In the latter case the computations could still be organized by the use of contiguous arrays, but the arrays will necessarily be short. Therefore the iterative methods for solving band systems could be implemented on a high level of performance on vector processors, while the speed of computations is normally not very high when direct methods are selected. In this way the iterative methods might perform better on vector processors even in the case where the direct methods perform better than the iterative methods when the same systems are solved on scalar machines.

The matrices that appear after the discretization of partial differential equations are not only band; furthermore these normally contain many zero diagonals within the band. This special kind of sparsity can easily be exploited (without a great reduction of the speed of computations). If this is done, then the efficiency is, as a rule, increased considerably. The iterative methods are clearly more attractive in this situation also, because while fill-ins do appear when direct methods are in use and when the sparsity is exploited, no fill-in appears when an iterative method is chosen and when the sparsity is exploited.

The use of modules that perform efficiently basic linear algebra operations plays a crucial role in many modern programs for solving systems of linear algebraic equations. Such modules, basic linear algebra subprograms (**BLAS**'s), were first proposed by **Lawson et al.** [4] and then by **Dongarra et al.** [2]. The latter set of modules is often called **BLAS 2**. For the modules in **BLAS 2** it is typical that: the matrix-vector multipliers for dense matrices are much more efficient than the matrix-vector multipliers for band matrices. Consider a matrix of order 1024 with bandwidth 65. If the matrix is considered as dense (i.e. if the bandedness of the matrix is ignored), then the speed of computations, measured in **MFLOPS**, is about **160**. If the bandedness is exploited, then the speed of computations is only about **30 MFLOPS** (in both cases it is assumed that a **CRAY M-XP** computer is used). This is so because the computations in the **BLAS 2** modules are organized by columns. While this is a good strategy when the matrix is dense, the situation changes when the matrix is band. In the latter case the arrays involved in the computations (the parts of the columns within the band) are normally short. If a direct method is to be used, then probably one can not do anything better. However, if iterative methods are to be applied, then it may be better to organize the computations by diagonals in order to achieve computations with both long and contiguous arrays. This technique is described in **Zlatev** [6] (for general band matrices) and in **Zlatev et al.** [7] (for symmetric matrices). The speed of computations achieved by the modules developed by the use of the techniques described in [6,7] is much higher than the corresponding modules in **BLAS 2** and it is comparable with the speed of computations of the modules for dense matrices in **BLAS 2** (which is an excellent result; no more could be achieved). Even when the sparsity is exploited the speed of computations remains very high (because indirect addresses are used only in outer loops that do not vectorize anyway). These statements are illustrated by many numerical experiments in **Zlatev** [6] and **Zlatev et al.** [7].

The organization of the calculations by diagonals has an additional advantage when band matrices are to be handled. It is very easy to exploit, in an efficient way, all processors in parallel when a multiprocessor, such as, for example, **CRAY X-MP48**, is in use. This is demonstrated in **Zlatev et al.** [7].

In **Section 2** the efficiency of the use of iterative methods for solving systems with symmetric and positive matrices together with the modules developed in **Zlatev et al.** [7] will be demonstrate by numerical experiments. In **Section 3** some preliminary results obtained in the solution of systems with general band matrices will be presented. In **Section 4** the exploitation of some special properties of the matrices under consideration, tri-diagonal and five-diagonal matrices, will shortly be discussed. Some conclusions will be drawn in the last section, **Section 5**. Before the start of the discussion it should be mentioned that calculations by diagonals have also been used, in another context, by **Madsen et al.** [5].

2 Symmetric and Positive Definite Matrices

The use of a conjugate gradient method in the solution of systems of linear algebraic equations with band, symmetric and positive definite matrices is discussed in **Zlatev et al.** [7]. The particular conjugate gradient algorithm chosen in [7] is that used in **Dubois et al.** [3]. Preconditioning can also be applied with this algorithm, but no preconditioning is used in this study (see also [3] and [7]). The speeds of computations on an **AMDAHL VP1100** are presented in **Table 1** (the case where the sparsity the matrix bands is not exploited) and in **Table 2** (the case where the sparsity of the matrix band is exploited).

The matrices used in the experiments are obtained as follows. Assume that NBANDL and NBANDU are the bandwidths of the lower and upper bands respectively. Assume also that NBANDL is already chosen. Set NBANDU = NBANDL, NBAND = NBANDL+NBANDU+1 and N = $(\text{NBANDL})^2$. Let δ be an arbitrary real number. Consider a matrix of order N such that

(a) all its elements on the main diagonal are equal to four,

(b) the elements in the diagonals whose first elements are a_{12} and a_{1k} with k = NBANDU are equal to $-1-\delta$,

(c) the elements in the diagonals whose first elements are a_{21} and a_{k1} with k = NBANDL are equal to $-1+\delta$,

(d) all other elements are equal to zero. The matrices so defined are in general neither symmetric nor positive definite. However, if $\delta = 0$ is selected, then the matrices become both symmetric and positive definite, and such matrices are used in this section.

N	NBAND	LINPACK	BLAS 2	DEPTH 0
100	21	1	8	45
400	41	2	15	79
900	61	3	20	93
1600	81	4	24	103
2500	101	4	29	107
3600	121	5	34	111
4900	141	6	38	113
6400	161	7	42	114
8100	181	7	45	114
10000	201	8	49	115

DEPTH 2	DEPTH 4	DEPTH 8	DEPTH 16
43	43	45	43
102	120	125	126
125	152	163	166
137	163	188	196
150	175	199	216
155	181	207	222
154	183	214	229
156	185	217	234
158	185	216	234
159	187	219	239

Table 1
The speeds of computations obtained in solving systems of linear algebraic equations with symmetric and positive definite matrices when the sparsity of the matrix bands is not exploited.

N	NBAND	LINPACK	BLAS 2	DEPTH0	DEPTH2	DEPTH4	DEPTH8	DEPTH16
100	21	1	8	35	32	32	31	30
400	41	2	15	79	85	85	84	82
900	61	3	20	97	112	114	114	111
1600	81	4	24	113	127	129	127	124
2500	101	4	29	120	142	143	142	138
3600	121	5	34	125	143	147	146	142
4900	141	6	38	132	146	153	152	149
6400	161	7	42	130	149	152	152	152
8100	181	7	45	133	152	153	152	155
10000	201	8	49	135	152	155	156	156

Table 2
The speeds of computations obtained in solving systems of linear algebraic equations with symmetric and positive definite matrices when the sparsity of the matrix bands is exploited.

It is immediately seen that unrolling with depth greater than two is very efficient when the matrix band is dense, while such unrolling has practically no effect when the matrix band is sparse and its sparsity is exploited. In the latter situation this is so, because the matrices used in these experiments contain two non-zero diagonals only (in the upper band as well as in the lower band). If the number of non-zero diagonals in the upper band (and due to the symmetry also in the lower band) is greater than two, then unrolling with depth greater than two will become more efficient also when the sparsity is exploited.

Assume that no unrolling is used. Then the speed of computations is greater when the sparsity is exploited. If only the kernels for performing matrix-vector multiplications are used (with no unrolling), then the speed of computations is higher (about 10%) when the sparsity is not exploited, because indirect addresses are used when the sparsity is exploited. If systems of linear algebraic equations are solved, however, not only matrix-vector multiplications are to be carried out. Some operations involving vectors are also to be performed. These operations are carried out very efficiently on vector machines. While the matrix-vector multiplications are normally the dominating part of the computational work when the sparsity is not exploited, this is not so when the sparsity is exploited. In the latter situation the relative part of the other operations is often rather considerable and this explains why the speed of the sparse version is greater. If unrolling is performed, then this advantage is lost. Now the speed of the matrix-vector multiplier becomes greater than the speed of the other operations. Because of the use of indirect addresses, the speed of computations for the sparse matrix-vector multiplier is smaller than that of the dense matrix-vector multiplier (the other operations being performed with the same speed in both versions). This explains why the speed of computations of the unrolled sparse version is slower than the speed of computations of the unrolled dense version.

The high speeds of computations, achieved when calculations by diagonals are carried out, do not guarantee better performance of the iterative methods. The computing times should also be compared. This is done in **Table 3**. It is seen that the subroutines based on calculations by diagonals perform considerably better than the subroutines based on direct methods (the **LINPACK**, [1], subroutines **SPBFA** and **SPBSL**) and the subroutines based on the **BLAS 2** matrix-vector multiplier. It should be mentioned here that the same iterative method is used in the solver in which matrix-vector products are calculated by diagonals and in the solver based on the **BLAS 2** matrix-vector multiplier. Therefore the numbers of iterations spent are the same for the **BLAS 2** solver and the diagonal solver (for the runs of the results of which are shown in Table 1 - Table 3 the number of iterations varies between 23 and 174).

N	NBAND	LINPACK	BLAS 2	SPARSITY NOT EXPLOITED	SPARSITY EXPLOITED
100	21	0.02	0.02	<0.01	<0.01
400	41	0.10	0.12	0.01 (8.3%)	<0.01
900	61	0.35	0.41	0.05 (12.2%)	0.01 (2.4%)
1600	81	0.83	0.94	0.12 (12.8%)	0.02 (2.1%)
2500	101	1.62	1.89	0.25 (13.2%)	0.04 (2.1%)
3600	121	2.79	3.30	0.50 (15.1%)	0.06 (1.8%)
4900	141	4.45	5.43	0.89 (16.4%)	0.10 (1.8%)
6400	161	6.63	7.68	1.38 (18.0%)	0.13 (1.7%)
8100	181	9.48	11.30	2.19 (19.4%)	0.19 (1.7%)
10000	201	13.00	14.86	3.04 (20.5%)	0.24 (1.6%)

Table 3

The computing times, measured in seconds, on AMDAHL VP1100. The results obtained when unrolling with depth 16 is applied are given in the case where the sparsity is not exploited. When the sparsity is exploited, the results given in this table are obtained using unrolling with depth 4. The percentages are obtained by using the BLAS 2 computing times.

Consider the case where the sparsity of the matrix band is not exploited. The comparison of the results obtained by the **LINPACK** solver with the results obtained by the solver in which the computations are organized by diagonals (and unrolling with depth 16) shows that the speed of computations in the latter case is about 40 times greater, while the computing time is reduced by a factor 5-6. This means that on a scalar machine, where the speed of computations will, roughly speaking, be the same in both situations, one should expect the **LINPACK** solver to perform better with regard to the computing time needed. However, in the situation where the sparsity is exploited, the performance of the **LINPACK** solver will still be considerably slower than the diagonal solver when a scalar computer is in use in connection with matrices of this class.

Unrolling with depth 16 gives best results when an **AMDAHL VP1100** computer is used (and when the sparsity of the matrix band is not exploited). On **CRAY X-MP48** best results are achieved when unrolling with depth 8 is applied. Unrolling with depth 4 gives best results when an **ISP SPERRY** is used. If the sparsity is exploited, then unrolling has no great effect for this class of matrices. There are two non-zero diagonals on each side of the main diagonal and, therefore, unrolling with depth greater than two should practically has the same effect as unrolling with depth two (on some computers the compilers try automatically to unroll with depth two).

3 Unsymmetric Matrices

The use of iterative methods in the case where the systems solved are neither symmetric nor positive definite is connected with many difficulties. In this case the choice of a proper preconditioner will very often be the most important step in the solution. Some results obtained with a conjugate gradients technique, applied to the residual vectors, will be presented. No preconditioning is used (and this leads to performing many iterations), but some work is now being carried out to construct a good preconditioner for the method used in this section.

N	NBAND	LINPACK	BLAS 2	DEPTH 0	DEPTH 2	DEPTH 4	DEPTH 8	DEPTH 16
100	21	4	16	45	45	40	39	40
400	41	9	30	79	130	112	114	115
900	61	14	43	91	131	147	157	159
1600	81	18	56	105	135	162	182	190
2500	101	23	66	105	144	174	197	208

Table 4
The speeds of computations obtained in solving systems of linear algebraic equations with unsymmetric matrices when the sparsity of the matrix bands is not exploited.

N	NBAND	LINPACK	BLAS 2	DEPTH 0	DEPTH 2	DEPTH 4	DEPTH 8	DEPTH 16
100	21	4	16	33	31	30	30	27
400	41	9	30	77	77	78	78	74
900	61	14	43	97	103	108	109	106
1600	81	18	56	111	118	123	122	121
2500	101	23	66	119	134	136	135	132

Table 5
The speeds of computations obtained in solving systems of linear algebraic equations with unsymmetric matrices when the sparsity of the matrix bands is exploited.

N	NBAND	LINPACK	BLAS 2	SPARSITY NOT EXPLOITED	SPARSITY EXPLOITED
100	21	0.01	0.03	0.01 (33.3%)	0.01 (33.3%)
400	41	0.08	0.17	0.04 (23.5%)	0.02 (11.8%)
900	61	0.24	0.51	0.13 (25.5%)	0.04 (7.8%)
1600	81	0.58	1.14	0.32 (28.1%)	0.08 (7.0%)
2500	101	1.13	2.19	0.66 (30.1%)	0.14 (6.4%)

Table 6

The computing times, measured in seconds, on AMDAHL VP1100. The results obtained when unrolling with depth 16 is applied are given in the case where the sparsity is not exploited. When the sparsity is exploited, the results given in this table are obtained using unrolling with depth 4. The matrices are obtained by using $\delta = 2.0$. The percentages are obtained by using the BLAS 2 computing times.

Again the solvers based on calculations by diagonals are compared with direct solvers (the **LINPACK**, [1], subroutines **SGBFA** and **SGBSL**) and with a subroutine based on the same iterative method and the **BLAS 2** kernel for performing a matrix-vector multiplication.

The same remarks as in the previous section could be made for the speeds of computations of the dense and the sparse versions of the code (in the case where unrolling is not used as well as in the case where unrolling is in use).

It is seen that the **LINPACK** solver for unsymmetric matrices performs considerably better than the **LINPACK** solver for symmetric and positive definite matrices. The same is true when the speeds of computations of the **BLAS 2** solvers (for unsymmetric matrices and for symmetric and positive definite matrices) are compared. This explains, together with the fact that much more iterations are needed now to obtain the accuracy required, why the computing times obtained when the calculations are organized by diagonals are not so good (compared with those obtained in the previous section). The **LINPACK** solver will perform much better on a scalar machine. As already mentioned, some kind of preconditioning seems to be necessary in this situation.

4 Special Matrices

The improvement achieved by unrolling is caused by the reduction of the sweeps in the outer loops. Of course, it will be even better if the double loops can be replaced by single loops. This can be done sometimes. In many applications the matrices involved in the computations are very special: the number of their non-zero diagonals is very small (say, three or five). In such a case the codes used to obtain the results in the previous two sections can be simplified so that all double loops are replaced by single loops. The implementation of this idea is straightforward. Kernels prepared in this way have been applied in some of the models at the Danish Air Pollution Laboratory. The numerical results indicate that such kernels are very useful, especially when the problems treated are very large.

5 Conclusions and Plans for Future Work

The storage scheme should be selected carefully when a vector processor is to be used in the iterative solution of large systems of linear algebraic equations with band matrices. The matrix elements should be stored by diagonals and all computations should be organized by diagonals in order to utilize completely the power of the vector machine used.

The use of iterative methods in the solution of linear algebraic equations could result in many (or even too many) iterations. Therefore preconditioning should be used when this happens. If a vector processor is in use, then not only should the number of iterations be decreased by applying the preconditioner chosen, but it is also desirable to keep the high speed of convergence. These two requirements work in opposite directions (as often happens in practice) and, therefore a compromise is to be found. The search for a good preconditioner, especially for systems of linear algebraic equations with general band matrices is still continuing.

Acknowledgements. *The work of the author has been partly supported by the Danish Natural Science Research Council. The author should like to thank very much for this support.*

References

[1] J. J. Dongarra, J. R. Bunch, C. B. Moler and G. W. Stewart: *"LINPACK: Users guide"*, SIAM Publications, Philadelphia, 1979.

[2] J. J. Dongarra, J. J. Du Croz, S. Hammarling and R. J. Hanson: *"A proposal for an extended set of Fortran basic linear algebra subprograms"*, ACM SIGNUM Newsletter, 20(1985), pp. 1-18.

[3] P. F. Dubois, A. Greenbaum and G. H. Rodrigue: *"Approximating the inverse of a matrix for the use of iterative algorithms on vector processors"*, Computing, 22(1979), pp. 308-323.

[4] C. Lawson, R. Hanson, D. Kincaid and F. Krogh: *"Basic linear algebra subprograms for Fortran usage"*, ACM Trans. Math. Software, 5(1979), pp. 308-323.

[5] N. K. Madsen, G. H. Rodrigue and J. I. Karush: *"Matrix multiplication by diagonals on a vector parallel processor"*, Information Processing Letters, 5(1976), pp. 41-45.

[6] Z. Zlatev: *"Computations with large and band matrices on vector processors"*, In:"Advances in Parallel Processing" (D.J.Evans, ed.), JAY Press, London, 1988, to appear.

[7] Z. Zlatev, Ph. Vu, J. Wasniewski and K. Schaumburg: *"Computations with symmetric, positive definite and band matrices on a parallel vector processor"*, Parallel Computing, 6(1988), to appear.

List of Participants

⋆ member of the Organizing Committee

G. Ammar	– Northern Illinois University, USA
T. Antoulas	– ETH, Zürich, Switzerland
Z. Bai	– New York University, USA
J. Barry	– Australian Nuclear Science & Technology Organisation, Australia
S. Basu	– Stevens Institute of Technology, USA
C. Beattie	– Virginia Polytechnic Institute and State University, USA
M. Bellanger	– T.R.T., Le Plessis Robinson, France
M. Berry	– University of Illinois, USA
R. Bitmead	– Australian National University, Australia
A. Björck	– Linköping University, Sweden
R. Brent	– Australian National University, Australia
A. Bultheel	– Katholieke Universiteit Leuven, Belgium
J. Bunch	– University of California, San Diego, USA
J-L. Calvet	– LAAS du CNRS, Toulouse, France
J-P. Charlier	– Philips Research Laboratory, Belgium
V. Cherkassky	– University of Minnesota, USA
P. Comon	– Thomson Sintra, Cagnes sur Mer, France
J. Cuppen	– Philips & Picker Medical Systems, The Netherlands
P. d'Almeida	– University do Porto, Portugal
E. de Doncker	– Katholieke Universiteit Leuven, Belgium
P. de Groen	– Vrije Universiteit Brussel, Belgium
P. Delsarte	– Philips Research Laboratory, Belgium
B. De Moor	– Katholieke Universiteit Leuven, Belgium
J. de Pillis	– University of California, Riverside, USA
E. Deprettere	– Delft University of Technology, The Netherlands
M. Dracopoulos	– National Technical University, Athens, Greece
A. Edelman	– M. I. T., USA
M. Eiermann	– Universität Karlsruhe, Federal Republic of Germany
A. Ephremides	– University of Maryland, USA
M. Ewerbring	– Cornell University, USA
D. Farrier	– University of Southampton, UK
V. Fernando	– Numerical Algorithms Group Ltd, Oxford, UK
L. Foster	– San Jose State University, USA

Y. Genin	– Philips Research Laboratory, Belgium
G. Golub⋆	– Stanford University, USA (Co-Director)
W. Govaerts	– Rijksuniversiteit Gent, Belgium
M. Gover	– University of Bradford, UK
C. Gueguen	– Ecole Nationale Supérieure des Télécommunications, Paris, France
G. Gulak	– University of Toronto, Canada
K. Gustafsson	– Lund Institute of Technology, Sweden
S. Hammarling⋆	– Numerical Algorithms Group Ltd, Oxford, UK
M. Hanke	– Universität Karlsruhe, Federal Republic of Germany
I. Ipsen	– Yale University, USA
E. Jessup	– Yale University, USA
T. Kailath	– Stanford University, USA
A. Kiper	– Middle East Technical University, Turkey
M. Köksal	– Gaziantep University, Turkey
F. Luk	– Cornell University, USA
W. Madych	– University of Connecticut, USA
P. Manneback	– Laboratoire de Recherches Solvay, Brussels, Belgium
J. Mc Whirter	– Royal Signals and Radar Establishment, UK
V. Mehrmann	– Universität Bielefeld, Federal Republic of Germany
J. Meinguet	– Université Catholique de Louvain, Belgium
G. Meurant	– Centre d'Etudes de Limeil-Valenton, France
D. Miller	– Wright State University, USA
G. Miminis	– Memorial University of Newfoundland, Canada
J. Modi	– Cambridge University, UK
M. Moonen	– Katholieke Universiteit Leuven, Belgium
H. Nelis	– Delft University of Technology, The Netherlands
M. Neumann	– University of Connecticut, USA
N. Nichols	– University of Reading, UK
L. Pacquet	– Université de Mons, Belgium
C. Paige	– McGill University, Canada
R. Plemmons	– North Carolina State University, USA
J. Pryce	– Royal Military College of Science, Shrivenham, UK
L. Reichel	– Bergen Scientific Centre IBM, Norway
D. Roose	– Katholieke Universiteit Leuven, Belgium
T. Sarkar	– Syracuse University, USA

D. Schimmel	– Cornell University, USA
R. Schreiber	– Rensselaer Polytechnic Institute, USA
H. Schwarz	– University of Zurich, Switzerland
G. Shroff	– Rensselaer Polytechnic Institute, USA
D. Slock	– Rijksuniversiteit Gent, Belgium
G. Sonnevend	– Universität Würzburg, Federal Republic of Germany
D. Sorensen⋆	– Argonne National Laboratory, USA
J. Sorensen	– Technical University of Denmark, Denmark
G. Starke	– Universität Karlsruhe, Federal Republic of Germany
N. Tsao	– Wayne State University, USA
M. Van Barel	– Katholieke Universiteit Leuven, Belgium
R. van de Geijn	– The University of Texas at Austin, USA
J. Vandewalle⋆	– Katholieke Universiteit Leuven, Belgium
M. van Dijke	– Utrecht University, The Netherlands
P. Van Dooren⋆	– Philips Research Laboratory, Belgium (Director)
S. Van Huffel	– Katholieke Universiteit Leuven, Belgium
T. Varvarigou	– National Technical University, Athens, Greece
B. Verdonck	– Universitaire Instellingen Antwerpen, Belgium
M. Verhaegen	– Delft University of Technology, The Netherlands
E. Verriest	– Georgia Institute of Technology, USA
M. Vespucci	– Istituto Universitario di Bergamo, Italy
U. von Matt	– ETH, Zürich, Switzerland
D. Watkins	– Washington State University, USA
L. Watson	– Virginia Polytechnic Institute & State University, USA
B. Yang	– Ruhr-Universität Bochum, Federal Republic of Germany
H. Zhang	– CNET/RPE/ETP, Issy-les-Moulineaux, France
Z. Zlatev	– Air Pollution Laboratory, Denmark

NATO ASI Series F

Including Special Programmes on Sensory Systems for Robotic Control (ROB) and on Advanced Educational Technology (AET)

Vol. 1: Issues in Acoustic Signal – Image Processing and Recognition. Edited by C. H. Chen. VIII, 333 pages. 1983.

Vol. 2: Image Sequence Processing and Dynamic Scene Analysis. Edited by T. S. Huang. IX, 749 pages. 1983.

Vol. 3: Electronic Systems Effectiveness and Life Cycle Costing. Edited by J. K. Skwirzynski. XVII, 732 pages. 1983.

Vol. 4: Pictorial Data Analysis. Edited by R. M. Haralick. VIII, 468 pages. 1983.

Vol. 5: International Calibration Study of Traffic Conflict Techniques. Edited by E. Asmussen. VII, 229 pages. 1984.

Vol. 6: Information Technology and the Computer Network. Edited by K. G. Beauchamp. VIII, 271 pages. 1984.

Vol. 7: High-Speed Computation. Edited by J. S. Kowalik. IX, 441 pages. 1984.

Vol. 8: Program Transformation and Programming Environments. Report on a Workshop directed by F. L. Bauer and H. Remus. Edited by P. Pepper. XIV, 378 pages. 1984.

Vol. 9: Computer Aided Analysis and Optimization of Mechanical System Dynamics. Edited by E. J. Haug. XXII, 700 pages. 1984.

Vol. 10: Simulation and Model-Based Methodologies: An Integrative View. Edited by T. I. Ören, B. P. Zeigler, M. S. Elzas. XIII, 651 pages. 1984.

Vol. 11: Robotics and Artificial Intelligence. Edited by M. Brady, L. A. Gerhardt, H. F. Davidson. XVII, 693 pages. 1984.

Vol. 12: Combinatorial Algorithms on Words. Edited by A. Apostolico, Z. Galil. VIII, 361 pages. 1985.

Vol. 13: Logics and Models of Concurrent Systems. Edited by K. R. Apt. VIII, 498 pages. 1985.

Vol. 14: Control Flow and Data Flow: Concepts of Distributed Programming. Edited by M. Broy. VIII, 525 pages. 1985.

Vol. 15: Computational Mathematical Programming. Edited by K. Schittkowski. VIII, 451 pages. 1985.

Vol. 16: New Systems and Architectures for Automatic Speech Recognition and Synthesis. Edited by R. De Mori, C.Y. Suen. XIII, 630 pages. 1985.

Vol. 17: Fundamental Algorithms for Computer Graphics. Edited by R.A. Earnshaw. XVI, 1042 pages. 1985.

Vol. 18: Computer Architectures for Spatially Distributed Data. Edited by H. Freeman and G.G. Pieroni. VIII, 391 pages. 1985.

Vol. 19: Pictorial Information Systems in Medicine. Edited by K. H. Höhne. XII, 525 pages. 1986.

Vol. 20: Disordered Systems and Biological Organization. Edited by E. Bienenstock, F. Fogelman Soulié, G. Weisbuch. XXI, 405 pages.1986.

Vol. 21: Intelligent Decision Support in Process Environments. Edited by E. Hollnagel, G. Mancini, D.D. Woods. XV, 524 pages. 1986.

NATO ASI Series F

Vol. 22: Software System Design Methods. The Challenge of Advanced Computing Technology. Edited by J. K. Skwirzynski. XIII, 747 pages. 1986.

Vol. 23: Designing Computer-Based Learning Materials. Edited by H. Weinstock and A. Bork. IX, 285 pages. 1986.

Vol. 24: Database Machines. Modern Trends and Applications. Edited by A.K. Sood and A.H. Qureshi. VIII, 570 pages. 1986.

Vol. 25: Pyramidal Systems for Computer Vision. Edited by V. Cantoni and S. Levialdi. VIII, 392 pages. 1986. *(ROB)*

Vol. 26: Modelling and Analysis in Arms Control. Edited by R. Avenhaus, R.K. Huber and J.D. Kettelle. VIII, 488 pages. 1986.

Vol. 27: Computer Aided Optimal Design: Structural and Mechanical Systems. Edited by C.A. Mota Soares. XIII, 1029 pages. 1987.

Vol. 28: Distributed Operating Systems. Theory und Practice. Edited by Y. Paker, J.-P. Banatre and M. Bozyiğit. X, 379 pages. 1987.

Vol. 29: Languages for Sensor-Based Control in Robotics. Edited by U. Rembold and K. Hörmann. IX, 625 pages. 1987. *(ROB)*

Vol. 30: Pattern Recognition Theory and Applications. Edited by P.A. Devijver and J. Kittler. XI, 543 pages. 1987.

Vol. 31: Decision Support Systems: Theory and Application. Edited by C.W. Holsapple and A.B. Whinston. X, 500 pages. 1987.

Vol. 32: Information Systems: Failure Analysis. Edited by J.A. Wise and A. Debons. XV, 338 pages. 1987.

Vol. 33: Machine Intelligence and Knowledge Engineering for Robotic Applications. Edited by A.K.C. Wong and A. Pugh. XIV, 486 pages. 1987. *(ROB)*

Vol. 34: Modelling, Robustness and Sensitivity Reduction in Control Systems. Edited by R.F. Curtain. IX, 492 pages. 1987.

Vol. 35: Expert Judgment and Expert Systems. Edited by J. L. Mumpower, L. D. Phillips, O. Renn and V.R.R. Uppuluri. VIII, 361 pages. 1987.

Vol. 36: Logic of Programming and Calculi of Discrete Design. Edited by M. Broy. VII, 415 pages. 1987.

Vol. 37: Dynamics of Infinite Dimensional Systems. Edited by S.-N. Chow and J.K. Hale. IX, 514 pages. 1987.

Vol. 38: Flow Control of Congested Networks. Edited by A.R. Odoni, L. Bianco and G. Szegö. XII, 355 pages. 1987

Vol. 39: Mathematics and Computer Science in Medical Imaging. Edited by M.,A. Viergever and A. Todd-Pokropek. VIII, 546 pages. 1988.

Vol. 40: Theoretical Foundations of Computer Graphics and CAD. Edited by R.A. Earnshaw. XX, 1246 pages. 1988.

Vol. 41: Neural Computers. Edited by R. Eckmiller and Ch. v. d. Malsburg. XIII, 566 pages. 1988.

NATO ASI Series F

NATO ASI Series F

Vol. 66: Sensor-Based Robots: Algorithms and Architectures. Edited by C. S. G. Lee. X, 285 pages. 1991. *(ROB)*

Vol. 67: Designing Hypermedia for Learning. Edited by D. H. Jonassen and H. Mandl. XXV, 457 pages. 1990. *(AET)*

Vol. 68: Neuro-Computing. Algorithms, Architectures and Applications. Edited by F. Fogelman Soulié and J. Hérault. XI, 455 pages. 1990.

Vol. 69: Real-Time Integration Methods for Mechanical System Simulation. Edited by E. J. Haug and R. C. Deyo. VIII, 352 pages. 1991.

Vol. 70: Numerical Linear Algebra, Digital Signal Processing and Parallel Algorithms. Edited by G. H. Golub and P. Van Dooren. XIII, 729 pages. 1991.

MIX
Papier aus verantwortungsvollen Quellen
Paper from responsible sources
FSC® C105338

If you have any concerns about our products,
you can contact us on
ProductSafety@springernature.com

In case Publisher is established outside the EU,
the EU authorized representative is:
Springer Nature Customer Service Center GmbH
Europaplatz 3, 69115 Heidelberg, Germany

Printed by Libri Plureos GmbH
in Hamburg, Germany